Geophysics: Processes and Properties of Earth

Geophysics: Processes and Properties of Earth

Editor: Cortez Ford

R CALLISTO REFERENCE

www.callistoreference.com

Callisto Reference,
118-35 Queens Blvd., Suite 400,
Forest Hills, NY 11375, USA

Visit us on the World Wide Web at:
www.callistoreference.com

ISBN: 978-1-64116-080-3 (Hardback)

Cataloging-in-Publication Data

Geophysics : processes and properties of earth / edited by Cortez Ford.
    p. cm.
Includes bibliographical references and index.
ISBN 978-1-64116-080-3
1. Geophysics. 2. Earth sciences. I. Ford, Cortez.
QE501 .G46 2019
550--dc23

Table of Contents

Chapter 24   **A Study of Earthquakes in Bangladesh and the Data Analysis of the Earthquakes that were generated In Bangladesh and Its' Very Close Regions for the Last Forty Years (1976-2016)**...194
            Md. Abdullah Al zaman and Nusrath Jahan Monira

Chapter 25   **Preliminary Magnetostratigraphic and Isotopic Dating of the Ngwa Formation (Dschang Western Cameroon)**..199
            Benammi M, Hell JV, Bessong M, Nolla D, Solé J and Brunet M

            Permissions

            List of Contributors

            Index

Preface

This book aims to highlight the current researches and provides a platform to further the scope of innovations in this area. This book is a product of the combined efforts of many researchers and scientists, after going through thorough studies and analysis from different parts of the world. The objective of this book is to provide the readers with the latest information of the field.

Geophysics is a branch of natural science concerned with the scientific analysis of the physical properties and processes of Earth and its surrounding atmosphere or space. The principles of geophysics are used to explore and study geological phenomena such as tectonic shifts, volcanism, rock formation, etc. Applications of geophysics can be found in the areas of natural hazard mitigation, mineral resource exploration and environmental protection. Chapters in this book are compiled comprehensively to provide detailed information about the upcoming concepts and theories of geophysics while also discussing modern techniques and principles. The book also attempts to present studies that have transformed this discipline and aided its advancement. Those in search of information to further their knowledge will be greatly assisted by this book.

I would like to express my sincere thanks to the authors for their dedicated efforts in the completion of this book. I acknowledge the efforts of the publisher for providing constant support. Lastly, I would like to thank my family for their support in all academic endeavors.

Editor

The Geology, Geochemistry and Petrogenetic Studies of the Precambrian Basement Rocks around Iworoko, Are and Afao Area, Southwestern Nigeria

Olusiji Samuel Ayodele*

Department of Geology, Ekiti State University, P.M.B. 5363, Ado-Ekiti, Nigeria

Abstract

The geology, geochemistry and petrogenetic studies of the Precambrian basement rocks around Iworoko, Are and Afao Ekiti were carried out to determine their geochemical and petrogenetic characteristics. Three lithologies including migmatite-gneiss, granite gneiss and banded gneiss with a pegmatite dyke that occurred as an intrusion were recognized in the study area. A total of seventeen rock samples were collected from the study area which were described based on their field relationships. Ten fresh rock samples were later selected for geochemical analysis. The result of the geochemical analysis revealed that silica (SiO_2) is the most abundant major oxide when compared with other oxides present in all the rock samples analyzed with an average percentage composition of 66.31%. The average percentage composition of other oxides present in all the rock samples are as follows; (16.41%) Al_2O_3, (3.67%) Fe_2O_3, (0.25%) CaO, (4.28%) K_2O, (3.53%) Na_2O, (1.75%) MgO, (0.78%) P_2O_5, (0.54%) TiO_2 and (0.061%) MnO. The results of the trace and rare earth elements analyses revealed that Barium (Ba) is the most abundant with an average value of 328.7 ppm compared to other trace and rare earth elements present in the rock samples. The high concentration of barium in the migmatite-gneisses of the study area revealed the radioactive nature of this lithology. Petrological and chemical data suggests a sedimentary protolith, probably greywacke for the migmatite gneiss, gneiss and banded gneiss in the study area which may have been derived from a continental environment.

Introduction

Nigeria lies to the rest of the West Africa craton in the region of late Precambrian to Early Paleozoic orogenesis. The basement complex is made up of Precambrian rocks and the schist belts infolded in them. The Precambrian rocks of southwestern Nigeria is part of the Precambrian Basement complex of Nigeria, the Basement complex itself is made up of Gneiss-migmatite complex and the Pan African older Granite rocks. However, the lithologies in the study area include; migmatite-gneiss, granite-gneiss and banded gneiss. These rocks have undergone polycyclic deformation thereby causing the deformation of both the macro and micro structures. Secondary structures in rocks that can be used as clues to determine the geologic history of an area include; joints, folds, fractures and foliations etc. Some of these structures are not deformational but were formed at the same time the rocks were emplaced. A lot of literatures abound on the study of basement geology of Nigeria and its associated structures which include Geochemical dispersion of gold in stream sediments in Paleoproterozoic Nyong series, southern Cameroon undertaken by Mubfu and Nforba [1] in an attempt to explore for gold using stream sediments collected in the Ngo Vayang area of southern Cameroon, and revealed that the Au-Hf element association from the R-mode factor analysis indicated gold mineralization while U-Th-Pb-W, Nb-Ta-Co-V, Au-Hf-Cu associations reflected lithologic controls. Akintola et al., [2] carried out the petrography and stream sediment geochemistry of Ede and its environs in order to identify the rock units with their mineralogical appraisal and to determine the concentration and distribution of major and trace elements in the stream sediments with a view to elucidate the mineral potentials of the study area. Emmanuel et al., [3] also carried out geochemical investigation of the southern part of Ilesha with the aim of clarifying the potential source of mineralization in the area. Geologic mapping of the area revealed that the area is made up of different lithologies such as undifferenced schists, gneisses and migmatites with pegmatites, schists and epidiorite complex, quartzite and quartz schist. Sixty-one soil samples were collected and analyzed for elements such as Pb, Fe, Ni, Cd, Cr, Cu, Zn and Mn, using multivariate analysis to obtain the coefficient of principal components. The elemental association ratio revealed high metallic concentrations which led to the mineralization trend in southern Ilesa. According to Okunlola and Akinola, the Precambrian pegmatites of Ijero area occur as steeply dipping intrusives into the other rocks of migmatite-gneiss complex and the schistose rocks. Thin section studies revealed that the

pegmatite samples contain mainly quartz (37%), plagioclase (12%), and microcline (7%) with accessories such as biotite, hornblende and tourmaline. Geochemical analysis of muscovite extracts from the pegmatite using ICP-MS analytical method showed that the pegmatites is siliceous with average SiO_2 content of 71.79%. Trace element analysis using variation plots of Ta/ (Ta+Nb) versus Mn/ (Mn+Fe) revealed that the muscovite-quartz-microcline pegmatites of Ijero area is of rare element class, lepidolite subtype. Odeyem et al., [4] suggested that almost all the foliation exhibited by the rocks of southwestern Nigeria excluding the intrusive are tectonic in origin because pre-existing primary structures have been obliterated by subsequent deformation while Anifowose et al., [5] noted that joints ranging from minor to major ones are found in all the rock types, some of which are filled with quartz, feldspars or a combination of both. They lie generally in the NE-SW direction.

The study area lies in Irepodun/Ifelodun area, north east of Ekiti State, and falls within latitudes $07^0\ 41^1$ N - $07^0\ 43^1$ N and longitudes $05^0\ 15^1$ E – $05^0\ 18^1$ E respectively (Figure 1). These include Iworoko, Are and Afao-Ekiti, with an area coverage of about 242.6 km^2. The study area is generally accessible and motorable with availability of tarred and untarred roads which linked the towns and villages together in the area. Most of the outcrops are located in the thick forests in which the use of cutlass was essential in creating footpaths. The major towns around the study area include Ilokun, Ilupeju etc while the localities within Iworoko, Are and Afao-Ekiti were named after the features or events which occurred within each town amongst these include Ori oke Adura, Oke ode, oke iyanu etc. The settlement pattern in the study area is the nucleated type and the major occupation of the inhabitants are farming and trading.

*Corresponding author: Olusiji Samuel Ayodele, Departmennt of Geology, Ekiti State University, P.M.B. 5363, Ado- Ekiti, Nigeria, E-mail: samuelayodeleolusiji@yhoo.com

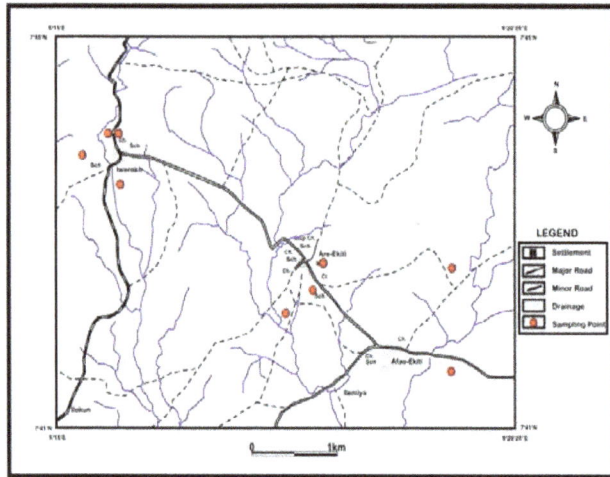

Figure 1: Location map of the study area showing rock sampling points.

Geologic Setting of the Study Area

The study area is underlain by the Precambrian Basement Complex of Southwestern Nigeria, which is also part of the Basement Complex rocks of Nigeria. The basement complex is one of the three major litho-petrological components that make up the geology of Nigeria. The Nigerian basement complex forms a part of the Pan – African mobile belt and lies between the West African and Congo cratons and south of the Tuareg shield [6]. It is intruded by the Mesozoic Calc-alkaline ring complexes (Younger granites) of the Jos plateau and is unconformably overlain by Cretaceous and younger sediments. The Nigerian basement complex was affected by the 600 Ma Pan African orogeny and it occupies the reactivated region which resulted from plate collision between the passive continental margin of the West African craton and the active Pharusian continental margin [7]. The Basement rocks are believed to be the results of at least four major orogenic cycles of deformation, metamorphism and remobilization corresponding to the Liberian (2,700 Ma), the Eburrnean (2500 Ma), the kibaran (1,100 Ma), and the Pan-African cycles (600 Ma). The first three cycles were characterized by intense deformation and isoclinals folding accompanied by regional metamorphism, which was further followed by extensive migmatization. The Pan-African deformation was accompanied by a regional meta-induced syntectonic granites and homogenous gneisses. Late tectonic emplacement of granites and granodiorites and associated contact metamorphism accompanied the end stages of this last deformation. The end of the orogeny was marked by faulting and fracturing [8]. Anifowose [9] was of the opinion that the granitic emplacement was probably controlled by fractures within the basement, and also showed outcrop pattern indicating that the older granite cut across all other structures with sharp and chilled contact. Within the basement complex of Nigeria, four major petro-lithological units are distinguishable namely [7];

- The Migmatite – Gneiss-Quartzite Complex
- The schist belts
- The Pan African granitoids
- Under formed acid and basis dykes.

Local geology of the study area

The study area is dominated by granite-gneiss, banded-gneiss and migmatite-gneiss respectively. These rocks are mostly metamorphic and belong to the Precambrian basement rocks of southwestern Nigeria, which itself is part of the basement rocks of Nigeria (Figure 2). The various lithologic units in the area are discussed below, while their field description and relationships are presented in Table 1.

Migmatite-Gneiss-Quartzite complex

The migmatite-gneiss-quartzite complex occurs as ridges and highlands at Iworoko and Afao-Ekiti, but as valleys at Are Ekiti. Most of these rocks have been subjected to mechanical and chemical weathering. The migmatite-gneiss-quartzitecomplex is presumably the oldest group of rocks in the study area and the most widespread. Its occurrence is not restricted as it is found in the entire area of study. The texture of the migmatite-gneiss-quartzite complex in the study area varies from fine grained to medium grained while the structures observed on the outcrop include folds, joints, cracks and veins. The dominant mineral assemblages of this rock include quartz, mica, plagioclase, and hornblende. The field relationships between the migmatite-gneiss-quartzite complex rocks and other surrounding rocks could not be easily ascertained as a result of thick vegetation.

Granite-gneiss

The granite-gneiss in the study area occurs as a hilly outcrop especially around Afao and Iworoko Ekiti. The texture of these rocks varies from fine to medium grained. The dominant mineral assemblages found in the granite-gneiss include quartz, feldspar, and mica, and the structures found on the outcrop are microfolds, veinlets and quartz veins. The surface of the outcrop has been subjected to exfoliation as a result of physical weathering. The granite-gneiss has a clear and sharp contact with the surrounding migmatite-gneiss-quartzite complex.

Banded gneiss

The banded gneiss extends across some parts of the study area particularly Iworoko and Afao. In these areas, the rocks are medium to coarse grained texturally and occur as a hill in the study area. Also, the rock is characterized by banding in which there is mineral alignment. The banding varies from about 1 centimeter to 3 centimeters in the banded varieties found in Are and Afao Ekiti. The rock is typified by clear fine banding which shows alternating white and black (hornblende,

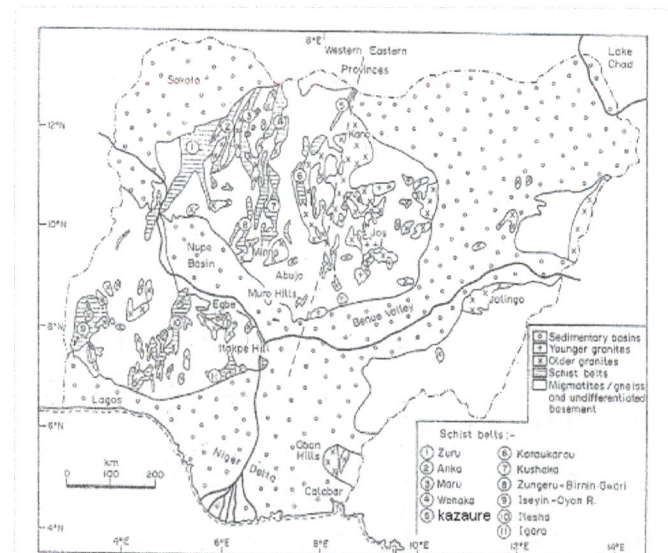

Figure 2: Generalized Geological Map of the Crystalline Rocks of Nigeria (After Rahaman [11]).

S/N	Location	Longitude	Latitude	Lithology	Texture	Structure	Dip Angles
1	Ori-Oke Adura Iworoko	07° 43 494′	05° 15 756′	Granite Gneiss	Medium to fine grained	Cracks, Joints, Dykes, e.t.c.	
2	Behind Iworoko mosque	07° 44 018′	05° 15 738′	Granite Gneiss	Fine to medium grained	Veins, Veinlets, Folds e.t.c.	
3	Behind Iworoko Grammer School	07° 43 794′	05° 15 315′	Migmatite Gneiss	Medium to fine grained	Quartz Vein, Cracks, Folds, Dykes, Exfoliation	
4	Eniafe road	07° 44 020′	05° 15 615′	Granite Gneiss	Medium to coarse grain	Folds, veinlets, quartz veins	
5	Off Iworoko Are road	07° 44 058′	05° 15 425′	Granite Gneiss	Fine to medium grained	veins, dykes.	
6	Aba Sunday Are	07° 43 375′	05° 17 114′	Migmatite gneiss	Medium to coarse grained	Exfoliation, veinlet, fold, veins	
7	Behind OBA's Palace Are	07° 43 691′	05° 18 147′	Banded gneiss	Medium to coarse grained	Solution hole, banding, lineation, exfoliation	42°
8	Oke Ode Are	07° 42 174′	05° 17 698′	Migmatite gneiss	Medium to coarse grained	Exfoliation, lineation, solution hole	42°
9	Oke Isoro	07° 42 158′	05° 17 410′	Migmatite gneiss	Medium to coarse grained	Fold, Dyke, Veins, Xenolith	
10	Off Isara road	07° 42 410′	05° 18 020′	Migmatite Gneiss	Fine to medium grained	Fold, Dyke, Veinlets,	
11	Isara Are	07° 42 310′	05° 17 512′	Migmatite Gneiss	Fine to medium grained	Solution hole, fold, veinlets	
12	Afao round about	07° 41 647′	05° 19 593′	Granite Gneiss	Medium to coarse grained	folds, dyke, vein, solution hole	
13	Behind Oke iyanu Afao	07° 41 578′	05° 19 669′	Granite Gneiss	Medium to coarse grained	Folds, vein, cracks, solution hole	
14	Oke Iyanu Afao	07° 41 249′	05° 19 405′	Granite Gneiss	Coarse to fine grained	Folds, fracture, solution hole	
15	Igbemo road, Afao	07° 43 158′	05° 20 684′	Granite Gneiss	Coarse to fine grained	Vein, solution hole.	
16	Off Igbimo road, Afao	07° 43 206′	05° 20 409′	Granite gneiss	Fine to coarse grained	Dyke, vein, solution hole	
17	Ilemiya road, Afao	07° 42 640′	05° 19 680′	Migmatite gneiss	Medium to fine grained	Dyke, cracks, vein.	

Table 1: Field data.

biotite) layers. Structures observed on this outcrop include folds, veins and pegmatite dyke, the pegmatite shows no zoning thus making it simple in nature with a width of about 32 centimeters. The present state of the rock is as a result of weathering.

Method of Study

The method of study adopted for the different aspects of this work include field and laboratory operations. The field operations involve sample collection and the sampling method used is the grid-controlled type. In this method, the rock samples were picked at a sampling density of one sample per 4 km^2. A total of seventeen rock samples were collected from the study area which were lithologically described based on their field relationships after which were labeled properly and put in a sample bag for onward transmission to the laboratory. Ten (10) fresh samples were selected from the whole samples and were prepared for geochemical analysis for major, trace and rare earth elements determination (Table 2). The rocks were initially crushed, using the jaw crusher into fragments and were pulverized later using the agate mortar. The four acid method of digestion was adopted in preparing the samples for geochemical analysis using 0.5 g of each sample and the analysis were carried out using the Atomic Absorption Spectrophotometer (AAS) at the Central Laboratory, University of Ibadan, Nigeria.

Discussion

The systematic geologic mapping, geochemical and petrogenetic studies of the Precambrian basement rocks in Iworoko, Are and Afao-Ekiti has been carried out and the overall results showed that SiO$_2$ (silica) is by far the most abundant mineral in all the rock types with the highest percentage present in the migmatites (66.2%) and the lowest in the banded gneisses (63.0%), while TiO$_2$ is the least abundant mineral with an average composition less than 1% in all the rock samples. The ferromagnesian compounds (FeO and MgO) have varying abundances in the rock samples. Fe$_2$O$_3$ value is higher in the migmatite gneisses(4.65%) while the lowest percentage is present in banded gneisses(2.67%), and the granite gneiss has percentage of 3.05%; MgO

is mostly abundant in the granite gneiss(3.03%) because it contains more of mafic minerals than the other rocks. The banded gneisses have 1.81% while the migmatite gneisses have the lowest value of MgO (1.80%). From this it shows these rock samples (migmatites and granite gneisses) must have been emplaced in a continental environment as explained by the plots of TiO$_2$- K$_2$O- P$_2$O$_5$ (Figure 3). The results of the trace and rare earth elements show that Ba is the most abundant trace elements in the migmatite-gneiss. This indicated that the migmatite gneiss in the study area harbors some radioactive materials which are worth investigating. However, no mineral of economic importance has been reported in the area lately, but the migmatite gneisses in the study area could serve as a potential source of mineralization, especially radioactive minerals if well exploited. Also, the migmatite-gneiss in the study area could be chemically comparable to the metasediments of the south western Nigeria [10,11]. The plot of SiO$_2$ versus CaO for the samples fell within the field of Francisian greywacke (Figure 4) while the Na$_2$O/Al$_2$O$_3$ against K$_2$O/Al$_2$O$_3$ for the discrimination of sedimentary/metasedimentary and igneous series (Figure 5) by Garrels and Mackenzie [10] showed that majority of the samples plotted were within the sedimentary field [12]. Also, The TiO$_2$-K$_2$O-P$_2$O$_5$ plot (Figure 3) confirmed the continental nature of the sediments [13]. The range and mean concentration (ppm) of the trace and rare earth elements present (Table 3 and Figure 6) indicated that Barium (Ba) has the range of 36 – 458 ppm with an average value of 328.7 ppm. Migmatite gneiss has the highest concentration of Ba 458 ppm while the pegmatite has the lowest concentration of Ba 36 ppm. Chromium content ranges from 22 – 136 ppm with an average value of 70.8 ppm which is quite high. This high concentration in the samples could be responsible for the presence of some ultramafic bodies in the migmatite-gneiss. The pegmatite also contains fairly equal amounts of Cu, Ga, Cr, As and Pb compared to the migmatite-gneisses.

Conclusion and Recommendation

The geochemical and geological studies (Figure 7) of the basement rocks in Iworoko, Are and Afao Ekiti, south western Nigeria has

Major elements	Are-Ekiti	Are-Ekiti	Iworoko-Ekiti	Afao-Ekiti	Iworoko-Ekiti	Iworoko-Ekiti	Are-Ekiti	Are-Ekiti	Iworoko-Ekiti	Afao-Ekiti
SiO_2	74.97	65.46	64.85	66.36	60.30	70.07	65.20	64.30	69.90	61.66
Al_2O_5	15.11	17.00	15.20	15.85	14.83	14.16	17.10	17.40	15.99	21.49
Fe_2O_3	0.40	7.60	7.20	4.67	4.77	3.39	3.05	3.15	0.27	2.18
TiO_3	0.01	1.08	0.76	0.58	0.74	0.01	0.53	0.75	0.01	0.95
CaO	0.27	1.72	2.90	3.02	4.75	0.09	3.08	5.88	0.26	3.68
P_2O_5	0.22	0.06	0.17	1.21	1.36	0.29	1.02	0.95	0.96	1.24
K_2O	3.53	4.25	2.82	2.91	3.47	8.42	3.75	1.75	8.30	3.56
MnO	0.04	0.02	0.15	0.10	0.09	0.06	0.05	0.03	0.05	0.02
MgO	0.11	1.40	2.45	2.32	4.35	0.10	3.03	2.20	0.10	1.41
Na_2O	5.28	2.87	3.15	3.08	4.06	2.76	3.28	3.76	3.57	3.45
Total	99.94	101.46	99.65	100.1	98.72	99.35	100.09	100.1	99.41	99.64

Table 2: Major elements (wt %).

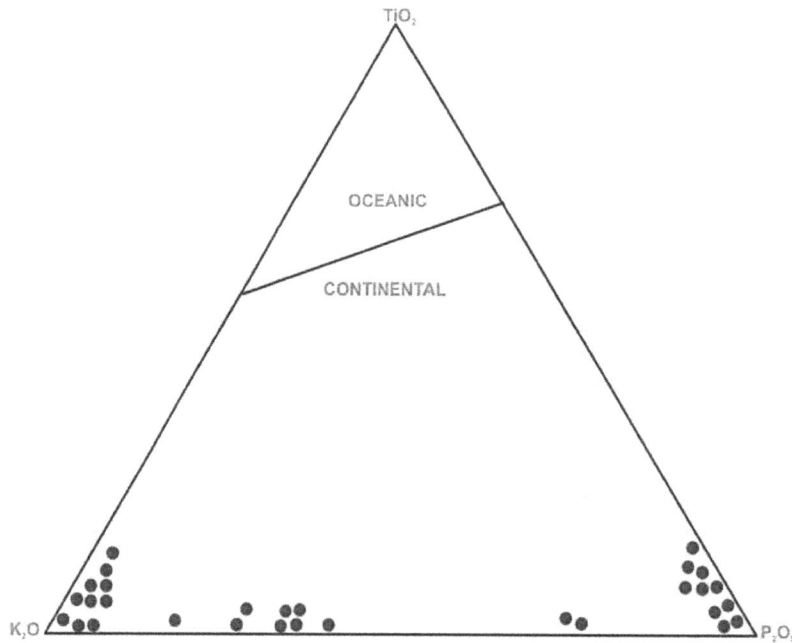

Figure 3: TiO$_2$-K$_2$0-P$_2$0$_5$ plot of the rocks from Iworoko, Are and Afao Ekiti (Pearce et al., [13]).

Oxides	SiO_2	Al_2O_3	Fe_2O_3	TiO_2	CaO	P_2O_5	K_2O	MnO	MgO	Na_2O
Migmatite gneiss	66.2	15.5	4.65	0.53	2.12	0.68	5.0	0.078	1.80	3.24
Granite gneiss	65.20	17.10	3.05	0.53	3.08	1.02	3.75	0.05	3.03	3.28

Table 3: Average concentration of major elements (wt%).

Trace and rare earth elements	Are-Ekiti	Are-Ekiti	Iworoko-Ekiti	Afao-Ekiti	Iworoko-Ekiti	Iworoko-Ekiti	Are-Ekiti	Are-Ekiti	Iworoko-Ekiti	Afao-Ekiti
Ba	36	372	382	316	458	346	358	446	237	336
Ga	39	17	38	44	39	37	34	37	26	29
Cu	14	45	25	30	25	49	29	33	15	29
Cr	52	131	109	96	48	136	42	46	22	26
As	62	57	4	6	61	52	4	3	6	3
Pb	61	88	15	18	36	28	246	78	60	86

Table 4: Trace and rare-earth elements (ppm).

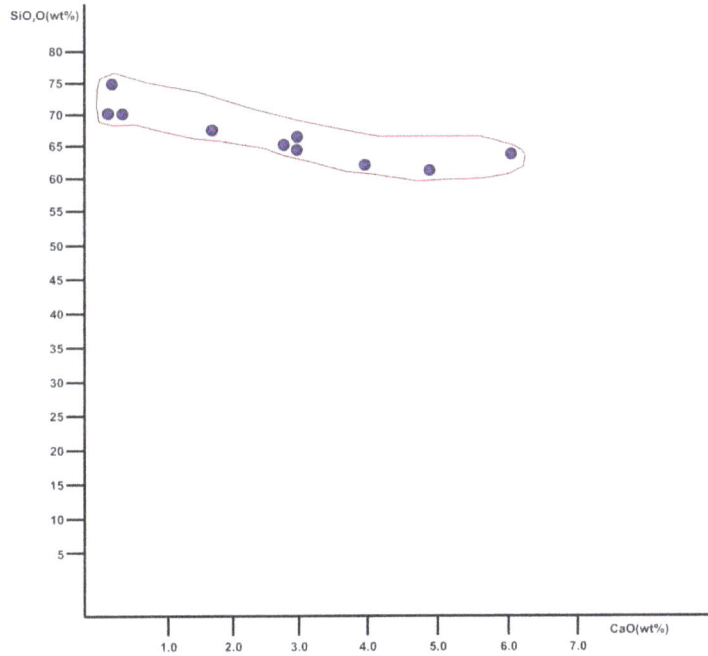

Figure 4: SiO_2-CaO Plot for Iworoko Are and Afao Ekiti Gneisses (Field of Francisian Greywacke after Brown et al., [12]).

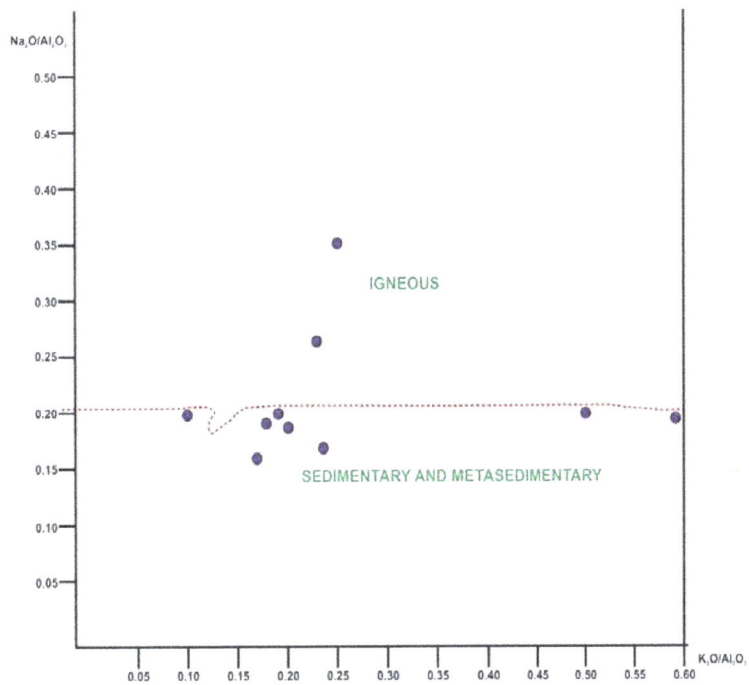

Figure 5: Na_2O/Al_2O_3 Vs K_2O/Al_2O_3 Variation diagram fields of igneous and sedimentary/metasedimentry rocks (After Garrels and Mackenzie [10]).

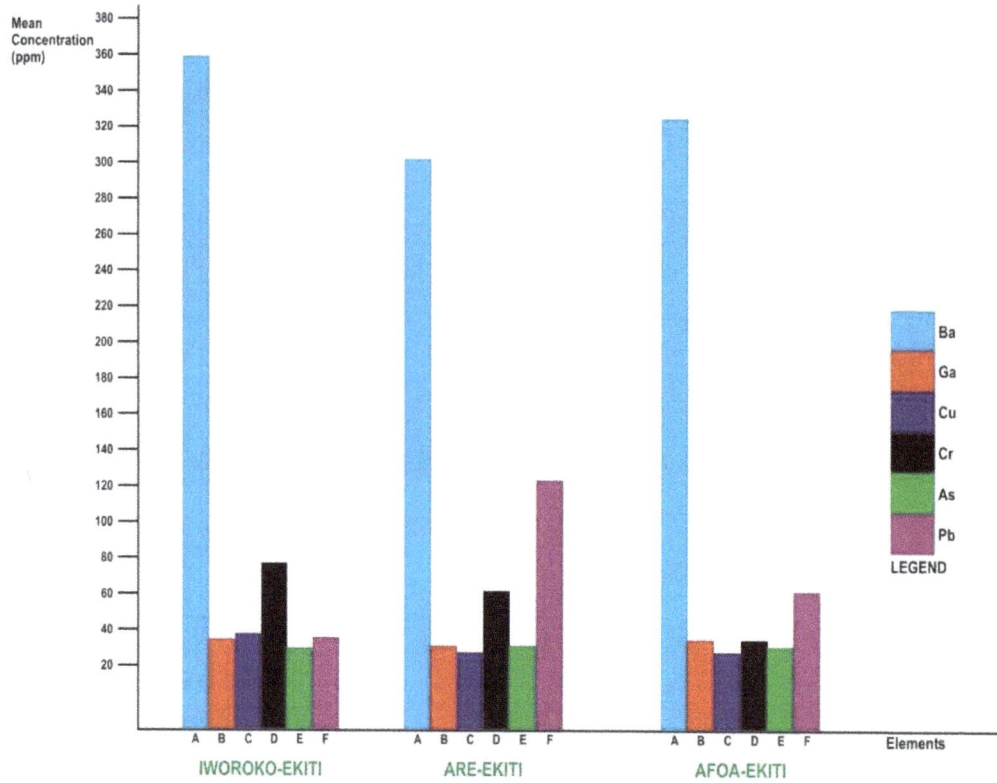

Figure 6: Histogram of mean concentration of the trace and rare-earth elements in the study area.

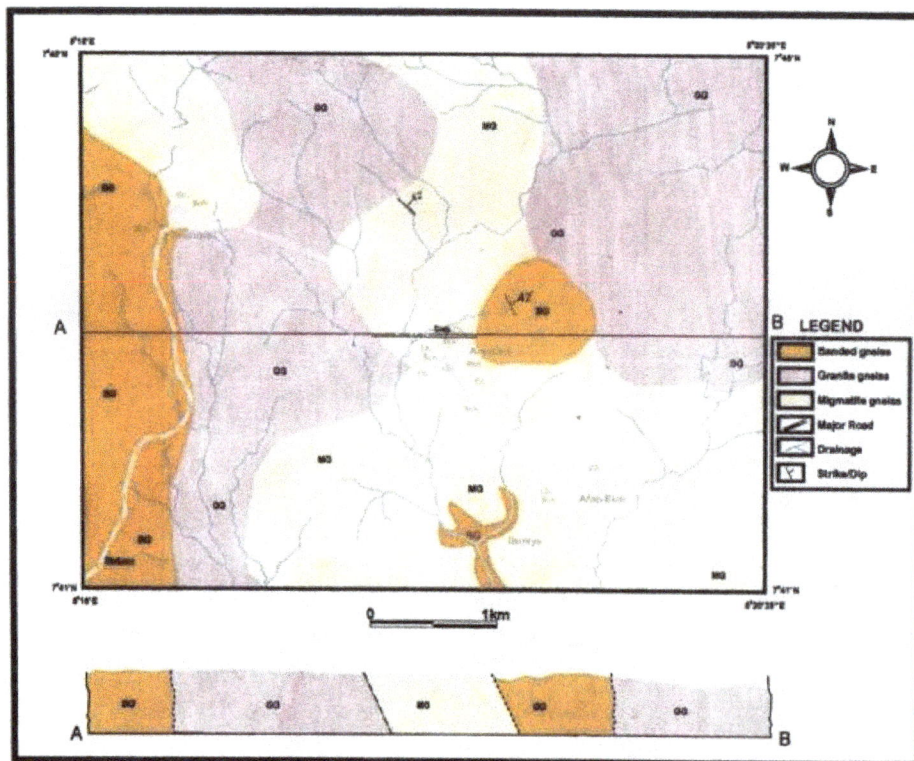

Figure 7: Geologic and cross-section map of the study area.

been studied. Through the results acquired from this analysis, it can be concluded that some of the trace and rare earth elements detected are radioactive in nature (Table 4). Petrological and geochemical data interpreted strongly indicated a sedimentary protolith, probably greywacke for the migmatite gneisses, granite gneiss and banded gneisses of the Iworoko, Are and Afao Ektit area. The greywacke sediments may have been derived from continental source. It is recommended that the inhabitants should be quarantined against possible exposure to radioactive elements which could cause infant mortality, mutations, and carcinogenic radiations emanating from the area. Furthermore, the level of radioactivity of the migmatite gneiss in the area should be ascertained by carrying out radioactive survey program of the entire area.

Aknowledgement

I hereby acknowledge the efforts of Akinniyi Richard Akinwale, who assisted me during the geologic field mapping exercise of this research and whose immense contribution to this work is highly appreciated. Thank you very much.

Refreneces

1. Mumbfu EM, Forba NMT, Cheo ES (2014) Geochemical dispersion of gold in stream sediment in Paleoproterozoic Nyong Series, Southern Cameroon. Science Research 2: 155-165.

2. Akintola AI, Ikhane PR, Bankole SI, Adeokurolere DN (2013) Petrography and Stream sediment geochemistry of Ede and its environs, southwestern Nigeria. International Research Journal of Geology and Mining 3: 2276-6618.

3. Emmanuel A, Samuel A, Folami L, Bankole D, Taye A, et al. (2011) Applications of the principal component analysis on geochemical data; A case study in the basement complex of southern Ilesa area, Nigeria. Arab Journal of Geoscience 4: 239-247.

4. Odeyemi IB, Anifowose AYB, Asiwaju-Bello YA (1999) Multi-technique graphical analysis of fractures from remotely-sensed images of basement region of Nigeria. Journal of Mining and Geology 35: 9-21.

5. Anifowose AYB, Borode AM (2007) Photogeological study of the fold structure in Okemesi area, southwestern Nigeria. Journal of Mining and Geology 43: 125-130.

6. Black R (1980) Precambrian of West African Episodes. International Journal of Geosciences 4: 3-8.

7. Dada SS (2006) Crust forming ages and Proterozoic crustal evolution in Nigeria, a reappraisal of current interpretations. *Precambrian Research* 87: 65-74.

8. Gandu AH, Ojo SB, Ajakaye DE (1986) A gravity study of the Precambrian in the Malufashi area of Kaduna State, Nigeria. Tectonophysics 126: 181-194.

9. Anifowose AYB (2004) An integrated Remote Sensing Analysis of the Ifewara-Zungeru Megalinear in Nigeria. Ph.D Thesis, Federal University of Technology, Akure, Nigeria.

10. Garrels RM, Mackenzie FT (1971) Evolution of sedimentary rocks. W.W. Norton and Company, Incorporated New York.

11. Rahaman MA, Ocan O (1978) On relationships in the Precambrian Migmatite-gneiss of Nigeria. Niger Journal of Mineral Geology 15: 23-32.

12. Brown EH, Babcock RS, Clark MD, Livingstone DE (1979) Geology of the older Precambrian rocks of the Grand Canyon 1. Petrology and Structure of the Vishnu Complex. Precambrian Research 8: 219-241.

13. Pearce M, Gorman BE, Birkett TC (1975) The relationship between major element chemistry and tectonic environment of basic and intermediate volcanic rocks. Journal of Earth Science letters 36: 121-132.

2

Spatial Confirmation of Major Lineament and Groundwater Exploration using Ground Magnetic Method near Mecheri Village, Salem District of Tamil Nadu, India

Muthamilselvan A*, Srimadhi K, Nandhini R, Pavithra P, Balamurugan T and Vasuki V

Centre for Remote Sensing, Bharathidasan University, Trichy, India

Abstract

Geophysical methods are widely used in various applications, especially to know about the subsurface features of the Earth. In the present study, the ground magnetic method has been done to map the NE-SW trending major fault traversing Mecheri block, Salem district for the spatial correlation study and also to locate possible groundwater potential zones with in the study area. The instrument used is Proton Precession Magnetometer-600 which produces a weak magnetic field that is picked up by an inductor, amplified electronically, and fed to a digital frequency counter whose output is typically scaled and displayed directly as field strength. The survey is done across the major fault marked by the GSI geologist with 100 m spacing in profile direction and 30 m sample spacing along the profile and for each profile line, 14 to 15 samples have been collected along with coordinates, time and magnetic value with all the necessary precautions. Then the data is processed and diurnal corrections were made for the interpretation using geophysical software. After the necessary corrections, profiles, contours and maps were generated for quantitative and qualitative analysis which includes magnetic contour map, total magnetic intensity, reduction to pole, analytical signal, upward and downward continuation, horizontal and vertical derivative, and radially average power spectrum. Based on the visual interpretation and interpreter's knowledge, it was identified that the major NE-SW trending magnetic break is present at the southeastern corner of the map which is spatially correlated with the major fault marked by the GSI geologist. Apart from that there are two magnetic highs were notice in the southwestern part of the map which is mainly due to presence of isolated granite and syenite bodies. A small another magnetic break in the E-W direction has also been noticed. Intersection point of the NE-SW and NW-SE fault zones are favorable zone for groundwater potential zone. Other than this, the magnetic anomaly depth has been inferred from the radially average spectrum method shows the anomaly at 11 m, 21 m and 51 m depth.

Keywords: Exploration geophysics; Lithology; Subsurface geology; Geomorphology

Introduction

Geophysics is the branch of Earth science that uses physical measurement and mathematical models to develop and understanding of Earth interior. Exploration geophysics can be used to directly to detect the target style of mineralization via measuring its physical properties directly. The exploration geophysics use physical method at the surface of the Earth to measure physical properties of the subsurface along with the anomalies in those properties.

A magnetic survey is a powerful tool for delineating the lithology and subsurface structure of buried basement terrain. Such a survey maps the variation of the geomagnetic field, which occurs due to changes in the percentage of magnetite in the rock. It reflects the variations in the distribution and type of magnetic minerals below the Earth's surface [1]. Magnetic minerals can be mapped from the surface to greater depths in the rock property, of the rock. Sedimentary formations are usually nonmagnetic and, consequently, have little effect, whereas mafic and ultramafic igneous rocks exhibit a greater variation and are useful in exploring the bedrock geology concealed below cover formations [1].

Magnetic survey used to investigate subsurface geology on the basis of magnetic anomalies resulted from magnetic properties of the underlying rocks [2]. It is also used to map lithological boundaries between magnetically contrasting litho units including faults [3]. A magnetic anomaly originates as a result of magnetization contrast between rocks which shows different magnetic properties. Most rocks contain some magnetite, hematite or other magnetic material which will produce disturbances in the local magnetic field. Because of this, most soils and man - made objects that contain nickel or iron have magnetic properties detectable by a sensitive magnetometer, because they create local or regional magnetic anomalies in the earth's main field. Anomalies are revealed by systematic measurement of the variation in magnetic field strength with position. Folami and Ojo [4] expressed their opinion about magnetic methods which are sensitive to susceptibility within the subsurface geology and so ideal for exploring in the basement complex regions which make this method suitable for this research work. Total magnetic intensity which traverses over an area can aid understanding geological information and, in the case of iron ore deposits, can indicate very clearly their locations. The main objective of this study is to delineate the trend of major fault marked by the geologist from GSI for spatial confirmation, to delineate subsurface structures and to study about the groundwater potentialities of the study area. To attain these goals, a ground

*Corresponding author: Muthamilselvan A, Assistant Professor, Centre for Remote Sensing, Bharathidasan University, Trichy-23, India
E-mail: thamil1978@gmail.com

magnetic survey area was conducted Mecheri village with the help of PPM-600 magnetometer. The magnetic survey data were subjected to a quantitative interpretation that involved some geophysical processing and interpretational techniques, which include (1) reducing the total intensity magnetic data to the north magnetic pole; (2) isolating the magnetic data into their residual, and regional, components, using Fast Fourier Transformation (FFT) techniques; and (3) power spectrum.

The magnetic method involves the study of lateral changes in magnetic field caused by variations in magnetization due to differing magnetic properties of rocks. This method is relevant to groundwater exploration in hard rock terrains, faults and basic dyke intrusive associated with prominent magnetic signatures [5-8].

Geologically the study area comprises the various geological formations from Archean to early paleozoic period. The most of the study area is covered by hornblende biotite gneiss followed by syenite and granite. Small amount of area is covered by dunite, pyroxene, calc. granulite, meta-limestone, fuchsite quartzite and amphibolites [9]. The amphibolite is found in the areas near Doramangalam, Avadathur and Jalakandapuram. The Fuchsite quartzite is found near Ramaswamymalai. Kumaramangalam area shows the occurrence of Calc granulite and Limestone. The Pakkanadu area shows occurrence of Dunite, Peridotite and Pyroxene [10]. The Granite is formed between the period of late Proterozoic and early Palaeozoic as intrusive which is noticed near Kumarapalayam, Kaveripatti, Thevur, Mothaiyanoor and Devanagoundanur. The Syenite occurs along koratti shear zone (Pakkanadu, Pulampatti, Vanavasi and Koonandiyur) as a dyke [11]. However, the Mecheri area comprises mainly Hornblende biotite gneiss and at places Calc granulite with lot of Pyrite specks and Malachite staining were noticed.

Materials and Methods

Spatial confirmation and depth extension survey against the major fault marked by the GSI scientist has been carried with the help of Magnetometer near Mecheri (Figure 1). Magnetic survey in Mecheri block has been carried out in the NW-SE profile direction with 100 m spacing and 30 m sample spacing along the profile approximately [12]. Traverse was planned perpendicular direction to the major fault which is orienting in the NE-SW direction. There are five profiles covered over an area of 0.21 sq km has been taken up for the present study. For each profile line, 14 to 15 samples have been collected along with co-ordinates, time and magnetic value with all necessary precautions. Then, necessary corrections were made and prepared various images and contours for qualitative and quantitative study. Further, these images visually interpreted and digitized available magnetic breaks. Intersection of magnetic breaks and faults are marked for groundwater potential zones [13]. The magnetic breaks further spatially correlated with the existing lineament (major fault) for validation and confirmation. Power spectrum is also generated to know the depth of the magnetic anomalies noticed in this study area.

Instrument used

The instrument used in this study is Proton Precession Magnetometer-600. The proton magnetometer, also known as the proton precession magnetometer (PPM), uses the principle of Earth's field nuclear magnetic resonance (EFNMR) to measure very small variations in the Earth's magnetic field, allowing ferrous objects on land and at sea to be detected [14]. The principle of the instrument is that, a direct current flowing in a solenoid creates a strong magnetic field around a hydrogen-rich fluid (kerosene and water are popular, and even water can be used), causing some of the protons to align themselves with that field. The current is then interrupted, and as protons realign themselves with the ambient magnetic field, they process at a frequency that is directly proportional to the magnetic field [15]. This produces a weak rotating magnetic field that is picked up by a (sometimes separate) inductor, amplified electronically, and fed to a digital frequency counter whose output is typically scaled and displayed directly as field strength or output as digital data.

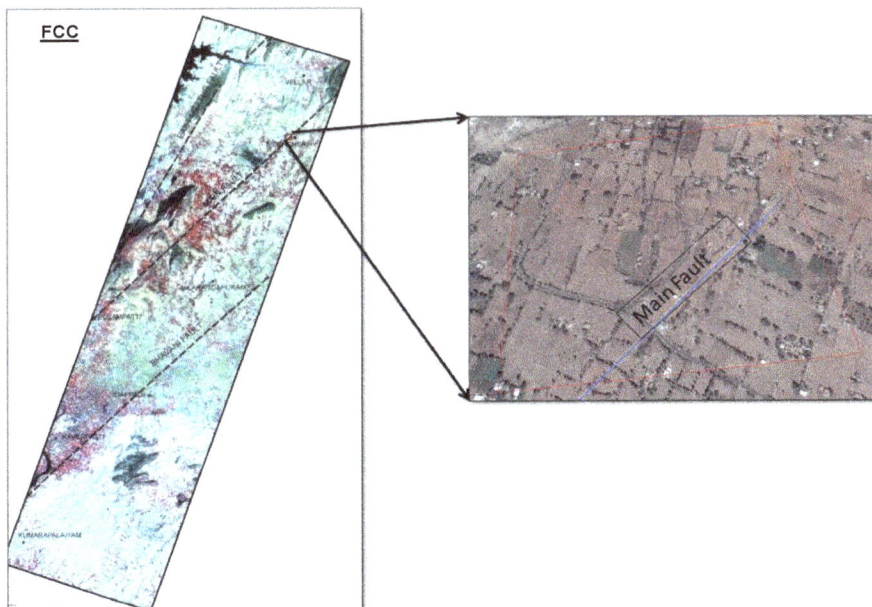

Figure 1: Spatial confirmation and depth extension survey with the help of Magnetometer near Mecheri.

Data preprocessing and analysis

The magnetic data collected in the field is raw data which needs to be corrected to simplify the interpretation. The following corrections are done to the collected data using Oasis Motanj software.

Diurnal correction

Diurnal changes are the daily changes in the magnetic field are related to the rotation of the earth; the changes may range from few Gammas to 100 Gammas or more. A continuous record of the changes in diurnal variations may be obtained using magnetic recording instrument [16]. If that is not available, the field magnetometer may be used to read the variations by repeating the observations over a base station several time during the course of day's survey.

The diurnal changes have been done by taking the readings one hour once at the base station and correcting the field magnetic values by subtracting the base station magnetic value [17]. After the corrections, magnetic profiles were generated (Figure 2).

Total magnetic intensity

Total intensity is the measurement from the magnetometer after a model of earth's normal magnetic field is removed. It is generally a reflection of the average magnetic susceptibility of broad, large-scale geologic features.

The values that are corrected are plotted as contour lines with an interval of 10 m. The values vary from -38 to 140 gammas [18]. Magnetic gradient has been noticed in the SW part of the Mecheri block where the main fault traversed in the NE-SW direction. The magnetic gradient trend has spatial correlation with the main fault is confirmed. Apart from this, another magnetic break has been recorded in the NW-SE direction (Figure 3). The magnetic high noticed in the SW corner of the images is due to the intrusion of syenite body which is confirmed during the field work.

Reduction to pole

Reduction to pole uses mathematical filtering methodology to calculate the magnetic anomaly that would be observed at pole i = ± 90°. It is Process by which effects of inclination (9.1) and declination (-1.5) are removed from the data [19]. The data are mathematically

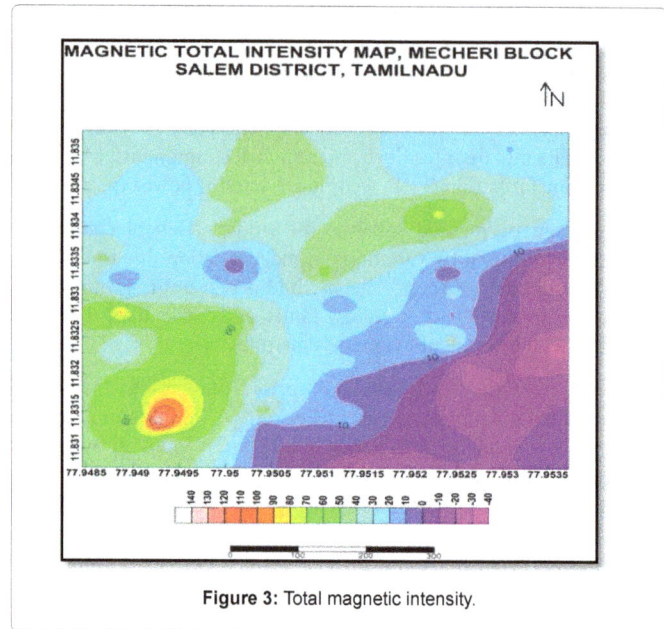

Figure 2: Profile lines.

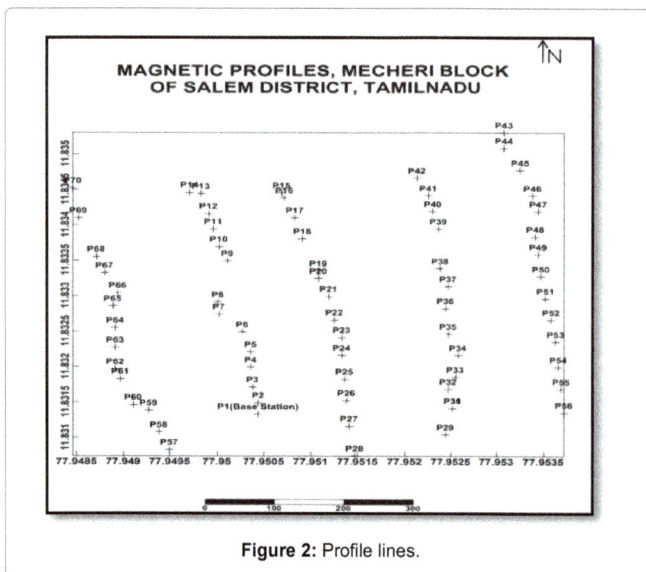

Figure 3: Total magnetic intensity.

transformed to measurements over the same geologic structure, but at the magnetic pole where the inducing field is vertical. So, reduction to pole has been done which shows the exact position of magnetic anomalies in the survey area. The RTP data for our study area shows subsurface occurrence of dykes that is exposed in the surface in E-W direction (Figure 4).

Analytical signal

The analytic signal or total gradient is formed through the combination of the horizontal and vertical gradients of the magnetic anomaly. The analytic signal has a form over causative body that depends on the locations of the body (horizontal coordinate and depth) but not on its magnetization direction Figure 5. The analytical signal image shows magnetic high in S-W part and low in eastern parts of the block.

Upward and downward continuation

Upward continuation predicts the magnetic field at a higher elevation and emphasizes the longer spatial wavelengths. The upward continuation has been done for 100 m (Figure 6) and 200 m (Figure 7) in both the continuation the anomaly is seen prominent i.e. it is not diluted. The upward continuation output shows magnetic anomaly in south and north sides with a break in between [20]. Downward continuation is a mathematical procedure that computes magnetic field at a lower level. This process will emphasize shorter wavelengths, but can be unstable and produce artifacts. The downward continuation has been done for 25 m (Figure 8). The downward continuation output shows magnetic anomaly in parts of south-west and north-east with breaks in W-E and NE-SW.

Horizontal and vertical derivative

Second order vertical derivative is nothing but the change in magnetic intensity in vertical direction. Vertical derivative magnetic map (Figure 9) shows low value in south and east parts and at the center in W-E direction and high values in south-west and north parts of block. Horizontal derivative magnetic map (Figure 10) indicates magnetic anomaly in south and in center part along the break.

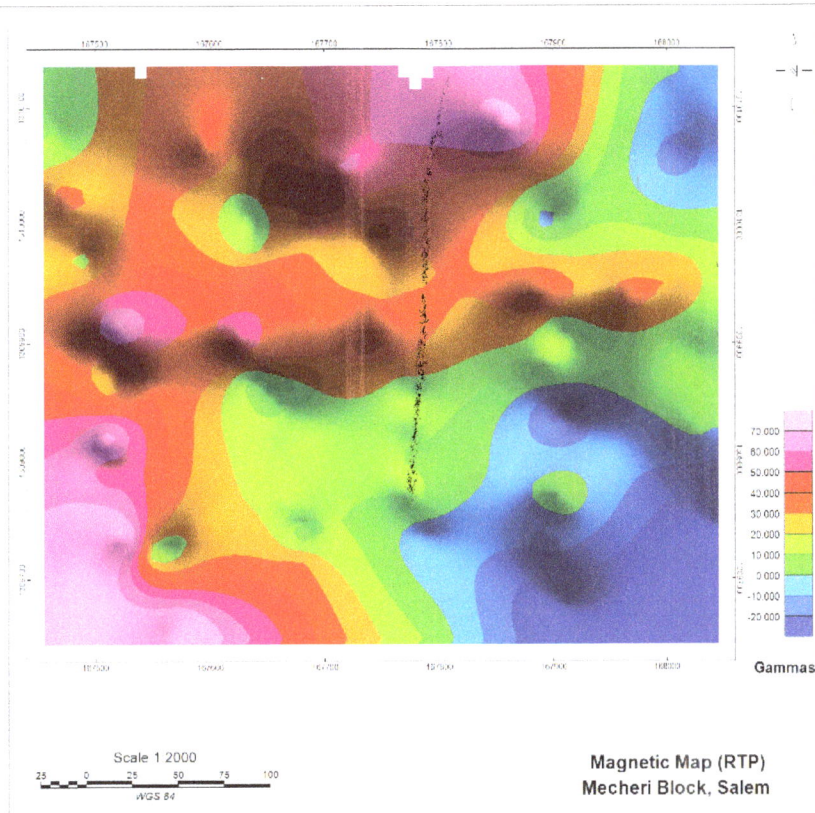

Figure 4: Reduction to pole image.

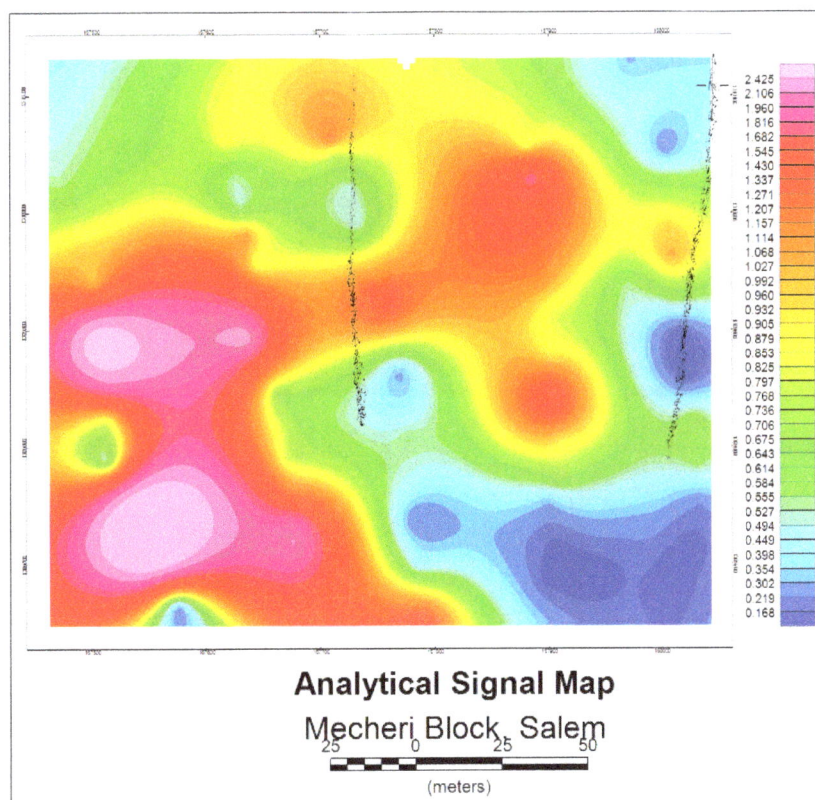

Figure 5: Analytical signal map.

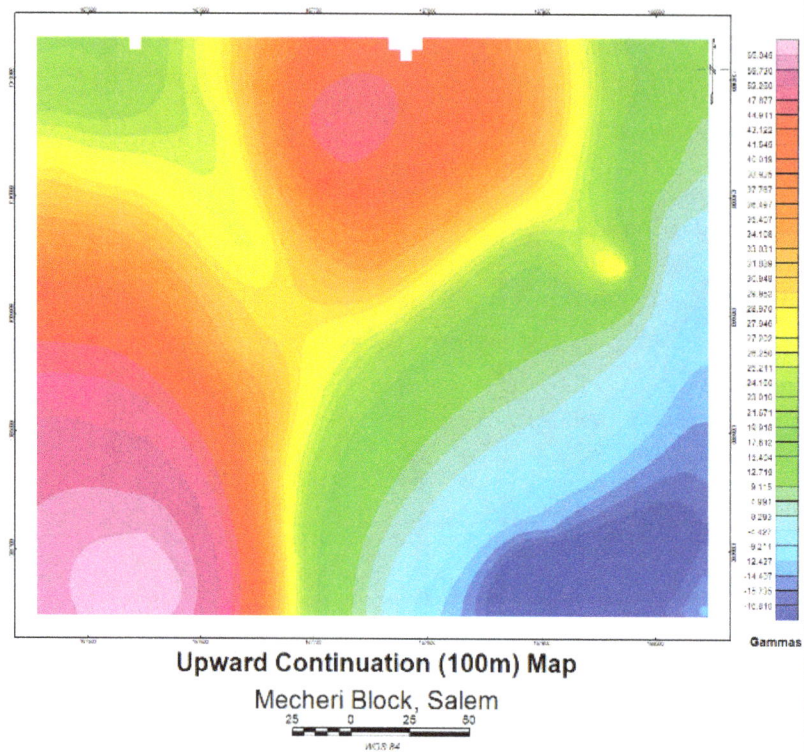

Figure 6: Upward continuation (100 m) map.

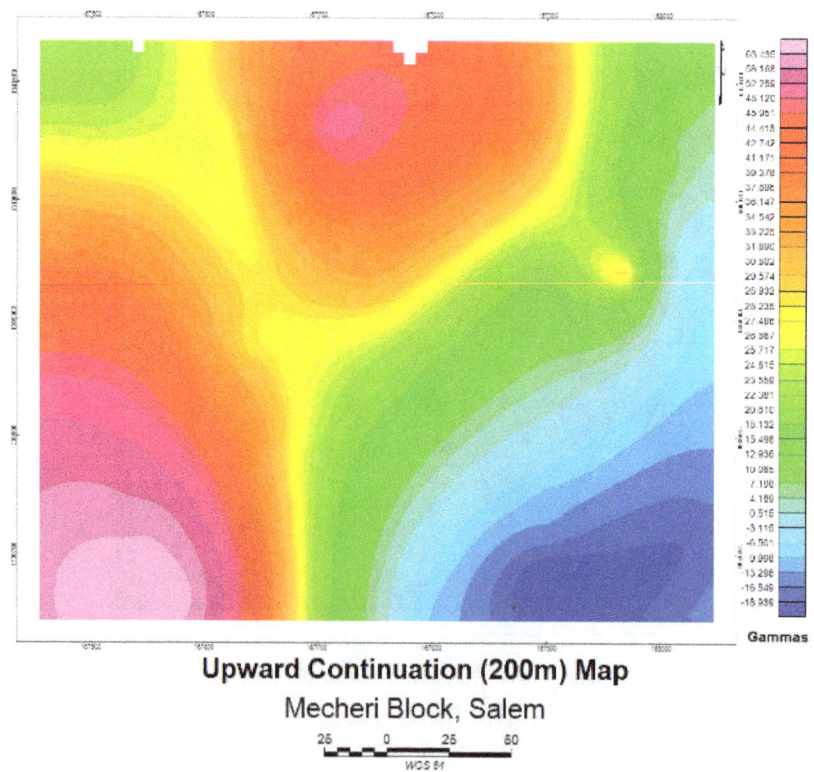

Figure 7: Upward continuation (200 m) map.

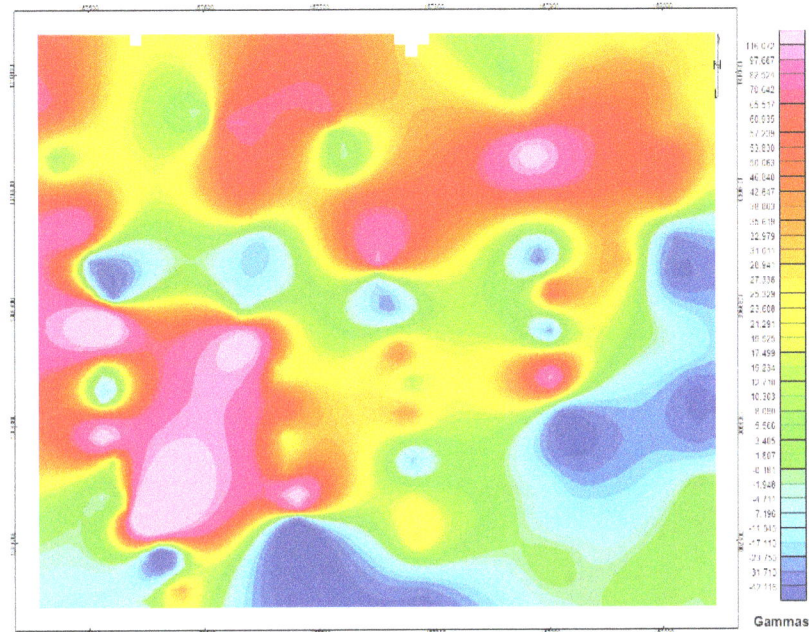

Figure 8: Downward continuation (25 m) map.

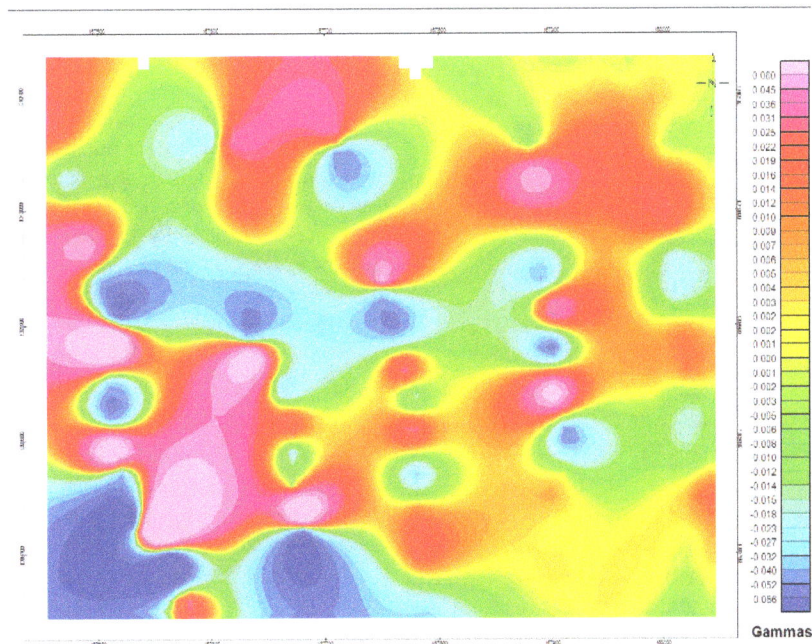

Figure 9: Vertical derivative map.

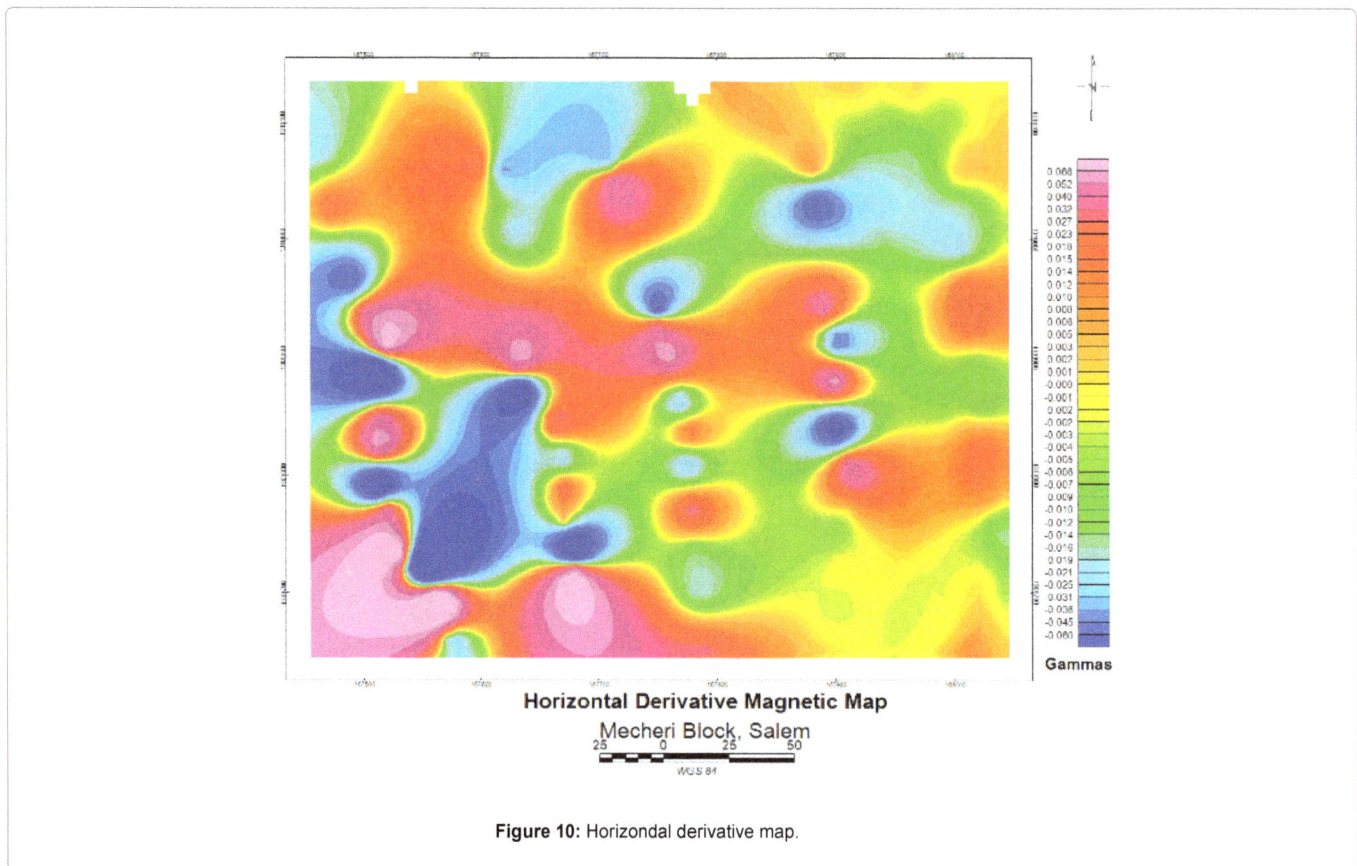

Horizontal Derivative Magnetic Map
Mecheri Block, Salem

Figure 10: Horizondal derivative map.

Radially average power spectrum

It is a technique done to approximately locate the depth of magnetic anomaly. Based on this, the slope is determined and from the slope the depth is estimated using the formula $\left[h = \frac{s}{4\pi}\right]$. For the study area this spectrum shows anomaly at 11 m, 21 m and 51 m (Figure 11).

Groundwater potential zone

Magnetic method is one of the best and simple methods for groundwater exploration related studies in the hard rock terrain. Mecheri area comprises two major magnetic breaks which are trending in NE-SW and NW-SE directions (Figure 12). The NE-SW break spatially well correlated with the major fault (lineament) marked by the GSI scientists. Intersection of these magnetic breaks was noticed about 500 m west of Mecheri village (Figure 13). Intersection of lineaments is good for groundwater exploration, because, in that particular zone structural porosity and permeability will be very high which can act as hard rock aquifer for groundwater.

During the data collection, it was also noticed that the area falls in the northwestern side of the NE-SW magnetic breaks are having good amount of groundwater and lot of agricultural activities going on when compared to the northeastern side of the magnetic breaks [21]. In addition, Stanley reservoir is located about 9 Km along the direction of NW-SE magnetic breaks which is not exposed on the surface. This lineament must be the reason for good groundwater potential in the western part of the major fault.

The area falls on the eastern side of the major fault can be artificially recharged through the direct and indirect methods of artificial recharge. In this regards, further detailed study about lithology, geomorphology, land use land cover soil type etc. are required to suggest suitable method.

Conclusion

The present study is taken up for the spatial confirmation of lineament and groundwater potential in Mecheri block using ground magnetic survey. Magnetic data has been collected across the main fault near Mecheri block recorded by GSI scientists. Data has been corrected and processed to simplify the interpretation. Total Magnetic Intensity map shows the gradient of magnetic values that leads to identification of two breaks in NE-SW and NW-SE directions. The RTP images shows the magnetic anomaly in NNW-SSE direction of the block which is not exposed in the surface. Analytical signal image for the block shows high magnetic value in southern and western parts and low magnetic values in the east.

The upward continuation has been done at 100 m and 200 m. In both the images the breaks are seen prominent in NW-SE and NE-SW direction indicates its depth persistence. The downward continuation has been done for 25 m. The magnetic high values are observed on SW and NE sides of the magnetic breaks. The second order vertical and horizontal derivative has been done which shows that the magnetic low values in south and east parts and magnetic high values in north and SW parts which may be due to presence of syenite or granite dykes in the subsurface. The radially average power spectrum shows the anomaly may occur at 11 m, 21 m and 51 m depth in Mecheri block.

The study carried out in this area so far suggest that the magnetic methods are best suitable method for locating subsurface fault, fractures and shear. This method is also suitable for spatial confirmation and

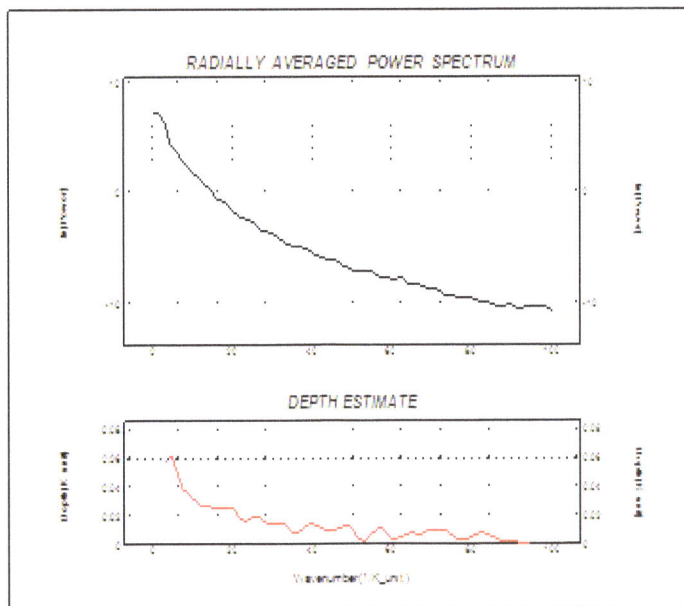

Figure 11: Radially average spectrum.

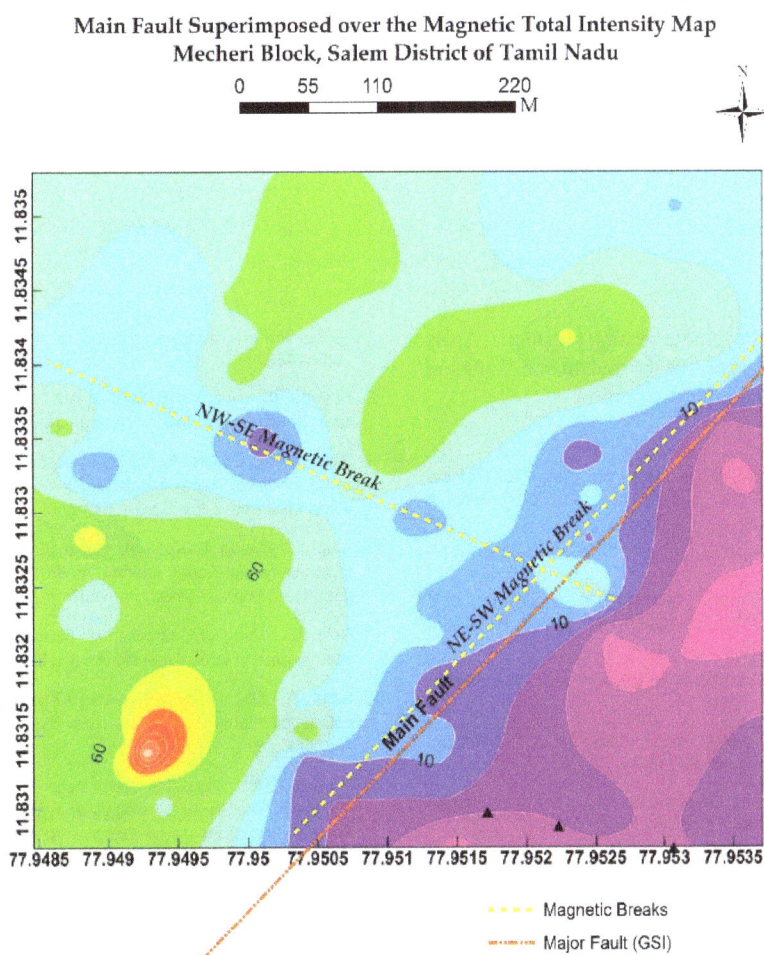

Figure 12: Major Fault (GSI) superimposed over total magnetic map.

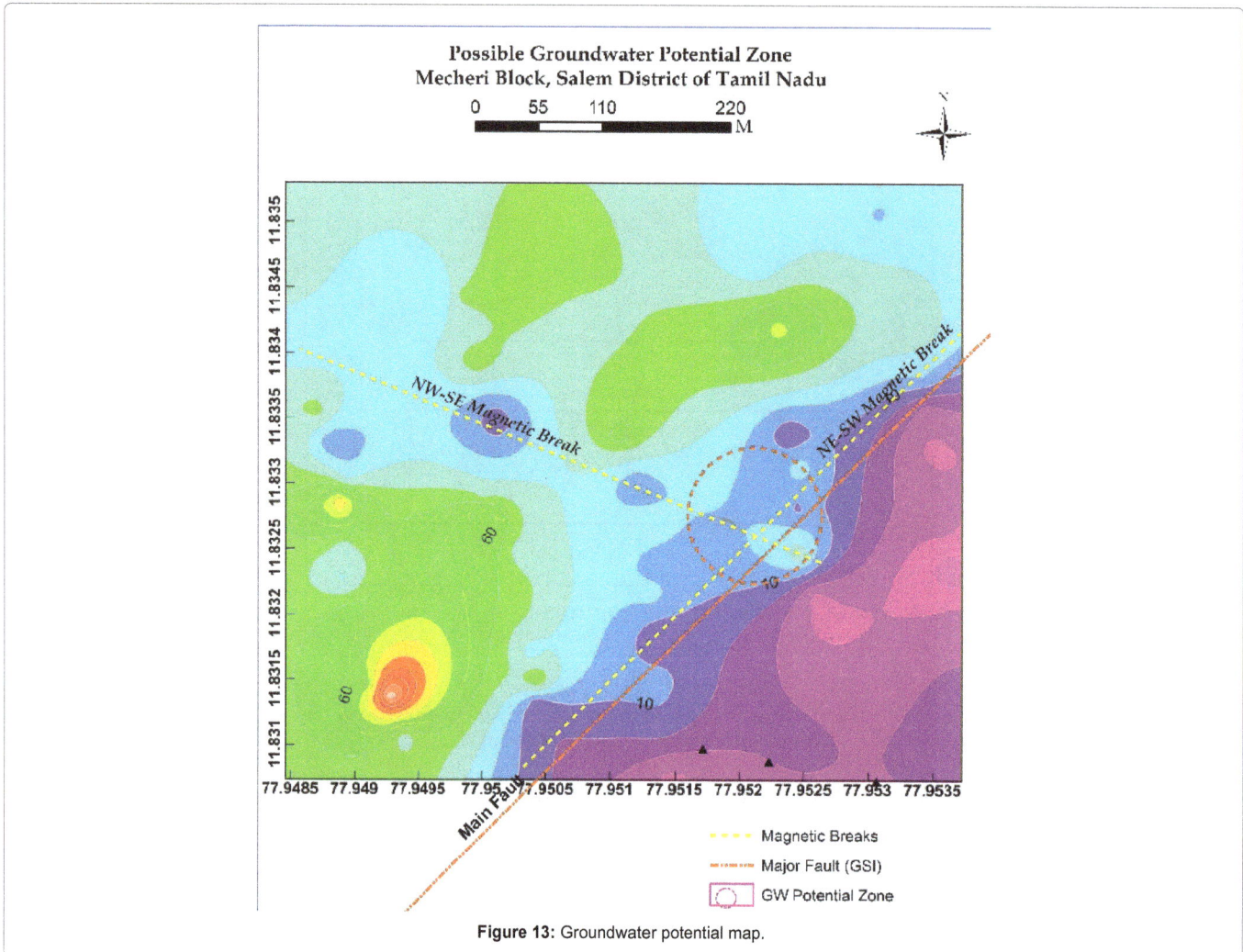

Figure 13: Groundwater potential map.

ground validation study done by the remote sensing technique. Apart from that the intersection of magnetic breaks and major faults are assumed to be good zone for groundwater exploration.

Acknowledgement

I am sincerely thankful to all the faculty members from Centre for Remote Sensing, Bharathidasan University for their support to my various research activities.

References

1. Mekonnen TK (2004) Interpretation and geodatabase of Dykes using aeromagnetic data of Zimbabwe and Mozambique. M.Sc. Thesis, ITC, Delft, the Netherlands p: 80.

2. Kearey P, Brooks M, Hill I (2002) An Introduction to geophysical exploration, Third Edition. Blackwell Science Ltd p168.

3. Telford WM, Geldart LP, Sheriff RE (2001) Applied Geophysics, 3rd Edition, Cambridge University Press, Cambridge p632-638.

4. Folami SL, Ojo JS (1991) Gravity and magnetic investigations over marble deposits in the Igara area Bendel State. J Min Geol 27: 49-54.

5. Birch FS (1984) Bedrock depth estimates from ground magnetometer profiles. Groundwat 22: 427-432.

6. Chandra PC, Reddy PHP, Singh SC (1994) Geophysical studies for groundwater exploration in Kasai and Subarnarekha River Basins UNDP project CGWB, Min. of Water Resources, Govt of India. Technical report.

7. Kumar D, Murthy NSK, Nayak GK, Ahmed S (2006) Utility of magnetic data in

delineation of groundwater potential zones in hard rock terrain. Curr Sci 91: 1456-1458.

8. Kayode JS, Nyabese P, Adelusi AO (2010) Ground magnetic study of ilesa east, Southwestern Nigeria. African J Environ Sci Technol 4: 122-131.

9. Batayneh A, Ghrefat H (2012) Lineament characterization and their tectonic significance using gravity data and field studies in the Al-JufrArea, Southeastern Jordan Plateau. J Ear Sci 23: 873-880.

10. Alaa A, Masoud A, Katsuaki Koike (2011) Auto-detection and integration of tectonically significant lineaments from SRTM DEM and remotely-sensed geophysical data. J Photogramm Rem Sens 818-832.

11. Reid AB, Allsop JM, Granser H, Millettg AJ, Somerton IW (1990) Magnetic interpretation in three dimensions using Euler deconvolution. Geophy 55: 80-91.

12. Salem A, Williams Sand, Fairhead JD (2007) Tilt-depth method: A simple depth estimation method using first-order magnetic derivatives. The Leading Edge 26: 1502-1505.

13. Oruç B, Selim HH (2011) Interpretation of magnetic data in the Sinop area of Mid Black Sea, Turkey, using tilt derivative, Euler deconvolution and Discrete Wavelet Transform. J Appl Geophy 74:194-204.

14. Chernicoff CJ, Richards JP, Zappetini EO (2002) Crustal lineament control on magmatism and mineralization in Northwestern Argentina: Geological, Geophysical. Ore Geol Rev 21: 127-155.

15. Radhakrishna Murthy IV, Bangarubabu S (2009) Magnetic anomalies across Bastar craton and Pranhita-Godavari basin in south of central India. J Ear Sys Sci 118: 81-87.

16. Koike K, Nagano S, Ohmi M (1995) Lineament analysis of satellite images using a segment tracing algorithm (STA). Comp Geosci 21: 1091-1104.

17. Chandra S, Krishnamurthy NS, Ahmed S (2006) Integrated studies for characterization of lineaments used to locate groundwater potential zones in a hard rock. Hydrogeol J 14: 767-776.

18. Shu-Kun H, Coppens D, Chuen-Tien S (1998) Depth to magnetic source using the generalized analytic signal. Geophy 63: 1947-1957.

19. Hildenbrand TG, Berger, Jachens RC, Ludington S (2000) Regional crustal structures and their relationship to the distribution of ore deposits in the Western United States, Based on magnetic and gravity data. Geol 95: 1583-1603.

20. Venkatesan N, Babu NK, Sukumar M (2014) A review of various lineament detection techniques for high resolution Satellite Images. Int J Adv Res Comp Sci Software Eng 4: 72-78.

21. Roest WR, Verhoefs J, Pilkington M (1992) Magnetic interpretation using the 3-D analytic signal. Geophy 57: 116-125.

3

Performing Spiking and Predictive Deconvolution on 2D Land Data (PSTM)

Mohamed Mhmod*, Liu Hai Yan, Liu Cai and Feng Xuan

College of Geo-Exploration Science and Technology, Jilin University, Changchun 130026, China

Abstract

In this paper we are performing spiking and predictive deconvolution on land 2D data, final (PSTM) data. For spiking deconvolution we are going to test the effect of operator length (operator length with n=operator length (where (n=240, 128, 40, 10) ms), and effect of percent prewhitening (0% and 1%) for each value of n. While for predictive deconvolution we will test effect of operator length ((operator length with n=operator length (where (n=240, 128, 40, 10)) ms) and lag (α=0 ms, α=1 ms, α=2 ms) for each value of n. The data used in this paper 2d data (PSTM), first we will apply spiking deconvolution on our data, for spiking deconvolution, the Deconvolution which in this case the desired output is zero-lag spike. While for predictive deconvolution, the Deconvolution which in this case the desired output is a lagged version of the input. The later lags are used as the cross-correlation of the input and desired output. The standard equations are solved for the predictive operator. In final comparison for each test were made on our data. The effect of parameters on spiking deconvolution was also published.

Keywords: Spiking deconvolution; Predictive deconvolution; Surface consistent deconvolution; Prewhitening

Introduction

The definition of deconvolution is a filtering process that removes a wavelet from the recorded seismic trace [1] and is this done by reversing the process of convolution [2]. The commonest ways that perform deconvolution, by designing a Wiener filter to transform one wavelet into another wavelet in a least-squares sense [3]. It is often applied at least once to marine seismic data. The attenuation of short-period multiples (most notably reverberations from relatively flat, shallow water-bottom) can be achieved with predictive deconvolution [4]. The periodicity of the multiples is exploited to design an operator, which identifies and removes the predictable part of the wavelet, leaving only its non-predictable part (signal) [5].

Algorithm principle

The spiking deconvolution in seismic data processing is routinely applied to compress the source wavelet included in the seismic traces to improve temporal resolution. The general form of the matrix equation for a filter of length n is represented in equation (1), [6]:

$$\begin{pmatrix} r_0 & r_1 & r_2 \dots r_{n-1} \\ r_1 & r_0 & r_1 \dots r_{n-2} \\ r_2 & r_1 & r_0 \dots r_{n-3} \\ \vdots & \vdots & \vdots & \ddots & \vdots \\ r_{n-1} & r_{n-2} & r_{n-3} \dots r_0 \end{pmatrix} \begin{pmatrix} a_0 \\ a_1 \\ a_2 \\ \vdots \\ a_{n-1} \end{pmatrix} = \begin{pmatrix} g_0 \\ g_1 \\ g_2 \\ \vdots \\ g_{n-1} \end{pmatrix} \qquad (1)$$

Here, r_i, a_i, and g_i, i=0,1,2,.....n-1 are the autocorrelation lags of the input wavelet, Winer coefficients, and the Crosscorrelation lags of the desired output with the input wavelet respectively.

If the desired output is zero delay spike, it is call spiking deconvolution (equation 2):

$$\begin{pmatrix} r_0 & r_1 & r_2 \dots r_{n-1} \\ r_1 & r_0 & r_1 \dots r_{n-2} \\ r_2 & r_1 & r_0 \dots r_{n-3} \\ \vdots & \vdots & \vdots & \ddots & \vdots \\ r_{n-1} & r_{n-2} & r_{n-3} \dots r_0 \end{pmatrix} \begin{pmatrix} a_0 \\ a_1 \\ a_2 \\ \vdots \\ a_{n-1} \end{pmatrix} = \begin{pmatrix} 1 \\ 0 \\ 0 \\ \vdots \\ 0 \end{pmatrix} \qquad (2)$$

The Equation (2) was scaled by $(1/x_0)$. The least squares inverse

filter has the same form as the matrix in equation (2). Therefore, spiking deconvolution is mathematically identical to least squares inverse filter. A distinction, however, is made in practice between the two types of filtering. The autocorrelation matrix on the left side of equation (2) is computed from the input seismogram, in the case of spiking deconvolution (statistical deconvolution), whereas it is computed directly from the known source wavelet in case of least squares inverse filtering. If the input wavelet is not a minimum phase, spiking deconvolution cannot convert it to a perfect zero-lag spike. Although the amplitude spectrum is virtually flat, the phase spectrum of the output is not a minimum phase. The spiking deconvolution operator is the inverse of the minimum-phase equivalent of the input wavelet. This wavelet may or may not be minimum phase.

There is always noise in the seismogram and its additive in both time and frequency domain. An artificial level of white noise is introduced before deconvolution [7]. This is called prewhitening.

If the percent prewhitening is given by a scalar, $0 \le \varepsilon < 1$, then the normal equations (2) are modified as in equation (3):

$$\begin{pmatrix} \beta r_0 & r_1 & r_2 \dots r_{n-1} \\ r_1 & \beta r_0 & r_1 \dots r_{n-2} \\ r_2 & r_1 & \beta r_0 \dots r_{n-3} \\ \vdots & \vdots & \vdots & \ddots & \vdots \\ r_{n-1} & r_{n-2} & r_{n-3} \dots \beta r_0 \end{pmatrix} \begin{pmatrix} a_0 \\ a_1 \\ a_2 \\ \vdots \\ a_{n-1} \end{pmatrix} = \begin{pmatrix} 1 \\ 0 \\ 0 \\ \vdots \\ 0 \end{pmatrix}, \qquad (3)$$

where $\beta = 1 + \varepsilon$. Adding a constant εr_0 to the zero lag of the autocorrelation function is the same as adding white noise to the spectrum, with its total energy equal to that constant.

---

*Corresponding author: Mohamed Mhmod, College of Geo-Exploration Science and Technology, Jilin University, Changchun 130026, China
E-mail: baveciwan-23@hotmail.com

The predicative deconvolution desired output, a time –advance from of input series suggests a predication processes. Given input x (t), we want to predict its value at some full time (t+α), where α is predication lag [8-12]. Wiener show that the filter used to estimate (x+α) can be computed by using a special form of the matrix equation (4) [6].

$$
\begin{bmatrix}
r_0 & r_1 & r_2 & \cdots & r_{n-1} \\
r_1 & r_0 & r_1 & & r_{n-2} \\
r_2 & r_1 & r_0 & & r_{n-3} \\
, & , & , & & , \\
, & , & , & & , \\
, & , & , & & , \\
r_{n-1} & r_{n-2} & r_{n-3} & & r_0
\end{bmatrix}
\begin{bmatrix}
a_0 \\ a_1 \\ a_2 \\ , \\ , \\ , \\ a_{n-1}
\end{bmatrix}
=
\begin{bmatrix}
g_0 \\ g_1 \\ g_2 \\ , \\ , \\ , \\ g_{n-1}
\end{bmatrix}
\qquad (4)
$$

Where Here r_i, a_i, and g_i, i=0, 1, 2 ….n, n-1 are the autocorrelation lags of the input wavelet, the Wiener filter coefficients, and the cross correlation lags of the desired output .with the input wavelet respectively. Since the desired output x (t+α) is the time-advance version of the input x (t), we need to specialize the right side of equation (4) for the predication problem. Consider a Five-point input time series x (t): (x_0, x_1, x_2, x_3, x_4), and set α=2. The designed may be carried out using equation (5) and applied on input series as shown in Figure 1.

$$
\begin{bmatrix}
r_0 & r_1 & r_2 & \cdots & r_{n-1} \\
r_1 & r_0 & r_1 & & r_{n-2} \\
r_2 & r_1 & r_0 & & r_{n-3} \\
, & , & , & & , \\
, & , & , & & , \\
, & , & , & & , \\
r_{n-1} & r_{n-2} & r_{n-3} & & r_0
\end{bmatrix}
\begin{bmatrix}
a_0 \\ a_1 \\ a_2 \\ , \\ , \\ , \\ a_{n-1}
\end{bmatrix}
=
\begin{bmatrix}
r_\alpha \\ r_{\alpha+1} \\ r_{\alpha+2} \\ , \\ , \\ , \\ r_{\alpha+n-1}
\end{bmatrix}
\qquad (5)
$$

Work method

This Process is executed by following steps; first we are going to apply spiking deconvolution. CMP sort is required. For Design window, is entry trace, operator length (240, 128, 40, 10) ms, the percent prewhitening (0%, 1%) for each value of n. Also using amplitude scaling (Mean scale) with applying signal band pass filter (Low Truncation Frequency 10 HZ, Low Cut Frequency 15 HZ, High Cut Frequency 200, High Truncation Frequency 250 HZ). Figure 2 shows the flow were used. The execution parameters, shot _sequence number (0-1856), receiver _sequence _ number (0-59423), channel _number (1-40), CMP _no (0-7478), inline: 2 Xline: 7479[0-7478], input traces (59424).

The spiking deconvolution algorithm is applied to original data. Figure 1a and the results are shown in Figures 3 and 4. With different

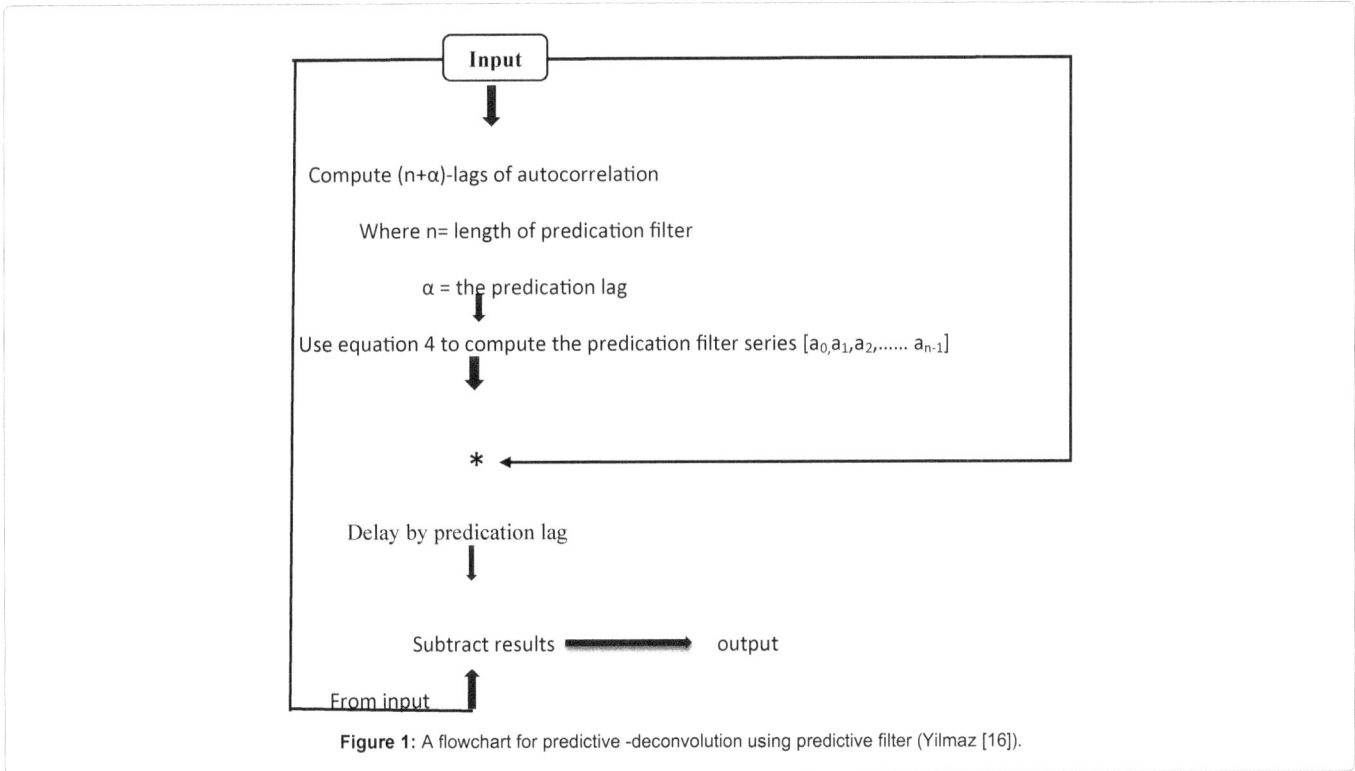

Figure 1: A flowchart for predictive -deconvolution using predictive filter (Yilmaz [16]).

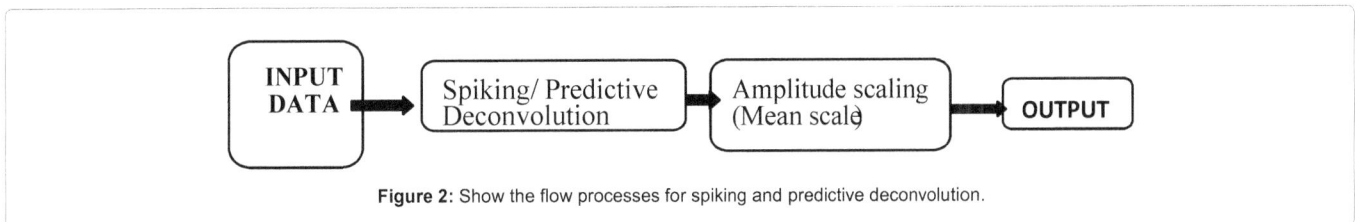

Figure 2: Show the flow processes for spiking and predictive deconvolution.

(a)

(b)

(c)

(d)

(e)

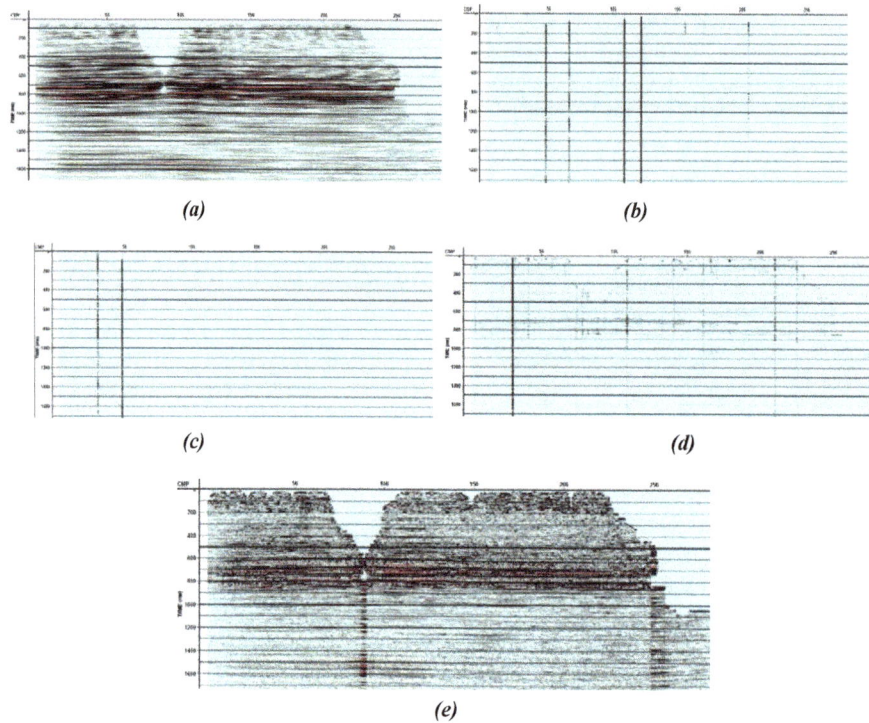

Figure 3: (a) 2d land PSTM Data, CMP sort, before applying spiking-deconvolution, (b) after applying spiking-Deconvolution with Operator-length 240 ms, (c) after applying spiking Deconvolution with Operator-length 128 ms, (d) after applying spiking-Deconvolution with Operator-length 40 ms, (e) after applying spiking-Deconvolution with Operator length 10 ms and the, percent prewhitening for all value of operator-length (0%).

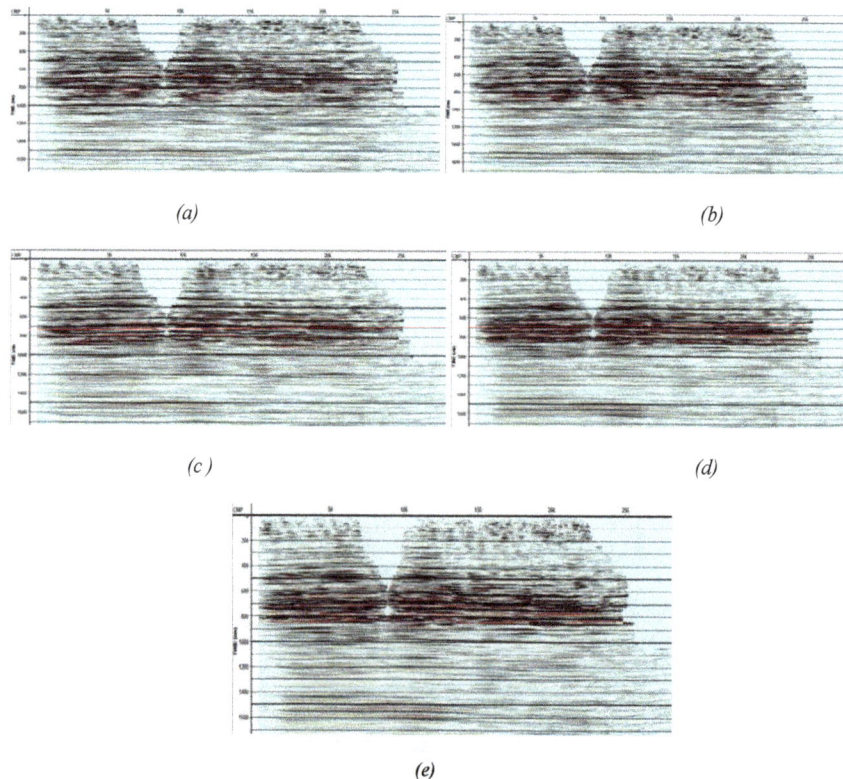

(a)

(b)

(c)

(d)

(e)

Figure 4: (a) 2d land final PSTM Data CMP sort, before applying spiking deconvolution, (b) after applying spiking Deconvolution with Operator length 240 ms, (c) after applying spiking Deconvolution with Operator length 128 ms, (d) after applying spiking Deconvolution with Operator length 40 ms, (e) after applying spiking Deconvolution with Operator length 10 ms and the, percent prewhitening for all value of operator length (1%).

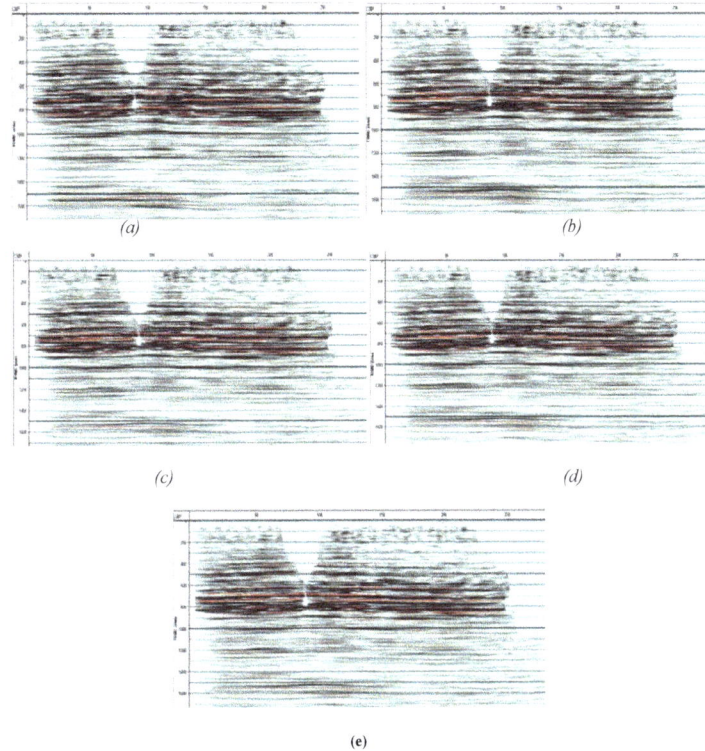

Figure 5: (a) 2d land final PSTM Data CMP sort, before applying predictive- deconvolution, (b) after applying predictive-Deconvolution with Operator-length 240 ms, (c) after applying predictive-Deconvolution with Operator- length 128 ms, (d) after applying predictive-Deconvolution with Operator-length 40 ms, (e) after applying predictive -Deconvolution with Operator-length 10 ms and the, lag for all value of operator-length(α=0 ms), and the percent prewhitening for all value of operator length (1%).

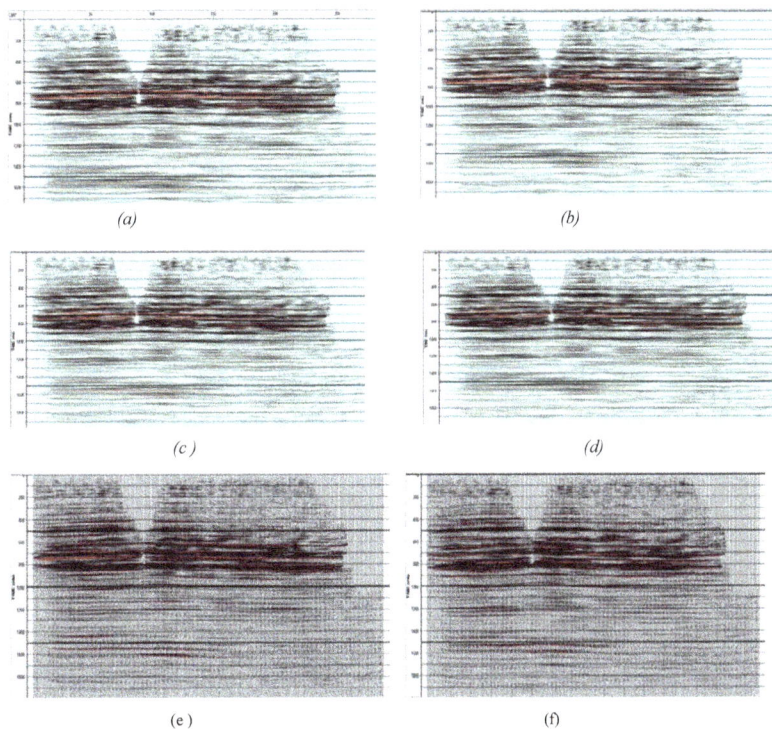

Figure 6: (a) before applying predictive-deconvolution, after applying predictive-Deconvolution with Operator-length 240 ms, (b) after applying predictive-Deconvolution with Operator-length 128 ms, (c) after applying predictive-Deconvolution with Operator-length 40 ms, (d) after applying predictive-Deconvolution with Operator- length 10 ms and the, lag for all value of operator-length (α=1ms), and the percent prewhitening for all value of operator- length (1%). (e) Final pstm, (f) data after applying predictive-Deconvolution.

operator lengths. Wiener deconvolution causes boosting in random noise, which is standard since it generally boosts the amplitudes of high frequency noise in the data [13-15]. For this reason, a conventional Wiener deconvolution process is generally followed by a band-pass filter to suppress this boosted high frequency noise. The band pass filter parameters are low truncation frequency is 10 Hz, low cut frequency is 15 Hz, high cut frequency is 200 Hz, and high truncation frequency is 250 Hz.

Second predictive deconvolution, this step creates a new data set with applied trace by trace predictive, No sort is required. For Design window, is entry trace, operator length (240, 128, 40) ms, the predication lag is unity and equal to ($\alpha=0$ ms, $\alpha=1$ ms, $\alpha=2$ ms) sampling rate, the percent prewhitening (1%). Also using amplitude scaling (Mean scale) with applying signal band pass filter (Low Truncation Frequency 10 Hz, Low Cut Frequency 15 Hz, High Cut Frequency 200 Hz, High Truncation Frequency 250 Hz). Figure 2 show the flow was used.

We also applied predictive -Deconvolution with same parameters and $\alpha=2$ ms, and due gives exactly same results, we won't display it.

Comparing the seismic data before and after deconvolution, we can see that the deconvolved seismic data shows a significant improvement in vertical resolution and enhanced reflections which correspond to the geology. Such high resolution reflection detail is a desirable feature for seismic interpretation [16,17]. This example indicates that suitable parameters can properly enhance the resolution of the seismic data.

There are many papers done this kind of work but each of them just only took one parameter, or used for other kind of geophysics methods such as Multichannel Wiener deconvolution of vertical seismic profiles [18] and predictive deconvolution in seismic data processing in Atala prospect of rivers State, Nigeria [5]. The second paper was used same methods as in this paper but didn't show exactly the effect of predication lag due used only one value.

Conclusions

The spiking-deconvolution operator is the inverse of the (minimum phase) equivalent of the input wavelet. This wavelet may or may not be minimum phase. When the source signature is known a designature process can be applied as an alternative or a complement to this step. In our cause we had a different approach. First we applied a trace by trace spiking deconvolution, and the Deconvolution which In this case desired output (zero-lag spike). We tested operator length with n= operator length (where n=240, 128, 40, 20, 10) ms, and the percent prewhitening (1%). Then we test effect of operator length and lags for predicative deconvolution and we found that for spiking deconvolution when operator length was 10 ms and prewhitening (0%) give perfect results (Figure 3e), and for predicative deconvolution Changing the

predication lag doesn't effect, while applying predictive Deconvolution give better results and no matter the value of α (lag) (Figures 5 and 6). The standard equations are solved for the predictive operator in final comparison for each test was made on our data.

Acknowledgements

We thank CREWES Project from University of Calgary for giving out permission to use their raw data.

References

1. Claerbout JF (1976) Fundamentals of geophysical data processing.

2. Leinbach J (1995) Wiener spiking deconvolution and minimum-phase wavelets: A tutorial. The Leading Edge 14: 189-192.

3. De Monvel JB, S Le Calvez, Ulfendahl M (2001) Image restoration for confocal microscopy: improving the limits of deconvolution, with application to the visualization of the mammalian hearing organ. Biophys J 80: 2455-2470.

4. Schoenberger M (1985) Seismic Deconvolution Workshop Sponsored by the SEG Research Committee, July 18-19, 1984, Vail, Colorado. Geophysics 50: 715-715.

5. Egbai J, Atakpo E (2012) Predictive deconvolution in seismic data processing in Atala prospect of rivers State, Nigeria. Advances in Applied Science Research 3: 520-529.

6. Robinson EA, Treitel S (1980) Geophysical signal analysis, Prentice-Hall New Jersey.

7. White RE (1984) Signal and noise estimation from seismic reflection data using spectral coherence methods. Proceedings of the IEEE 72: 1340-1356.

8. Gibson B, Larner K (1984) Predictive deconvolution and the zero-phase source. Geophysics 49: 379-397.

9. Lines L, Ulrych T (1977) The Old and the New in Seismic Deconvolution and Wavelet Estimation*. Geophysical Prospecting 25: 512-540.

10. Margrave GF, Gibson PC (2005) The Gabor transform, pseudodifferential operators, and seismic deconvolution. Integrated Computer-Aided Engineering 12: 43-55.

11. Margrave GF, Lamoureux MP (2010) Nonstationary predictive deconvolution, GeoCanada.

12. McCann D, Forster A (1990) Reconnaissance geophysical methods in landslide investigations. Engineering Geology 29: 59-78.

13. Mendel JM, Kormylo J (1978) Single-channel white-noise estimators for deconvolution. Geophysics 43: 102-124.

14. Peacock K, Treitel S (1969) Predictive deconvolution: Theory and practice. Geophysics 34: 155-169.

15. Perez MA, Henley DC (2000) Multiple attenuation via predictive deconvolution in the radial domain. CREWES Project 12th Annual Research Report.

16. Yilmaz O (1987) Seismic Data Processing, SEG, Tulsa. Practical 3D Refraction Statics 211.

17. Yilmaz ÖGA, Claerbout JF (1980) Prestack partial migration. Geophysics 45: 1753-1779.

18. Haldorsen JB, Miller DE, Walsh JJ (1994) Multichannel Wiener deconvolution of vertical seismic profiles. Geophysics 59: 1500-1511.

4

Uplift History of Syenite Rocks of the Sushina Hill, Tamar Porapahar Shear Zone (TPSZ), Purulia: Constraints from Fission-track Ages of Two Cogenetic Minerals

Amal Kumar Ghosh*, Virendra Kumar Sharma and Rajeev Kumar Singh

Bhagwant University, Ajmer, Rajasthan-305004, India

Abstract

Fission-track ages of cogenetic minerals namely apatite, and zircon from syenite rocks of the Sushina hill, Tamar Porapahar Shear Zone (TPSZ), coupled with the corresponding closure temperatures of the minerals have been used to reveal the uplift history of syenite rocks. Offset of their FT ages indicates that the samples uplifted at the rate of 9.97 m/Ma during the period 535 Ma-970 Ma.

Keywords: Metamorphic minerals; Cogenetic minerals; Syenite rocks; Uranium samples

Introduction

It is widely accepted that radiometric ages determined on metamorphic minerals from orogenic belts reflect their cooling history rather than their primary crystallization [1-5]. Cooling histories obtained using different radiometric techniques on cogenetic minerals from a single sample often include zircon and apatite fission track ages as low-temperature bounds to a temperature interval of several hundred degrees [6-15].

Fission-track ages were determined on cogenetic minerals from syenite rocks of the Sushina hill, TPSZ. Sushina hill in TPSZ lies within Singhbhum Group (SG) of rocks. The North Singhbhum Mobile Belt (NSMB) in its northern margin has a tectonic boundary with the Chhotanagpur Gneissic Complex (CGC) along the TPSZ [16-22]. Singhbhum Shear Zone (SSZ) is located 40 km south of the TPSZ. At places, TPSZ passes either through rocks of SG or CGC. Singhbhum orogenic cycle experienced three major phases of deformation. CGC underwent four phases of deformation [23-28]. TPSZ witnessed reactivation near 500 Ma due to overthrusting and suffered rapid exhumation near 600 Ma [29-34]. TPSZ was thus affected by the intense deformation in the area adjoining this shear zone. With this background of geological events, it was thought that fission track dating on cogenetic minerals might help unravel the uplift history of syenite rocks of the Sushina hill.

Experimental Procedure

The samples for this study were processed in the laboratory of the Geological Survey of India, Kolkata, after obtaining permission from the Director General, GSI, Kolkata, West Bengal. The samples were prepared using standard separation, grinding and polishing techniques [35]. All the samples were prepared for the external detector method. AFT mounts were etched with 70% HNO_3 at room temperature for 30 s. Zircons were mounted in PFA Teflon. Zircons were etched in KOH-NaOH eutectic etchant [36-39] at 215°C on Spinot digital hot plate for ~8 hrs. The sample was placed in 48% HF for 2 hrs to clean up grains. After etching, mica sheets were firmly attached on the sample mounts. The samples were irradiated in the thermal facilities of FRMII at Garching, Germany together with dosimeter glass IRMM-540R (15 ppm). Mica sheets were etched using 48% HF at room temperature for 19 min [40-43]. The fission tracks were counted under a total magnification of 1000X. The calibrated area of one grid is 0.64×10⁻⁶ cm². Durango apatites were used as the age standard mineral, which was provided by Prof. Barry Paul Kohn, University of Melbourne,

Australia. FT age of zircon was determined using equation without zeta value. Zeta calibration was not performed on zircon because of the unavailability of age standard minerals [44-48].

Interpretation and Results

Fission track age determinations were made on 15apatite, and 18 zircon separates from syenite rocks from the Sushina hill shown in Figure 1. The mean fission track ages of apatite and zircon are 535.25 Ma, and 970 Ma respectively as shown in Table 1.

The uplift rate has been calculated according to the equation:

$$\text{Uplift rate} = \frac{\text{Cooling rate}}{\text{Geothermal gradient}} \qquad (1)$$

Where, Cooling rate = $(T_1\text{-}T_2) \div (A_1\text{-}A_2)$

T_1, T_2 = Closure temperatures of cogenetic minerals

A_1, A_2 = Mean FT ages of cogenetic minerals

Average geothermal gradient of the order of 30°C/km has been adopted. Closure temperatures for apatite and zircon have been adopted 110°C and 240°C respectively.

FT age of apatite sample namely SAP has been calculated according to the equation [49]:

$$T = 1/\lambda_D \ln\{1 + \lambda_D Z.G\rho_d p_s / \rho_i\} \qquad (2)$$

Where, is the surface density of etched spontaneous fission racks, ρ_i is the surface density of etched induced fission tracks, and G is the integrated geometry factor of etched surface. $\lambda_d = 1.55 \times 10^{-10}\ yr^{-1}$ = total decay constant of ^{238}U, Z = calibration factor based on EDM of fission-track age standards. ρ_d = induced fission-track density for a uranium standard corresponding to the sample position during neutron irradiation.

*Corresponding author: Amal Kumar Ghosh, Ph.D. Student, Bhagwant University, Ajmer, Rajasthan-305004, India, E-mail: ghosh.971@gmail.com

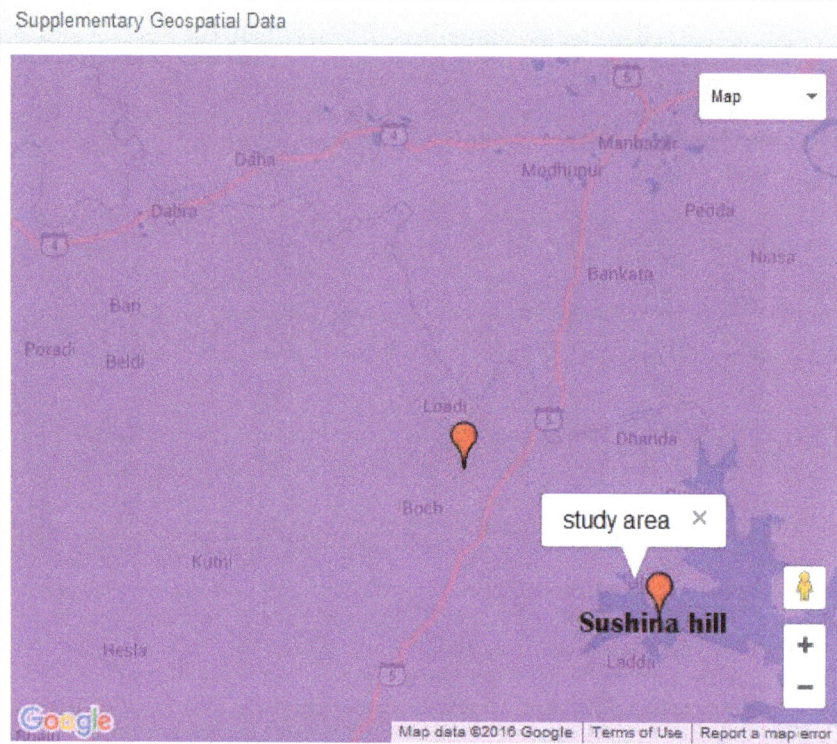

Figure 1: Location map of the Sushina hill.

Sample name	Rock type and Mineral	$P_d \times 10^6$	N_d	$P_s \times 10^6$	Ns	$P_i \times 10^6$	N_i	P{x}2(%)	U(ppm)	Mean age ± 1δ (Ma)	No. of grain
SAP	Syenite, Apatite	1.925	1232	0.325	156	0.146	70	0.33	0.68	535.25 ± 14.52	15
SZR	Syenite, Zircon	1.82	1168	17.66	452	1.773	244	15.39	41.65	970 ± 7.94	18

Table 1: Results of AFT analyses : ages calculated using dosimeter glass IRMM-540R with 15 ppm U, zeta = 250, irradiated at FRMII, calibrated by traditional zeta approach and external detector method for apatite sample SAP, N=Number of grains, ρ – track densities given in 10^6 tr cm^{-2}, ρ_d – dosimeter track density, N_d – number of tracks counted on dosimeter, $\rho_s(\rho_i)$ – spontaneous (induced) track densities, $N_s(N_i)$ – number of counted spontaneous (induced) tracks, P(^{2}x) – probability for obtaining ^{2}x value for n degrees of freedom, where n=no. of grain – 1, Neutron flux(Φ) for zircon sample SZR = 1.75×10^15 neutron/cm^2.

Region	Closure temperatures	Time Span (Ma)	Cooling rate (°C/Ma)	Uplift rate (m/Ma)
Sushina hill in TPSZ	110°C for apatite 240°C for zircon	535-970	0.299	9.97

Table 2: Cooling and uplift rate of the Sushina hill.

FT age of zircon sample namely SZR has been calculated according to the equation without zeta value [50]:

$$T = 1/\lambda_d \ln\left(1 + 9.25 \times 10^{-18} p_s / \rho_i.\Phi\right) \text{years} \qquad (3)$$

Where, Φ = Neutron flux

The large age errors, e.g. 14.52% are found in sample SAP. As already known, low Uranium samples present a problem because of low induced track densities [51-55]. The P (X^2) test was performed to measure the uranium variation in the samples. A value of P(X^2) larger than 5% means that the grains are assumed to be a single age. Sample SAP failed the X^2 test, which may indicate bimodal distributions for the sample.

By applying apatite fission track analysis, the possible uplift rate of the Sushina hill was attempted to be revealed, which could be reflected by an offset age of two cogenetic minerals (Table 2).

Conclusion

The largest age error (14.52%) occurs in sample SAP. This high error is most likely due to a very low uranium concentration (0.68 ppm). As already known, low uranium samples place limits on how robust the ages could be. In low uranium samples, an exact match between the areas counted in the grains and the mica is often hard to achieve. An adjustment by eye is difficult and subjective because the outline of the induced tracks on the mica does not reflect the shape of the analyzed grain. In this study, closure temperatures for apatite and zircon have been adopted 110°C and 240°C respectively. In reality, the closure temperature concept cannot be straight forward applied.

For determining zircon FT age, zeta calibration was not performed. It places limit on precise calculation of FT age. Syenite rocks of the Sushina hill in TPSZ uplifted at the rate of 9.97 m/Ma in the range from 535 Ma-970 Ma.

Acknowledgement

I thank Prof. Barry Paul Kohn, "University of Melbourne", Australia for his overall support and advice throughout my time as a Ph.D. student. He also sent me age standard minerals (Durango apatite) for zeta calibration, free of charge. Without his contribution, my work would have never been possible.

I thank Prof. Richard Ketcham "University of Texas", U.S.A., who kindly reviewed my AFT models and provided valuable guidance for the improvement of the models.

I am highly indebted to the Director General of Geological Survey of India, 27, J.L. Nehru Road, Kolkata – 700 016, for his kind permission to perform my work in the laboratory of G.S.I, Kolkata.

I thank Mr. Partha Nag, Officer-in-Charge, WBMTDC, Purulia for his dedicated help with the field work.

I thank Mr. T. Ray Barman, Ex-Scientist, G.S.I, Kolkata, for his constructive advice.

Many thanks are due to the entire family of G.S.I., Kolkata, for their help and encouragement.

I thank the entire team of FRMII, Garching, Germany for providing me with use of the irradiation facility, free of charge.

References

1. Acharya A, Basu SK, Bhaduri SK, Chaudhury BK, Ray S, et al. (2006) Proterozoic Rock Suites along South Purulia Shear Zone, Eastern India; Evidence for Rift-Related Setting. Geological Society of India 68: 1069-1086.

2. Basu SK (1993) Alkaline-Carbonatite Complex in Precambrian of South Purulia Shear Zone, Eastern India; Its Characteristics and Minerals Potentialities. Indian Minerals 47: 179-194.

3. Belton DX, Brown RW, Gleadow AJW, Kohn BP (2002) Fission Track Dating of Phosphate minerals and the Thermochronology of Apatite 16: 579-630.

4. Brown R, Gallagher K, Johnson C (1998) Fission Track Analysis and its Applications to Geological Problems. Annual Review Earth Planet Science 26: 519-572.

5. Carter A, Clift PD, Dorr N, Gee DG, Liskar F, et al. (2012) Late Mesozoic – Cenozoic exhumation history of northern Svalbard and its regional significance; constraints from apatite fission track analysis. Tectonophysics 514-517.

6. Dodge FCW, Ross DC (1971) Coexisting hornblendes and biotites from granite rocks near the San Andreas fault, California. Journal of Geology 79: 158-172.

7. Fayon A, Peacock SM, Stump E, Reynolds SJ (2000) Fission track analysis of the footwall of the Catalina detachment fault, Arizona: Tectonic denudation, magmatism, and erosion. J Geophys Res 105: 11047-11062.

8. Feinstein S, Kohn BP, Eyal M (1989) Significance of combined vitrinite reflectance and fission-track studies in evaluating thermal history of sedimentary basins: an example from southern Israel. In: Naeser ND, McCulloh TH (eds.) Thermal History of Sedimentary Basins: Methods and Case Histories. Springer-Verlag, Berlin pp: 197-216.

9. Fitzgerald PG (1994) Thermochronological constraints on the post-Paleozoic tectonic evolution of the central Transantarctic Mountains, Antarctica. Tectonics 13: 818-836.

10. Fitzgerald P, Gleadow AJW (1990) New approaches in fission track geochronology as a tectonic tool: Examples from the Transantarctic Mountains. Nucl Tracks Radiat Meas 17: 351-357.

11. Fitzgerald PG, Sandiford M, Barrett PJ, Gleadow AJW (1986) Asymmetric extension in the Transantarctic Mountains and Ross Embayment. Earth Planet Sci Lett 86: 67-78.

12. Fitzgerald PG, Fryxel JE, Wernicke BP (1991) Miocene crustal extension and uplift in southeastern Nevada: Constraints from fission track analysis. Geology 19: 1013-1016.

13. Fitzgerald PG, Reynolds SJ, Stump E, Foster DA, Gleadow AJW (1993) Thermochronologic evidence for timing of denudation and rate of crustal extension of the South Mountain metamorphic core complex and Sierra Estrella, Arizona. Nucl Tracks 21: 555-563.

14. Fitzgerald PG, Sorkhabi RB, Redfield TF, Stump E (1995) Uplift and denudation of the central Alaska Range: a case study in the use of apatite fission track thermochronology to determine absolute uplift parameters. J Geophys Res 100: 20175-20191.

15. Fitzgerald PG, Munoz JA, Coney PJ, Baldwin SL (1999) Asymmetric exhumation across the Pyrenean orogen: implications for the tectonic evolution of a collisional orogen. Earth Planet Sci Lett 173: 157-170.

16. Fleischer RL, Hart HR (1972) Fission track dating: Techniques and problems. In: Bishop WW, Miller DA, Cole S (eds.) Calibration of hominid evolution. Scottish Academic Press, Edinburgh pp: 135-170.

17. Fleischer RL, Price PB (1964) Techniques for geological dating of minerals by chemical etching of fission fragment tracks. Geochim Cosmochim Acta 28: 1705-1715.

18. Fleischer RL, Price PB, Symes EM (1964) On the origin of anomalous etch figures in minerals. Am Mineral 49: 794-800.

19. Fleischer RL, Price PB, Walker RM (1965a) Effects of temperature, pressure and ionization of the formation and stability of fission tracks in minerals and glasses. J Geophys Res 70: 1497-1502.

20. Fleischer RL, Price PB, Walker RM (1965b) The ion explosion spike mechanism for formation of charged particle tracks in solids. J Appl Phys 36: 3645-3652.

21. Fleischer RL, Price PB, Walker RM (1975) Nuclear Tracks in Solids. University of California Press, Berkeley.

22. Fletcher JM, Kohn BP, Foster DA, Gleadow AJW (2000) Heterogeneous Neogene cooling and uplift of the Los Cabos block, southern Baja California: Evidence from fission track thermochronology. Geology 28: 107-110.

23. Foster DA, Gleadow AJW (1992a) Reactivated tectonic boundaries and implications for the reconstruction of southeastern Australia and northern Victoria Land, Antarctica. Geology 20: 267-270.

24. Foster DA, Gleadow AJW (1992b) The morphotectonic evolution of rift-margin mountains in central Kenya: constraints from apatite fission track analyses. Earth Planet Sci Lett 113: 157-171.

25. Foster DA, Gleadow AJW (1993) Episodic denudation in East Africa - a legacy of intracontinental tectonism. Geophys Res Lett 20: 2395-2398.

26. Foster DA, Gleadow AJW (1996) Structural framework and denudation history of the flanks of the Kenya and Anza Rifts, East Africa. Tectonics 15: 258-271.

27. Foster DA, John BE (1999) Quantifying tectonic exhumation in an extensional orogen with thermochronology: Examples from the southern Basin and Range Province. In: Ring U, Brandon MT, Lister G, Willett SD (eds.) Exhumation Processes: Normal Faulting, Ductile Flow, and Erosion, Geol Soc London Special Publ 154: 343-364.

28. Foster DA, Raza A (2002) Low-temperature thermochronological record of exhumation of the Bitterroot metamorphic core complex, northern Cordilleran Orogen. Tectonophysics 349: 23-36.

29. Foster DA, Miller DS, Miller CF (1991) Tertiary extension in the Old Woman Mountains area, California: evidence from apatite fission track analysis. Tectonics 10: 875-886.

30. Foster DA, Gleadow AJW, Reynolds SJ, Fitzgerald PG (1993) The denudation of metamorphic core complexes and the reconstruction of the Transition Zone, west-central Arizona: constraints from apatite fission-track thermochronology. J Geophys Res 98: 2167-2185.

31. Foster DA, Gleadow AJW, Mortimer G (1994) Rapid Pliocene exhumation in the Karakoram, revealed by fission-track thermochronology of the K2 gneiss. Geology 22: 19-22.

32. Fuchs LH (1962) Occurrence of whitlockite in chondritic meteorites. Science 137: 425-426.

33. Galbraith RF (1990) The radial plot: graphical assessment of spread in ages. Nucl Tracks Radiat Meas 17: 207-214.

34. Galbraith RF, Green PF (1990) Estimating the component ages in a finite mixture. Nucl Tracks Radiat Meas 17: 197-206.

35. Galbraith RF, Laslett GM (1993) Statistical models for mixed fission track ages. Nucl Tracks Radiat Meas 21: 459-480.

36. Gallagher K (1995) Evolving temperature histories from apatite fission-track data. Earth Planet Sci Lett 136: 421-435.

37. Gallagher K, Brown RW (1997) The onshore record of passive margin evolution. J Geol Soc London 154: 451-457.

38. Gallagher K, Brown RW (1999a) Denudation and uplift at passive margins: the record on the Atlantic Margin of southern Africa. Phil Trans Roy Soc London A 357: 835-859.

39. Naeser CW (1967) The use of apatite and sphene for fission-track age determinations. Geol Soc America Bull 78: 1523-1526.

40. O'Sullivan PB, Tagami T (2005) Fundamental of Fission – Track Thermochronology 58: 19-47.

41. Turner DL, Forbes RB, Naeser CW (1973) Radiometric ages of Kodiak Seamount and Giacomini Guyot, Gulf of Alaska: implications for Circum-Pacific tectonics. Science 182: 579-581.

42. Turner DL, Frizzell VA, Triplehorn DM, Naeser CW (1983) Radiometric dating for ash partings in coal of the Eocene Puget Group, Washington: Implications for paleobotanical stages. Geology 11: 527-531.

43. van der Beek PA (1997) Flank uplift and topography at the central Baikal Rift (SE Siberia): A test of kinematic models for continental extension. Tectonics 16: 122-136.

44. van der Beek PA, Cloetingh S, Andriessen PAM (1994) Mechanisms of extensional basin formation and vertical motions at rift flanks: Constraints from tectonic modelling and fission track thermochronology. Earth Planet Sci Lett 121: 417-433.

45. van der Beek PA, Andriessen PAM, Cloetingh S (1995) Morpho-tectonic evolution of rifted continental margins: inferences from a coupled tectonic-surface processes model and fission-track thermochronology. Tectonics 14: 406-421.

46. van der Beek PA, Delvaux D, Andriessen PAM, Levi KG (1996) Early Cretaceous denudation related to convergent tectonics in the Baikal region, SE Siberia. J Geol Soc London 153: 515-523.

47. Villa F, Grivet M, Rebetez M, Dubois C, Chambaudet A, et al. (2000) Damage morphology of Kr tracks in apatite: dependence on thermal annealing. Nucl Instr Meth B 168: 72-77.

48. Vineyard GH (1976) Thermal spikes and activated processes. Radiat Effects 29: 245-248.

49. Wagner GA (1968) Fission track dating of apatites. Earth Planet Sci Lett 4: 411-415.

50. Wagner GA (1969) Traces of the spontaneous fission of 238 Urans as a means of dating of apatite and a contribution to the geochronology of the Odenwald. N Jahrb Mineral Abh 110: 252-286.

51. Wagner GA (1977) Fission track dating of apatite and titanite from the Ries: A Contribution to the age and thermal history. Geol Bavarica 75: 349-354.

52. Wagner GA, Reimer GM (1972) Fission track tectonics: The tectonic interpretation of fission track apatite ages. Earth Planet Sci Lett 14: 263-268.

53. Wagner GA, Storzer D (1970) The interpretation of fission track ages (fission track ages) using the example of natural glasses, apatite and zircon. Eclogae Geol Helv 63: 335-344.

54. Wagner GA, Storzer D (1972) Fission track length reductions in minerals and the thermal history of rocks. Trans Am Nucl Soc 15: 127-128.

55. Wagner GA, Storzer D (1977) Fission track dating of meteorite impacts. Meteoritics 12: 368-369.

5

Ground Vibrations and Air Blast Effects Induced by Blasting in Open Pit Mines: Case of Metlaoui Mining Basin, Southwestern Tunisia

Monia Aloui[1]*, Yannick Bleuzen[2], Elhoucine Essefi[3] and Chedly Abbes[1]

[1]Faculty of Sciences of Sfax, Road Soukra, Sfax 3038, Tunisia
[2]Society of Mining and Industrial Engineering, ZA Brushes 2, 9 Rue de la Hyssop 37270 Larcay, Tunisia
[3]National Engineering School of Sfax, University of Sfax, Tunisia

Abstract

Blasting is widely used in quarries and mining production processes. It is the beneficial industrial technology which provides achievement of expected results in a short period of time with relatively low cost. Never one less, blasts produce undesirable vibrations and sounds. This current investigation reports the ground motion and airblast over pressure measurements around the open pit mine nearby Metlaoui village (south-western Tunisia). An empirical relation proposed by USBM was utilized to calculate K (site constant) and n (slope of the PPV vs scaled distance) and verifies whether the recorded shots overcame the thresholds indicated by DIN and USBM. The site parameters obtained provide us with the propagation equation of the blast induced seismic waves in the study site. Also, an overview of frequency for the study area revealed the dominance of low frequencies (>40 Hz). Such values can cause damage to the nearby structures when a specific PPV value is reached by blasting. Moreover, it has been demonstrated in this study that all over pressure magnitudes are less than 134 dB, which is the safe limit of air blast level.

Keywords: Rock blasting; Peak particle velocity; Open pit mine; Ground vibration: Air blast

Introduction

On surface mining, blasting technique may be considered as the most economical method used for fragmenting rocks masses. Nonetheless, only 20-30% of the used energy is served for rocks fragmenting and displacing, while the rest is wasted in the form of ground vibration, air blast, noise and fly-rocks [1]. Both ground vibration and air blast are matter of great concern as they would result in damage to the existing surface structures and nuisances to the inhabitants in the vicinity of mines, which are exceedingly approaching near populated areas. In order to analyze the vibration-related problems, the combined effect of several factors such as site characteristics, propagation of surface, the body waves in the ground, and response of structures should be taken into consideration. The best approach for estimating the charge weight, that at given distance produces vibrations below the limit of safety, is using instrumentation on blasts to determine the constants of the actual blasting conditions. Furthermore, the effective control of problems related to vibration requires the development of a reliable vibration monitoring system and an assessment of attenuation characteristics of different vibrations [2].

Ground motion, which represents the major important effect of blasting, requires some regulations in relation with its structural damage. These regulations are mainly based on peak particle velocity (PPV) due to blasting operations [3]. Many scientists have investigated the PPV [4-7]. The United States Bureau of Mines (USBM) proposed the first significant PPV predictor equation.

There are regulatory limitations on blasting vibrations in relation to the maximum charge and the distance to a concerned location. Several established criteria of damage (USBM, DIN 4150) associated to the potential damage of blasting, can be used to confirm the design. Moreover, the relatively small charge per delay shall not result in potential damage.

In this work, the recorded measurements of ground vibration in Metlaoui open pit mine of phosphate were analyzed according to several established criteria of damage in order to study the ground vibrations characteristics resulting from production blasts and to evaluate the vibrations impact on the nearby structures.

Study area

The study site is an opencast mine of phosphate located in southwestern Tunisia, about 5 Km north of Metlaoui (Figure 1). The phosphate deposit of Metlaoui, belonging to dissymmetrical synclinal with layers dips not exceeding 10 degrees [8]. It has been exploited using opencast method since 1999. It contains considerable reserves of phosphate that are covered by the Eocene limestone deposits of the Kef Eddour Formation. This Eocene deposits constitute hard materials and, thus, its fragmentation must be carried out using explosive in order to facilitate the extraction of phosphates by opencast.

The sedimentary record of the Metlaoui phosphatic deposit is composed of various lithostratigraphic successions. The sedimentary rocks outcropping are composed by phosphatic Chouabine formation (Ypresian) and the studied limestone of the Kef Eddour formation (Early Lutetian), which is fragmented by explosive. The Chouabine Formation containing a significant amount of phosphates is overlain by limestone of the Kef Eddour formation [9,10].

Methods

Peak particle velocity (PPV)

The ground vibration intensity was measured in terms of peak particle velocity (PPV) to evaluate its potential damage. Peak particle velocity, which corresponds to an indicator of a structural damage, largely depending on the maximum charge, the distance between the blast and measuring point and the characteristics of the medium [11]. At areas with unknown transmission characteristics, seismic constants

---

*Corresponding author: Monia Aloui, Faculty of Sciences of Sfax, Road Soukra, Sfax 3038, Tunisia, E-mail: moniaaloui@yahoo.fr

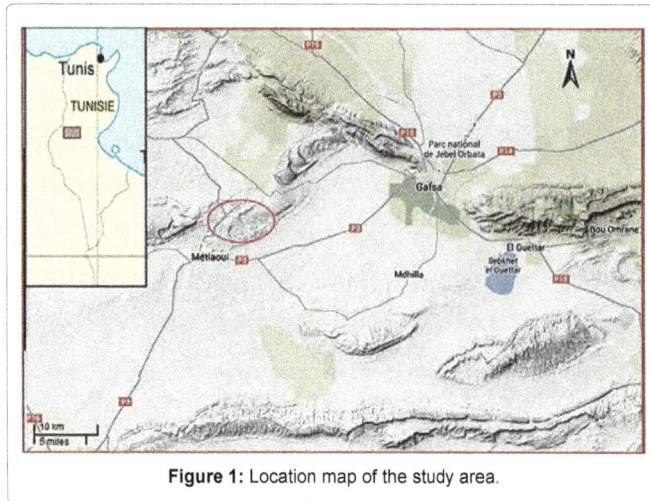

Figure 1: Location map of the study area.

of sites are determined by monitoring the ground vibrations at variable distances for identifying the blasting parameters [12-16]. The PPV is related to the charge weight and distance by the prediction equation (1) proposed by the USBM:

$$V \ = \ K \times \left(\frac{D}{\sqrt{W}} \right)^{n} \tag{1}$$

Where:

V: is the peak particle velocity (mm/s)

D: is the distance between the center of blasting site and measuring unit (m)

W: is the charge per delay (kg)

$D / \sqrt{W}$: is the scaled distance SD (m/kg$^{1/2}$)

K and n are site constants which vary from site to another; K represents the line intercept at SD=1 on the log-log graph. It is the initial energy transferred from the explosive to the surrounding rocks. Finally, n is a slope factor that induces the attenuation rate of the PPV caused by the geometric spreading and the influence of rocks characteristics. The variation in site constants can be attributed to different rock-mass properties at different sites.

Air blast over pressure

Air blast represents an undesirable and unavoidable output of blasting technique. The air blast damage and annoyance may be influenced by numerous factors such as blast design, weather, field characteristics, and human response. Air blast disturbances propagate as compression wave in air. Under specific weather conditions and poor blast designs, air blast can travel for long distances [11]. The over pressure may be expressed in Pascal (Pa) or with decibels (dB). The dB is calculated by the following formula (2):

$$P_{dB} = \ 20 \ log \ \left(P / P_{0} \right) \tag{2}$$

Where P is the measured pressure. P_{0} is the reference pressure of $2 \ 10^{-5}$ Pa.

Air blast is an atmospheric pressure waves emanating from explosion in air. This wave comprises (Figure 2):

- The audible part of the airblast (acoustic). It is characterized by higher frequency from 20 to 20,000 Hz.

- The sub-audible part of the airblast (infrasound) having a low frequency content below 20 Hz.

Unlike the audible air blast (Acoustic), which is classified as noise, the air blast at frequencies below 20 Hz is called concussion. These are classified as an "over pressure" when air blast pressure exceeds atmospheric pressure. Airblast overpressure exerts a force on structures and in turn causes a secondary and audible rattle within a structure. It is very often confused with vibrations transmitted by the ground [17].

Blast vibration regulatory limits

Ground vibrations related to rock blasting result in several deleterious problems for mining, construction, quarry, and pipelines. Worldwide, numerous researches were established for providing the damage criteria such as the energy, the energy ratio, the displacement, the velocity or the acceleration. It has been demonstrated that the particle velocity of ground motion near structures is an effective criterion of damage evaluation. The United States Bureau of Mine (USBM) RI 8507 reveals that peak particle velocity corresponds to the best description of the single ground vibration [18]. The safe maximum velocity of particles around 2 in/sec (50 mm/s) is recommended for all buildings [19,20]. During the last two decades, the PPV and frequency have been used as a damage criteria; and some standards from USBM and DIN (Deutsches Institut für Normung) were developed (Table 1). Most modern blasting seismographs must express the vibration data according to the USBM limiting criterion. In general, at lower frequencies, the ground vibration should not exceed 12.7 mm/s, but at higher frequencies, the limit can increase to 50 mm/s [21].

Another blasting level set of smooth criteria was recommended by the USBM – RI 8507 (Table 1). For each blast, it requires a frequency analysis of blast-generated ground vibration wave as well as the measurements of the particle velocity. This method would be the best means to evaluate the potential damage to residential structures as well as human annoyance. Any seismic record for any component (longitudinal, transverse, vertical) for the particle velocity at a particular predominant frequency below any part of the solid line graph of Figure 3 may be considered safe. On the other hand, any value above any part of the solid line graph increases the possibility of human annoyance and residential damage.

Figure 3 shows the German Standard DIN 4150 [22] criteria. DIN 4150 suggests three lines of time-dependent vibration limits for different types of structures [11,12]. The first line is utilized for buildings used for commercial and industrial purposes. The second line is applied for dwellings and buildings of similar design and/or use. The third line is used for structures, which are not included in those listed in lines 1 and 2, due to their particular sensitivity to vibration; and are of great intrinsic values, e.g., building that are under a preservation order. Line 1 is close to the upper boundary of vibration limits in the 4-100 Hz frequency range [23]. The low-frequency portion of the seismic waves thereby plays an important role. Furthermore, the potential damage of blast-induced vibrations on structures is conditioned by the particle velocity and the low-frequency portion of seismic waves. The low-

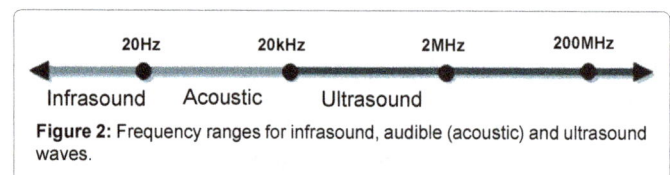

Figure 2: Frequency ranges for infrasound, audible (acoustic) and ultrasound waves.

US Bureau of Mines RI 8507			DIN 4150-3			
Structure	PPV (mm/s)		Structure	PPV (mm/s)		
	<40 Hz	≥40 Hz		10 Hz	10-50 Hz	50-100 Hz
Modern homes dry-	18,75	50	Industrial buildings	20	20-40	40-50
wall interiors			Residential buildings	5	5-15	15-20
Older homes	12,50	50	More sensitive buildings than above	3	3-8	8-10

Table 1: Safe level blasting criteria: thresholds of PPV values at different frequencies (USBM and DIN 4150) [18,22,25].

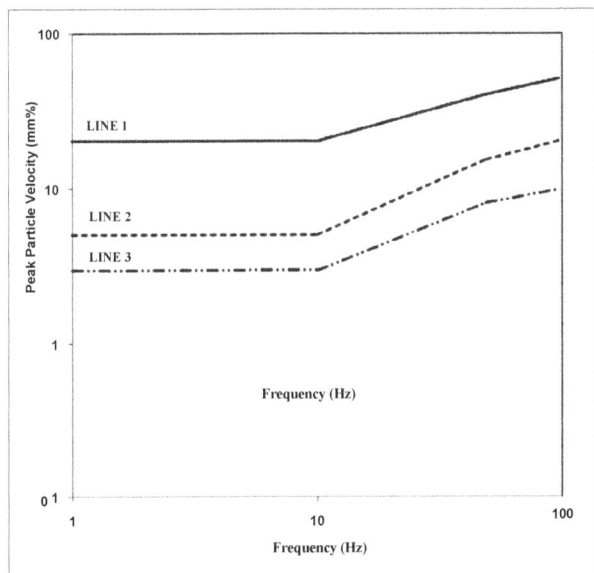

Limit of the potential damage for:

LINE 1: Industrial Building

LINE 2: Residential Building

LINE 3: Sensitive Building

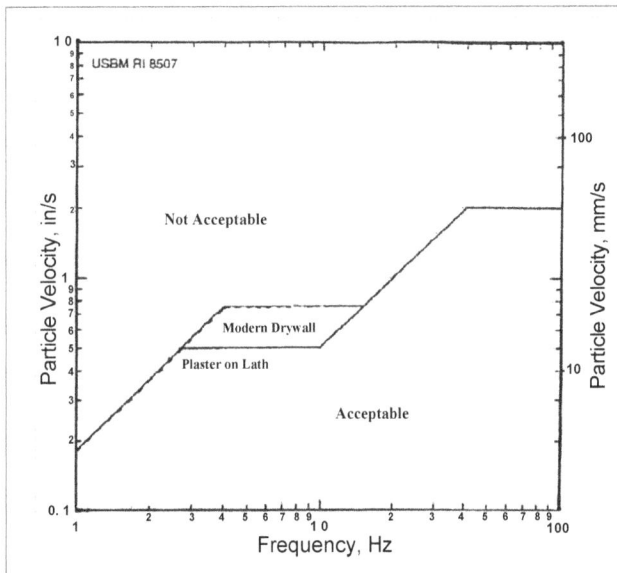

The safe limit for

---- Modern Homes

——— Older Homes

Figure 3: Recommended safe levels for blasting vibration by DIN 4150 (a) and USBM (b) criteria.

frequency range (<40 Hz) potential damage is considerably higher than that of the high- frequency range (>40 Hz). This is due to the resonance effects at the natural frequency range (5-16 Hz).

Air blast regulatory limits

The Bureau of Mines Report of Investigations 8485 (1980), "Structure Response and Damage Produced by Air blast From Surface Mining" generally recommends a maximum safe overpressure of 0.014 psi (134 dB) for air blast recorded at residential structures [24]. This set of criteria was based on the major superficial type of damage affecting residential-type structures. Any criterion of the sets represents the safe maximum air blast levels although the best is recommended to be the 134 dB at 20 Hz high pass system. These recommended levels provide 95-99% of a non-damage probability and 95-90% of annoyance acceptability (Table 2).

Air blast can be affected considerably by surface winds and climatic conditions such as temperature inversions (increase of temperature with altitude). Under these conditions, the peak over pressure can increase by a factor of 5-10, requiring therefore the adoption of certain precautions. Indeed, high air blast over pressure could cause structural damage, while, those produced by routine blasting operations under normal atmospheric conditions are not likely to do so [25-27].

Measurements and methods

During the present study, the blasting parameters of three shots with nonel initiation systems were carefully recorded in Metlaoui quarry. Blasting was performed within the Eocene limestone. We have used five seismographs; three of them are Deltaseis model and the two other are Instantel model. The seismographs have also attached microphones which can measure sound and airblast overpressure (Figure 4).

In order to monitor the ground vibration and air blast over pressure generated from blast, seismographs were installed at predetermined distances from the blast site to the town of Metlaoui (Figure 5). The threshold of release of each seismograph is selected according to its location.

Results and Discussion

Each seismograph consists of 3-axis velocity transducer, an air over-pressure transducer, and a data acquisition and storage device. Figure 6 displays a typical printout of seismograph obtained for each blast. It contains information about the graph such as the duration, the acoustic scale, the seismic scale and the interval between the two lines of time. Also, it gives details about amplitudes, frequencies and Fourier analysis.

Overpressure unit		Description
Db	**Pa**	
180	206842,71	⇦ Structural damage
170	6550,01	⇦ Most windows break
160	2068,42	
150	655	⇦ Some windows break
140	206,84	⇦ OSHA* maximum for impulsive sound
130	65,50	⇦ USBM TPR 78 maximum
		⇦ USBM TPR 78 safe level
120	20,68	⇦ Threshold of pain for continuous sound
110	6,55	⇦ Complaints likely
100	20,684	⇦ OSHA maximum for 15 minutes
90	0,65	
80	0,20	⇦ OSHA maximum for 8 hours

*Occupational Safety and Health Administration.

Table 2: Overpressure unit conversion (dB and psi) and effects on human annoyance and structural damage [18].

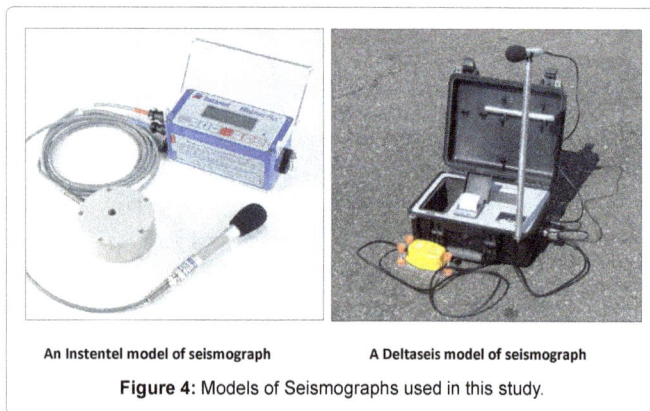

An Instentel model of seismograph   A Deltaseis model of seismograph

Figure 4: Models of Seismographs used in this study.

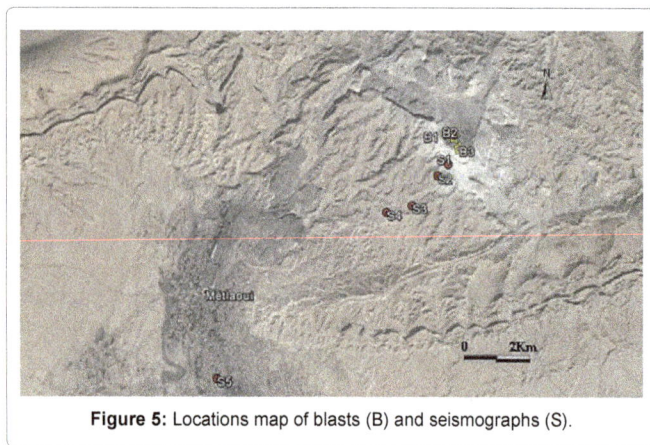

Figure 5: Locations map of blasts (B) and seismographs (S).

The records of the seismographs were printed including full waveforms, summary of peak values of ground motion as well as air blast over pressure. Table 3 shows the data for each blast, including geophone, SD, PPV (mm/s) for longitudinal (L), transverse (T), and Vertical (V) components of ground vibration along with their frequencies (Hz), and their vector sum resultant (PPVR), air blast (sound) over pressure in Decibels (dB).

Vibration predictor equation

In order to establish a relationship between PPV and scaled distance

for this site, regression analysis was carried out by using all data pairs. The determined equation for this site can be used to determine the permissible explosive charge at any distance for a specific scaled distance value. Also, this equation can be used to determine a compliant or safe value of the distance for a given charge weight of explosive as well as a specified scaled distance. The determined equation for this site study is given below.

The ground vibration data including PPV and scaled distance for various blasts was analyzed to understand the effect of ground vibrations induced by blasting in the open pit mine of phosphate in Metlaoui. The following predictor equation in terms of scaled distance and PPV is:

$$V_{PPV} = 1508 \times \left(\frac{D}{\sqrt{W}} \right)^{-1.73} \tag{3}$$

$R^2 = 0.847$

The empirical factors k and n were determined by regression analysis as 1508 and -1.73, respectively.

In order to develop a statistically reliable curve, relationship between blast-induced vibrations scaled distance were established on a log-log diagram (Figure 7).

The relationship shows that the PPV decreases when maximum charges per delay in the holes are decreased or else if the distance is increased. In order to find maximum charge per hole with respect to the structures at fixed distances, the stipulated PPV as per the requirements of the status is to be maintained. This case analysis is site specific and cannot be adopted for general use in other mines.

Evaluation of damage risk

The measured magnitudes of PPV and frequency of the shots were evaluated taking into account several established damage criteria (USBM and DIM 4150) used in mining and geotechnical/structure engineering. According to USBM damage criteria (Table 1 and Figure

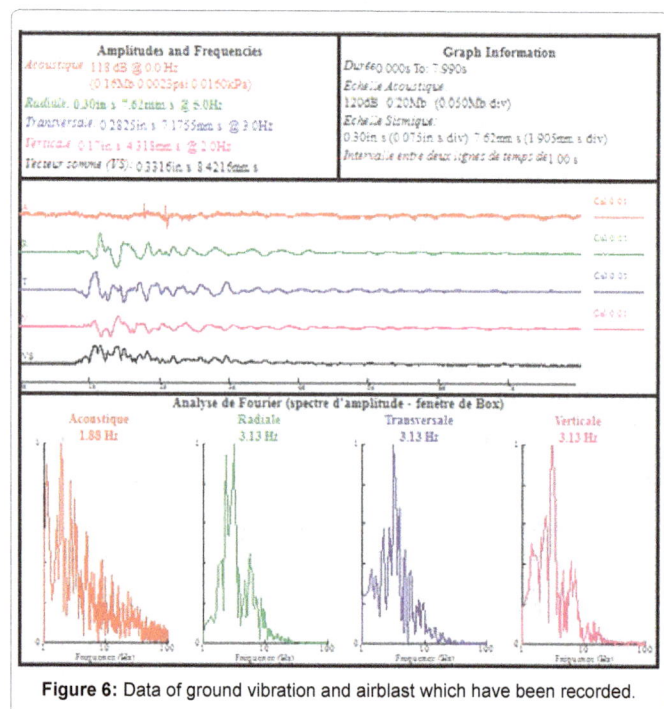

Figure 6: Data of ground vibration and airblast which have been recorded.

N Blast NO	Seismograph	PPV$_{MAX}$	Particle velocity (mm/s)			Ground vibration Frequency (Hz)	Charge per delay W (kg)	Distance D (m)	Scaled distance SD (m/kg 0.5)	Air blast (dB)	Air blast frequency (Hz)
			Longitudinal Peak (PVL)	Vertical peak (PVV)	Transverse peak (PVT)						
B1	S1	14,2	11,1	7	14,2	5	390	434	22	117	2,25
	S2	1,7	0,7	1,6	1,7	2,25	390	610	57,8	108	10,63
	S3	-	-	-	-	-	-	1141	-	-	-
	S4	-	-	-	-	-	-	1429	-	-	-
	S5	-	-	-	-	-	-	4867	-	-	-
B2	S1	7,6	7,6	4,3	7,2	3,13	275	391	23,6	118	1,88
	S2	2,1	2,1	1,7	1,3	2,5	275	592	35,7	111	1,5
	S3	1,1	0,8	1	1,1	2,5	275	1144	69	108	2,4
	S4	1	0,6	0,6	1	2	275	1452	87,6	100	8,1
	S5	-	-	-	-	-	275	4898	-	-	-
B3	S1	13,6	9,5	5,5	13,6	5,63	300	280	16,2	110	1,38
	S2	1,7	1,7	0,9	1,7	2	300	503	29,1	106	6
	S3	0,9	0,6	0,6	0,9	2	300	1069	61,7	108	5
	S4	0,6	0,5	0,5	0,6	1,63	300	1404	81,1	115	3,88
	S5	-	-	-	-	-	300	4850	-	-	-

Table 3: Results of ground vibration and air blast measurements.

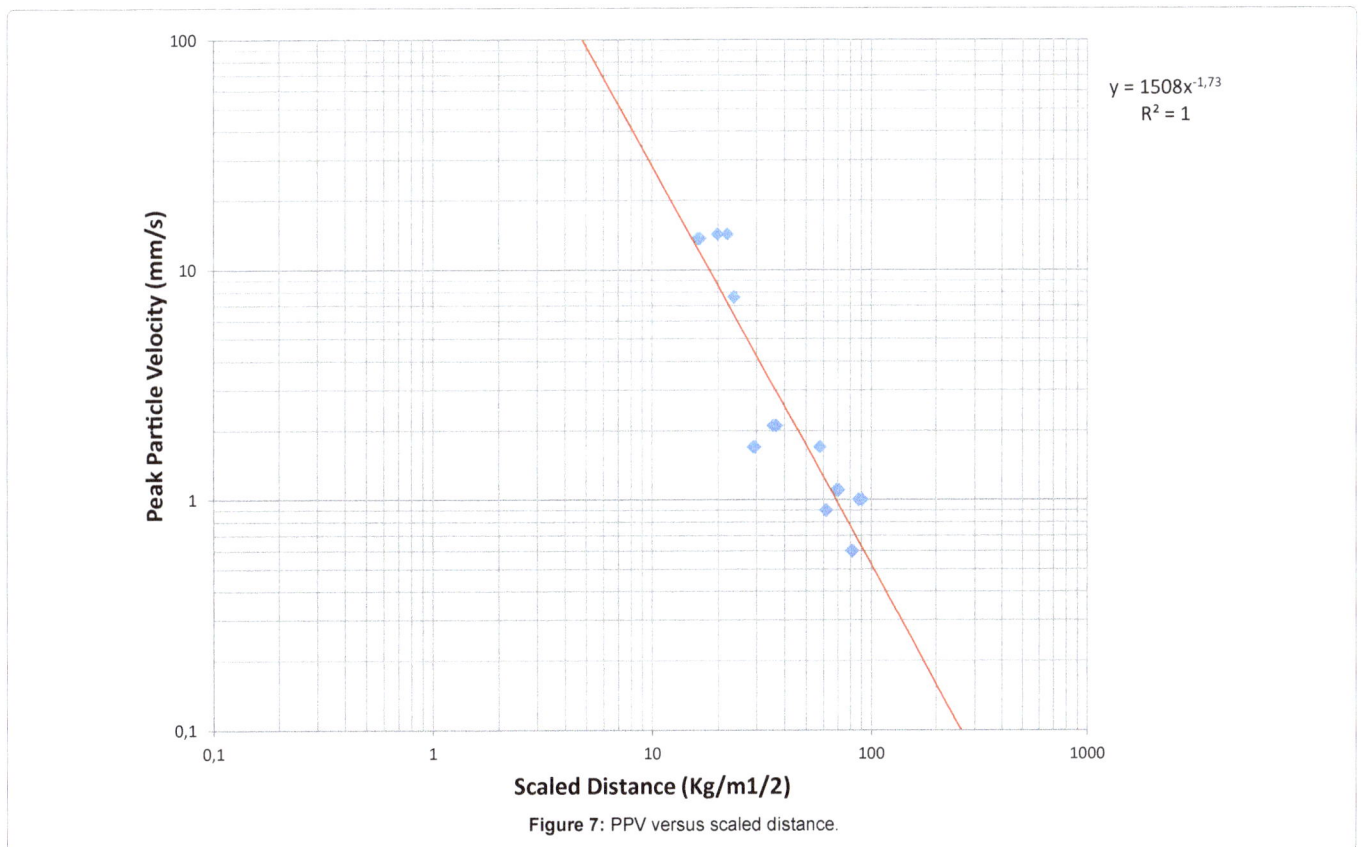

Figure 7: PPV versus scaled distance.

8) for the recorded ground vibration frequencies data, which are less than 40 Hz in all blast PPV are below the safe limits for moderns homes (PPV=18.5 mm/s) but, some of them exceeded the safe limits for older homes (PPV=12.50 mm/s). Moreover, according to DIN damage criteria (Table 1), PPV and frequency measurements are in the class of Residential and More sensitive buildings (PPV=5 mm/s and F=10 Hz). It was observed that in the plot of PPV versus frequency (Figure 9), the majority of the shots were below permissible limits described in

the damage criteria corresponding for residential and more sensitive buildings.Nevertheless, some of measured PPV exceed the safe limits, which may have a damage risk.

Based on USBM and DIN 4150 criteria, PPV and frequency measurements are in the class of modern homes and residential buildings, respectively. It was observed that in the plot of PPV versus frequency, somerecorded ground vibration datawere above permissible

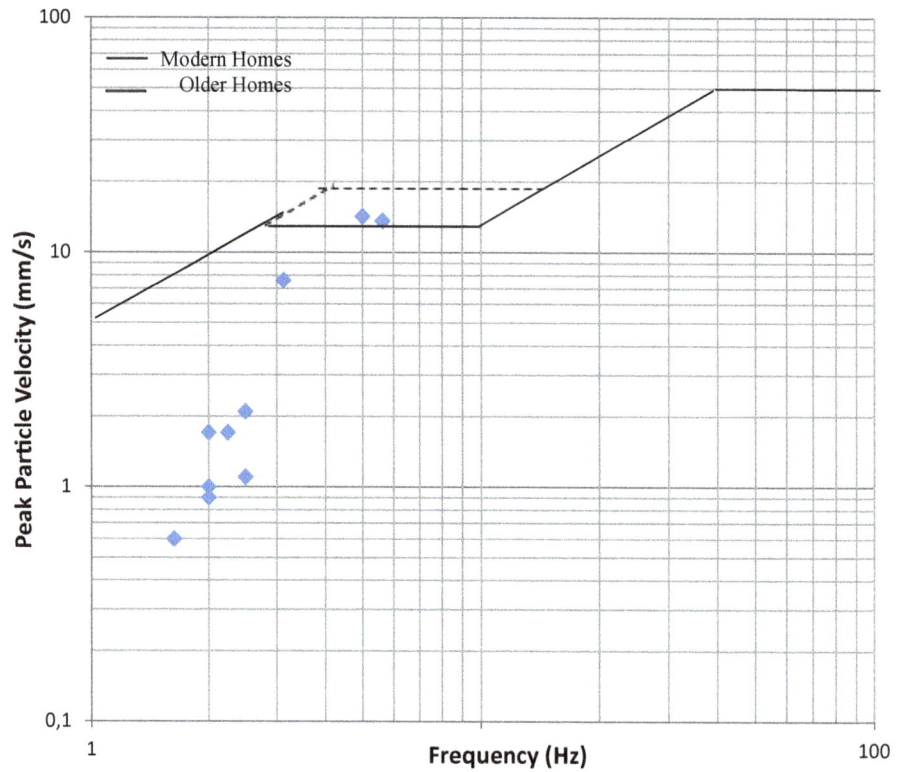

Figure 8: Evaluation of damage risk of the shots according to USBM criterion.

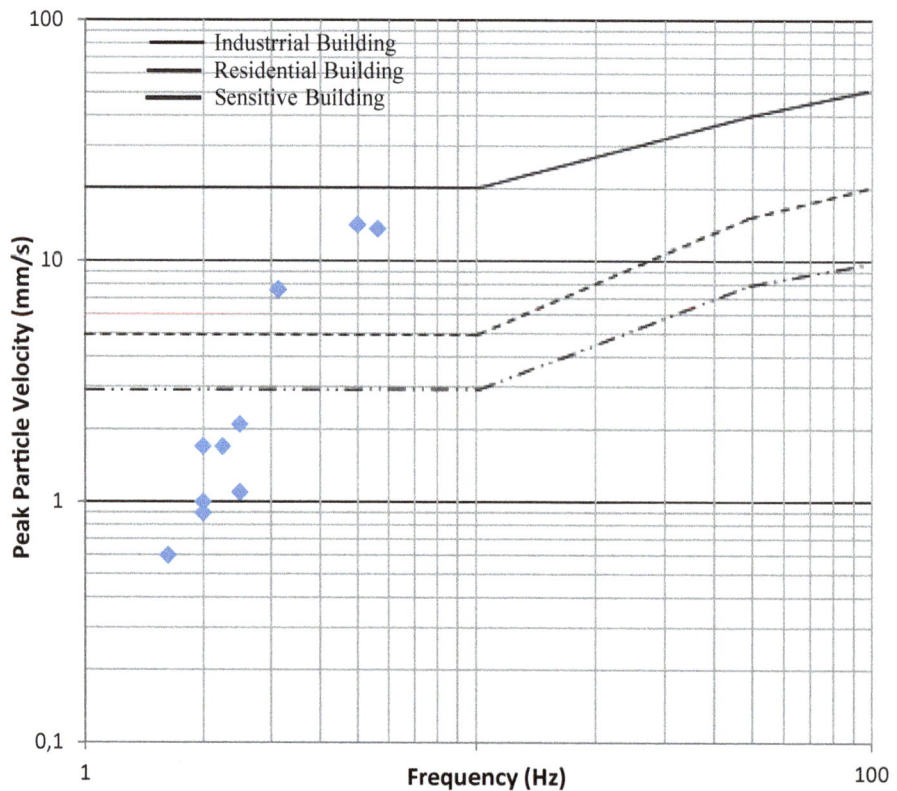

Figure 9: Evaluation of damage risk of the shots according to DIN 4150 criterion DIN (Deutsches Institute for Norming) standards.

limits described in both of the damage criteria,thereforeit may be increase the probability of structural damage.

Air blast results

The over pressure values, which range from 100 and 118 dB (Table 3), decrease with increasing scaled distance. This over pressure level is characterized by rattling windows and banging sound causing fear and annoyance but not damaging structures. However, considering the air blast over pressure levels, the building and the instrumentation are marginally safe, which is within the USBM limits (Table 2). Hence, some changes in the current blasting practice have to be made to make sure that the over-pressure is below the safety margin by greater magnitude. This will provide additional security against unpredictable atmospheric adverse conditions. In addition, it is recommended to measure the wind speed and direction before blasting to avoid blasting if the wind direction is towards the town of Metlaoui.

Frequency analysis

Ground vibration frequency: It was observed that the majority of blast frequencies were less than 5 Hz (Table 3). Below 5 Hz, the structures are generally solicited indirectly by the vibrations. The foundation soil may be mobilized especially if unconsolidated, heterogeneous, sloping and / or the foundations do not redistribute properly the stresses. In fact, it is the case of soil that characterizes the Metlaoui area which consists of sandy clays and conglomerates (lower Mio-Plio-Pleistocene) also silt and gypseous clays (lower Holocene, higher Pleistocene) [8].

The measured event frequencies of blast induced ground vibrations represent high-potential damage risk due to resonance effects. However, the frequency interval of 1-5 Hz that has higher damage risk constitutes the majority of all shots. These low frequencies are very critical to residential structures because they are in the range of their natural frequencies. The measured values frequency are near the natural frequency of residential structures (<20 Hz) is the most dangerous because it causes amplification of ground vibration.

Among the potential causes that can explain the records of lower frequencies, we mention the fracturating that characterizes the study area. Generally, the frequencies are higher in the non-fractured hard rocks. They decrease when the rock becomes increasingly fractured. The soil acts as a filter, imposes its own frequency and absorbs rapidly the high frequencies.

The distribution of recorded frequency values is given in Table 3.

Air blast frequency: All frequencies of induced air blast monitoring recorded were less than 20 Hz (Table 3), which increases risk of damage. In fact, air blast is considered as an ever annoying phenomenon in Metlaoui Mine and mostly propagates in low frequencies (<20 Hz), and causes perceptible rattling of windows easily in the building.

Conclusion

To better understand the ground vibration it is required to get the optimum blast result. The present study corresponds to the application of regression technique in blasting operation considering two variables (the distance and the maximum charge per delay), to find out the maximum PPV representing safe limit with minimum spoil to environment.

As a result of the particle velocity measurements in certain number of explosions by means of seismic devices and according to the ground vibration propagation equations, the site parameters obtained provide

us with the propagation equation of the blast induced seismic waves in that area. The propagation equation, then, is used to keep the particle velocity within certain limits through alterations to the explosive parameters. The site parameters were determined based on the particle velocity data measured in the site; and in accordance with the equation given by USBM, which has been the most widely used one among international propagation equations. Frequency values of ground vibrations were below 40 Hz, which are considered as low frequency according to the international standards. Such values can cause damage to the nearby structures when a specific PPV value is reached by blasting. In addition, it was thought that such low frequencies could lead to human response at this site.

The Control of blasting done by recording airblast clearly shows that the quality of blasts taken can affect buildings, airblast over pressure frequencies are usually less than 20 Hz. On the other hand, it has been demonstrated that all over pressure magnitudes are less than 134 dB, which is the safe limit of air blast level.

References

1. Uysal O, Erarslan K, Cebi MA, Akcakoca H (2008) Effect of barrier holes on blast induced vibration. Int J rock mech mining sci 45: 712-719.

2. Rai R, Singh TN (2004) A new predictor for ground vibration prediction and its comparison with other predictors. Ind J Eng Mat Sci 11: 178-184.

3. Adhikari GR (2005) Role of blast design parameters on ground vibration and correlation of vibration level to blasting damage to surface structures. S&T Project Report: MT/134/02.

4. Görgülü K, Arpaz E, Demirci A, Koçaslan A, Dilmaç MK, et al. (2013) Investigation of blast-induced ground vibrations in the Tülü boron open pit mine. Bulletin of Engineering Geology and the Environment 72: 555-564.

5. Ozer U, Kahriman A, Aksoy M, Adiguzel D, Karadogan A (2008) The analysis of ground vibrations induced by bench blasting at Akyol quarry and practical blasting charts. Environmental geology 54: 737-743.

6. Azizabadi HRM, Mansouri H, Fouché O (2014) Coupling of two methods, waveform superposition and numerical, to model blast vibration effect on slope stability in jointed rock masses. Computers and Geotechnics 61: 42-49.

7. Saadat M, Khandelwal M, Monjezi M (2014) An ANN-based approach to predict blast-induced ground vibration of Gol-E-Gohar iron ore mine, Iran. Journal of Rock Mechanics and Geotechnical Engineering 6: 67-76.

8. Zargouni F (1985) Tectonics of the southern Atlas of Tunisia: Evolution geometric and kinematic shear zone structures. Thesis University of Louis Pasteur 302.

9. Zouari H (1995) Geodynamic Evolution of the Atlas central- southern Tunisia: stratigraphy, geometric analysis, kinematic and tectonic- sedimentary. Thèse Doc Etat Univ Tunis11: 278.

10. Burollet PF (1956) Contribution to the study of stratigraphic central Tunisia. Ann Mines et Géol 18: 1-345.

11. Elsemain IA (2000) Measurement and analysis of the effect of ground vibrations induced by blasting at the limestone quarries of the Egyptian cement company. College of Engineering, Assiut University, ASIUT EGYPT.

12. AK H, Iphar M, Yavuz M, Konuk A (2008) Evaluation of ground vibration effect of blasting operation in a magnesite mine. Soil Dynamics and Earthquake Engineering 29: 669-676.

13. Adhikari GR, Singh MM (1989) Influence of rock properties on blast – induced vibration. Mining Science and Technology 8: 297-300.

14. Giraudi A, Cardu M, Kecojevic V (2009) An assessment of blasting vibrations: a case study on quarry operation. American Journal of Environmental Sciences 5: 468-474.

15. Simangunsong GM, Wahyudi S (2015) Effect of bedding plane on prediction blast-induced ground vibration in open pit coal mines International. Journal of Rock Mechanics and Mining Sciences 79: 1-8.

16. Kumar R, Choudhury D, Bhargava K (2016) Determination of blast-induced ground vibration equations for rocks using mechanical and geological properties. Journal of Rock Mechanics and Geotechnical Engineering.

17. Faramarzi F, Ebrahimi F, Mohammad A, Mansouri H (2014) Simultaneous investigation of blast induced ground vibration and air blast effects on safety level of structures and human in surface blasting. International Journal of Mining Science and Technology 24: 663-669.

18. Nicholls HR, Johnson CF, Duvall WI (1971) 'Blasting vibrations and their effects on structures', U.S. Department of Interior, Bureau of Mines Bulletin 656.

19. Brinkmann JR (1987) The control of ground vibration from colliery basting during the undermining of residential areas. J S Inst Min Metall 87: 53-61.

20. Edwards M, Rudenko PGD (2011) Site attenuation and vibramap Study for Hanson aggregates east Crabtree quarry Raleigh, Wake country, North Carolina. Vibra-Tech Engineers report.

21. Rorke AJ (2011) Blasting impact assessment for the proposed new largo colliery based on new largo mine plan 6. AJR_NL001_2011 Rev 4.

22. DIN 4150 (1986) 3 Structural vibration-Effects of vibration on structures.

23. Svinkin MR (2007) Assessment of safe ground and structure vibrations from blasting. European Federation of Explosives Engineers, ISBN 978 -0-9550290-1-1.

24. Nicholson RF (2005) Determination of blast vibrations using peak particle velocity at Bengal Quarry, in St Ann, Jamaica. Master's thesis. Lulea University of Technology.

25. Siskind DE, Stagg MS, Kopp JW, Dowding CH (1980) 'Structure response and damage produced by ground vibrations from surface blasting', RI 8507, U.S. Bureau of Mines, Washington, D.C.

26. Egan J, Kermode J, Skyrman M, Turner LL (2001) Ground vibration monitoring for construction blasting in urban areas. California Department of Transportation. Division of structural foundations. Technical report. F-00-OR-10.

27. Abdel-rasoul IE (2000) Measurement and analysis of the effect of ground vibrations induced by blasting at the limestone quarries of the Egyptian Cement Company. ICEHM 54-71.

6

Application of Vertical Electrical Soundings (VES) for Delineating Sub Surface Lithology for Foundations

Sandeep Meshram* and Sunil P Khadse

College of engineering-PUNE, Maharastra, India

Abstract

In today's world, field investigation is a necessity to get a detailed overview of any area for civil engineering construction purpose. Precise determination of engineering geological properties is essential to plan for a proper design and successful construction for any civil engineering structure. The traditionally practiced conventional methods for the same are invasive, costly and time consuming. Electrical Resistivity Survey is an attractive tool for delineating subsurface geology without soil disturbance. Reliable correlation between electrical resistivity values and other field geological parameters can help in successful interpretation of the engineering properties and behavior of soil in evaluation of difficult terrains e.g. for obtaining hard rock position, obtaining continuity of rock strata and for knowing the position of various sub – stratifications. This has led to develop and put in practice the geophysical method of sub-surface investigation for a more precise, economical and fast assessment of large areas like the present study area of Cauvery-Vaigai link canal project. The present paper presents the results of the use of Vertical Electrical Sounding coupled with Hydrogeological studies of the Cauvery-Vaigai-Gundar (CVG) link canal area. On the basis of generalized Vertical Electrical Soundings (VES) log it is seen that the aquifers in tertiary rocks are deep – seated, the depth varying from 80 to 300 m from the ground level. It was observed that the open wells are tapping unconfined aquifer system is highly weathered rock, moderately weathered/ fractured hard rock and highly fractured hard rock. The bore wells are tapping the confined aquifer system 40 m Below Ground Level (BGL) in highly fractured hard rock. The aquifers in the bore wells were reported to be in between 40 to 120 m BGL. Most of the bore wells are dry at present due to depletion of aquifer owing to very little precipitation and over exploitation. From resistivity survey it is inferred that where ever highly fractured hard rock are expected along the link, abundant quantity of ground water may be present there in that fractured zones. The result of resistivity values of different rock types of the area show resistivity of less than 20 ohm-m in the weathered soil, while highly fractured rocks show its range between 100 to 250 ohm-m, moderately fractured rocks shows range between 250 and 500 ohm-m and the massive crystalline rocks exhibit the range of more than 500 ohm-m. The water table shows periodical variations. During the excavation of the canal, the study of water table condition will give an idea about the chances of striking ground water to cut and cross drainage portions as ground water poses a serious hazard during the excavation works.

Introduction

Geophysical survey is necessary to ascertain subsurface geological and hydrogeological conditions and helps to delineate regional hydrogeological features, even pinpoint locations for drilling of boreholes. Geophysical data provides information on local geological environment such as type and extent of surface material, extent an degree of weathered mantle, the nature and extent of underlying bedrock, the structural elements etc. that influence ground water occurrence and movement. The important geophysical method involved in ground water exploration is Geo - Physical prospecting.

The Pre Quaternary biostratigraphy and lithostatigraphy and environment of deposition and the associated tectonism in the Cauvery basin is well documented specially the Creataceous formations [1-4]. The previous studies on Pre Quaternary sediments have only emphasized the depositional pattern and history in response to the basin tectonics and environment and not related it to the source region climate and tectonism. There is dearth of literature on the Quaternary geology of the Cauvery delta. The first comprehensive geological study of Cauvery river basin was done by Blanford of GSI in 1862. Most of the studies in this region is concentrated on Late Jurrasic to Mio-Pliocene sediments, with major emphasis on the Cretaceous sedimentary rocks [2,5,6]. They have built a detailed biostratigraphy for this period of sediments. Coming to the Quaternary the literature is almost missing except for some work done on sedimentology and geochemistry of surface sediments and channel suspended sediments more so to assess the solute load, nutrient and pollutant transfer by the river [7-10].

The river Cauvery branches off into a number of distributory rivers with its apex located at Grand Anicut (East of Trichy) and forms a triangular shaped delta. While the main Cauvery River flows in the North-eastern rim of the delta as river Coleroon, all its distributaries stand as paleochannels over which the rivers like Vennar, Vettar, Arasalar, Kudamurthi flow as misfit rivers [11]. Such abandoned nature of distributory channels and fluvial activity only in northern rim can suggest that entire Cauvery delta is an abandoned one. There has been lot of debate on the Quaternay tectonics in the Cauvery delta and associated changes in the landforms. Based on the morphological and land sat imagery study suggest northward migration of river Cauvery in the delta from the initial south-east course from Trichy to Bay of Bengal through Vedaranyam to present North-east course along Coleroon [12,13]. In the main Cauvery delta, the present day Cauvery flows in the northwestern rim of the delta as river Colleroon and all the distributary system of rivers, stand exposed as palaeo drainages over which the present day Vennar – Palamcauvery schemes of drainages are misfit. Occurrence of palaeo distributary system in the south with present day Cauvery (Colleroon) in the northern rim was attributed to anticlock wise rotational migration of Cauvery in its deltaic region related to

*Corresponding author: Sandeep Meshram, Ph.D, College of engineering-PUNE Maharashtra, India, E-mail: mailsandeep2010@gmail.com

the tectonic upliftment of Mio pliocene sandstone in Mannargudi - Pattukkottai region [11].

Electrical resistivity in support of Geological mapping has been well studied by Dale Rucker et al., [14] wherein the resistivity data was correlated to geological maps and to the data obtained by bore hole logging. Schepers et al., [15] have emphasized that Different geophysical methods have been used successfully to solve a number of special geotechnical problems.

The proposed work of Geophysical resistivity studies was undertaken to know the subsurface Lithology, texture, structure, and mechanical behavior of rock material respectively, and its application to the Project of Interlinking of Indian Rivers. The outcome of the study will be socially relevant because the Rivers have played a central role in Economic and Cultural development of any country but in India it is more than life, here Rivers is divine, River is Goddess.

Study Area

The Cauvery - Vaigai - Gundar link project cover parts of Karur, Thiruchirapalli, Pudukkottai, Shivaganga, Ramanathapuram and Viruthnagar districts of Tamil Nadu. The Study area falling in parts of Survey of India degree sheets No. 58 I, 58 J, 58 M and 58 N bounded by latitudes 90 10' 00" to 110 00' 00" and longitudes 780 05'00" to 790 00' 00" covering an area of 18,000 sq. km The link canal will off take from a proposed regulator located on the right bank of river Cauvery at the location upstream of the proposed Kattalai barrage. The off take falls within Krishnarayapuram taluk of Karur district in Tamilnadu (Annexure 1).

The rivers that would be crossed by the canal are Napalli, Koraiyar, Kondar, Vellar, Pambanar, Virisalar, Sarugani, Vaigai, Gridhamal, and Gundar. Command area of the link canal lie between north latitudes 9° 10' 00" to 11° 00' 00" and east longitudes 78° 05' 00" to 79° 0' 00". The entire link is enclosed in the parts of Survey of India Toposheets bearing Nos 58 J and 58 K of scale 1: 250000 in the districts of Karur, Tiruchchirappalli, Pudukkottai, Sivaganga, Ramanathapuram and Virudhunagar in Tamil Nadu.

Geology of the area

Precambrian crystalline rocks cover 80 percent of the terrain and Paleozoic sedimentary rocks cover the eastern costal terrain and the river valley account for the rest. In the deeply eroded Precambrian terrain rocks of the Khondalite and Charnockite Groups and migmatites derived from them are extensively traced within this west array of crystalline rocks, igneous emplacement of anorthosited, granites, ultramafic bodies and basic sills and dykes are defined. The geological setup of the Cauvery-Vaigai-Gundar link is as follows:

a) Quaternary sediments

b) Tertiary sediments

c) Granites

d) Migmatic rocks

e) Charnockites

Hydrogeology of the area

The hydrogeology parameters include noting the depth to water table, outcrops at surface, road cutting, pits, tunnels and well sections. The observations were carried out in the month of April to May. The water table ranges from 5 to 25 m BGL along the link. The link canal mostly passes through the draught prone area where rainfall is very scanty and hence at most places the water table is 15 m Below Ground Level (BGL). Most of the open wells are 10 to 12 m deep was found to be dry.

Materials and Methods

Profiling: The resistivity profiling is a technique to locate lateral variations in resistivity at a constant depth. Here the electrode configuration remains the same for the entire area and it is moved as a whole along a traverse, generally normal to the strike of the rock formations. This resistivity profiling is used as a semi-reconnaissance tool in the search of vertical structures with large resistivity contrasts such as faults, geological contacts, dykes, shear zones and veins. Electrode separation for profiling is chosen after conducting a number of vertical electrical soundings in an area. Based on these soundings, an optimum spacing is chosen for the required depth of investigation. Interpretation of resistivity profiling data is usually qualitative in nature. If we run a number of profiles in an area, we can even expect to find somewhat similar results and by joining these anomalies it is possible to support or otherwise the lineaments shown by the remote sensing method.

The sub-surface layers were interpreted using the graphs prepared from resistivity measurements coupled with geological and hydro geological observations. The sub-surface layers are classified on the basis of degree of weathering and fracturing, which are the essential engineering properties from the point of view of excavation of the material.

Vertical Electrical Sounding (VES): As part of the present investigations, detailed geophysical investigation has been carried out in the Cauvery- Vaigai Link Canal Project using the Schlumberger array at the selected locations. Current electrode spacing (AB/2) was gradually increased up to 100 m for delineation of deeper structures. Electrodes were spread in the direction parallel to strike direction. Over layered earth structures (1-D situation), variation in apparent resistivity with current electrode separations is quite smooth. Further, this variation is also smooth when the direction of spread is parallel to the strike and erratic when the direction of spread is perpendicular to the strike for 2-D situation. In the present study, a rather smooth variation in apparent resistivity is observed up to large electrode separations in the strike direction. Therefore, we assume that in such situation 1-D interpretation will yield significant subsurface features for recommending appropriate locations for civil engineering structures.

The vertical electrical soundings, otherwise known as electric drilling, depth sounding or depth probing, is used to determine the resistivity variation with depth and provide the information about the vertical distribution of fresh, brackish and saline water bodies if any and their aerial extent. A VES is typically carried out in Schlumberger array, where the potential electrodes are placed in fixed position with a short separation and the current electrodes are placed symmetrically on the outer sides of the potential electrodes (Figure 1). After each resistivity measurement the current electrodes are moved further away from the centre of the array. In this way the current is stepwise made to flow through deeper and deeper parts of the ground. The positions of the current electrodes are typically logarithmically distributed with at least 10 positions per decade. For large distances between the current electrodes, the distance of the potential electrodes is increased to ensure that the measured voltage is above the noise level and the detection level in the instrument. With the expansion of electrical array, the depth of penetration of electrical current is increased there by detailed

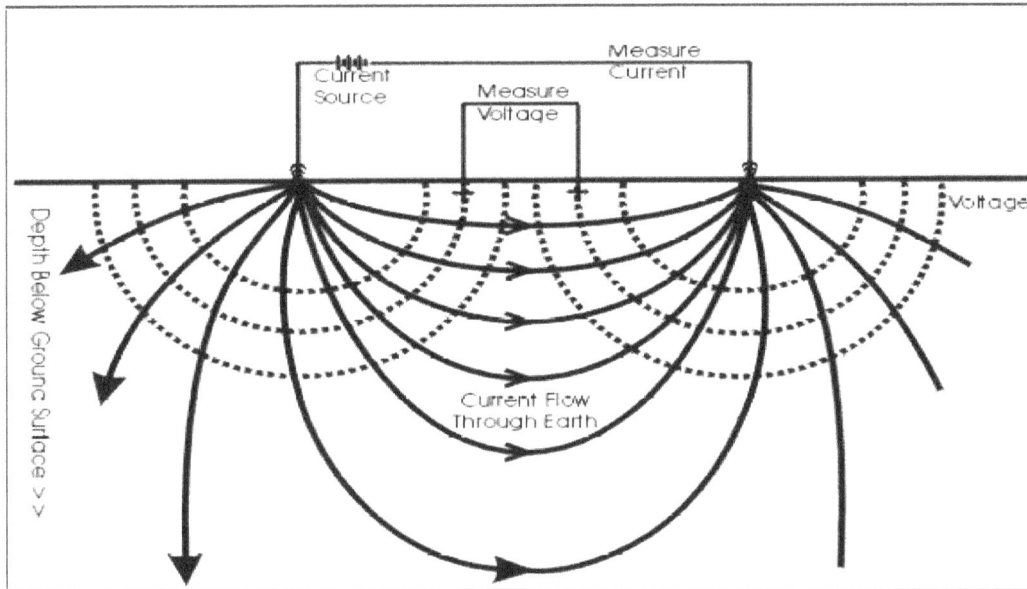

Figure 1: Schlumberger electrode arrangement used in present study.

Rock/Material	Almost Dry	Saturated with water
Quartzite	$4.4 \times 10^3 – 2 \times 10^8$	50–500
Granite	$10^3 – 10^8$	50–300
Limestone	$600 – 10^7$	50–1000
Basalt	$4 \times 10^4 – 1.3 \times 10^8$	10–50
Gneiss	$6.8 \times 10^4 – 3 \times 10^8$	50–350
Sand	$150 – 2 \times 10^3$	10–100

Table 1: Ranges of resistivity values (Ω.m) of common rocks/materials [16].

information and the vertical succession of various conductive zones, their individual true resistivity and thickness can be obtained. Some of the most common electrode arrays are Wenner, Schlumberger, pole-pole, pole-dipole and dipole-dipole array.

In reality, the subsurface ground does not conform to the homogenous medium and hence the resistivity obtained is 'apparent resistivity' (pa), rather than the 'true resistivity' (pt). Apparent resistivity values calculate from measured potential differences can be interpreted in terms of overburden thickness, water table depth, and the depths an thicknesses of subsurface strata. The apparent resistivity is obtained as the product of measured resistance(R) and geometric factor K, which depends on the geometric spread of electrodes.

pa= (AV/I) X K where AV is the measured potential, I is the transmitted current, and K, the geometrical factor .

The apparent resistivity values along the y-axis can be plotted against logarithmic value of current electrode half spacing along the x axis to obtain apparent resistivity curves, which is also known as field curves. Further these field curves are interpreted by different techniques like curve matching, curve breaks etc to get the different layer parameters, such as, depth to water table, depth to basement topography, thickness of weathered layers, detection of fissures, fractures as well as fault zone and also the quality of ground water in terms of dissolved salt content. In the present study the curve matching techniques are used to interpret the field curves obtained. VES data together with the subsequent borehole information has helped us to compile the hydrogeological potential of the areas along the link

Singhal and Gupta et al., [16] have given a range of resistivity values

of common rock/ materials under dry and saturated conditions and is shown in Table 1.

In crystalline terrains, the maximum depth of penetration needed is to reach the compact rock (bed rock), usually having infinite resistivity. Infinitely, resistive basement is reflected in the sounding curves as an asymptote at the extreme right hand part, making an angle of 45° with the x–axis on the double log sheet. Survey will be stopped as soon as this asymptotic part of the curve is obtained. In resistivity survey, Vertical Electrical Soundings (VES) were conducted at 269 predetermined locations using Schlumberger electrode configuration, at the cut and Cross Drainage portions. The soundings were separated by a distance 1 km except at Cross Drainage portions, where it ranged from 100 to 200 meter. The maximum depth of penetration was 40 m at RD 105.000 km [17].

The interpreted resistivity data is correlated with the observed geological and hydro geological field data and the subsurface layers are classified into four categories as:

1. Soil

2. Highly weathered rock

3. Moderately weathered/fractured hard rock

4. Highly fractured hard rock

5. Hard rock

The top layer soil is mostly the deposited layer consisting of sand, gravels, pebbles etc. along with fine particles. The remaining layers are the highly weathered rock, moderately weathered and/or fact hard rock and hard rocks are the same type of rocks with differential weathering. Based on the resistivity, various layer boundaries are demarcated. The layer boundaries at each sounding is referred to the mean sea level in which the topographic level (NSL-Natural Soil Level) is as given by National Water Development Authority (NWDA) and the other layers are as per the depth of their occurrences. Sub surface lithological vertical cross sections are prepared for every 25 km length of the alignment. The layer continuity is drawn by interpolation of the boundaries of the same lithology between the successive sounding, which is prepared through computer graphics. The information in the vertical cross sections consists of surface topographic levels, lithological boundaries, water table sub-surface lithology, Canal Bed Level, Full Supply Level of the proposed alignment. Graphs showing pseudo resistivity section along the Cauvery-Vaigai-Gundar (CVG) Link Canal is enclosed

Results and Discussion

The sub-surface vertical layer inferred from electrical resistivity survey, geological and for each 25 km of the link canal alignment. The contoured apparent resistivity pseudo section was produced from plots of VES points against electrode spacing. The pseudo section was generated using IP12 WIN software. The curves were interpreted qualitatively through visual inspection and quantitatively using IP12 WIN software. Interpreted results were used to construct geo-electric section from the layered parameters. The field curves show three, four and five to six layers case. Most of the curves indicate multiple increase and decrease of resistivity with depths.

This graphical information gives an idea about the type of rock formation likely to be encountered at cut and Cross Drainage portions and the volume of the different type of material to be excavated along the alignment. From resistivity survey it is inferred that where ever highly fractured hard rock is expected along the link, abundant quantity of ground water may be present there in those fractured zones. The water table shows periodical variations. During the excavation of the canal, the study of water table condition will give an idea about the chances of striking ground water at cut and Cross Drainage portions as ground water poses a serious hazard during the excavation works. During the survey, over exploitation was observed at some places, where the shallow and deep aquifers are in good health. Some bore wells were also observed near the alignment, whose depths are ranging from 40 to 120 m BGL.

In the study area, shallow ground water aquifers occurs in alluvial plains, palaeochannels and the depth of the water table in the aquifers shows seasonal changes from about 2 to 7 m from the ground level. The aquifers in tertiary rocks are deep – seated, the depth wearying from 80 to 300 m from the ground level. It was observed that the open wells were tapping unconfined aquifer system in highly weathered rock, moderately weathered/fractured hard rock and highly fractured hard rock. The bore wells are tapping the confined aquifer system 40 m BGL in highly fractured hard rock. The aquifers in the bore wells were reported to be in between 40 to 120 m BGL. Most of the bore wells are dry at present due to depletion of aquifer owing to very little precipitation and over exploitation.

Analysis of four layer curve shows that, potential fracture zones occur in this type of resistivity profile. The first layer is top soil and second layer is saturated shallow aquifer, the third layer is basement rock and fourth is the fracture zone. Mostly in five layer cases, the fracture zone occurs as the fourth layer. In the six layer case, the first layer is top soil, the second layer is dry unsaturated zone, the third is saturated and the fourth layer is massive rock followed by fracture zone. In the seven layer case, the shallow saturated zone occurs as third layer. The fracture zone occurs as sixth layer. The above analysis show that fracture zone occurs in various depths in different type of resistivity profile. The areas where apparent resistivity are low in the fourth, fifth and sixth layers have high groundwater potential. Vertical Electrical Sounding (VES) locations show that most of the VES's near the lineament intersections has very good groundwater potential. It is can be concluded that the potential aquifers in the area are mainly the fracture zones, faults, joints and weathered column and it is under confined to semi-confined condition.

The top layer soil is mostly the deposited layer consisting of sand, gravels, pebbles etc. along with fine particles. The remaining layers are the highly weathered rock, moderately weathered and/Fractured hard rock and hard rocks are the same type of rocks with differential weathering. Based on the resistivity, various layer boundaries are demarcated. The layer boundaries at each sounding are referred to the mean sea level and the other layers are as per the depth of their occurrences. Sub surface litho logical vertical cross sections are prepared for 255 km length of the alignment. The main geological formations observed in the study area are metamorphic hard rock and the remaining stretch is dominated by sedimentary domain consisting of sand, clay and shale, capped by laterite. In metamorphic terrain, top soil and weathered migmatitic gneiss form the media. The sedimentary formations consisting of sand and clay exposed in canal route appear to have poor shear strength. In general, no adverse geological features are noticed along canal alignment, as seen by the present study.

Conclusion

The economy of any project depends upon the quality of the preliminary investigations carried out. The findings generated through this study will give a new insight to researchers in the interpretation of the foundation conditions and will facilitate the design of appropriate foundation systems. The results depicted in the vertical cross sections show the surface topographic levels, lithological boundaries, water table, sub-surface lithology, Canal Bed Level (CBL), and Full supply level (FSL) of the proposed alignment. The sub-surface section of the link is prepared from the data obtained by electrical resistivity survey and geological survey done by the author along the link canal alignment.

As seen from the present study the area is mainly underlain by Charnockite/Charnockite gneiss type formations. This rock type is generally massive and does not store or transmit considerable quantity of water except there are any fractures or joints in it. The Geomorphological map reveals the landforms in the study area which mainly are structural hills, pediplain, less dissected plateau, channel bar, water bodies and valley fill. Among these, Valley fills have the maximum groundwater potential. Slope map shows that steeper slopes in the southern and north eastern part of the study area. Drainage pattern of the basin is influenced by infiltration properties of soil; more is the drainage density lesser the infiltration. Interpretation of VES data gives the details of the underlying fracture zone and thickness of the potential fractured layers. Qualitative interpretation of VES data shows multi layered resistivity profiles for most of the locations (mare than four layers). The pseudosection of measured apparent resistivities as enclosed in Annexure 2 reveals that the values change both horizontally and vertically, although a general increase in the values with depth is expected. At some places, irregular topography between the top heterogeneous ground and the underlying material is observed which may be due to slight variation in lithology or to the weathering of rocks which gives higher resistivity values and indicates that soil is more compacted (128 ohm to 300 ohm) and has less moisture content than the rocks of surrounding areas. The study area has shallow ground water aquifers in alluvial plains and in palaeochannels of the rivers. The depth of the water table in the aquifers shows seasonal changes ranging from 2 to 7 m from the surface level. The aquifers present in the tertiary rocks are deep – seated, and their depth varying from 80 to 300 m from the ground level. It was observed that the open wells were tapping unconfined aquifer system in highly weathered rock, moderately weathered/fractured hard rock and highly fractured hard rock. In addition to this, the bore wells are tapping the confined aquifer system 40 m Below Ground Level in highly fractured hard rock. The aquifers in the bore wells are found to be in between 40 to120 m Below Ground Level. Most of the bore wells are dry at present due to depletion of aquifer owing to very little precipitation and over exploitation, but most of the time, under the sufficient rate of precipitation, the bore wells and dug wells yields good amount of water. The geomorphological characters of the landscape are helping for the additional recharge of the groundwater in this area.

Based on the above studies it is finally suggested that for a huge project like Interlinking of Rivers in India in the present study area of Cauvery-Vaigai-Gundar Link Canal Project, for construction purpose the material like soil/alluvium (including all types of gravel, pebbles, and boulders) should be excavated either manually or by machines. The highly weathered rock material should be excavated either manually or by machines. The material like moderately weathered hard rock is to be excavated either by chiseling or by blasting and the hard rock material should be excavated by blasting.

References

1. Mohan M, Kumar P, Narayanan (1977) Paleocene-Eocene boundary in Cauvery Basin. Journal Geological Society of India 18: 401-411.

2. Venkatachala BS, Sastri VV, Raju ATR, Sinha RN, Banerji RK (1977) Biostratigraphy and evolution of the Cauvery Basin. Journal Geological Society of India 18: 355-377.

3. Govindan A (1977) Upper Cretaceous and Lower Tertiary foraminiferal genus Bolikvinoides from the upper Cauvery Basin, South India and its palaeo-bio-geographical significance. Journal Geological Society of India 18: 459- 476.

4. Ramasamy SM, Saravanavel J, Selvakumar R (2006) Late Holocene Geomorphic Evolution of Cauvery Delta, Tamil Nadu. Journal of Geological Society of India 67: 649-657.

5. Nagendra A, Southworth JT (2002) Accessibility as determinant of landscape transformation in western Honduras; Linking pattern and process. Landscape Ecol 18: 141-158.

6. Madhavraju P, Ramasamy SM (2002) Sedimetological studies Pre-Cretaceous and Cretaceous sequence of key wells of Krishna Godavari basin. Petroliferous basins of India 1: 343.

7. Seralathan P (1987) Trace element geochemistry of modern deltaic sediments of the Cauvery River, east coast of India. Ind J Mar Sci 16: 235-239.

8. Seralathan P, Seetharamaswamy A (1979) Phosphorous distribution in modern deltaic sediments of the Cauvery river. Ind Jr Marine Sc 8: 130-136.

9. Vaithiyanathan P, Ramanathan AL, Subramanian V (1988) Chemical and sediment characteristics of the upper reaches of the Cauvery estuary, East Coast of India. Ind J Mar Sci 17: 114-120.

10. Vaithiyanathan P, Ramanathan AL, Subramanian V, Das BK (1994) Nature and transport of solute load in the Cauvery River basin, India. Science Direct Elsevier 28: 1585-1593.

11. Ramasamy SM (1991) A remote sensing study of river deltas of Tamil Nadu. Memoirs of Jr of Geol Soc of India 22: 75-89.

12. Ramasamy SM, Panchanathan S, Palanivelu R (1987) Pleistocene Earth Movements in Peninsular India- Evidences from Landsat MSS and Thematic Mapper Data. Proceedings of International Geo Science and Remote Sensing Symposium, Michigan University, Ann Arbor.

13. Meijerick AMJ (1971) Reconnaissance of Quaternary geology of Cauvery delta. Journal of Geological Society of India 2: 113-124.

14. Rucker DF, Noonan GE, Greenwood WJ (2010) Electrical resistivity in support of Geological mapping along the Panama Coast-Elsevier. Journal of Engineering Geology 117: 121-133.

15. Schepers R, Rafat G, Gelbke C, Lehman B (2001) Application of borehole logging, core imaging and tornography of geotechnical exploration. Intl J of Rock Mech and Mni Sci Elsevier 38: 867-876.

16. Singhal BBS, Gupta RP (1999) Applied Hydrology of Fractured Rocks. Springer Science and Business Media.

17. Ramasamy SM, Karthikeyan N (1998) Plieistocene/Holocene Graben along Pondicherry-Cumbum Valley, Tamil Nadu. Geocarto Imternational 13: 83-90.

7

Analysis of Hydrostatic Pressure Zones in Fabi Field, Onshore Niger Delta, Nigeria

Victor OM*, Ude AE and Valeria AC

Department of Geological Sciences, Nnamdi Azikiwe University, Awka, Nigeria

Abstract

The Analysis of hydrostatic pressure of Fabi Field onshore Niger delta was carried out to understand the subsurface depth pressure variations in the field. The well logs were analyzed to identify hydrostatic pressure zones through points of deviation from the compaction trends of the sediments and attribute crossplots while seismic inversion was carried out on the seismic data to obtain the lateral hydrostatic pressure variations in Fabi Field. The well logs exhibited normal compaction trends from depth level of 7200 ft to 8625 ft. At depth level of 8625 ft to 9000 ft, the velocity, density and resistivity logs deviated from the normal compaction trends. This point of deviations from the compaction trends of the sediments were identified as overpressure zones. Through crossplots of velocity and density logs, two overpressure generating mechanism of the Fabi Field such as under compaction and unloading were revealed. The under compaction was characterized by a linear drop in velocity and increase in density with depth while Unloading is characterized by abrupt decrease in velocity and at constant change in density. The results of acoustic impedance inversion revealed the lateral hydrostatic pressure variations of the Fabi Field. The onsets of over pressure zones are found around 1600 ms to about 1630 ms while very high overpressure gradients occur between 1950 ms to 2150 ms. These zones are characterized by very high acoustic impedance. The area extents of the positive anomalies (increase in acoustic impedance) are mostly consistent with the pressure while negative anomalies (low acoustic impedance) are interpreted as reservoir or sand zones.

Keywords: Well logs; Pressure zones; Acoustic impedance; Hydrostatic pressure; 3D Seismic; Wavelet analysis; Crossplots analysis

Introduction

In study of pore pressure, physical properties of subsurface formation with hydrostatic pressure are encountered worldwide during hydrocarbon exploration. Hydrostatic pressure as use here is simply the pore pressure at normal conditions. This occurs as a result of under-compaction due to rapid burial of the sediments. Rapid processes of deposition and sedimentation that builds up the basin have resulted into under-compaction at subsurface depths. If the loading process is rapid, fluid expulsion through compaction is seriously impeded, especially in fine-grained sediments with low permeability such as silts or clays. This confining layer that seal up the reservoir retard the escape of pore fluids at rates good enough to compensate for the rate of increase in vertical stress induced by the overlying beds and thus the pore fluid pressure begins to carry a large part of the load resulting to abnormal increase in pore fluid pressure. In many cases the shale pressure, and hence magnitude of overpressure, based on sedimentation rate is linked to the process of compaction disequilibrium although is empirical. This pressure tends to affect the seismic or sonic velocities where possible, thereby posing significant threats to drilling safety. The cost of mitigation is very high, to the tune of $1.08 billion per year world-wide. Dutta [1] discussed causes of overpressure and various ways to detect such a phenomenon. Boer et al. [2] presented a paper describing an approach used to estimate subsalt pore pressure in deep water Green Cayon area of Gulf of mexico. In their approach, they use a tomographyically derived 3D estimate of seismic velocity. The quality of seismic velocities allows an acurate velocity model of 3D salt distribution to be defined, the precise delineation which is essential for pore pressure prediction. Amonpantang [3] explained that methods for pore pressure estimation can give the approximate value of pore pressure but not the exact number. He reports that using well sonic data is the most suitable method for overpressure estimation. Real Formation Tester data is the measure of pore pressure that is used to compare with the predicted Pore pressure. Yu [4] reports that to meet today's challenge of high drilling cost and green environmental requirements and to obtain accurate and quantitative pore pressure information, pore pressure analysts are to know the abnormal pressure causes and building a good model. Solano et al. [5] reported that using the ratio between exponent and effective stress, pore pressure can be estimated more accurately than the standard exponent method for shaly formations. This approach is more objective for the definition of the normal compaction trend, because normal compaction trend is defined for the entire field rather than for individual wells. According to Standifird et al. [6], different measurements of the overpressure estimation which can be used to understand rock and fluid properties are based on three sources of data; Logging while drilling (LWD), wire line logging and seismic reflection surveys.

In this paper, we concentrate on the use of seismic and well log data in determination of overpressure zones. Seismic and well log data are very important in understanding of geopressured formation zones before and after drilling respectively. In prediction of abnormal pressure during drilling, the technique is based on mechanical drilling data (rock strength computed from rate of penetration (ROP), weight on bit (WOB) and torque). The rate of penetration is monitored to signal the penetration into overpressure zones, especially, if there is a transition zone. However, since is not easy to keep other drilling

Corresponding author: Victor OM, Department of Geological Sciences, Nnamdi Azikiwe University, Awka, Nigeria, E-mail: vctrmbah@yahoo.com

parameters constant, this method is not very reliable thus, is not considered in this research. In detection of abnormal pressure zones in Fabi Field, we construct rock model and predict overpressure by identifying the deviation points from their normal compaction trends of sediments from well logs and seismic inversion. Our objective is to identify hydrostatic pressure zones in the study area.

Location of the study area

The field name is Fabi. It is located some kilometres southwest of Port Harcourt province in Onshore Niger delta, Nigeria as shown in Figure 1. The area is situated on the continental margin of the Gulf of Guinea in West Africa at the south end of Nigeria. The Niger delta lies between Latitude 4° and 7° N and longitude 3° and 9° E bordering the Atlantic Ocean on the southern end of Nigeria. The northern boundary is the Benin flank and the North-eastern boundary is defined by cretaceous outcrop of Abakiliki High and southeastern end by Calabar flank. The province covers 300,000 Km and includes the geologic extent of the tertiary Niger delta (Akaka-Agbada) petroleum system [7-13].

Materials and Methods

Research data were provided by Shell Petroleum Development Corporation, Eastern division, Port-Harcourt in Rivers state, Nigeria. The data includes well logs and 3D seismic data, from Fabi Field, onshore Niger delta oil field.

The data were given with strict confidentiality for security reasons. The well logs were digitized in LAS format. Data interpretations and analysis were carried out using HRS program. Wire line log data comprise sonic log, density, resistivity log, caliper log, porosity log, and gamma ray log [14]. The inverse of the interval transit time of the sonic log were used to generate a compressional and shear wave velocity for each well. The well logs were marked with spurious events such as high frequency noise, and invasion problems. They were subjected to various corrections or editing so as to limit the possible interpretational error. The logs were despiked using media filtering, to ensure that they contain only appropriate range of values (Figure 2). The media filtering operation replaces the sample value at the centre of the operator. The longer the operator length, the smoother the log signatures. This process was largely experimental in order to isolate the best log operator length, we found that with an operator length 25, the logs were largely well smoothened. Figure 2 shows the difference between the non despiked well log using media filter and despiked well log respectively [15]. The data were subjected to wavelet analysis. Wavelet is defined by both amplitude and phase spectrum. We applied two wavelet extraction processes in the research. These include statistical and well log wavelet extraction. Statistical uses seismic traces alone to extract wavelet by Weiner- Levinson deconvolution process which uses autocorrelation function. The well log approach uses the log to determine the constant phase used in combination with statistical approach.

The main objective of wavelet extraction is to obtain qualitative seismic to well calibration. Before we apply seismic to well tie, an accurate depth time conversion was performed in order to make the vertical scale of the well acoustic impedance data match the vertical scale of the seismic data so as to allow spatial correlation [16]. This conversion was carried out using the sonic and the initial two ways travel time. The first sample provides the highest correlation coefficient between the synthetic and the observed trace. This is commonly known as seismic to well tie [17] . In this process, we manually stretch or squeeze the log and the seismic in order to improve the time correlation between the target logs and the seismic attributes. Once the needed bulk shift and stretches are applied, the well log depth to time map match the seismic times. This process simultaneously creates a composite trace from the seismic and synthetics seismogram from the log as shown in Figure 3.

With data properly corrected for error, the well logs were analyzed to determine the hydrostatic pressure trends through points of deviation from the compaction trends of the sediments. The deviations from the compaction trends of the sediments were identified as the point of the overpressure zones. In seismic analysis of overpressure zones, low frequency acoustic impedance model was created [18]. The

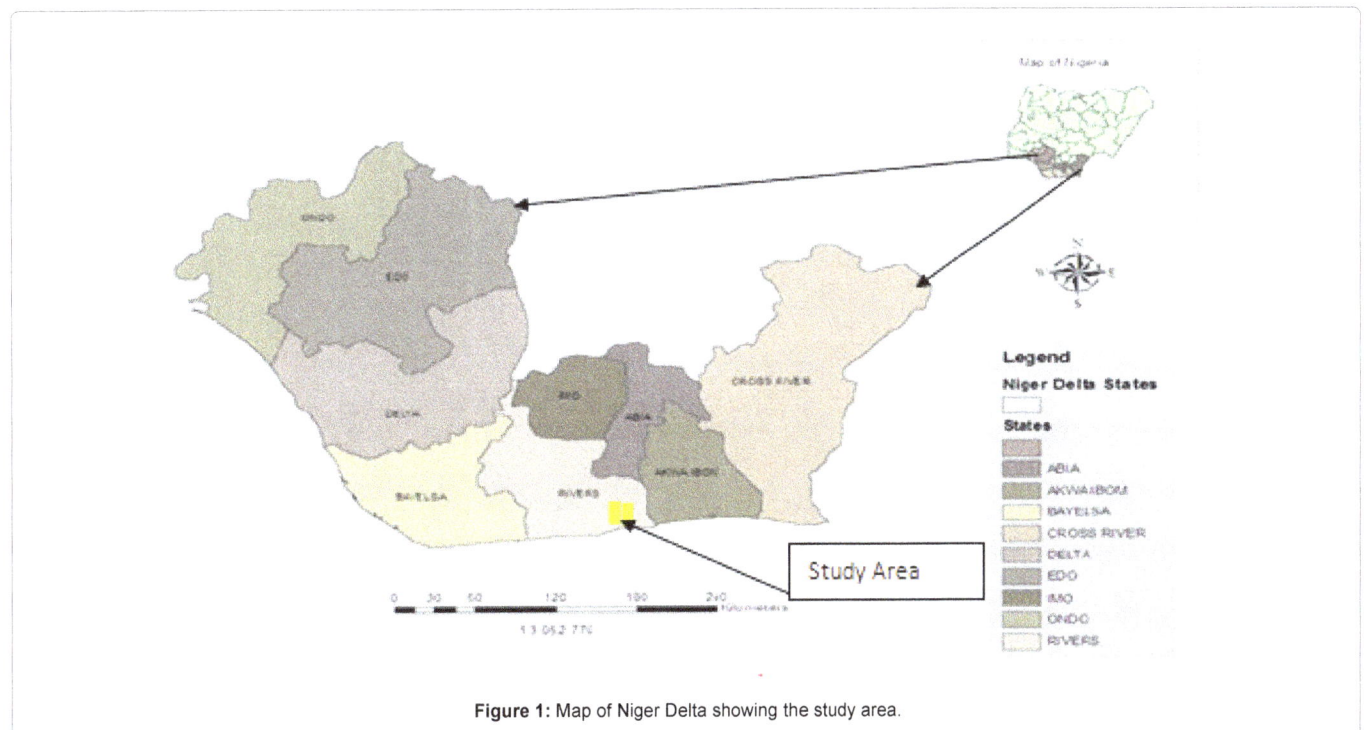

Figure 1: Map of Niger Delta showing the study area.

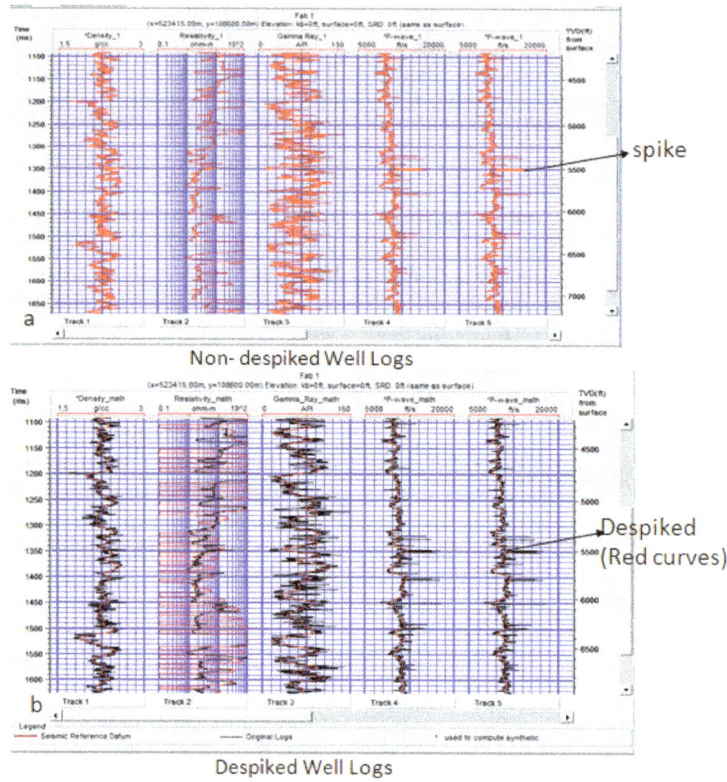

Figure 2: Well log correction showing (a) original well logs with spike and (b) despiked well logs (red curve).

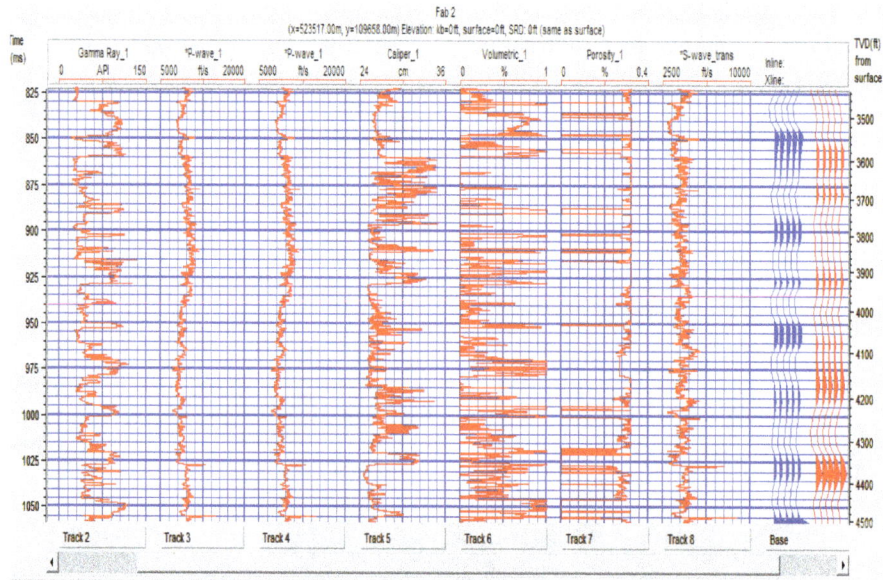

Figure 3: Seismic to well tie of (a) Fabi 1 and (b) Fabi 2.

model based inversion derived the impedance profile which best fit the modeled trace and the seismic trace in a least square sense using initial guess impedance guided by well logs and horizons. Basically this inversion process resolves the reflectivity from an objective function and compares its RMS amplitude with the assumed reflectivity size. The low frequency model was subjected to acoustic impedance inversion. The inverted seismic section in this case, provide means of detecting abnormal pressure regimes by looking at high acoustic impedance zones which may correspond to under compacted shale and hence overpressure zones.

Results and Interpretation

In well log identification of hydrostatic pressure, we exploited the deviation characteristics of rock property from normal compaction trends of both normal and inverted well logs. The pore pressure prediction points were picked from the clean shale depth intervals through the measured points in sands as shown in Figure 4. The well logs exhibited normal compaction trends from depth level of 7200 ft to 8625 ft. At depth level of 8625 ft to 9000 ft, the velocity, density and resistivity logs deviated from the normal compaction trends. This indicates the onset of the overpressure zones.

Sonic velocity increases with depth for a normal pressure zone. It has been observed to follow normal compaction trends. But when overpressure sets in, the velocity invariably slows down and is observed to fall below the normal compaction trend line (blue line). This is due to increase in porosity in the overpressure shale. The depth point where this change begins is known as the top of overpressure. The pink horizontal line highlights the top of overpressure and below it is the overpressure zone as shown in Figure 5. Bulk density measurement is characterized by normal increasing trend until the top of the overpressure zone is reached. In overpressure environments, the bulk density values are lower than normal trend value due to increase pore volume in this zone. However, as rock compacts, porosity is reduced in a normal pressure zones but increase in overpressure zones due to the inability of the pore fluid to escape in equilibrium with the rate of compaction but shale resistivity increases with depth, since porosity decreases due to compaction. In overpressure zones, the resistivity of shale departs from their normal trend, it invariably decreases. This is because of the relative change in the mineralogy whereby the illite part increases at the expense of the montmorillonite part of the clay structure, leading to high conductivity. The red marked horizons on the log suites as shown in Figure 6 identified the reservoir units in across the field identified as sands zones which indicate the measured pore pressure points while the remaining unmarked parts represent the shale

units which indicate the predicted pore pressure points with varying vertical depth of occurrence at far right end of the figure

To obtain these depth pressure variations, we introduced crossplots analysis of the predicted pore pressure and measured pore pressure points of study area. However, various mechanisms can cause hydrostatic pressure in rocks and their behavioral pattern differs. With the crossplots of velocity and density logs, two overpressure generating mechanism of the study area were revealed which include undercompaction and unloading mechanism.

The undercompaction is characterized by a linear drop in velocity and increase in density with depth. In this case, with increase in vertical stress, the pore fluids escape as rock pore spaces try to compact. If a layer of low permeability material prevents the escape of pore fluids at rates sufficient to keep up with the rate of increase in the vertical stress, the pore fluids begin to carry a large part of the load and the pore fluid pressure will increase as shown in Figure 7, identified via color codes at far right side of the figure. While unloading is characterized by abrupt decrease in velocity and at constant change in density. In this case, velocity-density plot recognize unloading, characterized by abrupt decrease in velocity at a constant density as shown in Figure 7. Fluid expansion unloading mechanism occurs due to processes like heating, clay dehydration and hydrocarbon maturation (source rock to oil and gas). It could also result when sediment under any given compaction condition has fluids injected into it from a more highly pressure zone. However, in the result of well log inversion, acoustic impedance has a very high degree of correlation to porosity as indicated by well log analysis (Figure 8). An inverted acoustic impedance model of the well log is shown in the red curve while the blue curve represents the original logs. High acoustic impedance represents points of the overpressure which were majorly observed in the shale zones. The horizons defined the net pay cutoffs which relatively aligned with the point of the low acoustic impedance values. The high acoustic impedance is associated with the right kick which probably represent shale zones with high overpressure gradients.

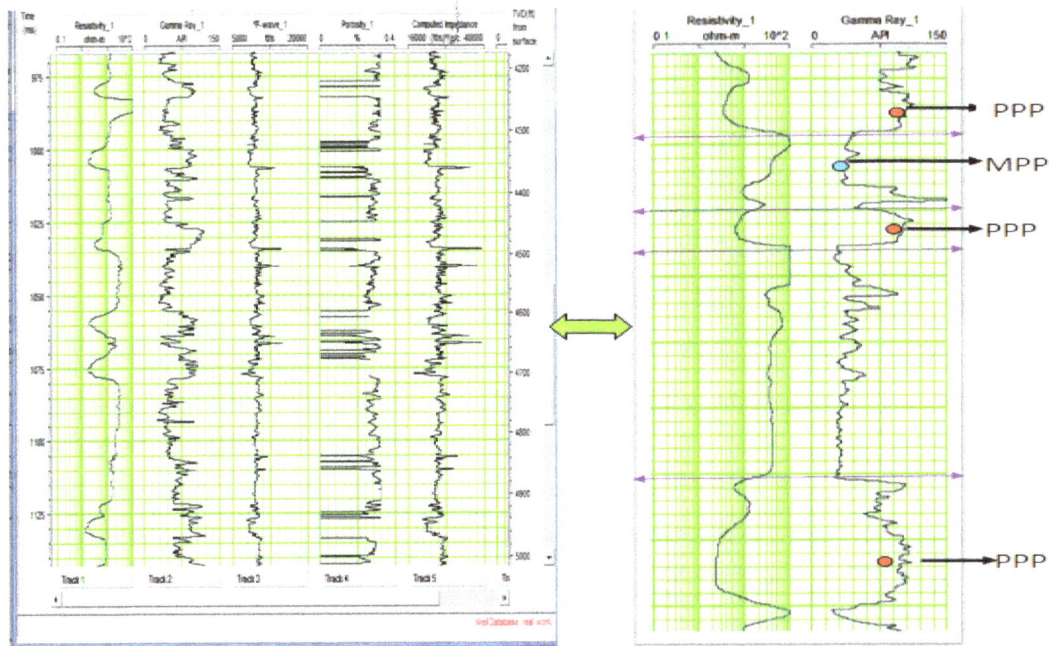

Figure 4: Predicted pore pressure (PPP) points picked from shale depth intervals (red dots) and measured pore pressure (MPP) points in the sands (blue dots).

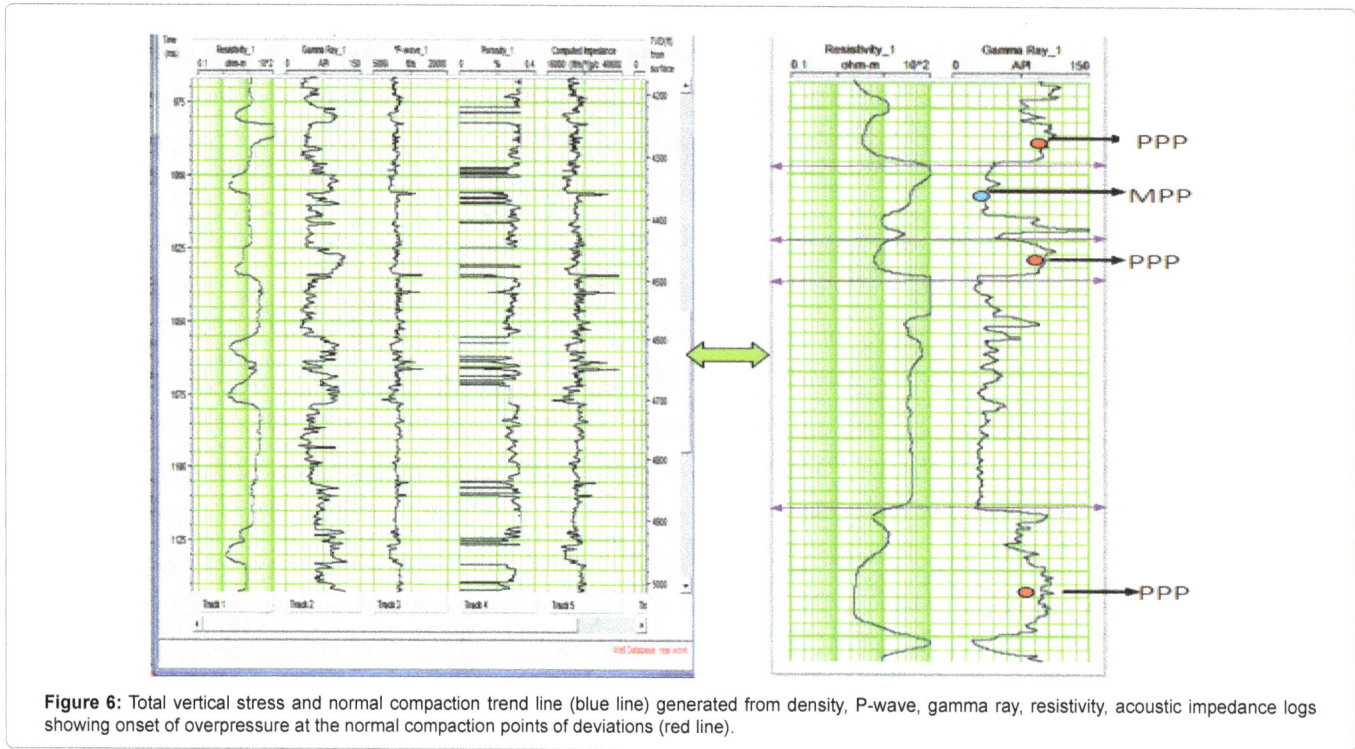

Figure 6: Total vertical stress and normal compaction trend line (blue line) generated from density, P-wave, gamma ray, resistivity, acoustic impedance logs showing onset of overpressure at the normal compaction points of deviations (red line).

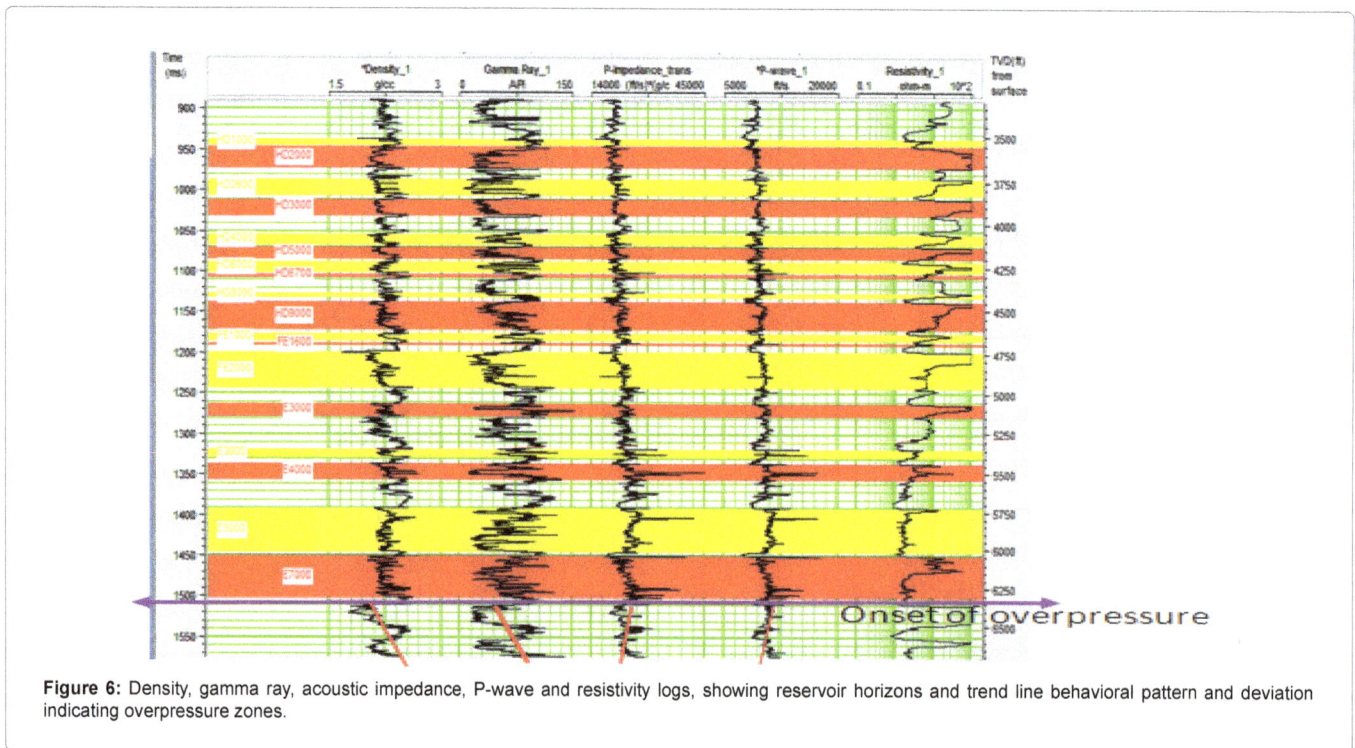

Figure 6: Density, gamma ray, acoustic impedance, P-wave and resistivity logs, showing reservoir horizons and trend line behavioral pattern and deviation indicating overpressure zones.

Well log acoustic impedance inversion presents a clearer picture of overpressure variations in a formation because it correlates the individual points of the pore pressure zones to the seismic data by matching the synthetic of the well logs (red) with that of seismic data (black) as shown in the far right side of Figure 8. To determine the formation overpressure variations, we present acoustic impedance inversion of the seismic data as shown in Figure 9. The results present the hydrostatic pressure zones

in the Fabi Field as defined by blue and pink color code which present high acoustic impedance zones in Fabi field. The onsets of over pressure zones are found around 1600 ms to about 1630 ms as indicated by blue color codes. However, very high overpressure gradients occur between 2000 ms to 2100 ms. These zones are characterized by very high acoustic impedance as shown by pink color codes identified via color bar. The result also show high sand/reservoir distributions which are associated

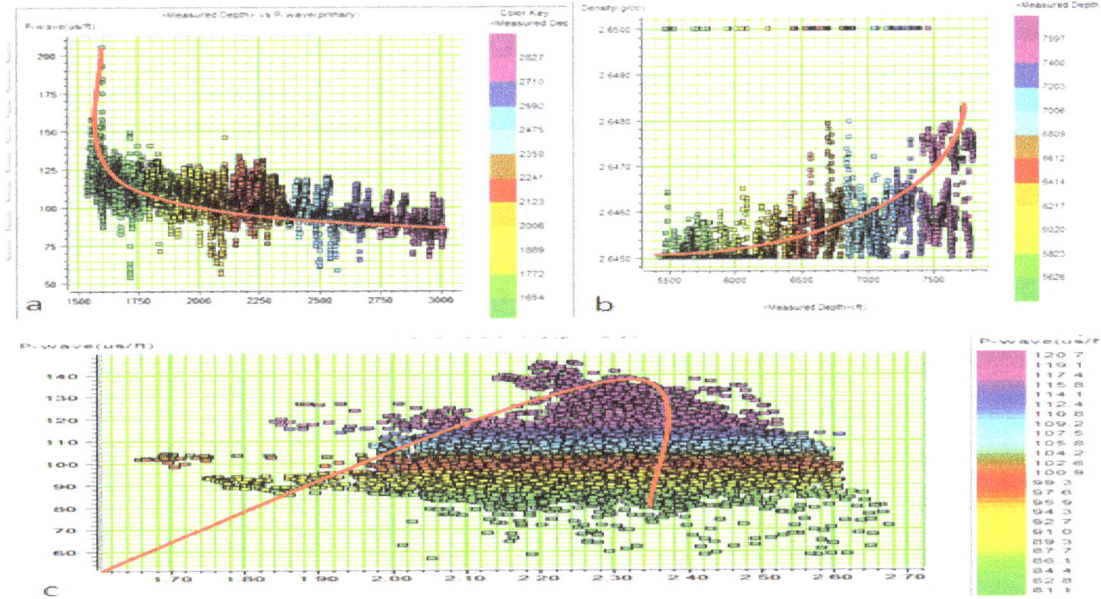

Figure 7: (a) Velocity-Measured depth and (b) Density-Measured depth crossplots recognizing decrease in velocity and increase in density with depth indicating over pressure (identified via color codes) (c) velocity-density plot recognizing Unloading, characterized by abrupt decrease in velocity at a constant density (identified via color codes).

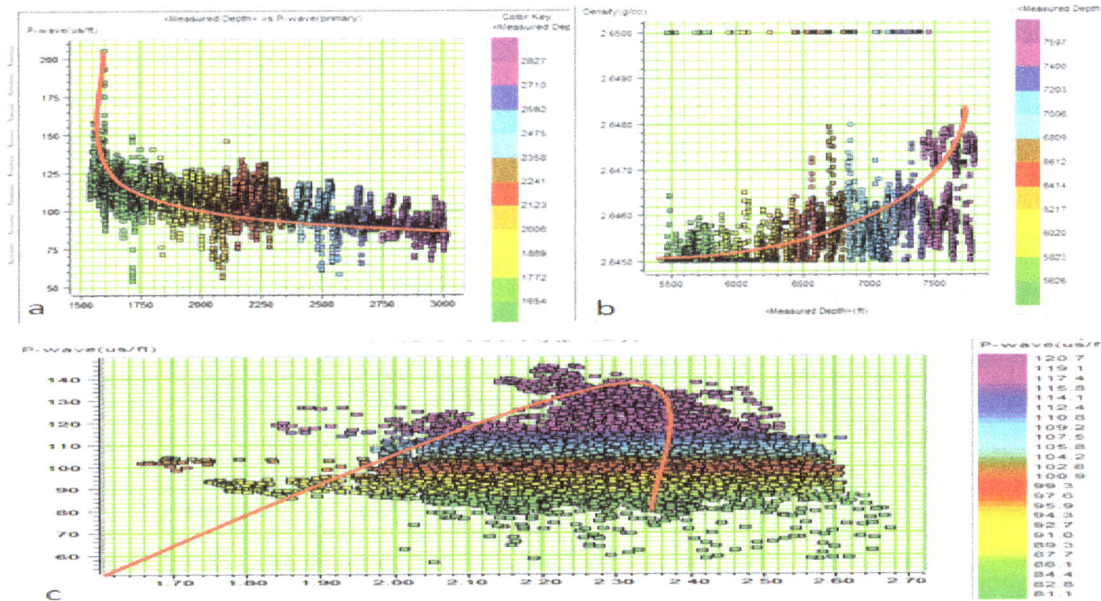

Figure 8: Inverted well log calibrated with synthetic seismic data, starting from 1500-2500 ms. Sands zones are characterized by low acoustic impedance defined by HD horizons indicating reservoir zones while the shale zones are indicated by high acoustic impedance.

with low acoustic impedance as indicated by green and yellow color codes. In acoustic impedance inversion, the area extents of the positive (increase in acoustic impedance) anomalies are mostly consistent with the pressure while low acoustic impedance is interpreted as reservoir or sand zones. However, the analysis of hydrostatic pressure using seismic inversion shows the spatial distribution of the overpressure zones in the study area. The pore pressure monitoring and overpressure detection technique proposed here are based on assumptions on an increase in

acoustic impedance techniques. Sand stones reservoir show a negative anomaly (decrease in acoustic impedance) while shale/clay zones are associated with positive anomaly (increase in acoustic impedance). But generally, velocity increases with depth under normal condition of sand homogeneity. Velocity tends to increase with increase in the compaction trends with depth but decreases when it encounters voids which are assumed to be overpressure zones. This assumption was used as a basis for overpressure estimation from well logs. Although we hold this

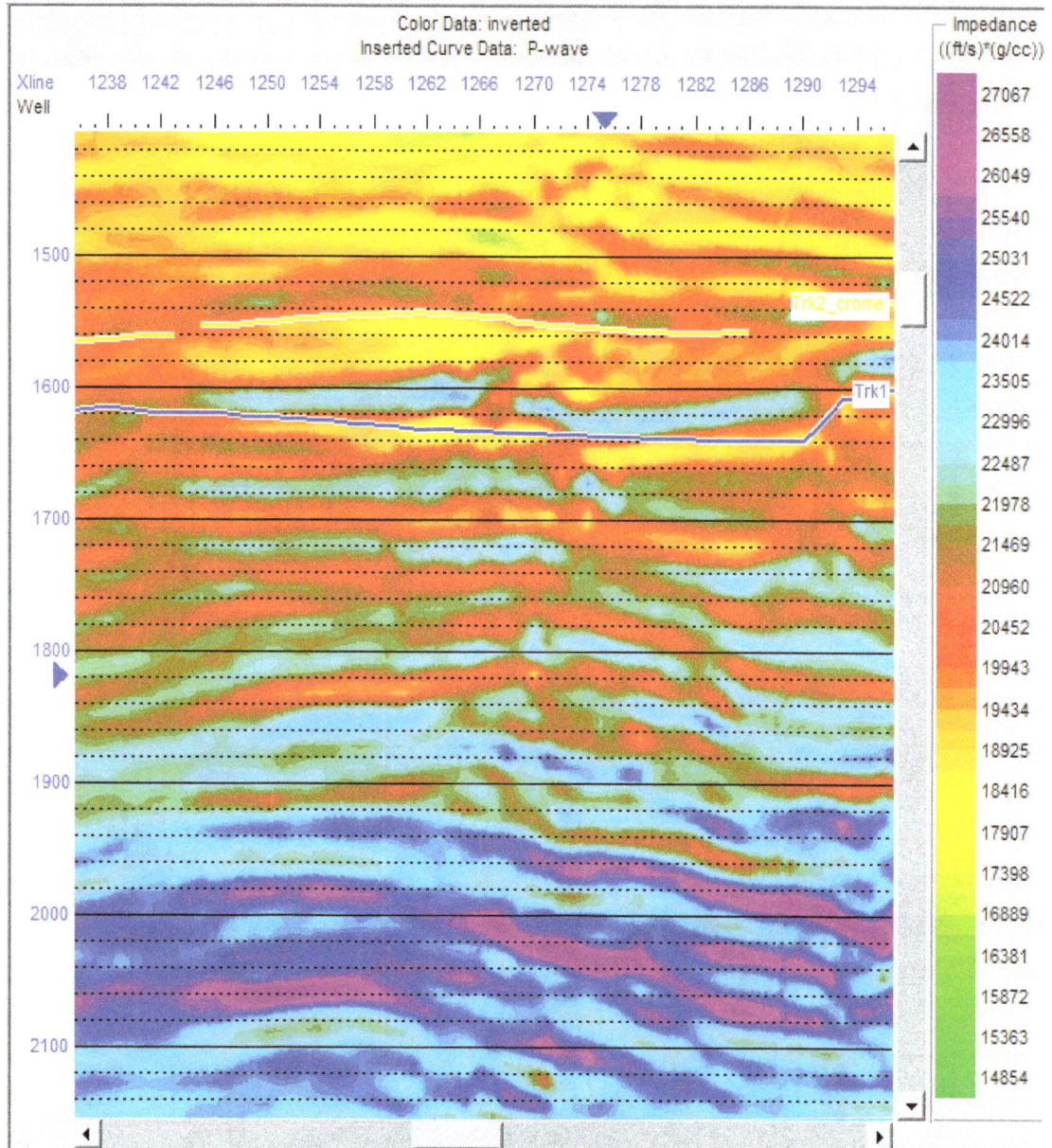

Figure 9: The inverted acoustic impedance volume showing overpressure variations in the field.

assumption not to be always true and this necessitate the use of acoustic impedance inversion in this research. Acoustic impedance uses the relationship between density and velocity to predict the pore pressure variations by conversion of the reflectivity data into quantitative rock properties that are descriptive of Formation.

Conclusion

Well logs and acoustic impedance inversion of Fabi Field, Onshore Niger delta has been studied to understand the hydrostatic pressure variations of the Field. The well logs exhibited normal compaction trends from depth level of 7200 ft to 8625 ft. At depth level of 8625 ft to 9000 ft, the velocity, density and resistivity logs deviated from the normal compaction trends. This indicates the onset of the overpressure zones because as rock compacts, porosity is reduced in a normal pressure zones but increase in overpressure zones due to the inability of the pore fluid to escape in equilibrium with the rate of compaction but shale resistivity increases with depth, since porosity decreases due to compaction. In overpressure zones, the resistivity of shale departs from their normal trend, it invariably decreases. However, with the crossplots of velocity and density logs, two overpressure generating mechanism such as undercompaction and unloading mechanism were revealed. Acoustic impedance inversion revealed the hydrostatic pressure variations of the Field. The onsets of over pressure zones are found around 1600 ms to about 1630 ms. However, very high overpressure gradients occur between 2000 ms to 2100 ms. These zones are characterized by very high acoustic impedance which was identified as high pressure zones.

References

1. Dutta NC (2002) Deep water geohazard prediction using Prestack inversion of large offset P-Wave data and rock model. Western Geco, Houston Texas, US. The leading Edge p: 193-198.

2. Boer LD, Sayers CM, Noeth S, Hawthorn A, Hooyman PJ, et al. (2011) Using tomographic seismic velocities to understand subsalt overpressure drilling risks in the Gulf of Mexico. Offshore Technology conference, Houston, Texas, USA.

3. Amonpantang P (2010) An overpressure investigation by sonic log and seismic data in Morgot Field, Gulf of Thailand. Bull Ear sci Thailand 3: 37-40.

4. Yu G (2008) Pore pressure prediction sees improvements. Hart energy publishing 1616S. Voss, Ste, 1000, Houston TX77057 USA. Geotrace.

5. Solano YP, Uribe R, Frydman M, Saavedra NF, Calderon ZH (2007) A modified approach to predict pore pressure using the d exponent method: An example from the carbonera formation, Colombia. CT&F- Ciencia, Tecnologia y Futuro 3: 103-111.

6. Standifird W, Paine K, Mathews M (2000) Nile delta study reports better models and technology required to improve exploration success. Knowledge systems, Petroleum Africa, Staford, Texax USA.

7. Tuttle MLW (1999) Tertiary Niger delta province, Nigeria, Cameroon, and Equatorial Guinea Africa, Central Region Energy Resources team, US Department of the interior. Geophysical Survey, USGS pp: 121-143.

8. Doust H, Omatsola E (1990) Edwards JD, Santogrossi PA (eds.) In: Niger Delta, Divergent/Passive Margin Basins. AAPG memoir 48: 239-248.

9. Stacher P (1995) Oti MN, Postma G (eds.) In: Present understanding of the Niger Delta hydrocarbon habitat. Geology of Deltas: Rotterdam, AA Balkema p: 257-267.

10. Evamy BD, Haremboure J, Kamerling P, Knaap WA, Molloy FA, et al. (1978) Hydrocarbon habitat of tertiary Niger Delta. AAPG Bull 62: 277-298.

11. Ejedawe JE (1981) Pattern of the incidence of oil reserves in the Niger Delta Basin. AAPG 66: 1574-1585.

12. Weber KJ (1971) Sedimentological aspects of oilfield in the Niger Delta. Geol Mynbouw 50: 559-576.

13. Weber KJ, Daukoru AA (1975) Geological aspect of the Niger Delta. 9th world petroleum congress Tokyo 2: 209-221.

14. Ekweozor CM, Daukoru EM (1994) Northern Delta Depobelt Portion of the Akata-Agbada petroleum System, Niger Delta, Nigeria p: 599-614.

15. Orife JM, Avbovbo AA (1982) Stratigraphic and non-conformity traps in Niger delta. AAPG Bull 66: 251-262.

16. Reijers TJA, Petters SW, Nwajide CS (1997) Selly RC (ed.) In: The Niger delta basin, Amsterdam, Elservier Science p: 151-172.

17. Schlumberger (2000) Pore pressure prediction: Interactive Petrophysics Technology and marketing support by schlumberger information solution.

18. Weber KJ (1987) Hydrocarbon distribution patterns in Nigerian growth fault structures controlled by structural style and stratigraphy. J Petrol Sci Eng 1: 91-104.

Mini Review Uranium-Thorium Decay Series in the Marine Environment of the Southern South China Sea

Yusoff AH and Mohamed CAR*

School of Environmental and Natural Resource Sciences, University Kebangsaan Malaysia, 43600 Bangi, Selangor, Malaysia

Abstract

The South China Sea (SCS) is divided into two parts namely northern SCS (nSCS) and southern SCS (sSCS). The sSCS is a semi-closed system that receives rapid large water flushing from the Western Pacific Ocean and the Java Sea during the northeast and southwest monsoon events. Major natural radionuclides in sSCS are expected to come from river water and terrestrial sediment discharge i.e., Mekong River, Chao Phraya River, Pahang River and Rajang River which contain high lithogenic and biogenic materials. A box model was developed to estimate the amount of ^{232}Th discharge from rivers to the sSCS basin. The result shows that the total flux of ^{232}Th entering into the sSCS was 140.3 × 10^3 Bq/km^2/yr, with the highest contribution from the Pahang River followed by the Rajang River, Mekong River and Chao Phraya River. The activity concentrations of natural radionuclides presented herein should be considered useful in order to understand the geochemical behavior of natural radionuclides in marginal sea areas. The review shows that publications on natural radionuclides are still limited; therefore further research needs to be done.

Keywords: Radionuclides; Southern South China Sea; Northeast monsoon; Mekong river; Marginal sea

Introduction

Components of the uranium-throium U-Th decay series have played an important role in the study of chemical oceanography in the southern South China Sea (sSCS) since the early 1990s. Natural radionuclides are widely used as a tool to investigate the oceanographic process that occurs in the sSCS region such as in the Gulf of Thailand [1,2], the Johor Straits, Malaysia [3], coastal Peninsular Malaysia [4-7], the Vietnam coast [8], the North coast of Java , Indonesia [9], Manila Bay [10,11] and the Sulu Sea [12].

The South China Sea (SCS) is divided into two parts; the northern South China Sea (nSCS) and the southern South China Sea (sSCS). The nSCS includes Taiwan, Hong Kong and the Republic of China, while countries in the sSCS basin comprise Malaysia, Singapore, Thailand, Indonesia, Brunei, Indonesia, Cambodia, Vietnam and the Philippines. While four rivers feed into the sSCS; the Mekong River, the Chao Phraya River, the Pahang River and the Rajang River (Figure 1). The Mekong River is the 12th longest river in the world and drains a catchment of 790 000 km2 and about 15 000 m^{3}s^{-1} as the 8th largest water discharge [13]. This river flows for 4909 km passing through six countries; China, Myanmar, Thailand, Laos, Cambodia, and Vietnam [14]. The Pahang River is located in the Pahang basin, and at 459 km is the longest river in Peninsular Malaysia. The source of the river is the Titiwangsa main range and is a main channel of water drainage to the sSCS region [15]. The longest river in East Malaysia is the Rajang River, located in northwest Borneo at about 563 km. In addition, sSCS has a unique geographical structure. Being a semi-enclosed ocean basin, there is a constant exchange of water between the sSCS and the surrounding ocean through the straits. Water is flushed from the western Pacific during northeast (NE) monsoon events and the basin receives a high input of natural radionuclide from land during southwest (SW) monsoon events [6,16]. Furthermore, the geological characteristic of the straits around the sSCS shows a unique structure (Figure 2). Mekong and Chao Phraya River which are located at the Indochina Peninsula consist mainly of Paleozoic-Mesozoic sedimentary rocks with minor intrusive and extrusive igneous rocks. While, the Pahang River which is located at the Malaysian Peninsula consists mainly of Paleozoic-Mesozoic granite and granodiorite rock [17]. The different geological characteristic surrounded the sSCS might contribute to the different behavior of radionuclides to the sSCS. Therefore, the aim of

this study is to review published reports of potential sources of natural radionuclide and develop a suitable simple model to estimate the flux of natural radionuclides entering into the sSCS. This review also reports on current levels of natural radionuclides in seawater, sediment and organisms in the sSCS area.

Uranium thorium (U-Th) decay series in marine environments

There are three major actinide nuclides with a very long half-life; ^{238}U (t$_{1/2}$ 4.5 × 10^9 years), ^{235}U (t$_{1/2}$ 1.4 × 10^6 years) and ^{232}Th (t$_{1/2}$ 7.0 × 10^8 years) as well as others described in other studies such as Cheng et al. and Loeff [18]. The decay process of these radionuclides produce three decay series as shown in Figure 3. The U-Th decay series begins with ^{238}U, ^{235}U and ^{232}Th, and ends with stable isotopes of lead [18-20]. In a closed system where the growth of the parent is balanced with the decay of the daughter, the ratio between parent and daughter is equal to unity (1). However, in a natural environment such as in marine, freshwater and terrestrial ecosystems disequilibrium will occur between parent and daughter [21-23].

Generally, the members of a U-Th decay series enter marine environments through four main pathways; in-situ production of a daughter nuclide, river transport, the diffusion process through sediment and from the atmosphere (Figure 4) [20].

Most natural radionuclides are commonly absorbed onto particles, deposited on seafloors and have their own parents. For example, ^{234}U, ^{226}Ra and ^{210}Po are produced from the decay of ^{238}U, ^{230}Th and ^{210}Pb, respectively. Meanwhile, natural radionuclides of riverine input such as ^{238}U, ^{232}Th and ^{228}Ra are carried in detrital particles from source during

***Corresponding author:** Mohamed CAR, Faculty of Science and Technology, School of Environmental and Natural Resource Sciences, University Kebangsaan Malaysia, 43600 Bangi, Selangor, Malaysia, E-mail: carmohd@ukm.edu.my

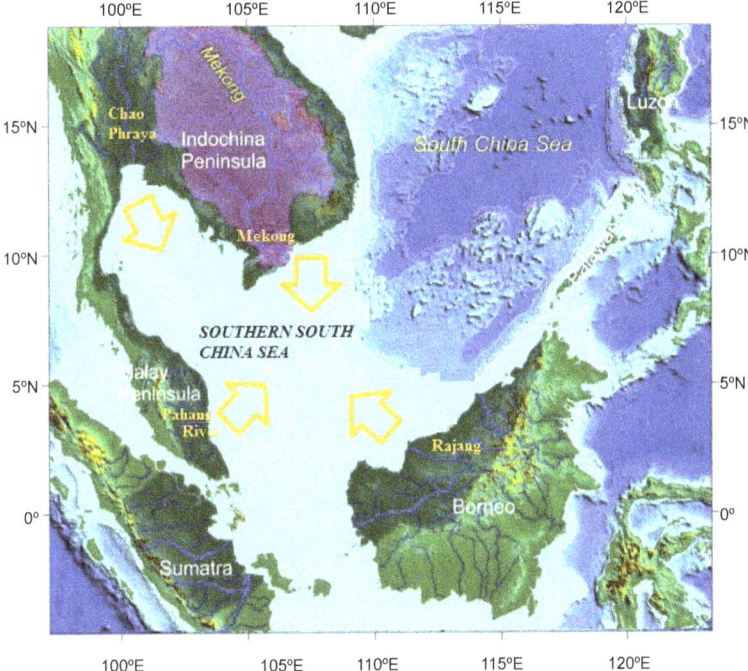

Figure 1: The four major rivers that feed into the southern South China Sea.

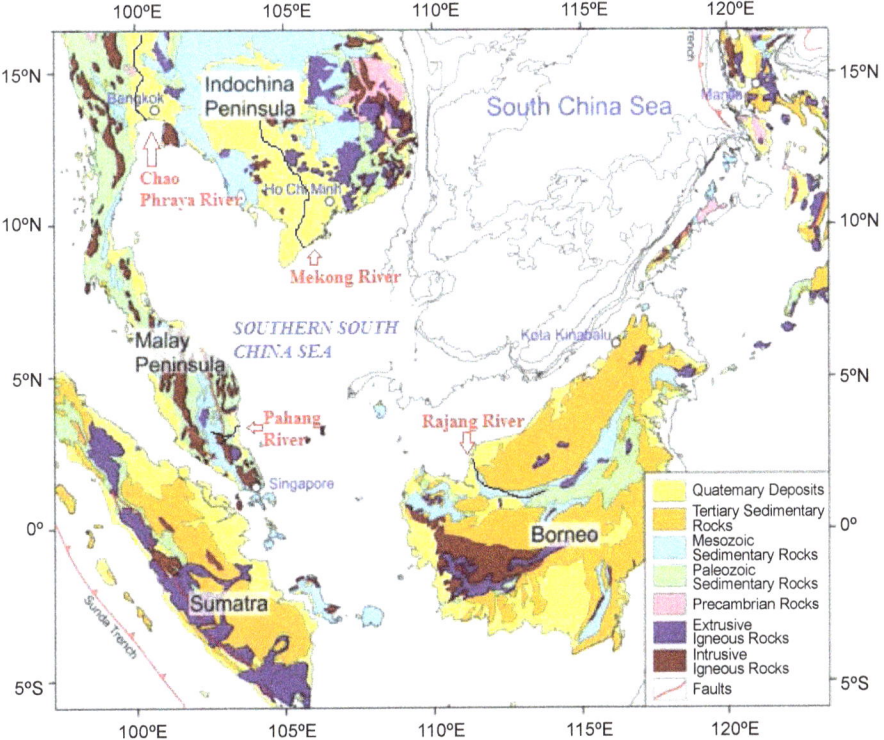

Figure 2: Geological characteristic of major rivers in the southern South China Sea [17].

river flow to the ocean [24]. In addition, the input of ^{210}Pb resulted from emissions of [222]Rn from rock and soil, and is also scavenged from the atmosphere by precipitation processes before being deposited onto the ocean surface [16].

The characteristics of each radionuclide in a uranium-decay series are unique and thus suitable for use as indicators of marine environmental conditions e.g., ^{210}Pb, [210]Po and [234]Th are used as tools to estimate biological productivity the water column [25-27]. The ratio

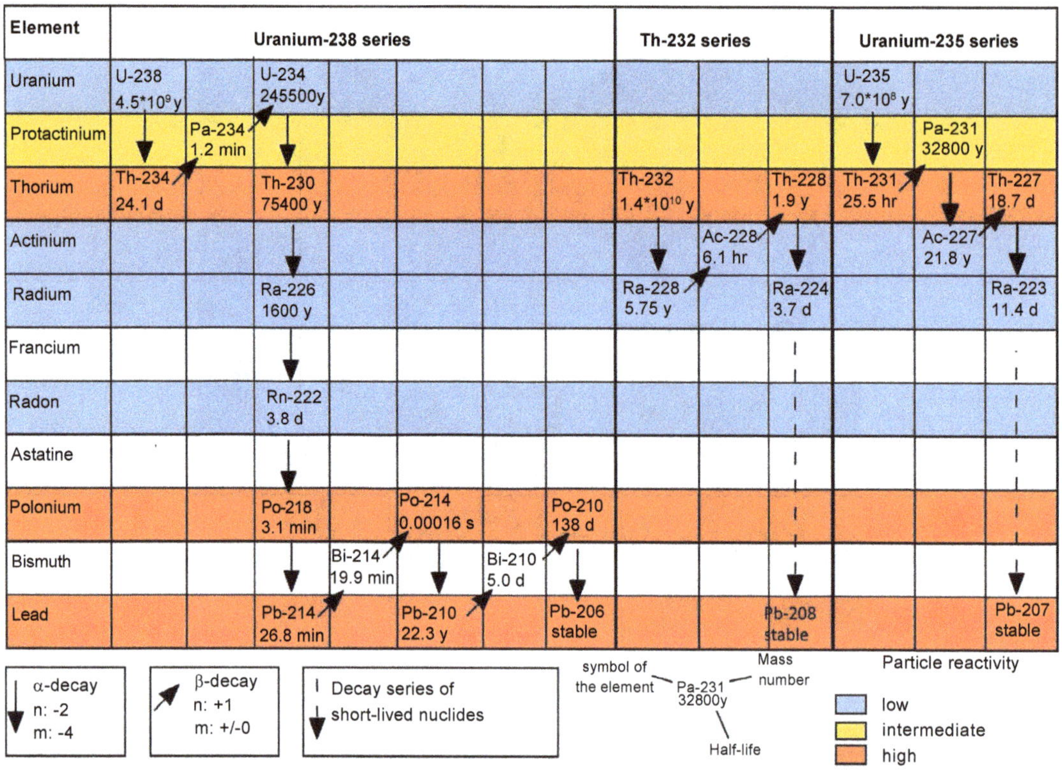

Figure 3: Natural uranium-thorium decay series, colored according to particle reactivity. The arrows represent decay with changes in atomic number (Z) and number of nucleons (N) indicated. All three series end with a stable lead isotope [18].

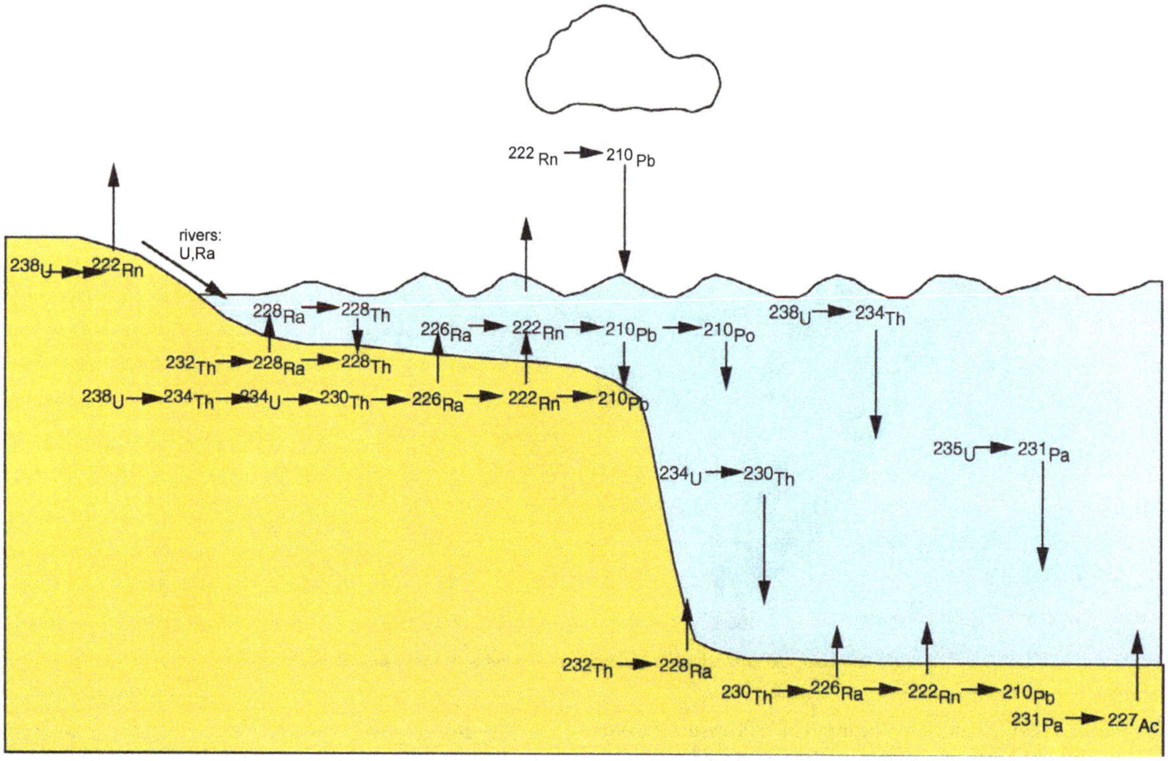

Figure 4: Schematic diagram showing oceanic cycles of selected members of the U and Th decay series. Horizontal arrows indicate radioactive decay characterized by a constant rate. Vertical arrows denote fluxes across the sediment-water interface and also removal from atmosphere (for ^{210}Pb) through chemical scavenging and particle settling [18,20].

of radionuclides such as [230]Th/[232]Th and [231]Pa/[230]Th has been used to determine pollution levels and biological productivity in open oceans [28], where both [231]Pa and [230]Th are also produced throughout the water column at a constant production ratio rate value of [231]Pa/[230]Th = 0.093, especially at in marginal seas.

Potential source and budget of radionuclides in the southern South China Sea

The sSCS has a unique geographical structure as it is a semi-enclosed ocean basin with water exchanges through the straits and bay. The area is dominated by the unique characteristic of an east Asian monsoon system with northeast winds during winter (November–February) and southwest winds during summer (June–August) [29,30]. Surface water current flows from north to south during the northeast monsoon and in the reverse direction during the southwest monsoon (Figure 5). This results in large amounts of water flushing both artificial and natural radionuclides from the western Pacific Ocean and the Java Sea into the sSCS during northeast and southwest monsoon events, resulting in a high deposition of radionuclides in this region [6]. Several studies have shown that monsoon events contribute significantly to high radionuclide activities in the sSCS basin. High [226]Ra and [228]Ra activity has been observed in surface sediment on the eastern coast of Peninsular Malaysia; the source of which are neighboring countries and the Western Pacific Ocean during northeast monsoon events [6]. Recent studies have also shown high [210]Po activity in Malaysian coastal waters as a result of haze event in dry season during southwest monsoon [16].

Besides monsoon phenomena, the SCS also have experienced with the typhoon events especially at the nSCS region [31]. However, the typhoon event is rare occurred in sSCS compared with nSCS because the sSCS located at the Equator region and always immune from the natural disaster [32]. Typhoon Vamei formed in the sSCS was a rare typhoon event that occurred at sSCS in year 2001 with a wind speed more than 36 m/s has significant affecting the atmospheric and oceanic conditions [33].

In addition to the monsoon, the sSCS also receives major radionuclides from the river either in particulate, dissolved or soil forms. The SCS receives about 1600 million tons per year (Mt/yr) of detrital sediments from numerous rivers including the Pearl, the Red, and the Mekong [29]. Terrigenous materials in the SCS are mainly generated through transport from the continent via large rivers (e.g., the Mekong River, the Rajang River) representing 8.4% of estimated global fluvial sediment discharge to the world's oceans [34]. In the sSCS, the highest sediment discharge is from the Mekong River (160 Mt/yr), while the Chao Phraya, Pahang and Rajang Rivers discharge 11 Mt/yr, 20.4 Mt/yr and 30 My/yr, respectively [17] (Table 1). Based on that scenario, the sSCS can be considered the largest sink for fluvial sediments in the SCS basin, and is thus expected to receive more natural radionuclides from its contributing rivers.

The riverine input of radionuclides such as [238]U, [232]Th and [228]Ra from major rivers to the sSCS is listed in Table 1. These radionuclides are carried in the particulate load of rivers where reactions occurring

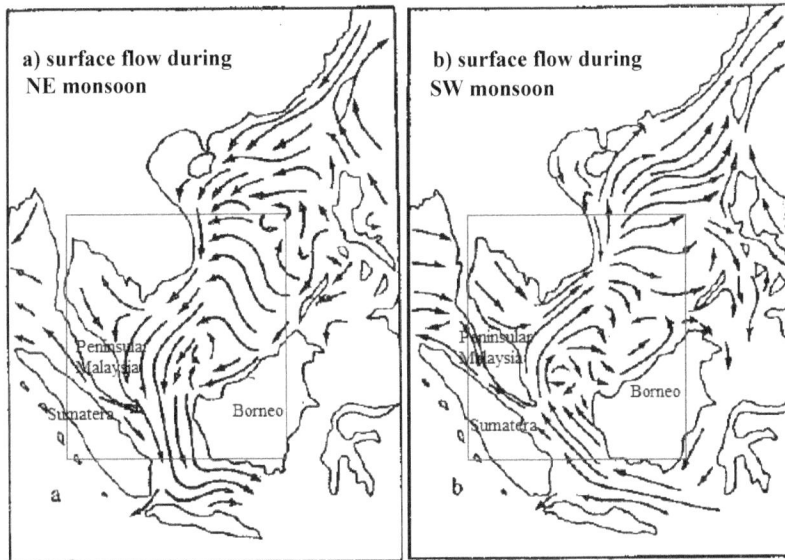

Figure 5: a) Surface current flows in the Southern South China Sea during the winter northeast monsoon, and b) Surface current flows during the summer southeast monsoon [17].

River (area)	Drainage area 10³ km²	Sediment discharge Mt/year	Radioactivities (Bq/kg)		
			[238]U	[232]Th	[228]Ra
Mekong (Vietnam)	790[h]	160[h]	n.a	50.7[a]	n.a
Chao Phraya (Thailand)	160[h]	11[h]	33.33[b]	48[b]	n.a
Pahang (Peninsular Malaysia)	19[i]	20.4[i]	n.a	102.5[c]	56[d]
Rajang (East Malaysia)	50[j]	30[j]	58[e]	27.8[f]	45[g]

Note: n.a = not available
Data sources: a) Huy and Luyen [53], b) Srisuksawad et al. [1] c) Mohamed et al. [6] d) Wan Mahmood and Yii [4] e) Yusoff and Mohamed [38], f) Wan Mahmood et al. [44], Yii et al. [43] h) (Milliman and Syvitski [13] I, Liu et al. [55] j) Staub et al. [54].

Table 1: Drainage area, suspended sediment discharge of major rivers flowing directly into the Southern South China Sea (sSCS) and radionuclide activity.

in the river may modify radionuclide fluxes to the oceans [24]. The drainage area, sediment load and radionuclide activity data were used to estimate radionuclide fluxes from river sediments (Equation 1) to the sSCS basin using the equation below:

$$F = C x S x 1 / A \qquad (1)$$

$$\Delta F = \left(C_{mk} \; x \; S_{mk} \; x \; 1 / A_{mk}\right) + \left(C_{cp} \; x \; S_{cp} x \; 1 / A_{cp}\right) + \\ \left(C_{ph} \; x \; S_{ph} \; x \; 1 / A_{ph}\right) + \left(C_{rj} x \; S_{rj} \; x \; 1 / A_{rj}\right) \qquad (2)$$

Where F is the fluxes of radionuclides, C is the radionuclides activity and A is the drainage area. While an equation 2 was apply to estimate the total fluxes of radionuclides contribute from four rivers into the sSCS. Where ΔF is the total fluxes of radionuclide entering into the sSCS, C_{mk}, C_{cp}, C_{ph} and C_{rj} are the activities of radionuclide in Mekong River, Chao Phraya River, Pahang River and Rajang River, respectively. Then S_{mk}, S_{cp}, S_{ph} and S_{rj} are refer to the amount of sediment loading contributed from Mekong, Chao Phraya, Pahang and Rajang Rivers, respectively. Finally, A_{mk}, A_{cp}, A_{ph} and A_{rj} are refer to the drainage area of Mekong, Chao Phraya, Pahang and Rajang Rivers, respectively.

A simple box model has been developed to estimate total fluxes of [232]Th discharge from the river to the sSCS (Figure 6). Total fluxes of [232]Th is 140.3×10^3 Bq/km²/yr with the highest amount contributed by the Pahang River (110.05×10^3 Bq/km²/yr) followed by the Rajang River (16.68×10^3 Bq/km²/yr), the Mekong River (10.27×10^3 Bq/km²/yr) and the Chao Phraya River (3.3×10^3 Bq/km²/yr).

High inputs from the Pahang River might be related to its geological formation. The Pahang basin's geographical location in Peninsular Malaysia has extensive granitic rock compared to other areas in Southeast Asia (Figure 7) [35]. Thorium is highly concentrated in granitic rock with a value range of 8-33 ppm [36]. During the weathering process, Th in the tetravalent state can be adsorbed onto surface minerals of granitic rock such as apatite, monazite and zircon then transported from rivers to the oceans [36,37]. The normal ratio of Th/U in most granitic rock ranges from 3.5 to 6.3 [36]while the value of Th/U in marine sediment on the Peninsular Malaysia coast ranges from 2.49 to7.66 with the average value being 3.38 [38]. The Chao Phraya and Mekong Rivers show an estimated value of 1.44 and

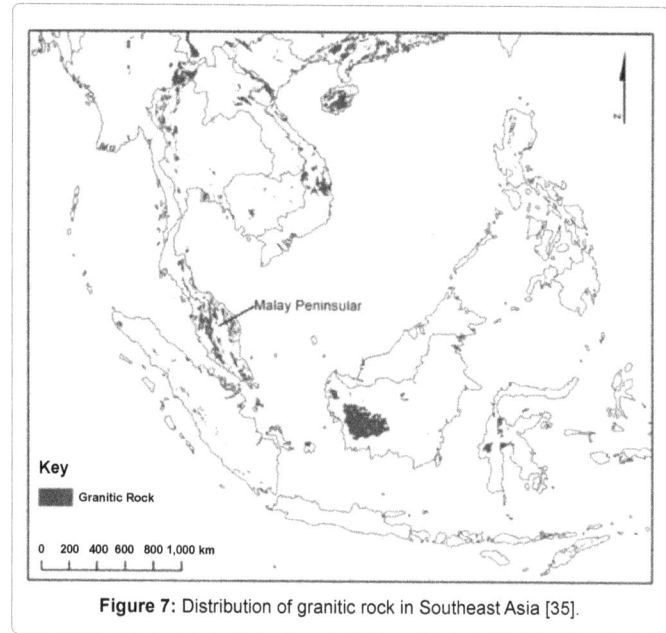

Figure 7: Distribution of granitic rock in Southeast Asia [35].

1.85, respectively [39,40] suggesting Th in marine sediment adjacent to Peninsular Malaysia is influenced by granitic rock.

Study of natural U-Th decay series in the southern South China Sea: Is there enough information?

Studies measuring natural radioactivity in various media around the sSCS have been conducted since the early 1990s. A number of expeditions such as Research on the Seas and Island (ROSES) organized by Universiti Sains Malaysia (USM) and the Prime Scientific Sailing Expedition (EPSP'09), which covered the west and east coasts of Peninsular Malaysia, the Exclusive Economic Zone (EEZ) of Sabah Sarawak, the Sulu Sea and the Celebes Sea have been held for Malaysian marine scientists. These expeditions have contributed new knowledge of radionuclide behavior in the sSCS near Malaysia [3,5,7,12,16,23,38,41-45].

Tables 2-4 presents the values of natural radionuclide in sediment, sea water and biota respectively as collected in the sSCS and adjacent sea areas. The value of [234]U and [238]U in sediment ranges from 6.83 Bq/kg to 65.3 Bq/kg and 5 Bq/kg to 91.5 Bq/kg, respectively [1,3,38,39]. Natural thorium isotopes such as [232]Th, [230]Th and [234]Th in sediments were detected in various areas in the sSCS such as the Johor Strait, the Vietnam coast and on the Malaysian coast with values ranging from 1.83 Bq/kg to 262.83 Bq/kg, 9 to 62.83 Bq/kg and n.d to 200 Bq/kg, respectively [1,3,6,44,46]. The values of [226]Ra and [228]Ra in sediment from the sSCS ranged from 2.9 Bq/kg to 64 Bg/kg and 23 to 130 Bq/kg, respectively [1,4,6,43]. In addition, high [232]Th activity was detected in sediment in the sSCS region (~262.83 Bq/kg) [3] which might due to marginal sea characteristics where the ocean receives more input of [232]Th from the straits surrounding the sea. Furthermore, a high organic matter contents in sediment also reported at the sSCS e.g., western coast of Borneo which can influence the concentration activity of natural radionuclides in sediments. A strong statistical correlation value between organic matter contents with natural radioisotopes of uranium such as [234]U (r^2=-0.959, p<0.01) and [238]U (r^2=-0.904, p<0.01) are well discussed by Yusof and Mohamed [38]. The adsorption of uranium with organic matters will occurred through the process of ion exchange for formation the stable of U(IV) in surface sediments [22,47].

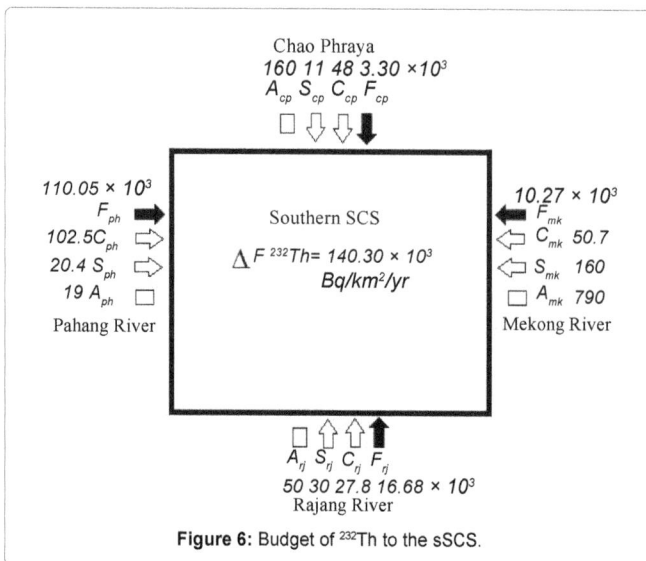

Figure 6: Budget of [232]Th to the sSCS.

Radionuclide	Samples	Technique	Range values (average) Bq/kg	Country (Area)	References
[234]U	Sediment core (n=22)	Alpha detector	21.12-65.3	Thailand gulf	[1]
	Sediment core (n=2)	Alpha detector	6.83-62.17	Malaysia (Borneo port)	[38]
[238]U	Sediment core (n=25)	INAA	30-91.5 (65.0)	Malaysia (Johor Strait)	[3]
	Sediment core (n=22)	Alpha detector	20.67-63.67 (33.33)	Thailand gulf	[1]
	Surface sediment (n=51)	ICP-MS	5-50 (23)	Thailand (eastern coast of gulf)	[39]
	Sediment core (n=2)	Alpha detector	8.5-58	Malaysia (Borneo port)	[38]
[226]Ra	Surface sediment (n=51)	HGPe detector	2.9-53.2	Thailand (eastern coast of gulf)	[39]
	Sediment core (n=16)	HPGe detector	16-46 (30)	Malaysia (EEZ of Peninsular)	[4]
	Surface sediment (n=31)	HPGe detector	16-21	Malaysia (SCS of Sabah and Sarawak)	[43]
	Surface sediment (n=15)	Alpha detector	8-48	Malaysia (East coast of peninsular)	[6]
	Surface sediment (n=18)	HPGe detector	13-64	Malaysia (West coast of peninsular)	[6]
[228]Ra	Sediment core (n=16)	HPGe detector	28-57 (56)	Malaysia (EEZ of Peninsular)	[4]
	Surface sediment (n=31)	HPGe detector	23-45	Malaysia (SCS of Sabah and Sarawak)	[43]
	Surface sediment (n=15)	HPGe detector	12-130	Malaysia (East coast of peninsular)	[6]
	Surface sediment (n=18)	HPGe detector	36-89	Malaysia (West coast of peninsular)	[6]
[230]Th	Sediment core (n=3)	Alpha detector	23.17-164	Malaysia (marine port of Sabah, Labuan and Klang)	[46]
	Surface sediment (n=31)	Alpha detector	9-12	Malaysia (SCS of Sabah and Sarawak)	[44]
	Sediment core (n=22)	Alpha detector	22.17-62.83	Thailand gulf	[1]
[232]Th	Surface sediment (n=31)	Alpha detector	6.8-27.8	Malaysia (SCS of Sabah and Sarawak)	[44]
	Sediment core (n=22)	Alpha detector	4-108 (48)	Thailand gulf	[1]
	Sediment core (n=25)	INAA	30.67-262.83 (86.67)	Malaysia (Johor Strait)	[3]
	Sediment core (n=3)	Alpha detector	1.83-153.67	Malaysia (marine port of Sabah, Labuan and Klang)	[46]
	Surface sediment (n=15)	Alpha detector	65-145	Malaysia (East coast of peninsular)	[6]
	Surface sediment (n=18)	Alpha detector	35-59	Malaysia (West coast of peninsular)	[6]
[234]Th	Sediment core (n=33)	HPGe detector	n.d-200	Vietnam (northern coast)	[55]
	Sediment core (n=25)	HPGe detector	n.a	Indonesia (north coast of Java Sea)	[9]
[210]Po	Sediment core (n=13)	Apha detector	2.33-141.4	Malaysia (SCS of Sabah)	[56]
[210]Pb	Sediment core (n=22)	Alpha detector	11.67-143.33	Thailand gulf	[1]
	Sediment core (n=10)	Alpha detector	22-110	Philippines (area affected by toxic harmful algal bloom (HAB))	[11]

Note: EEZ = exclusive economic zone; n.d = not detected (below detection limit); n.a = not available, HPGe= High purity vertical germanium detectors, INAA= Instrumental neutron activation analyses

Table 2: Range values of radionuclides in marine sediment in the southern South China Sea area.

The activity of isotopes in sediments in the sSCS is generally higher compared to other regions such as the Red Sea and the north eastern coast of India, thus further investigation is essential to determine the various sources of radionuclide entering the sSCS region.

In sea water, high values of [226]Ra and [228]Ra isotopes were detected in the EEZ of Peninsular Malaysia with values of up to 2.7×10^3 mBq/L and 3.5×10^3 mBq/L respectively as published by Amin et al. (Table 3) [48]. A high value of radium isotopes in sea water corresponds to the high activity value of radium isotopes detected in surface sediments from the east coast of Peninsular Malaysia [6]. In addition, oil and industry gas contribute up to almost 560 Bq/kg of [226]Ra from sludge discharge to the

marine environment [49]. Radium [210]Po isotope activity in sea water was higher in the sSCS compared to the entire South China Sea and western Pacific Ocean regions due to high radionuclide depositional fluxes from enhanced dry precipitation caused by haze events [16].

On the other hand in marine organisms, high [210]Po and [234]Th values were detected in green mussels collected from the Johor Strait with average values of 150 Bq/kg and 388 Bq/kg. It is plausible that the Straits of Johor might be liable to certain levels of radionuclide contamination from exiting natural radionuclides and anthropogenic activity inputs [50]. Furthermore, concentrations of [226]Ra in fish (0.80-2.13 Bq/kg) from the eastern coast of Peninsular Malaysia [4] also show high values

Radionuclide	Water Depth	Technique	Range values (average) mBq/L	Area	References
^{226}Ra	n.a	HPGe detector	n.d-2.7 x 10^3	EEZ of Peninsular Malaysia	[48]
	>1000 m	Gamma detector	0.94-3.66	Indonesian sea	[41]
	>4000 m	Gamma detector	1.08-3.73 (2.27)	Central of SCS	[41]
^{228}Ra	n.a	HPGe detector	n.d-3.5 x 10^3	EEZ of Peninsular Malaysia	[48]
	>1000 m	Gamma detector	0.21-1.81	Indonesian sea	[41]
	>4000 m	Gamma detector	0.35-2.48 (0.89)	Central of SCS	[41]
^{230}Th	>4000 m	TIMS	n.d-22.0 (13.0) 10^{-3}	Central of SCS	[12]
	>4000 m	Thermal ionization mass spectrometer	n.d-21.0 (8.2) 10^{-3}	Sulu Sea	[12]
^{232}Th	>4000 m	TIMS	0.65-2.73(2.33) x 10^{-3}	Central of SCS	[12]
^{210}Pb	<1000 m	Gross beta detector	0.22-0.96 (0.58)	SCS of Sabah and Sarawak, Sulu Sea and Celebes Sea	[16]
	>4000 m	Alpha detector	1.12-1.77	Central of SCS	[42]
^{210}Po	<1000 m	Alpha detector	1.52-8.98(4.10)	SCS of Sabah and Sarawak, Sulu Sea and Celebes Sea	[16]
	>4000 m	Alpha detector	1.17-1.30	Central of SCS	[42]

Note: n.a = not available, the value is below detection limit, however Amin et al. [48] stated the location is 50 km offshore distance, TIMS = Thermal ionization mass spectrometer

Table 3: Range values of radionuclides in seawater in the southern South China Sea area.

Radionuclide	Samples	Technique	Range values (average) Bq/kg	References	Area
^{226}Ra	Edible fish (n=9)	HPGe detector	0.7-4.5	[48]	EEZ of Peninsular Malaysia
	Mollusc (n=2)	HPGe detector	4.5-5.0		
	Crustaceans (n=4)	HPGe detector	1.2-3.9		
	Pelagic and demersal fish (n=16)	HPGe detector	0.80-2.13	[4]	
^{228}Ra	Edible fish (n=9)	HPGe detector	0.9-5.1	[48]	EEZ of Peninsular Malaysia
	Mollusc (n=2)	HPGe detector	3.8-4.0		
	Crustaceans (n=4)	HPGe detector	0.9-3.9		
	Pelagic and demersal fish (n=16)	HPGe detector	0.95-3.57	[4]	
^{210}Pb	Edible fish (n=16)	Alpha detector	0.65-23.10 (6.12)	[23]	West coast of Peninsular Malaysia
	Zooplankton	Alpha detector	364.67	[52]	East coast of Peninsular Malaysia
	Green mussel (n=9)	Alpha detector	68-257(150)	[50]	Johor strait
^{210}Po	Edible fish (n=16)	Alpha detector	0.47-68.10(18)	[23]	West coast of Peninsular Malaysia
	Zooplankton	Alpha detector	93.67	[52]	East coast of Peninsular Malaysia
^{234}Th	Green mussel (n=9)	Gross beta detector	236-641(388)	[50]	Johor strait

Table 4: Range values of radionuclides in organisms collected in the southern South China Sea area.

compare to other areas such as fish from Japan (^{226}Ra: 0.008 Bg/kg) and Puget Sound, USA (^{226}Ra: 0.003-0.75 Bq/kg) [4]. In addition, ^{210}Po and ^{210}Pb activity in the west coast of Peninsular Malaysia shows high values compared to ^{226}Ra [23,51]. Furthermore, high activity values of ^{210}Po and ^{210}Pb detected in zooplankton as published by Mohamed and Kuan [52] indicates a high input of these radionuclides in this area. However, there is limited data available on the accumulation of radionuclides in microorganisms such as bacteria and virus [53-57]. Therefore, it is strongly recommended that further and more complete research is undertaken to study bioaccumulation trends of radionuclides in organisms particularly in the sSCS [58].

Conclusion

This review discusses potential sources and budgets of natural radionuclides in the sSCS. The total flux of ^{232}Th discharge to the sSCS was successfully estimated from the box model with a value of 140.3 × 10^3 Bq/km^2/yr. The highest flux of ^{232}Th contributed to the sSCS is from the Pahang River with a value of 110.05 × 10^3 Bq/km^2/yr followed by

the Rajang River, Mekong River and Chao Phraya River. The activity values of natural radionuclides in organisms, sediment and seawater in the sSCS were also compiled and reviewed. Unfortunately, there is limited data available on the distribution and behavior of some natural radionuclides such as ^{231}Pa, ^{234}Th, ^{210}Bi, ^{7}Be and ^{10}Be in the sSCS. It is strongly recommended that further and more complete research is undertaken to study the behavior of these radionuclides in the sSCS.

Acknowledgements

The authors would like to thank the Ministry of Science, Technology and Innovation (MOSTI), for providing the research grant (04-01-02-SF0801). Thanks are also due to all the laboratory members and staff of Pusat Pengajian Sains Sekitaran dan Sumber Alam, Faculty of Science and Technology, UKM.

References

1. Srisuksawad K, Porntepkasemsan B, Nouchpramool S, Yamkate P, Carpenter R, et al. (1997) Radionuclide activities, geochemistry, and accumulation rates of sediments in the Gulf of Thailand. Continental Shelf Research 17: 925-965.

2. Cheevaporn V, Mokkonggpai P (1996) Pb-210 Radiometric dating of estuarine

sediments from the eastern coast of Thailand. Journal of Science Society, Thailand 22: 313-324.

3. Wood KH, Ahmad Z, Shazili NA, Yaakob R, Carpenter R (1997) Geochemistry of sediments in Johor Strait between Malaysia and Singapore. Continental Shelf Research 17: 1207-1228.

4. Wan Mahmood ZUY, Yii MW (2012) Marine radioactivity concentration in the Exclusive Economic Zone of Peninsular Malaysia : 226 Ra, 228 Ra and 228 Ra / 226 Ra. Journal of Radioanalytical and Nuclear Chemistry 292: 183-192.

5. Yii MW, Wan Mahmood ZU, Ahmad Z, Jaffary NA, Ishak K (2011) NORM activity concentration in sediment cores from the Peninsular Malaysia East Coast Exclusive Economic Zone. Journal of Radioanalytical and Nuclear Chemistry 289: 653-661.

6. Mohamed CAR, Wan Mahmood, Ahmad Z, Ishak AK (2010) Enrichment of natural radium isotopes in the southern South China Sea surface sediments. Coastal Marine Science 34: 165-171.

7. Mohamed CAR, Wan Mahmood ZUY, Ahmad Z (2008) Recent sedimentation of sediments in the coastal waters of Peninsular Malaysia. Pollution Research 27: 27-36.

8. Duong P, Tschurlovits M, Buchtela K (1996) Enrichment of radioactive materials in sand deposits of Vietnam as a result of mineral processing. Environment International 22: 271-274.

9. Boer W, van den Bergh GD, de Haas H, de Stigter HC, Gieles R, et al. (2006) Validation of accumulation rates in Teluk Banten (Indonesia) from commonly applied 210Pb models, using the 1883 Krakatau tephra as time marker. Marine Geology 227: 263-277.

10. Maria EJ (2009) Estimating sediment accumulation rates in Manila Bay, a marine pollution hot spot in the Seas of East Asia. Marine Pollution Bulletin 59: 164-174.

11. Sombrito EZ, Bulos AD, Sta Maria EJ, Honrado MC, Azanza RV, et al. (2004) Application of 210Pb-derived sedimentation rates and dinoflagellate cyst analyses in understanding Pyrodinium bahamense harmful algal blooms in Manila Bay and Malampaya Sound, Philippines. J Environ Radioact 76: 177-194.

12. Okubo A, Obata H, Gamo T, Minami H, Yamada M (2007) Scavenging of Th in the Sulu Sea. Deep Sea Research II 54: 50-59.

13. Milliman JD, Syvitski JPM (1992) Geomorphic/Tectonic Control of Sediment Discharge to the Ocean: The Importance of Small Mountainous Rivers1. The Journal of Geology 100: 525-544.

14. Pantulu VR (1986) The Mekong River system. In: Davies B, Walker K (eds.) The Ecology of River Systems. Dordrecht, The Netherlands: Dr. W. Junk Publishers 695-741.

15. Lun PI (2011) Hydrological Pattern of Pahang River Basin and Their Relation To Flood Historical Event. Jurnal e-Bangi 6: 29-37.

16. Sabuti AA, Mohamed CAR (2015) High 210 Po Activity Concentration in the Surface Water of Malaysian Seas Driven by the Dry Season of the Southwest Monsoon (June – August 2009). Estuaries and Coasts 38: 482-493.

17. Liu Z, Zhao Y, Colin C, Stattegger K, Wiesner MG, et al. (2015) Source-to-Sink transport processes of fluvial sediments in the South China Sea. Earth-Science Reviews 153: 238-273.

18. Loeff MMR (2015) Uranium-Thorium Decay Series in the Oceans: Overview. In Elias S (ed.) Earth Systems and Environmental Sciences, (Reference Module in Earth Systems and Environmental Sciences). Amsterdam: Elsevier 1-16.

19. Roy-Barman M, Jeandel C, Souhaut M, Rutgers M, Voege I, et al. (2005) The influence of particle composition on thorium scavenging in the NE Atlantic ocean (POMME experiment). Earth and Planetary Science Letters 240: 681-693.

20. Henderson G, Anderson R (2003) The U-series Toolbox for Paleoceanography. Reviews in Mineralogy and Geochemistry 52: 493-531.

21. Kronfeld J, Godfrey-Smith DI, Johannessen D, Zentilli M (2004) Uranium series isotopes in the Avon Valley, Nova Scotia. J Environ Radioact 73: 335-352.

22. Dawood YH (2010) Factors Controlling Uranium and Thorium Isotopic Composition of the Streambed Sediments of the River Nile, Egypt. JAKU: Earth Science 21: 77-103.

23. Mohamed CAR, Theng TL (2006) Activity concentration of Po-210 and Pb-210

24. Scott MR (1982) The chemistry of U- and Th-series nuclide in the rivers. In: Ivanovich M (ed.) Uranium Series Disequilibrium: Applications to Environmental Problems. New York: Oxford University Press 181-201.

25. Saili AB, Mohamed CAR (2014) Behavior of 210Po and 210Pb in shallow water region of Mersing estuary, Johor, Malaysia. Environment Asia 7: 7-18.

26. Theng TL, Mohamed CAR (2005) Activities of 210Po and 210Pb in the water column at Kuala Selangor, Malaysia. Journal of Environmental Radioactivity 80: 273-286.

27. Yang W (2006) Disequilibria between 210Po and 210Pb in surface waters of the southern South China Sea and their implications. Science in China Series D Earth Sciences 49: 103-112.

28. Miguel S, Bolívar JP, García-Tenorio R (2003) Mixing, sediment accumulation and focusing using 210Pb and 137Cs. Journal of Paleolimnology 29: 1-11.

29. Liu Z, Stattegger K (2014) South China Sea fluvial sediments: An introduction. Journal of Asian Earth Sciences 79: 507-508.

30. Wang B (2006) The Asian Monsoon, Netherlands: Springer Science & Business Media.

31. Ko DS, Shenn-Yu C, Chun-Chieh Wu, Lin II (2014) Impacts of Typhoon Megi (2010) on the South China Sea. Journal of Geophysical Research: Ocean 119: 4474-4489.

32. Tan F, Lim HS, Khiruddin A (2011) The Impact of the Typhoon to Peninsular Malaysia on Orographic Effects. In: IEEE Symposium on Business, Engineering and Industrial Applications (ISBEIA). Langkawi.

33. Aboobacker M, Pavel T, Vinod KK, Vethamony P (2013) Wind waves generated by Typhoon Vamei in the southern South China Sea. Geophysical Research Abstracts.

34. Milliman JD, Farnsworth K (2011) River Discharge to the Coastal Ocean: A Global Synthesis. New York: Cambridge University Press.

35. Chappell NA, Sherlock M, Bidin K, Macdonald R, Najman Y, et al. (2007) Runoff processes in Southeast Asia: Role of soil, regolith and rock type. In: Sawada H eds. Forest Environments in the Mekong River Basin. Springer 3-23.

36. Gascoyne M (1982) Geochemistry of the actinides and their daughters. In: Ivanovich M (ed.) Uranium Series Disequilibrium: Applications to Environmental Problems. New York: Oxford University Press 35-41.

37. Harmon R, Rosholt J (1982) Igneous rock. In: Ivanovich M, Harmon R (eds.) Uranium Series Disequilibrium: Applications to Earth, Marine and Environmental Sciences. Oxford UK: Clarendon Press 145-166.

38. Yusoff AH, Mohamed CAR (2015) Vertical Profiles of Natural Uranium Isotopes in Sediment Cores from Kota Kinabalu and Labuan Ports, Malaysia. EnvironmentAsia 8: 85-93.

39. Kritsananuwat R, Sahoo SK, Fukushi M, Pangza K, Chanyotha S (2015) Radiological risk assessment of ^{238}U, ^{232}Th and ^{40}K in Thailand coastal sediments at selected areas proposed for nuclear power plant sites. Journal of Radio analytical and Nuclear Chemistry 303: 325-334.

40. Huy NQ, Luyen T (2005) Study on external exposure doses from terrestrial radioactivity in Southern Vietnam. Radiation protection dosimetry 1: 1-6.

41. Nozaki Y, Yamamoto Y (2001) Radium 228 based nitrate fluxes in the eastern Indian Ocean and the South China Sea and a silicon-induced "alkalinity pump" hypothesis. Global Biogeochemical Cycles 15: 555-567.

42. Obata H, Nozaki Y, Dia Sotto Alibo, Yamamoto Y (2004) Dissolved Al, In, and Ce in the eastern Indian Ocean and the Southeast Asian Seas in comparison with the radionuclides 210Pb and 210Po. Geochimica et Cosmochimica Acta 68: 1035-1048.

43. Yii MW, Zaharudin A, Abdul-Kadir I (2009) Distribution of naturally occurring radionuclides activity concentration in East Malaysian marine sediment. Applied Radiation and Isotopes 67: 630-635.

44. Wan Mahmood ZUY, Ahmad Z, Izwan Abd Adziz M, Mohamed CAR, Ishak AK (2010a) Radioactivity distribution of thorium in sediment core of the Sabah-Sarawak coast. Journal of Radio analytical and Nuclear Chemistry 285: 365-372.

45. Wan Mahmood ZUY, Mohamed CAR, Yii MW, Ahmad Z, Ishak K, et al. (2010b) Vertical inventories and fluxes of 210 Pb, 228 Ra and at southern South China

Sea and Malacca Straits. Journal of Radioanalytical and Nuclear Chemistry 286: 107-113.

46. Yusoff AH, Sabuti AA, Mohamed CAR (2015) Natural uranium and thorium isotopes in sediment cores Off Malaysian Ports. Ocean Science Journal 50: 403-412.

47. Borovec Z, Kribek B, Tolar V (1979) Sorption of uranyl by humic acids. Chemical Geology 27: 39-46.

48. Amin YM, Mahat RH, Nor RM, Uddin KM, Ghazwa HT, et al. (2013) The presence of natural radioactivity and 137Cs in the South China Sea bordering peninsular Malaysia. Radiation protection dosimetry 156: 475-480.

49. Omar M, Ali HM, Abu MP, Kontol KM, Ahmad Z, et al. (2004) Distribution of radium in oil and gas industry wastes from Malaysia. Applied Radiation and Isotopes 60: 779-782.

50. Peng ML (2015) Radioactivity Levels of 234Th and 210Po in the Green Mussel (Perna viridis) at the Straits of Johor and the Estimated Accumulations to Human Body. In: Ahmad I, Syaizwan ZZ (eds.) ISIMBIOMAS 2015. Putrajaya: UPM press pp: 34-39.

51. Alam L, Mohamed CAR (2011) Natural radionuclide of Po210 in the edible seafood affected by coal-fired power plant industry in Kapar coastal area of Malaysia. Environmental Health 10: 43.

52. Mohamed CAR, Kuan PF (2005) Concentrations of 210Po and 210Pb in zooplankton at Pulau Redang, Terengganu, Malaysia. Journal of Biological Sciences 5: 312-314.

53. Huy NQ, Luyen TV (2006) Study on external exposure doses from terrestrial radioactivity in Southern Vietnam. Radiat Prot Dosimetry 118: 331-336.

54. Staub JR, Among HL, Gastaldo RA (2000) Seasonal sediment transport and deposition in the Rajang River delta, Sarawak, East Malaysia. Sedimentary Geology 133: 249-264.

55. Bergh GD, Boera W, Schaapveldb MAS, Ducc DM, van Weeringa TjCE (2007) Recent sedimentation and sediment accumulation rates of the Ba Lat prodelta (Red River, Vietnam). Journal of Asian Earth Sciences 29: 545-557.

56. Theng TL, Ahmad Z, Mohamed CAR (2003) Estimation of sedimentation rates using 210Pb and 210Po at the coastal water of Sabah, Malaysia. Journal of Radio analytical and Nuclear Chemistry 256: 115-120.

57. Liu Z, Wang H, Hantoro WS, Sathiamurthy E, Colin C (2012) Climatic and tectonic controls on chemical weathering in tropical Southeast Asia (Malay Peninsula, Borneo, and Sumatra). Chemical Geology 291: 1-12.

58. Mohamed CAR, Mohamed K, Ahmad Z (2006) Distribution of 234U and 238U in Sungai Selangor, Peninsular of Malaysia. Journal of Applied Sciences 6: 562-566.

9

Geophysical Investigation of Geothermal Potential of the Gilgil Area Nakuru County, Kenya Using Gravity

Nyakundi ER[1]*, Githiri JG[2] and Ambusso WJ[1]

[1]Kenyatta University, Department of Physics, Nairobi, Kenya
[2]Jomo-Kenyatta University of Agriculture and Technology, Department of Physics, Nairobi, Kenya

Abstract

In this study, gravity survey was used to investigate the geothermal potential field in Gilgil area Nakuru County, Kenya. The ground based CG-5 Autograv gravimeter was used to accurately measure gravity at each field station. A total of 147 gravity stations were established over an area of about 68 km^2 and gravity corrections done. The complete bouguer anomaly was computed and a contour map for the study area plotted using surfer 8.0 software. Qualitative interpretation of the map shows gravity highs in the study area which were interpreted as dense bodies within the subsurface. Five profiles along the gravity highs were drawn and oriented in the directions SW-NE, NW-SE and almost N-S. The regional trend of the profiles was subtracted from the observed data yielding the residual anomaly. 2D Euler deconvolution was done on the profile data and revealed subsurface faults and bodies at a depth range of 790m-4331m. Forward modelling of selected profiles using Grav 2DC software revealed presence of dense intrusive bodies on the northern and southern parts of the study area with the density contrast range of 0.25-0.28. These bodies were interpreted as intrusive dykes that have higher density than surrounding rocks. Such intrusive dykes may be geothermal heat sources.

Keywords: Geothermal potential; Gravitational field; Volcanism; Gilgil area

Introduction

Gilgil area is located between Naivasha and Nakuru in Nakuru County, Kenya. It lies 121 km north of Nairobi. Gilgil area is in the Kenyan rift where a number of geothermal fields lie. Preliminary surface investigations have been carried out in Suswa, Longonot, Olkaria, Eburru, Menengai, Bogoria, Baringo, Korosi, Silali and Emurangogolak geothermal fields [1]. Drilling has been done in Eburru and Olkaria. The present power station is in Olkaria. Thus this study was carried out to establish the potential of Gilgil area as a geothermal reservoir.

Gravity surveying has been done to gain information on geothermal potential areas. Gravity technique in geophysical exploration deals with measurements of changes in the Earth's gravitational field strength [2]. Gravity measurements and observations are done on the earth's surface. The gravimeter is an instrument used to measure changes in the Earth's gravitational field on the Earth's surface and records its values in milligals. It helps to find bodies within the subsurface of the earth which have greater or lesser density than the surrounding host rocks. Gravity can also constrain data during interpretation of other geophysical techniques such as seismic and magnetic.

Gravitational field is natural on the earth's surface similar to magnetic and radioactivity. It is a natural field technique that uses gravitational field of the earth. There is no energy required to be put into the subsurface to gain information [3]. It reveals change in these natural gravitational field that is attributed to economic feature of concern within the subsurface. This feature portrays a subsurface area of anomalous mass and causes localized change in gravity referred to as gravity anomaly.

Geology of Gilgil area

The geology of Gilgil area is as a result of volcanism and tectonic activities of the rift valley. The volcanism of the rift preceded and accompanied the rift tectonism. Gilgil area is dominated by quaternary volcanic ash and diatomaceous silts in the plain areas and some volcanic tuff, lava flow and diatomite deposits in the higher escarpments [4]. Alkaline volcanism composed of pumiceous pyroclastics, ashes, trachytes, ignimbrites, phonolites and phonolitic trachytes, tuffs, agglomerates and acid lava dominates Gilgil area. Also volcanic soil and diatomite deposits dominate the area with trona impregnated silts bordering Lake Elmenteita [4]. The area is also characterized by repeated volcanicity followed by movement. The eruptives in each episode start with basalt [5]. The southern part of Gilgil is within the Olkaria volcanic complex. Craters, fumaroles, hot springs and steam vents are found in several places within the Olkaria and Eburru area [5]. The earlier tectonic geology is reflected in the step-faults of Satima and Kinangop generating Kinangop plateau. Grid faulting generated Gilgil plateau while the Mau escarpment is as a result of fault flexures. The major fault escarpments influence topography of the rift floor that influences the drainage flow pattern [6].

Methodology

In gravity technique, the geology is examined on the foundations of changes in the Earth's gravitational field emerging from deviations of mass within the underlying rocks. The fundamental concept is the idea of a causative body, which is a rock of unusual density from the host masses (Figure 1). This causative body portrays a subsurface region of abnormal density and results in change in the Earth's gravitational

*Corresponding author: Nyakundi ER, Kenyatta University, Department of Physics, P.O BOX 43844-00100, Nairobi Kenya, E-mail: rayoraerick@yahoo.com

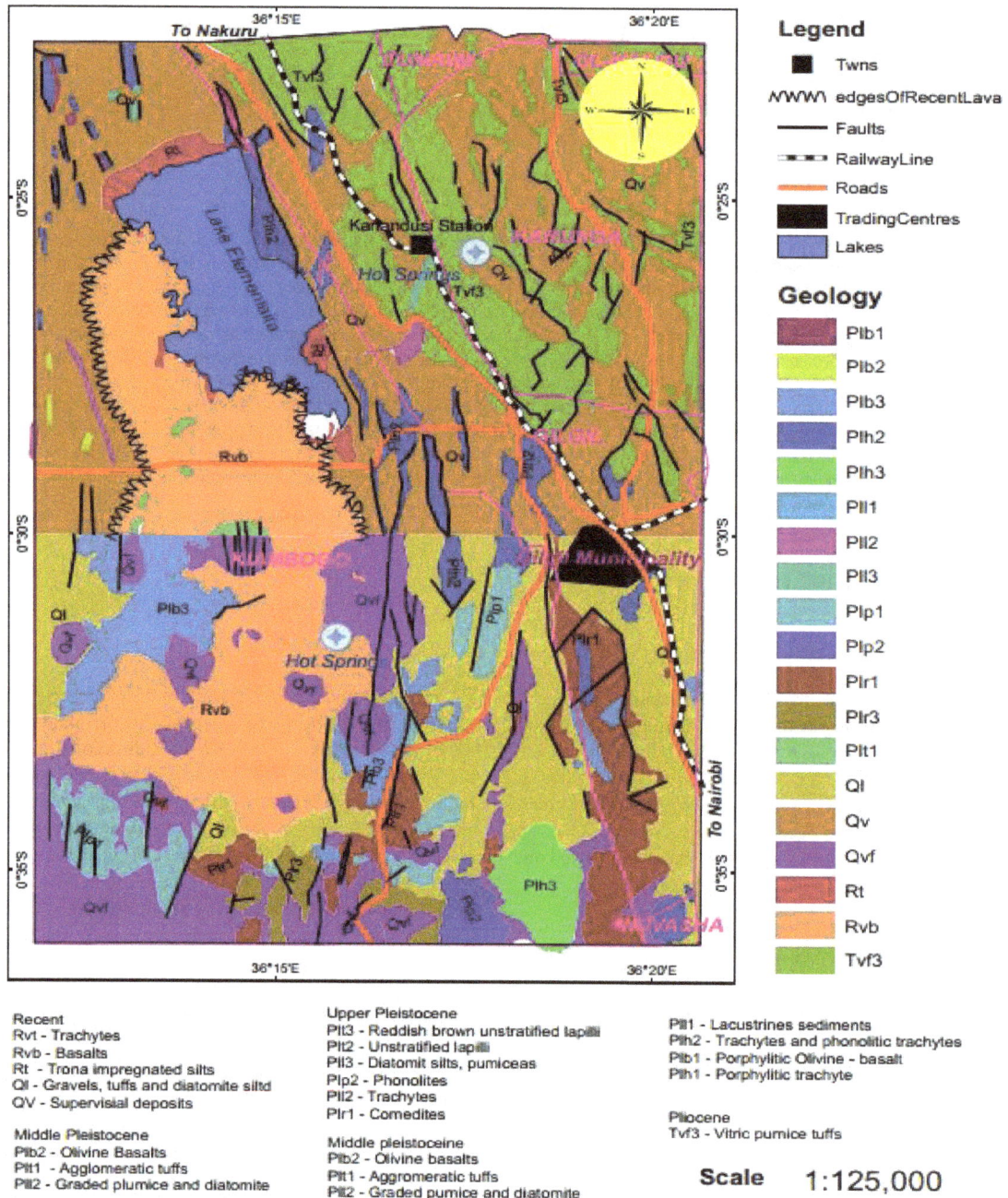

Figure 1: Map showing the geology of the study area.

field called gravity anomaly [7]. An area of approximately 68 km² was covered during this study. Data was gathered from 147 measurement points using CG-5 gravimeter. The base stations were formed for the purpose of drift corrections. Stations were spaced at 500 m apart. At each station the time, northing, easting, altitude and gravity value in milligals was recorded.

Regional density

The average density of rocks in the study area ρ_a was taken as 2.67g/

cm³ [4]. Density of an intruding body ρb ranges from 2.70 g/cm³-3.20 g/cm³ [7]. Density contrast is given by

$$\rho = \rho_b - \rho_a$$

Density contrast range was found to be 0.03 g/cm³-0.53 g/cm³ and was employed for modelling. Body of density contrast 0.03 g/cm³-0.53 g/cm³ is associated with heat source at its basin because it best forms at plumes and hotspots below the continent. Mostly forms as an extrusive rock such as lava flow but can also form as intrusive bodies like dike or sill [4]. During gravity forward modelling of this study, a density contrast of 0.25 cm³ and 0.28 cm³ produced the best fit between observed gravity anomaly and computed gravity anomaly.

The Bouguer anomaly map

Gravity anomalies are obtained after reductions have been done to the observed gravity data. If there were no mass distribution within the Earth's subsurface, the gravity anomaly would be zero.

The contour map in Figure 2 was generated from processed gravity data. This map shows contour intervals of 1 mgal with the highest value at −181 mgal and the least value at −211 mgal. To the Northeast, the map reveals gravity highs with few gravity lows. To the southeast, the map reveals gravity lows with a few gravity highs. To the northwest, the map reveals gravity lows with a small part of gravity high. An intruding rock has density ranging from 2.70 g/cm³-3.20 g/cm³ which is a gravity high [7]. Geothermal reservoir is associated with a gravity high because materials coming from the Earth's mantle are of higher density than materials found in the Earth's crust.

Euler deconvolution

Euler deconvolution technique provided automatic approximations of a causative body location and its depth within the Earth's subsurface. Therefore, Euler deconvolution located the boundary of the said resource and its depth from the surface. The most important outcome of Euler deconvolution is the description of trends and depths [8]. In this study, a structural index of 1.0 was used as it best delineates fractures and intruding dykes in the subsurface which are associated with heat sources.

Euler solutions along profile PP' as shown in Figure 3 suggest a causative body which occurs at maximum depth of 2053.74 m. It reveals a fault at 1000 m and 2000 m along the profile [9]. It also shows a causative body at 1000 m and 3000 m along the profile which has a material of higher density than the host rock.

Euler solutions along profile QQ' as shown in Figure 4 reveals a causative body which occurs at a maximum depth of 792.74 m. It has imaged a body of higher density than the surrounding rock at 1000 m

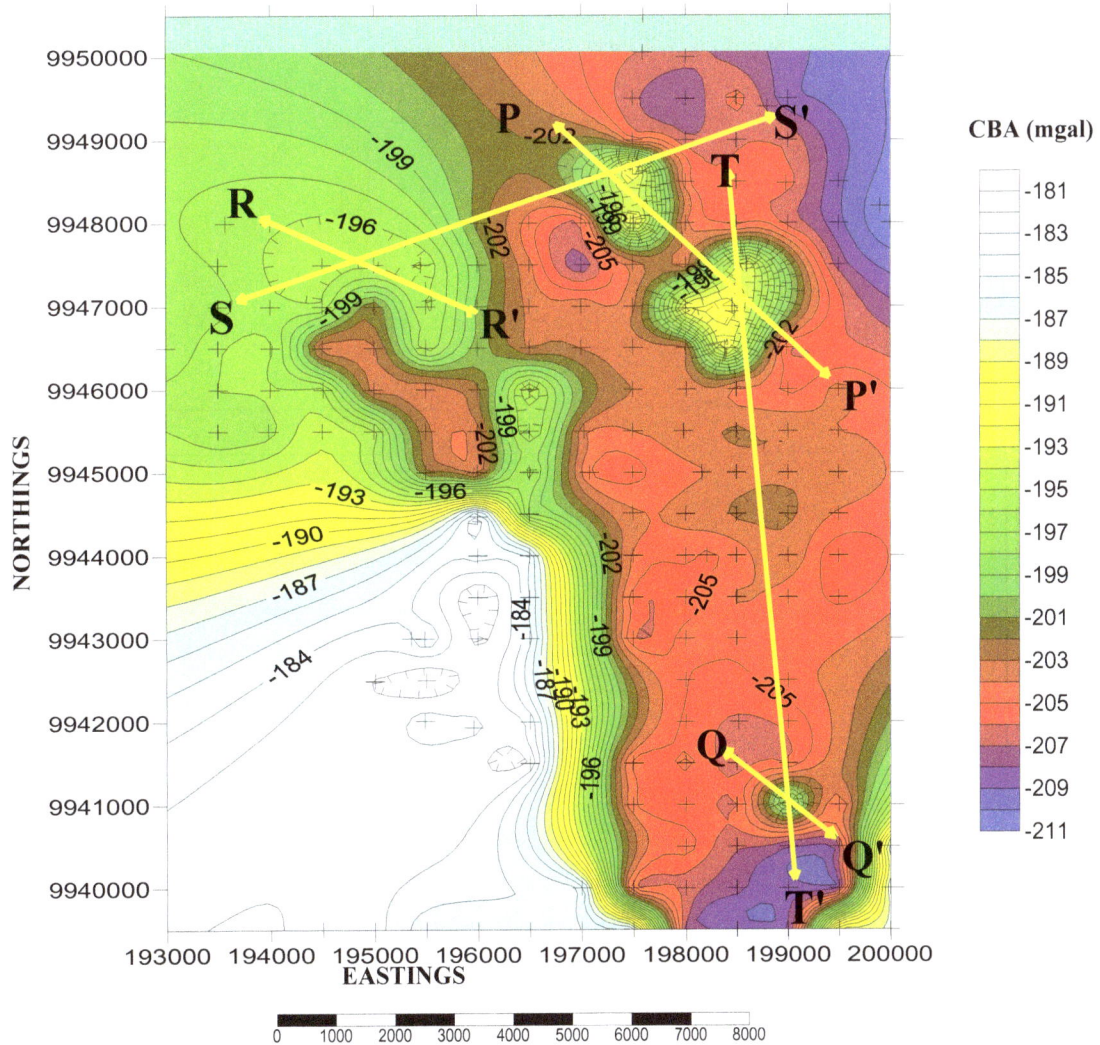

Figure 2: Contour map profiles for the Gilgil area.

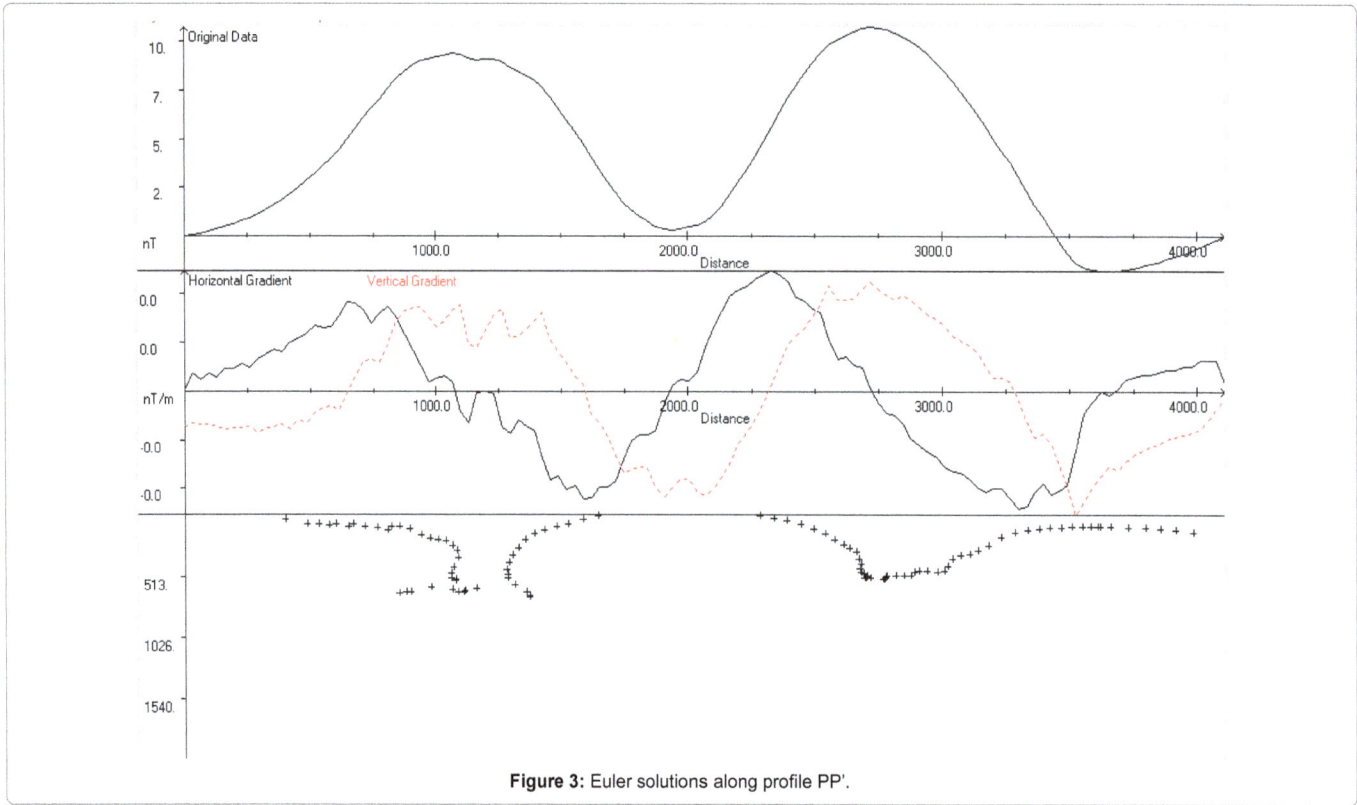

Figure 3: Euler solutions along profile PP'.

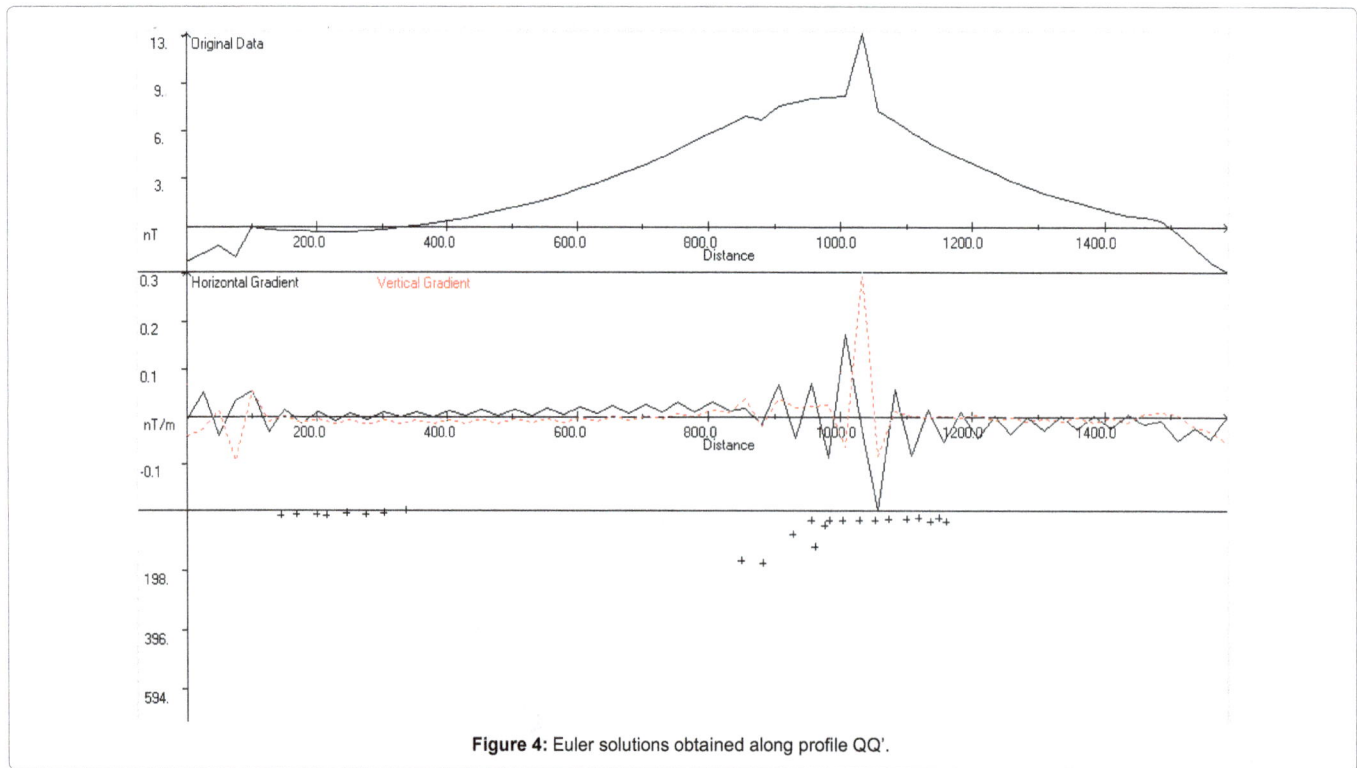

Figure 4: Euler solutions obtained along profile QQ'.

along the profile [10]. At 200 m along the profile, there is a shallow causative body.

Euler solutions along profile RR' as shown in Figure 5 shows a causative body which occurs at a maximum depth of 1194.21 m. The body occurs between 1200 m and 2000 m along the profile and has a higher density than the host rock. It also reveals a fault between 1200 m and 2000 m along the profile.

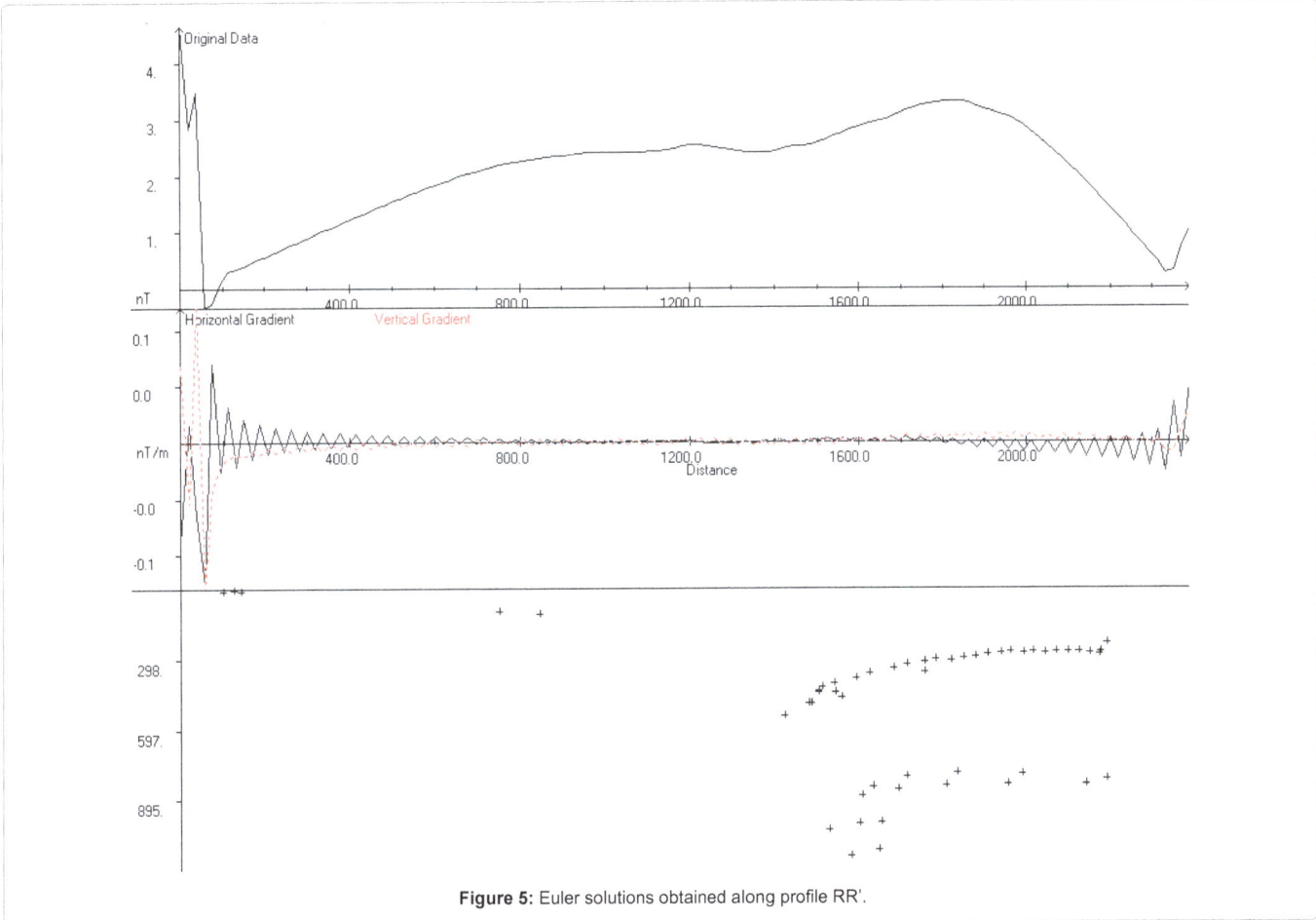

Figure 5: Euler solutions obtained along profile RR'.

Euler solutions along profile SS' as shown in Figure 6 shows an intrusive body which occurs at maximum depth of 2837.15 m. It has also imaged faults at about 2000 m and 4000 m along the profile which is filled by a material of higher density than the surrounding host rock.

Euler solutions along profile TT' as shown in Figure 7 reveals a causative body which occurs at a maximum depth of 4331.38 m. It has also imaged a fault at about 2000 m to 3000 m along the profile which has a higher density than the surrounding host rock.

Forward modeling

This was done using GRAV2DC software in surfer 8 computer programme. Modelling entailed construction of an appropriate model based on geological information of the study area [11]. The cross section data was transferred to GRAV2DC software for forward modelling. The parameters determined by Euler deconvolution acted as start-up parameters for the model bodies. The model's gravity anomaly was computed and compared to the observed anomaly. Features of the model were altered to increase the correspondence of observed anomaly and computed anomaly. In this interpretation, the depth and density contrast of a causative body was determined [12]. Models constructed are as shown in Figures 8-12.

Profile PP' is on the northern part of the study area as shown in Figure 2 and it cuts across a gravity high anomaly region trending in a NW-SE direction. Models on profile PP' as shown in Figure 8 reveals two subsurface intrusive bodies [13]. The first body has a density of

2.92 g/cm³ and imaged at a depth of 169.48 m while the second body has the same density of 2.92 g/cm³ and imaged at a depth of 159.51 m. This gravity high could be due to hot intrusive bodies of high density from the mantle which are probably feeding the hot spring in the area.

Profile QQ' is on the southern part of the study area and it cuts across a gravity high anomaly region trending in a NW-SE direction. Models on profile QQ' as shown in Figure 9 reveals an intrusive body of density 2.95 g/cm³ and imaged at a depth of 50.33 m. Presence of recent volcanic soil shows there was volcanic activity which deposited high density materials close to the surface hence the imaged body could be a cooling dyke injection [14].

Profile RR' is on the north western part of the study area and it cuts across a gravity high anomaly region trending in a NW-SE direction. Models on profile RR' as shown in Figure 10 shows an intrusive body of density 2.92 g/cm³ and imaged at a depth of 370.22 m. This was presumed to be a dense body imaged under a volcanoe which is probably a hot intruding dyke hence a heat source at the basin [15].

This model shown in Figure 11 was oriented in a SW-NE direction to constrain the density contrast and depth of profile RR' and PP'. It generated the same values as obtained in Figure 8 and Figure 10. The density contrast of the intruding bodies was found to be 0.25 g/cm³. The depth for body 1 was 365.99 m and body 2 was 169.48 m. It has imaged faults responsible for underground thermal movement and massive intrusions which could be heat sources [16].

The model fit shown in Figure 12 was oriented in a nearly N-S

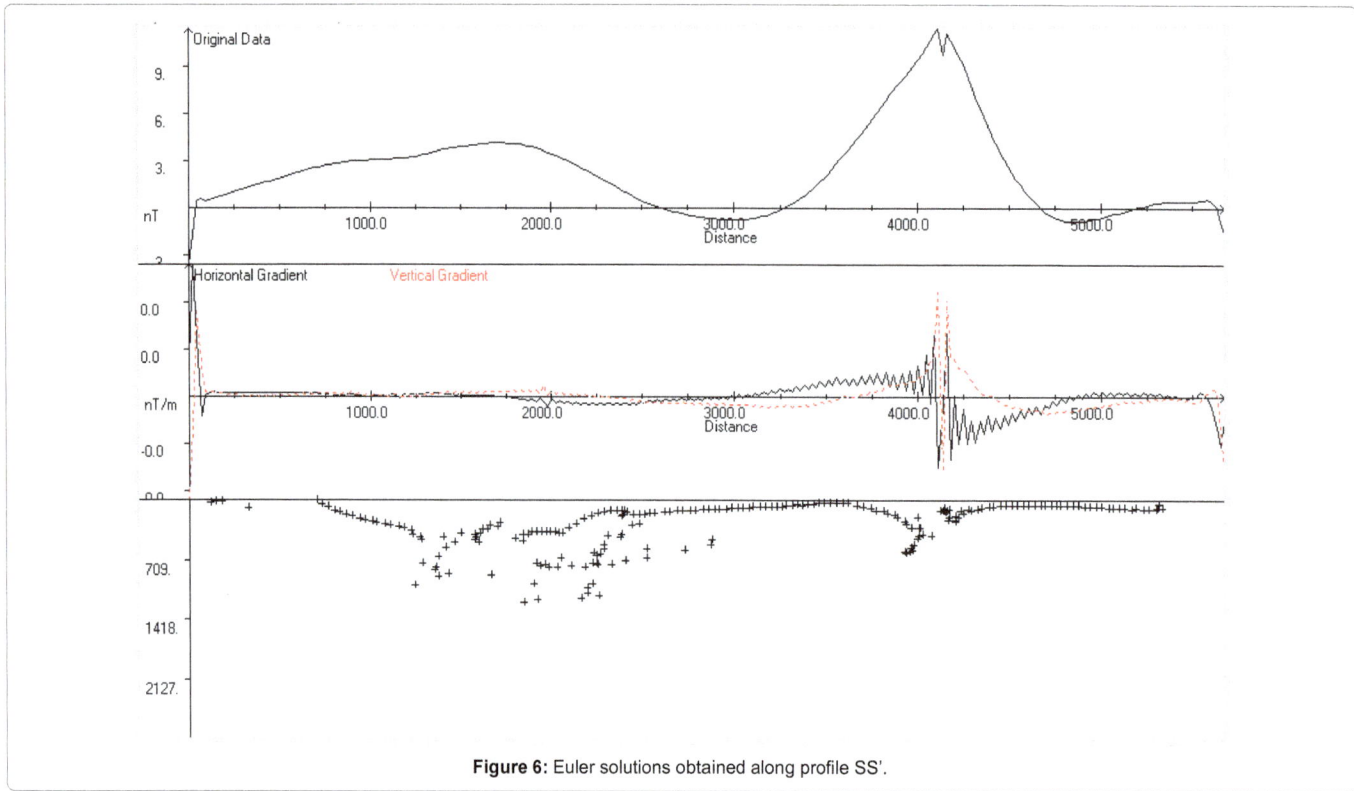

Figure 6: Euler solutions obtained along profile SS'.

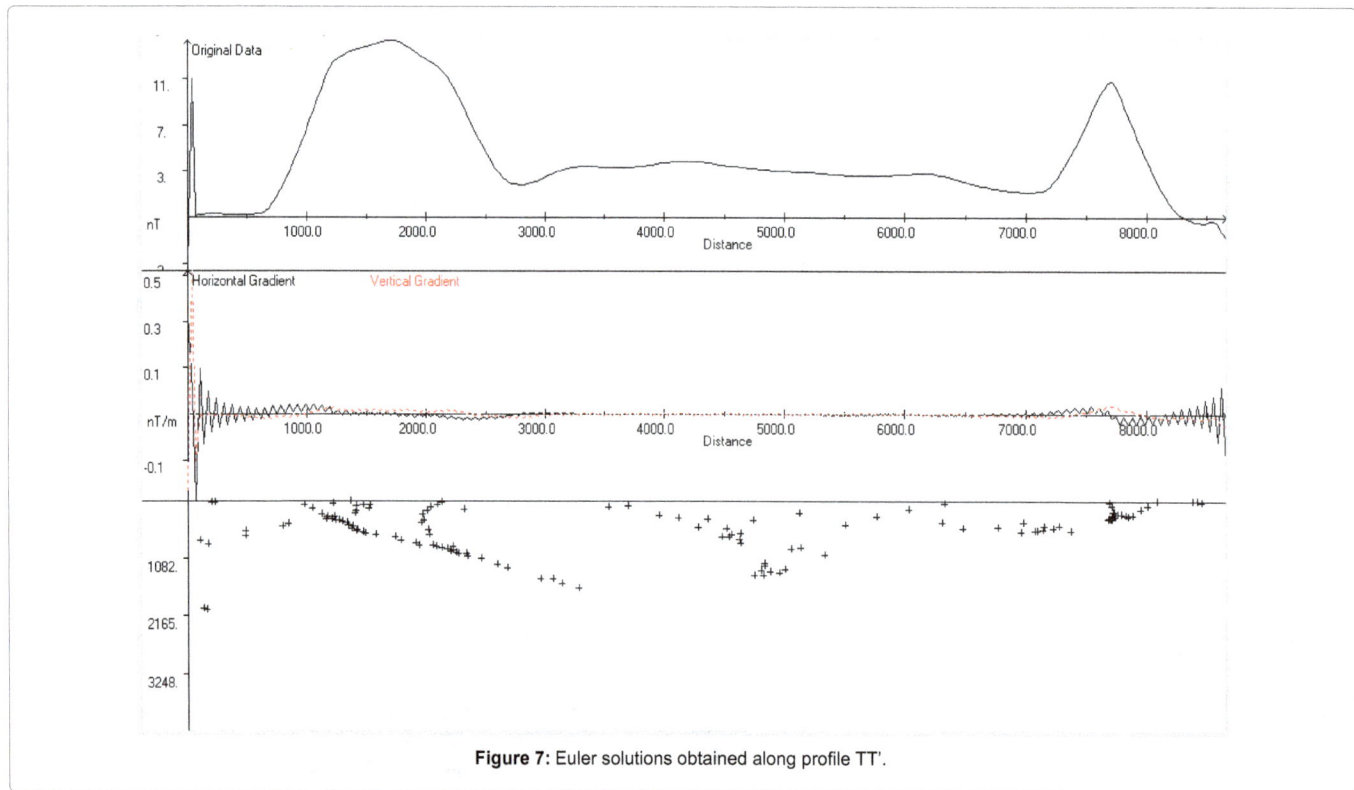

Figure 7: Euler solutions obtained along profile TT'.

direction to constrain the density contrast and depth of profile PP' and QQ'. It generated the same results as in Figures 8 and 9. The density contrast for body 1 was 0.25 g/cm³and body 2 was 0.28 g/cm³. The depth was 129.16 m for body 1 and 50.61 m for body 2. These bodies were interpreted to be dense intruding dykes into the subsurface which could be heat sources.

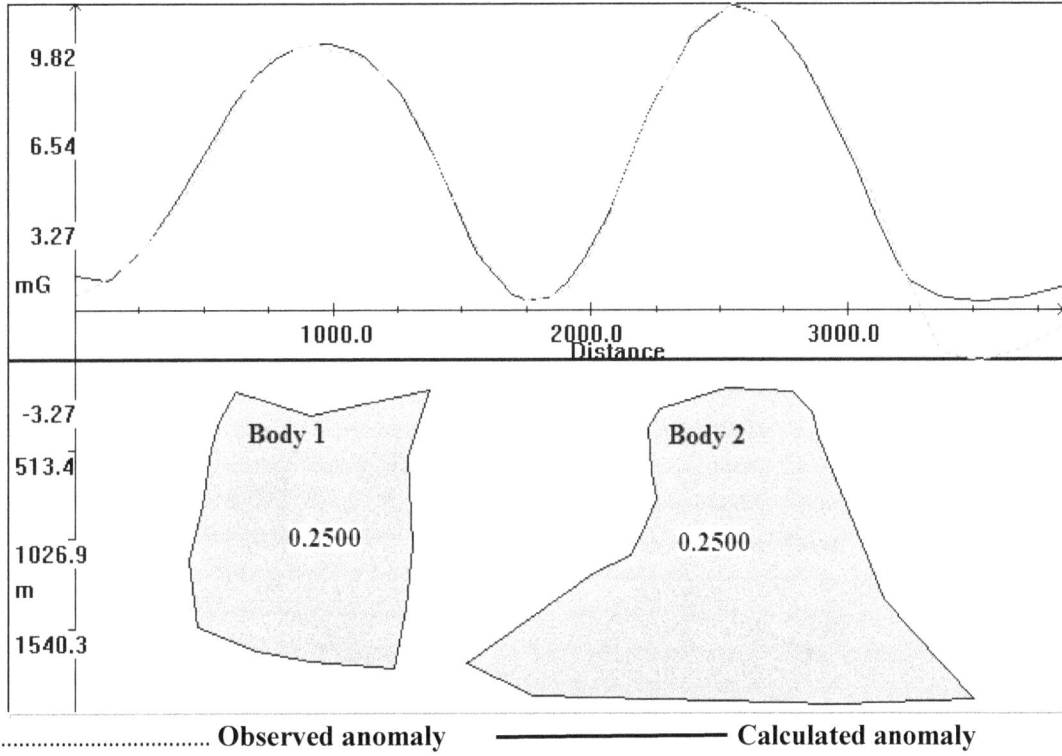

Figure 8: Model fit on residual bouguer anomaly profile PP'.

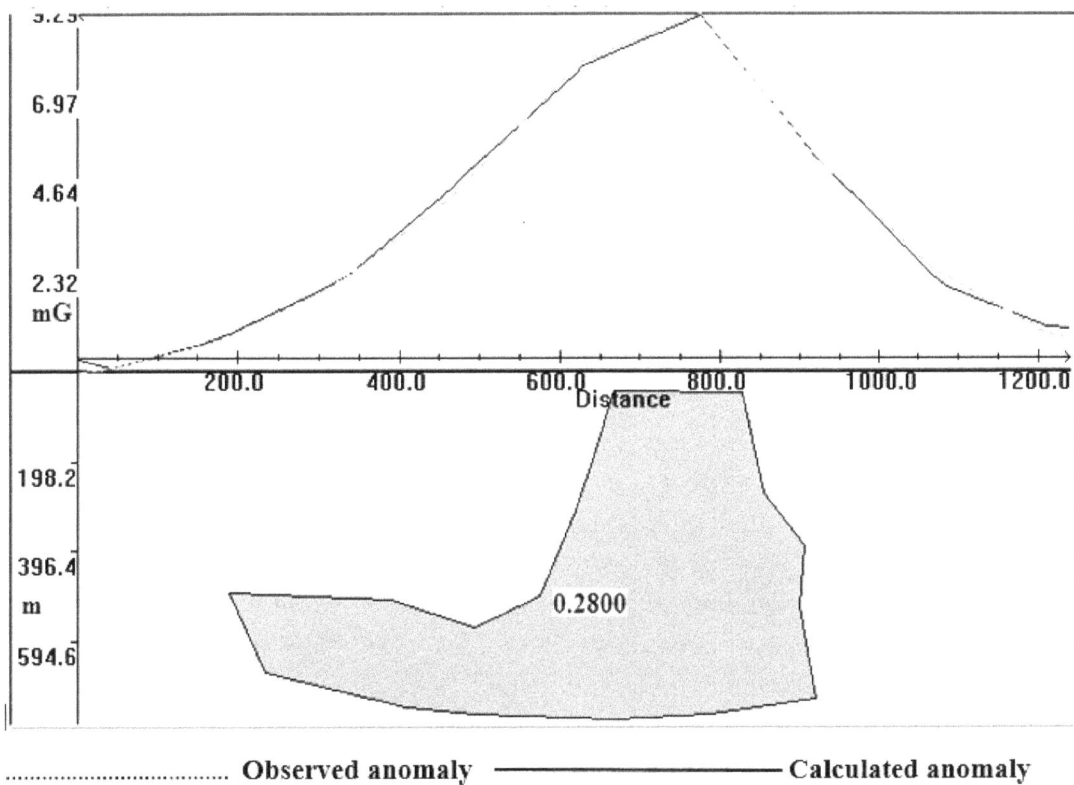

Figure 9: Model fit on residual bouguer anomaly profile QQ'.

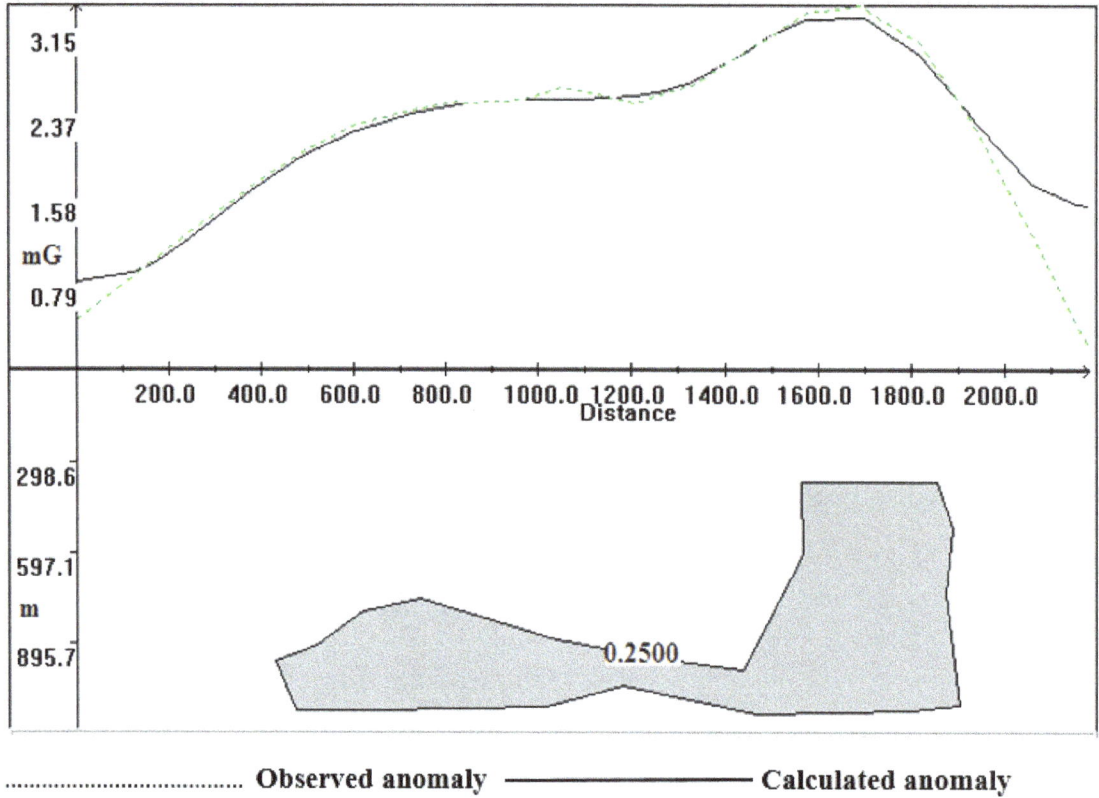

Figure 10: Model fit on residual bouguer anomaly profile RR'.

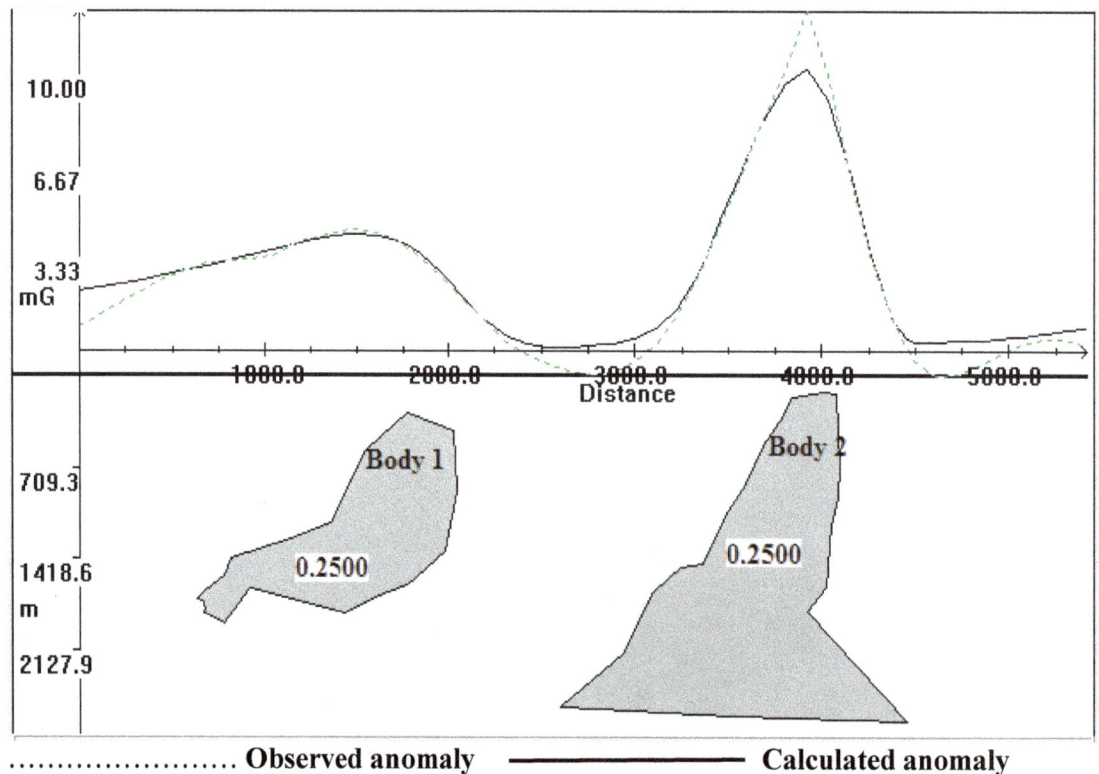

Figure 11: Model fit on residual bouguer anomaly profile SS'.

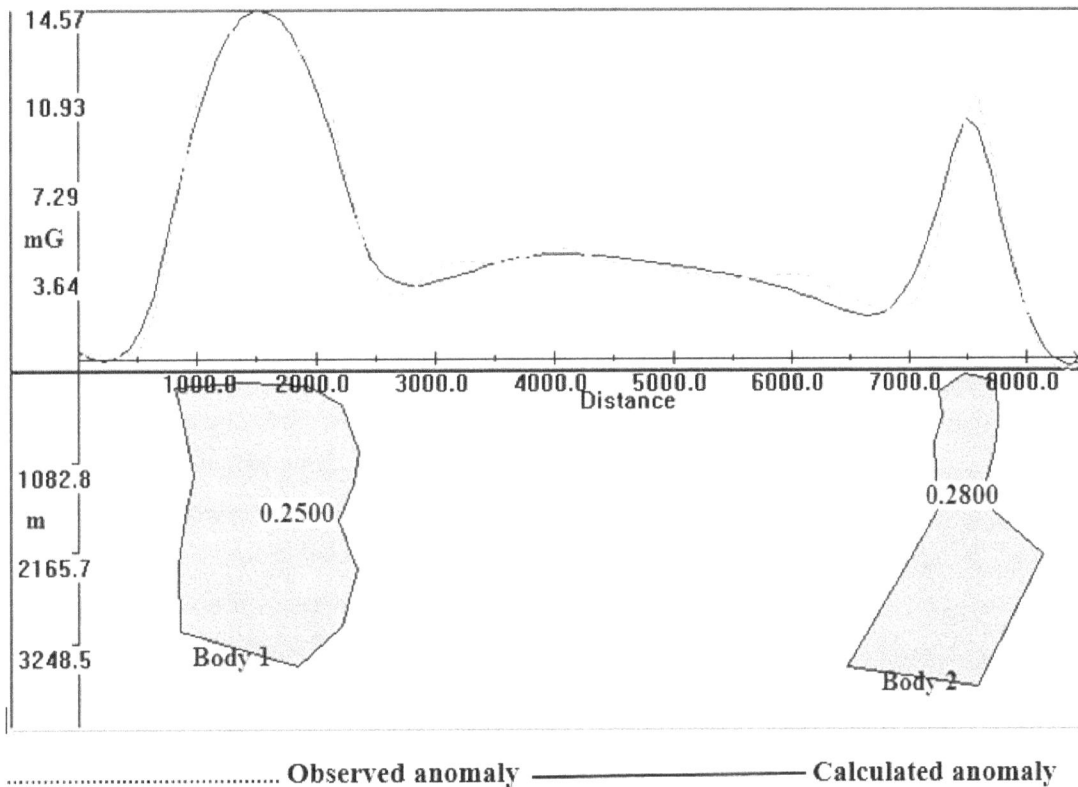

Figure 12: Model fit on residual bouguer anomaly profile TT'.

Discussion

Profile PP' is on the northern part of the study area as shown in Figure 2. It cuts across a gravity high anomaly region trending in a NW-SE direction. Models on profile PP' as shown in Figure 8 reveals two subsurface intrusive bodies. The first body has a density of 2.92 g/cm^3 and imaged at a depth of 169.48 m while the second body has the same density of 2.92 g/cm^3 and imaged at a depth of 159.51 m. This positive gravity anomaly could be a result of hot intrusive bodies of high density from the mantle under the volcanic complexes which are probably feeding the hot spring in the area hence there could be a heat source at the basin.

Profile QQ' is on the southern part of the study area as shown in Figure 2. It cuts across a gravity high anomaly region trending in a NW-SE direction. Models on profile QQ' as shown in Figure 9 reveals an intrusive body of density 2.95 g/cm^3 and imaged at a depth of 50.33 m. This was presumed to be due to phonolitic trachytes during the lower Pleistocene period in the study area. Presence of recent volcanic soil shows there was volcanic activity which deposited high density materials close to the surface. Probably there was a volcanic activity in the area which stopped hence the imaged body could be a cooling dyke injection.

Profile RR' is on the north western part of the study area as shown in Figure 2. It cuts across a gravity high anomaly region trending in a NW-SE direction. Models on profile RR' as shown in Figure 10 shows an intrusive body of density 2.92 g/cm^3 and imaged at a depth of 370.22 m. This was presumed to be a dense body imaged under a volcanoe which is probably a hot intruding dyke hence a heat source at the basin.

Profile SS' shown in Figure 2 was drawn to cut across profile RR' and profile PP' for the purpose of constraining the density contrast and depth of imaged body. Also profile TT' in Figure 2 was drawn to cut across profile PP' and QQ' for the purpose of constraining the density contrast and depth of imaged body. Profile SS' and profile TT' gave the same density contrast and depth as profile PP', profile QQ' and profile RR' as shown in Figure 11 and Figure 12. The two profiles have imaged faults and massive intrusions which could be heat sources. Due to hydrothermal activity and imaged fractures in the area, probably these hot intrusive bodies for profile PP' and RR' are responsible for the hot spring in the area.

Conclusion and Recommendation

The Gilgil prospect area is located in the Kenyan rift where a number of geothermal fields lie. It is characterized by major fracture lines, quiet volcanic craters, fumaroles and hot springs. Fractures resulting from extensional tectonics of continental rifting probably provides a good structural set up that allows water from the rift scarps to penetrate deep into the crust, towards the hot magmatic bodies as modelled under the volcanoes and normal faults conducting hot fluids from deep into possible geothermal reservoirs at shallower depth. Major fracture lines were imaged in this prospect area by Euler deconvolution. These faults are at different depths from the surface as shown by Figure 3 through Figure 7. There are those at the deep basement while others at the shallow subsurface. The deep faults transport thermal fluids from deep parts of the Earth to the subsurface. Also the shallow faults in the Earth's subsurface direct the flow of thermal fluids on the upper part of the basement. The top faults direct the flow of water from the rift scarps to the hot masses underground. This probably led to the trapping of a heat source in the area as evidenced by hot springs. The deep water circulation would therefore collect heat from the bodies and discharge it through hot springs along faults and fractures as observed in the study area.

Geothermal occurs along major fracture lines, inactive volcanic craters and where there are hot springs. Therefore the modelled bodies across the selected gravity profiles lie relatively at shallower depths as shown from the models in Figure 8 through Figure 12. Thus, the high heat flow observed in the area as evidenced by hot springs could be due to shallow dense intruding bodies within the rift floor faults. Gravity technique was able to locate these dense bodies within the Earth's subsurface as positive gravity anomalies. It is postulated that intrusives, in the form of dykes would be tapping heat from large magma bodies at few kilometres from the surface. Therefore this study was able to detect gravity highs that show evidence of a buried dense body compared to the surrounding rocks. The buried dense bodies were interpreted as intruding dyke injections within the subsurface which could be heat sources.

This gravity study was done to gather information on the possibility of geothermal occurrence in Gilgil area. It has provided information which will be used as a start point for future detailed geophysical study. Gravity technique is ambiguous and this implies that any anomaly could be a result of many possible sources. To reduce this ambiguity during interpretation, this study recommends the application of other geophysical techniques like seismic, magnetotelluric (MT) and magnetic for outcome comparisons. This ensures confirmation of this gravity results before drilling is done which is expensive.

Acknowledgement

Special thanks go to research supervisors, Dr. Willis Ambusso of Kenyatta University and Dr. John Githiri of Jomo Kenyatta University of Agriculture and Technology for their technical guidance, suggestions and encouragement during this research work. Thanks to the staff physics department of Kenyatta University for their valuable suggestions.

References

1. Clarke MCG (1990) Geological, volcanological and hydrogeological controls on the occurrence of geothermal activity in the area surrounding Lake Naivasha. Ministry of energy, Kenya 7-12.

2. Sharma p (2002) Environmental and engineering geophysics. Cambridge university press, Cambridge.

3. Dickerson PW (2004) Field geophysical training of astronauts in Taos region. University of Texas at Austin, Texas 278-281.

4. Mccall GJH (1967) Geology of the Nakuru-Thomson's fall-lake Hannington area. Report No.78, Geological Survey of Kenya.

5. Thompson AO, Dodson RG (1963) Geology of the Naivasha area. Report No.55, Geological Survey of Kenya.

6. Onywere SM, Mironga JM, Simiyu I (2012) Use of remote sensing data in evaluating the extent of anthropogenic activities and their impact on lake Naivasha Kenya. Open Environ Eng J 5: 9-18.

7. Kearey P, Brooks M, Hill I (2002) An Introduction to geophysical exploration. Blackwell science Ltd, Oxford.

8. Chenrai P, Meyers J, Charusiri P (2010) Euler deconvolution technique for gravity survey. J Appl Sci Res 6: 1891-1897.

9. Baker BH, Wohlenberg J (1971) Structural and evolution of the Kenya rift valley. Nature 229: 538-542.

10. Hirt C (2015) Gravity forward modeling. Curtain University, pert, WA, Australia. Encyclopedia of geodesy.

11. Lowrie W (1997) Fundamentals of geophysics. Cambridge university press, Cambridge.

12. Riaroh D, Okoth W (1994) Geothermal fields of the Kenya Rift. Tectonophys 236: 117-130.

13. Santos PA, Rivas AJ (2009) Gravity survey contribution to geothermal exploration in El Salvador: The cases of Berlin, Ahuachapán and San Vicente areas. United Nations University, Geothermal Training Programme.

14. Searle RC (1970) Evidence from gravity anomalies for the thinning of the lithosphere beneath the Rift valley in Kenya. Geophys J 21: 13-31.

15. Telford WM, Geldart LP, Sheriff RE (1990) Applied Geophysics. Cambridge University Press.

16. Yu G, He ZX, Hu ZZ, Porbergsdottir IM, Strack KM, et al. (2009) Geophysical exploration using MT and gravity techniques at Szentlorinc area in Hungary. Int Expos Ann Meet 4333-4338.

10

An Automatic Deconvolution Method for Modified Gaussian Model using the Exchange Monte Carlo Method: Application to Reflectance Spectra of Synthetic Clinopyroxene

Peng K Hong[1]*, Hideaki Miyamoto[1-3], Takafumi Niihara[1,4], Seiji Sugita[2], Kenji Nagata[5], James M Dohm[1] and Masato Okada[5]

[1]The University Museum, The University of Tokyo, 7-3-1 Hongo, Bunkyo-ku, Tokyo 113-0033, Japan
[2]Department of Earth and Planetary Science, The University of Tokyo, 7-3-1 Hongo, Bunkyo-ku, Tokyo 113-0033, Japan
[3]Planetary Science Institute, 1700 East Fort Lowell, Tucson, AZ 85719-2395, USA
[4]Lunar and Planetary Institute, Universities Space Research Association, 3600 Bay Area Boulevard, Houston, TX 77058, USA
[5]Department of Complexity Science and Engineering, The University of Tokyo, 5-1-5 Kashiwanoha, Kashiwa, Chiba 277-8561, Japan

Abstract

Deconvolution analysis of reflectance spectra has been a useful method to infer mineral composition and crystal structure. Many of the recent deconvolution analyses of reflectance spectra of major rock-forming minerals, such as olivine and pyroxene, have been based on a modified Gaussian Model (MGM). The numerical algorithm of the widely used MGM, however, utilizes the steepest descent method, which has a local minima problem. With inaccurate initial parameters, the steepest descent method converges into a local minimum, thus the analyzer must manually adjust initial parameters and calculate the model repeatedly to obtain the desired solution. In order to avoid the local minimum problem, we utilized Bayesian spectral deconvolution with the exchange Monte Carlo method, which is an improved algorithm of the Markov chain Monte Carlo method, aimed to both avoid local minima traps and remove the arbitrariness originated from initial parameters. We applied the model to visible to near infrared reflectance spectra of 31 synthetic clinopyroxene samples with wide ranging Mg, Fe and Ca compositions (solid solution). We obtained results consistent with the previous studies based on conventional MGM analyses, suggesting that the exchange Monte Carlo method can yield results consistent with the conventional MGM analyses purely based on the observed data. We also find that the center wavelengths of 1 μm absorption bands of high-Ca pyroxene samples have a linear dependence on Fe/Mg component. Both 1 μm and 2 μm absorption bands seem to follow approximation lines in the three-dimensional spaces of center wavelengths, Ca and Fe components. The successful application of the exchange Monte Carlo method to a wide range of clinopyroxenes would have a potential to expand the applicability of MGM to a variety of space/ground-based observations, especially when we cannot rely on prior information of the mineralogy.

Keywords: Pyroxene; Synthetic mineral; Reflection spectroscopy; Spectral deconvolution; Modified Gaussian model; Monte Carlo method

Introduction

Remote sensing of reflectance spectra of Earth and other planetary bodies can be useful for identifying mineral distribution on their surfaces, especially in remote regions that are exceedingly challenging to perform field-based investigation, and those planetary surfaces yet to have in situ observation, mapping, characterization, sampling, and analyses [1-8]. Many different factors, however, can influence the surface spectra, such as various alteration and weathering processes, and observational conditions [2,3,5,9-12]. Because a reflectance spectrum is a complex non-linear mixture of the above mentioned factors [13-16], it is highly challenging to segregate each factor and extract the true mineral spectra, based solely on remotely observed reflectance spectra, and thus confidence in the resulting signatures should be gained by comparing with the reference spectra obtained by field or laboratory measurements [17]. Though challenging, investigating the spectral change due to the variation of elemental composition should not be avoided, since it is one of the ultimate goals of remote sensing of reflectance spectra of planetary surfaces [2,18]. Compared to terrestrial surfaces, those of extraterrestrial bodies such as the Moon and asteroids are not covered by liquid water and vegetation, and have negligible to no atmosphere, and thus may be considered to be the best places to observe the true nature of mineral spectra [19-25]. Yet, there are many factors to contaminate reflectance spectra of such planetary surfaces such as regolith particles, space weathering, and horizontal and vertical mixing by impact cratering, thus identifying the variation of mineral distribution on extraterrestrial bodies is still difficult [10,22,26-30]. Therefore, studying the spectral change using simple pure minerals

that compose Earth and planetary surfaces is a critical foundational step to analyze reflectance spectra and is a prerequisite of reflectance spectroscopy, in light of continued application to planetary surfaces [8,31-35].

Deconvolution of reflectance spectra has been a common procedure for interpreting the experimental data of minerals and observation data of terrestrial and extraterrestrial surfaces. Among the most recognized spectral deconvolution methods in planetary science is the modified Gaussian model (MGM) [36,37]. In their paper, Sunshine et al. [36] showed that for 1 μm absorption band of orthopyroxene, Gaussian functions in the wavelength space fit better than Gaussians in the frequency space. The MGM express reflectance spectrum by the following function:

$$\log R = C(x) + \sum_{k=1}^{k} s_k . \exp\left(\left\{ \frac{-(x - \mu_k)^2}{2\sigma_k^2} \right\} \right)$$

where R is reflectance, x is wavelength, and s_k, μ_k and σ_k are the strength (amplitude), center (mean) and width (standard deviation)

*Corresponding author: Peng K Hong, The University Museum, The University of Tokyo, 7-3-1 Hongo, Bunkyo-ku, Tokyo 113-0033, Japan
E-mail: hong@um.u-tokyo.ac.jp

for k^{th} Gaussian function, respectively. K is the number of Gaussian functions. The function $C(x)$ is continuum of reflectance spectra:

$$C(x) = C_0 + \frac{C_1}{x}$$

where C_0 and C_1 are respectively intercept and slope of the continuum in frequency space.

The technical difficulty of applying the modified Gaussian model (MGM) is due to a local minima problem. The numerical algorithm of the widely used MGM utilizes the steepest descent method [36,38] or total inversion algorithm [37,39], both of which are not guaranteed to converge into a global solution. The analyzer can find an optimal solution only when the appropriate initial parameters are provided, based on preliminary knowledge of mineralogy [40,41]. With inaccurate initial parameters, however, these gradient descent methods converge into local minima, and thus the analyzer must manually adjust initial parameters and calculate the model iteratively to obtain the desired solution [16,42]. This would be a significant obstacle when one needs to automatically analyze large spectral databases obtained by space missions. In addition, preliminary knowledge of mineralogy may not always be available for space/ground-based observations, especially when the reflectance spectra are the only useful obtained data for interpreting the mineralogy of target bodies. Given the recent rapid increase of reflectance spectral data, automation of deconvolution analysis without requiring preliminary information on mineralogy is warranted. Makarewicz et al. [43] and Parente et al. [42] recently developed an algorithm to select initial band parameters automatically, based on inflection points of the derivatives of observed spectra. Although their algorithm does not depend on prior information, many spurious local minima and inflection points due to noise lead the authors to apply a smoothing filter, yielding arbitrariness on their analyses.

In order to overcome the local minima problem, the exchange Monte Carlo method, also known as parallel tempering [44], has been widely applied in the fields of physics, chemistry, biology, engineering and materials science [45]. Nagata et al. [46] developed a Bayesian spectral deconvolution model combined with the exchange Monte Carlo method with application to visible to near-infrared (Vis/NIR) reflectance spectra of fayalite and forsterite. This method is an improved algorithm of the Markov chain Monte Carlo method, aimed to avoid local minima traps [47] and to remove the arbitrariness originated from initial parameters. In order to solve the local minima problem, the simulated annealing scheme [48] has been incorporated into the model of Nagata et al. [46]. This algorithm introduces a pseudo-temperature and attempts to find the global minimum by heating and cooling the system. Nagata et al. [46] showed that the method can deconvolve reflectance spectral data of fayalite and forsterite into a few Gaussians with a continuum, purely based on the observed data, without requiring preliminary information of the band structure of olivine absorptions. In this paper, we report the applicability of the exchange Monte Carlo method to more complex rock-forming minerals (i.e., clinopyroxene). As described below, since the behavior of pyroxene spectra with the change of chemical composition is relatively well understood, the use of reflectance spectra of pyroxene minerals is suitable for testing the new spectral deconvolution method. Clinopyroxene (Cpx), with its general formula being $(M2)(M1)$ $(SiAl)_2O_6$, is one of the most important rock-forming mineral groups due to both its rich abundance on solid bodies in the solar system and distinguished absorption features [20,34,49]. Cpx includes a wide range solid solution of Mg, Fe and Ca compositions and has two crystal structures of C2/c and P2₁/c [50], which could reflect various physical and chemical processes

inside planetary bodies, such as the thermal history of magma [51]. Due to its wide range of chemical compositions, reflectance spectra of Cpx minerals vary significantly, and are generally grouped into three types: type-A, B and A/B [20,52,53]. Three major absorption bands are observed in type-B spectra, centered around 1.0, 1.2, and 2 μm, attributed to spin-allowed crystal field transitions of Fe cations in the octahedral (M1 and M2) sites [20]. These band centers are known to vary due to total iron and calcium content [19,52-55]. On the other hand, type-A spectra lack a strong 2 μm band, interpreted as a low Fe^{2+} content in the M2 site [53]. Type-A/B spectra are intermediate between type-A and B, although the boundaries are not well defined. MGM analyses have been performed to natural Cpx [36,53], synthetic Cpx [56], and mixtures of orthopyroxene-clinopyroxene [37,41].

Methods

Exchange Monte Carlo method

The technical details of the exchange Monte Carlo method is described elsewhere [46], thus we briefly summarize the key parameters of the model. In our study, the hyperparameters for Gamma and Gauss distributions used to yield probability densities of the parameters are those identified in Nagata et al. [46]: $\eta_a = 3.0$, $\lambda_a = 2.0$, $\nu_0 = 1.25$, $\xi_0 = 2.5$, $\eta_b = 5.0$ and $\lambda_b = 0.04$. The total number of temperatures L in our study was 80, and the inverse temperature β_l given by:

$$\beta_l = \begin{cases} 0 & (l=1) \\ 1.25^{l-L} & (l \neq 1) \end{cases}$$

Figure 1 shows typical examples of the evolution of root mean square (RMS) during the exchange Monte Carlo calculations for sample 088 using 4 Gaussian functions. Rapid decreases of RMS for the lowest temperature can be observed at about Monte Carlo steps = 300, 500, 700, 1500, and 6000, due to the parameter exchange between the lowest and middle temperatures. From Figure 1, it can be observed that the RMSs of higher temperatures are generally larger than those of lower temperatures, showing attempts to find better global minimum with wider fluctuations. On the other hand, for lower temperatures, each iteration attempts to find local minimum within a parameter range narrower than higher temperatures. After the Monte Carlo step exceeds 10^4, although the model still attempts to find better solution

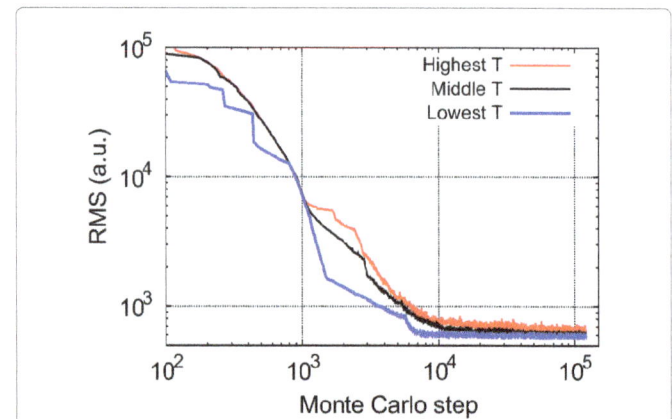

Figure 1: Evolution of root mean square (RMS) with every 10 Monte Carlo steps during the exchange Monte Carlo calculation for sample 088 with 4 Gaussian functions. Significant reduction of RMS for the lowest temperature can be observed approximately at Monte Carlo steps: 300, 500, 700, 1500, and 6000, which shows that the parameter exchange between the lowest and middle temperatures have occurred. The first 100,000 steps were used for the burn-in period and the last 20,000 steps for the expectation value calculations.

with certain fluctuations of parameters, the RMSs remain almost constant, and the expectation values of parameters converge to the best solution. Similar to Nagata et al. [46], the Monte Carlo calculation was iterated through 100,000 steps for the burn-in period and 20,000 steps for the expectation value calculations. Errors of band parameters are estimated from 2σ based on ten runs using a different series of random numbers.

The number of Gaussian functions, K, is an important parameter for deconvolution analysis. The model usually improves with more Gaussians. Though, too many Gaussians may cause overfitting, and the solution can be physically unrealistic. We performed spectral deconvolution using a various number of Gaussian functions. For one spectrum, we varied K from 3 to 10, thus the total number of free parameters ranges from 11 to 32, including the intercept and slope of the continuum. In order to select an optimal K for the deconvolution, we calculate the free energy, or stochastic complexity [57,58], which is an evaluation function for the model section problem [46]. With increasing K, we find that the free energy decreases, and when K is higher than about 5, it converges and fluctuates around a low value. Thus, we chose minimum Ks from the region where the free energies converge and interpret them as optimal Ks, listed in Table 1.

Spectral data

We used the visible to near-infrared spectra of synthetic clinopyroxene samples with a wide compositional range collected at the KECK/NASA Reflectance Experiment Laboratory (RELAB) at Brown University [59,60]. The sample IDs for the RELAB catalogue is summarized in Table 1. In order to compare our results with conventional MGM analyses, we collected 31 reflectance spectra that have been analyzed by a previous study [56]. The method of synthesis is detailed in Turnock et al. [61]. Individual synthetic pyroxene grains typically have 15-25 μm in size, however, these grains formed clumps [56]. Thus, samples are crushed and sieved at <45 μm. The spectra were measured at 5 nm intervals over the wavelength range of 0.3-2.6 μm. The incidence and emission angles were 30° and 0°, respectively [56,59,60]. We performed spectral deconvolution only over the wavelength range of 0.4-2.6 μm, since the standard deviations become larger near the shorter wavelengths [46]. The chemical compositions of the samples are measured with electron microprobe by Klima et al. [56]. The compositions of individual pyroxene samples are indicated in molar ratio with endmember compositions of enstatite (En: $Mg_2Si_2O_6$), ferrosilite (Fs: $Fe_2Si_2O_6$) and wollastonite (Wo: $Ca_2Si_2O_6$) and plot on a pyroxene quadrilateral (Table 1 and Figure 2). Minor compositions typical for natural CPx samples, such as Cr, Mn, Al and Fe^{3+}, are not observed [56]. Only the mineral structure of sample 088 is reported to be $P2_1/c$ [62], while the mineral structures for the remaining samples are not available. Based on the nomenclature of clinopyroxene [50], we assumed low-Ca pyroxene specimens with Wo < 20 to be pigeonite, high-Ca pyroxene with Wo > 45 and En > 25 diopside, and high-Ca

Sample ID	Mineral[a]	Composition[b]			1 μm band			1.2 μm band		
		En	Fs	Wo	Center (μm)	FWHM (μm)	Strength	Center (μm)	FWHM (μm)	Strength
9	Pigeonite	43	47	10	0.956 ± 0.0014	0.223 ± 0.004	-1.32 ± 0.03	1.229 ± 0.008	0.30 ± 0.03	-0.24 ± 0.03
11	Pigeonite	36	50	14	0.9579 ± 0.0003	0.2342 ± 0.0007	-2.040 ± 0.004	1.243 ± 0.0011	0.230 ± 0.003	-0.416 ± 0.0013
53	Pigeonite	23	70	8	0.9646 ± 0.0003	0.233 ± 0.0011	-2.042 ± 0.004	1.258 ± 0.0013	0.244 ± 0.003	-0.458 ± 0.005
88	Pigeonite	0	90	10	0.9801 ± 0.0004	0.2114 ± 0.0007	-1.334 ± 0.003	1.245 ± 0.002	0.351 ± 0.003	-0.3592 ± 0.0008
50	Augite	19	58	23	0.9808 ± 0.0003	0.235 ± 0.0013	-2.307 ± 0.007	1.278 ± 0.002	0.265 ± 0.003	-0.510 ± 0.005
51	Augite	39	34	27	0.991 ± 0.002	0.20 ± 0.010	-1.6 ± 0.11	1.26 ± 0.014	0.34 ± 0.07	-0.35 ± 0.08
54	Augite	6	70	23	0.988 ± 0.0013	0.250 ± 0.008	-2.50 ± 0.02	1.290 ± 0.007	0.30 ± 0.010	-0.70 ± 0.02
55	Augite	18	56	26	0.998 ± 0.002	0.201 ± 0.005	-1.9 ± 0.12	1.261 ± 0.003	0.331 ± 0.005	-0.475 ± 0.008
56	Augite	18	60	22	0.9715 ± 0.0004	0.241 ± 0.002	-1.32 ± 0.014	1.260 ± 0.002	0.257 ± 0.006	-0.32 ± 0.011
57	Augite	36	39	25	0.9915 ± 0.0004	0.1998 ± 0.0008	-1.414 ± 0.004	1.269 ± 0.004	0.28 ± 0.02	-0.182 ± 0.005
58	Augite	28	45	27	1.001 ± 0.004	0.198 ± 0.007	-1.3 ± 0.14	1.26 ± 0.011	0.28 ± 0.02	-0.29 ± 0.03
66	Augite	15	48	38	1.0128 ± 0.0003	0.1804 ± 0.0005	-1.43 ± 0.02	1.227 ± 0.004	0.386 ± 0.009	-0.443 ± 0.009
67	Augite	52	9	39	1.016 ± 0.003	0.17 ± 0.011	-1.0 ± 0.14	1.3 ± 0.13	0.3 ± 0.2	-0.2 ± 0.10
68	Augite	29	33	38	1.008 ± 0.006	0.20 ± 0.011	-1.8 ± 0.3	1.26 ± 0.010	0.33 ± 0.014	-0.38 ± 0.012
73	Augite	36	25	39	1.0099 ± 0.0003	0.163 ± 0.0010	-1.083 ± 0.007	1.12 ± 0.011	0.53 ± 0.014	-0.22 ± 0.010
74	Augite	24	37	39	1.0108 ± 0.0005	0.181 ± 0.002	-1.32 ± 0.03	1.204 ± 0.006	0.41 ± 0.010	-0.41 ± 0.013
85	Augite	0	61	39	1.024 ± 0.002	0.165 ± 0.007	-0.91 ± 0.05	1.23 ± 0.05	0.4 ± 0.13	-0.2 ± 0.13
87	Augite	0	71	29	1.000 ± 0.004	0.200 ± 0.007	-2.10 ± 0.06	1.245 ± 0.005	0.409 ± 0.006	-0.707 ± 0.009
33	Diopside	42	8	49	1.04 ± 0.03	0.5 ± 0.3	-0.3 ± 0.14	1.5 ± 0.11	0.8 ± 0.2	-0.1 ± 0.10
36	Diopside	27	24	49	1.03 ± 0.03	0.6 ± 0.5	-0.7 ± 0.3			
39	Diopside	29	22	49	1.04 ± 0.013	0.5 ± 0.4	-0.7 ± 0.3			
43	Diopside	45	6	49	1.08 ± 0.010	0.51 ± 0.04	-0.44 ± 0.07			
75	Diopside	46	9	45	1.03 ± 0.02	0.16 ± 0.02	-0.7 ± 0.13	1.2 ± 0.10	0.3 ± 0.2	-0.18 ± 0.07
77	Diopside	52	3	45	1.0193 ± 0.0002	0.161 ± 0.001	-0.701 ± 0.003	1.106 ± 0.0012	0.507 ± 0.004	-0.208 ± 0.002
79	Diopside	38	15	47	1.024 ± 0.0013	0.165 ± 0.003	-0.93 ± 0.03	1.05 ± 0.02	0.62 ± 0.03	-0.51 ± 0.04
37	Hedenbergite	16	35	49	1.04 ± 0.02	0.29 ± 0.09	-0.8 ± 0.4	1.3 ± 0.2	0.2 ± 0.3	-0.2 ± 0.3
70	Hedenbergite	14	41	45	1.028 ± 0.007	0.17 ± 0.010	-1.1 ± 0.10	1.23 ± 0.05	0.36 ± 0.07	-0.4 ± 0.2
71	Hedenbergite	23	31	46	1.025 ± 0.002	0.171 ± 0.002	-1.06 ± 0.02	1.24 ± 0.03	0.2 ± 0.10	-0.1 ± 0.11
76	Hedenbergite	18	35	46	1.03 ± 0.011	0.19 ± 0.04	-0.6 ± 0.2	1.2 ± 0.10	0.4 ± 0.11	-0.20 ± 0.07
82	Hedenbergite	1	50	49	1.06 ± 0.02	0.23 ± 0.09	-1.0 ± 0.3			
83	Hedenbergite	0	49	51	1.07 ± 0.02	0.25 ± 0.04	-0.4 ± 0.14	1.21 ± 0.02	0.26 ± 0.05	-0.7 ± 0.2

Sample ID	2 μm band			Continuum		
	Center (μm)	FWHM (μm)	Strength	C_0	C_1	K^c
9	2.046 ± 0.004	0.69 ± 0.02	-0.66 ± 0.02	-1.38 ± 0.02	-9.69E-3 ± 0.05	4
11	2.1228 ± 0.0003	0.6938 ± 0.0009	-1.171 ± 0.0014	-0.5536 ± 0.00010	-1.51E-4 ± 0.0007	4
53	2.1798 ± 0.0006	0.712 ± 0.002	-1.133 ± 0.002	-0.3729 ± 0.0002	-1.86E-3 ± 0.005	4
88	2.2022 ± 0.0002	0.6354 ± 0.0007	-0.565 ± 0.0010	-0.8344 ± 0.0002	-5.94E-2 ± 0.0010	4
50	2.2620 ± 0.0006	0.728 ± 0.002	-1.263 ± 0.003	-0.3828 ± 0.00014	-2.20E-4 ± 0.006	4
51	2.29 ± 0.04	0.73 ± 0.12	-0.90 ± 0.07	-0.393 ± 0.006	-4.85E-3 ± 0.12	5
54	2.2860 ± 0.0009	0.706 ± 0.004	-1.321 ± 0.006	-0.556 ± 0.003	-3.05E-3 ± 0.010	5
55	2.3060 ± 0.0005	0.731 ± 0.004	-1.072 ± 0.004	-0.264 ± 0.0014	-3.85E-3 ± 0.009	5
56	2.190 ± 0.0013	0.751 ± 0.009	-0.769 ± 0.008	-0.258 ± 0.004	-2.89E-3 ± 0.02	4
57	2.264 ± 0.0010	0.59 ± 0.011	-0.58 ± 0.010	-1.00 ± 0.02	-6.46E-2 ± 0.014	4
58	2.285 ± 0.0010	0.76 ± 0.02	-0.79 ± 0.02	-0.19 ± 0.02	-2.38E-2 ± 0.03	5
66	2.3139 ± 0.0009	0.665 ± 0.003	-0.804 ± 0.004	-0.24 ± 0.013	-4.77E-2 ± 0.02	5
67	2.328 ± 0.003	0.52 ± 0.02	-0.50 ± 0.02	-0.45 ± 0.05	-9.43E-2 ± 0.09	5
68	2.328 ± 0.0012	0.652 ± 0.008	-0.92 ± 0.011	-0.33 ± 0.03	-5.26E-2 ± 0.06	6
73	2.3078 ± 0.0006	0.494 ± 0.002	-0.427 ± 0.002	-0.952 ± 0.009	-4.66E-2 ± 0.02	4
74	2.312 ± 0.0013	0.614 ± 0.005	-0.682 ± 0.009	-0.50 ± 0.03	-6.63E-2 ± 0.05	5
85	2.290 ± 0.004	0.54 ± 0.02	-0.34 ± 0.02	-0.58 ± 0.05	-1.21E-1 ± 0.09	6
87	2.2882 ± 0.0005	0.638 ± 0.002	-1.058 ± 0.002	-0.4928 ± 0.0002	-7.09E-4 ± 0.002	5
33	2.33 ± 0.07	0.8 ± 0.3	-0.2 ± 0.10	-0.17 ± 0.07	-3.79E-2 ± 0.04	6
36	2.33 ± 0.06	0.2 ± 0.2	-0.07 ± 0.09	-0.86 ± 0.04	-4.20E-3 ± 0.11	6
39	2.30 ± 0.011	0.39 ± 0.03	-0.17 ± 0.010	-0.52 ± 0.07	-4.34E-2 ± 0.14	5
43	2.39 ± 0.013	0.26 ± 0.06	-0.111 ± 0.009	-0.29 ± 0.03	-2.62E-2 ± 0.04	5
75	2.33 ± 0.02	0.59 ± 0.06	-0.44 ± 0.03	-0.29 ± 0.05	-7.15E-2 ± 0.05	6
77	2.310 ± 0.0012	0.574 ± 0.004	-0.338 ± 0.002	-0.344 ± 0.002	-4.53E-4 ± 0.003	4
79	2.297 ± 0.002	0.47 ± 0.02	-0.42 ± 0.02	-1.01 ± 0.03	-8.65E-2 ± 0.09	5
37	2.3 ± 0.3	0.3 ± 0.7	-0.08 ± 0.09	-0.74 ± 0.05	-1.20E-1 ± 0.11	7
70	2.302 ± 0.002	0.58 ± 0.02	-0.63 ± 0.02	-0.32 ± 0.04	-3.74E-2 ± 0.09	6
71	2.302 ± 0.002	0.52 ± 0.013	-0.47 ± 0.010	-0.72 ± 0.03	-1.29E-1 ± 0.07	5
76	2.25 ± 0.02	0.36 ± 0.05	-0.11 ± 0.05	-1.55 ± 0.05	-2.05E-1 ± 0.2	5
82	2.297 ± 0.003	0.39 ± 0.05	-0.16 ± 0.02	-0.40 ± 0.04	-2.36E-2 ± 0.06	7
83	2.35 ± 0.02	0.4 ± 0.4	-0.05 ± 0.02	-0.48 ± 0.04	-9.79E-3 ± 0.06	10

[a]Based on Morimoto et al. [50].
[b]Data from Klima et al. [56]. En: Enstatite; Fs: Ferrosilite; Wo: Wollastonite.
[c]Number of Gaussian function.

Table 1: Compositions of synthetic clinopyroxene samples and results of spectral deconvolution calculated using the exchange Monte Carlo method. Errors are estimated from 2σ based on ten runs using different series of random numbers.

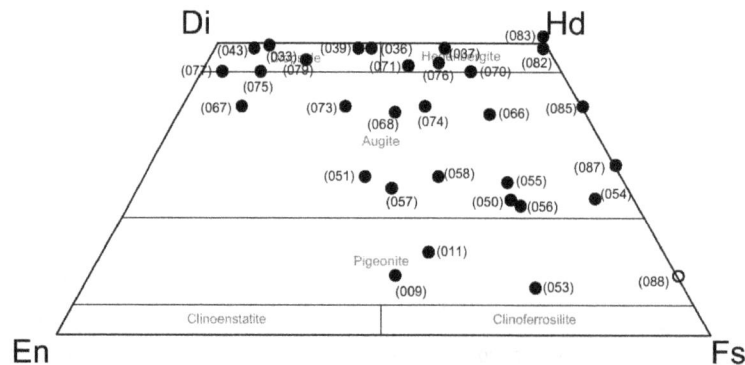

Figure 2: Clinopyroxene quadrilateral with the sample names. Compositions are measured with an electron probe microanalyzer (EPMA) by Klima et al. [56]. EPMA data for sample 088 is not available, thus it is plotted as an open circle based on the initial composition of the synthesis. The nomenclature of clinopyroxene is based on Morimoto et al. [50].

pyroxene with Wo > 45 and Fs > 25 hedenbergite. The remaining samples with intermediate Wo contents are assumed to be augite. Many of the samples locate in the "forbidden zone" where pyroxene exists merely as a metastable state under standard temperature and pressure [51]. These samples, however, were synthesized under a high-pressure up to 22.5 kbar [56], and thus are stable even under standard temperature and pressure for a geological timescale [63]. All of the reflectance spectra of synthetic pyroxene samples are shown in Figure 3. Spectra of pigeonite and augite are shown in Figures 3A-3C. The band minima of 1 μm absorptions for pigeonite and augite locate slightly shorter than 1 μm, similar to synthetic orthopyroxene [64]. Their band minima of 2 μm absorptions, caused by spin-allowed crystal field transition of Fe^{2+} in the M2 site [52,65], locate at longer than 2 μm similar to Fe-rich low-Ca pyroxene (orthopyroxene). Pigeonite and augite also show distinctive 1.2 μm absorption bands, which is attributed to spin-allowed crystal field transition of Fe^{2+} in the M1 site [66]. In Figures 3A-3C, the spectra of pigeonite and augite specimens, which are assigned to type-B spectra [52], show distinctive 1 and 2 μm absorption bands. On the other hand, some of the high-Ca pyroxene (i.e., diopside and hedenbergite) lack distinctive 2 μm absorption bands, as shown in Figures 3D and 3E. They are assigned to type-A pyroxene, in which the M2 sites are saturated cations other than Fe^{2+}, such as Ca^{2+} [52]. The spectrum of sample 083 has a broad 1 μm absorption band, probably due to a composite absorption by M1 bands near 1 and 1.2 μm [56]. This type of spectrum, exemplified by sample 083, may not be a suitable subject for MGM analyses, since the shape of 1 μm absorption band is far from Gaussian. However, since our objective in this paper is to test the applicability of the exchange Monte Carlo method and to compare the results with those of conventional MGM analyses, we included sample 083 in our analysis. Fitting additional small absorption bands centered near 0.50 and 0.55 μm in type-B spectra, which are associated with spin-forbidden crystal field transitions [66], is beyond the scope of this paper, since our focus was to evaluate the applicability of the model to two major absorption bands of clinopyroxene centered about 1 and 2 μm.

Results

Deconvolution with the exchange Monte Carlo method

Deconvolution results of pyroxene spectra are shown in Figure 4. The best optimized parameters for 1, 1.2, and 2 μm bands are summarized in Table 1. Errors estimated from 2σ based on ten runs using different series of random numbers are also shown. Most of the spectra are best fitted by 4 to 6 Gaussians with appropriate continuum.

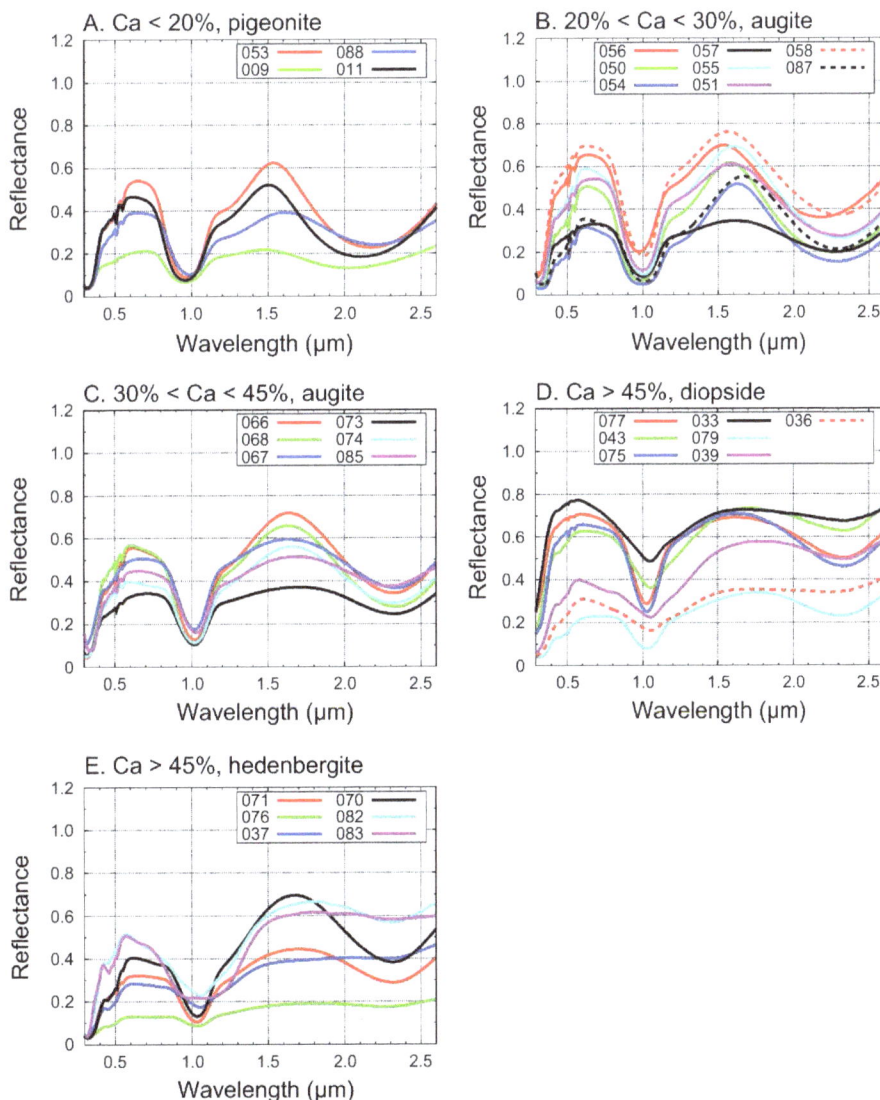

Figure 3: Visible to near infrared reflectance spectra of synthetic clinopyroxene. (3A) Pigeonite, (3B) Low-Ca augite, (3C) High-Ca augite, (3D) Diopside, (3E) Hedenbergite.

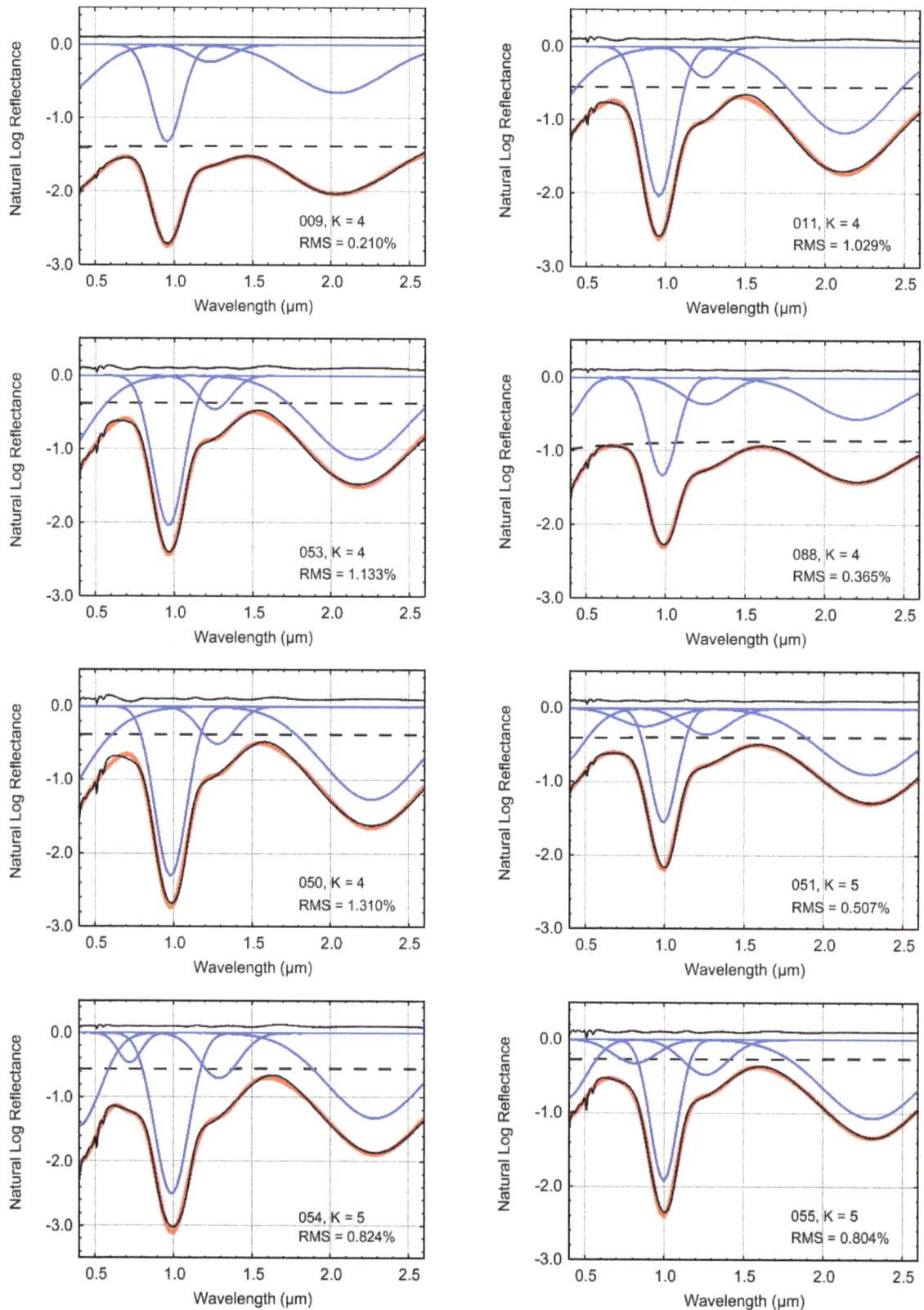

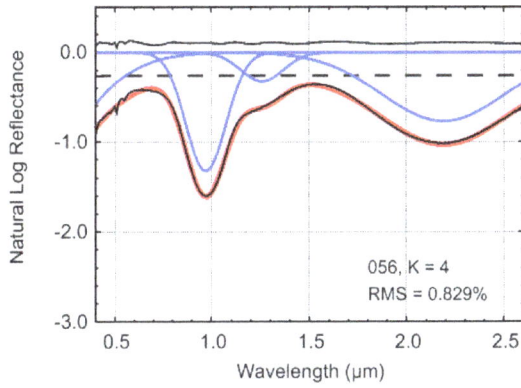

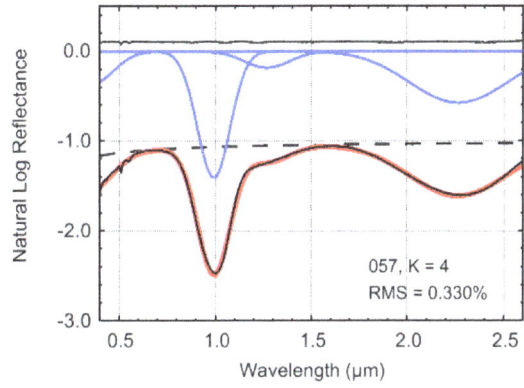

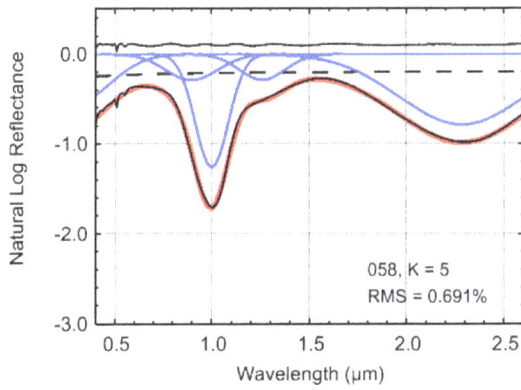

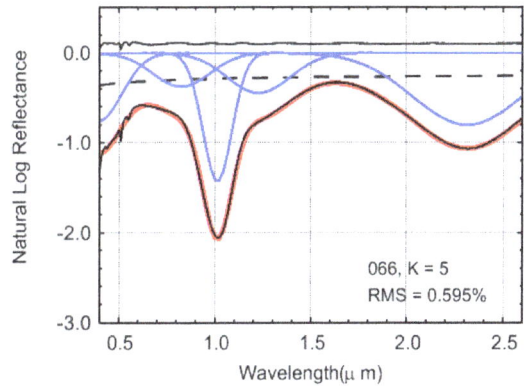

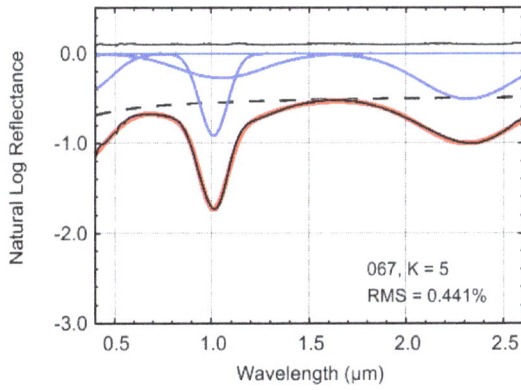

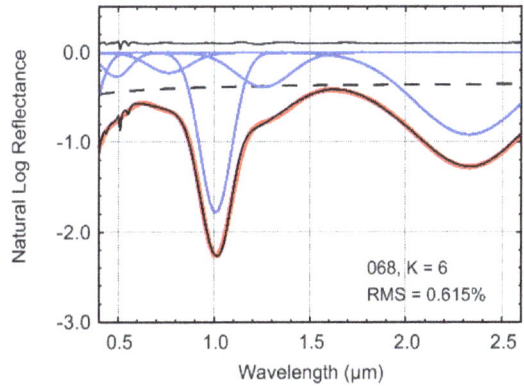

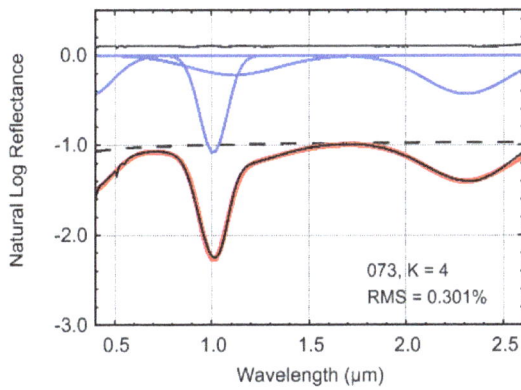

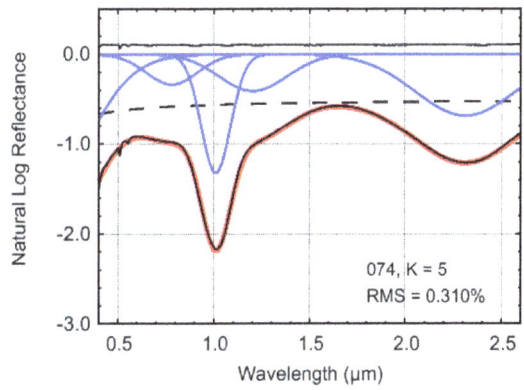

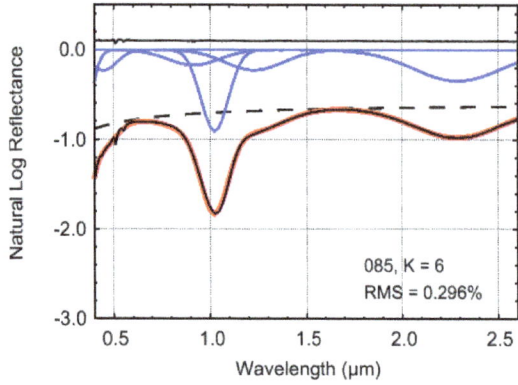

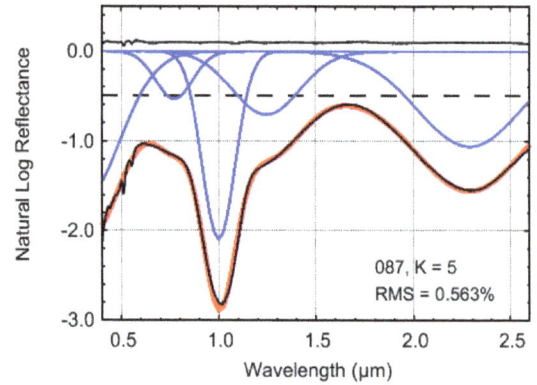

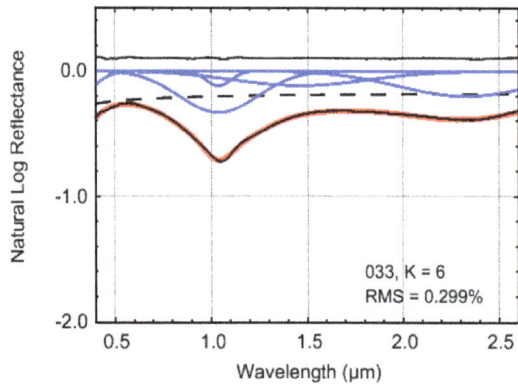

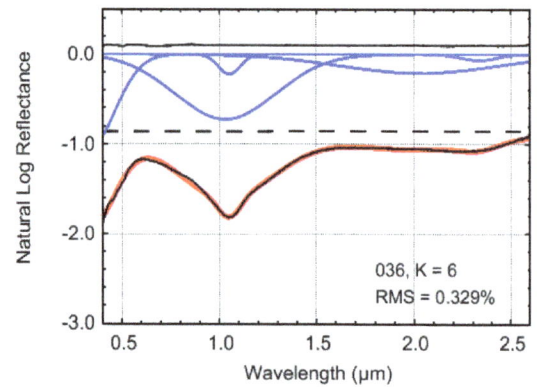

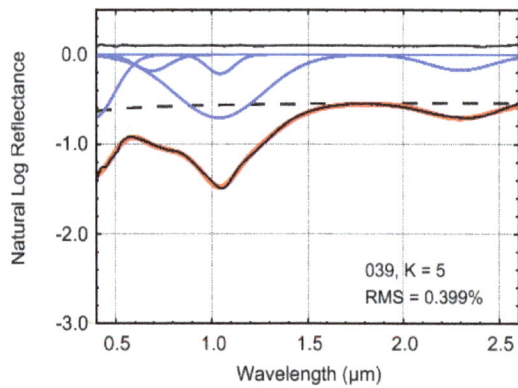

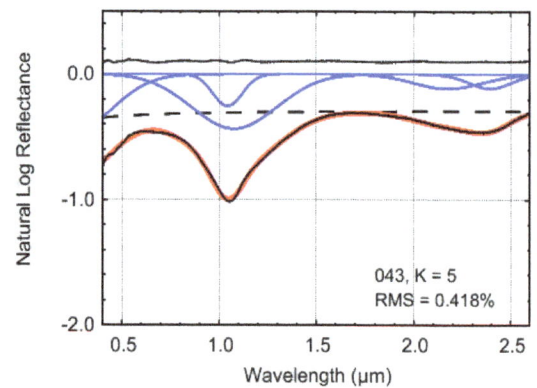

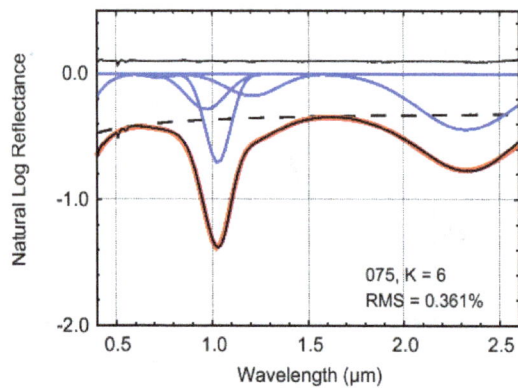

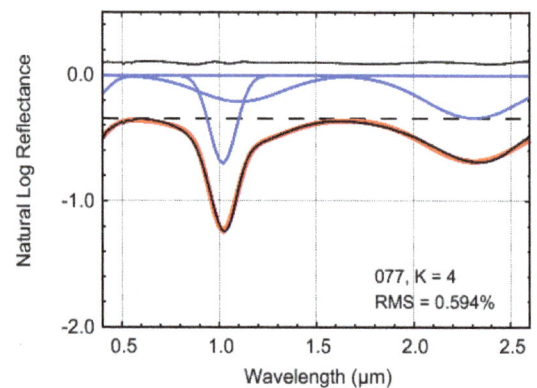

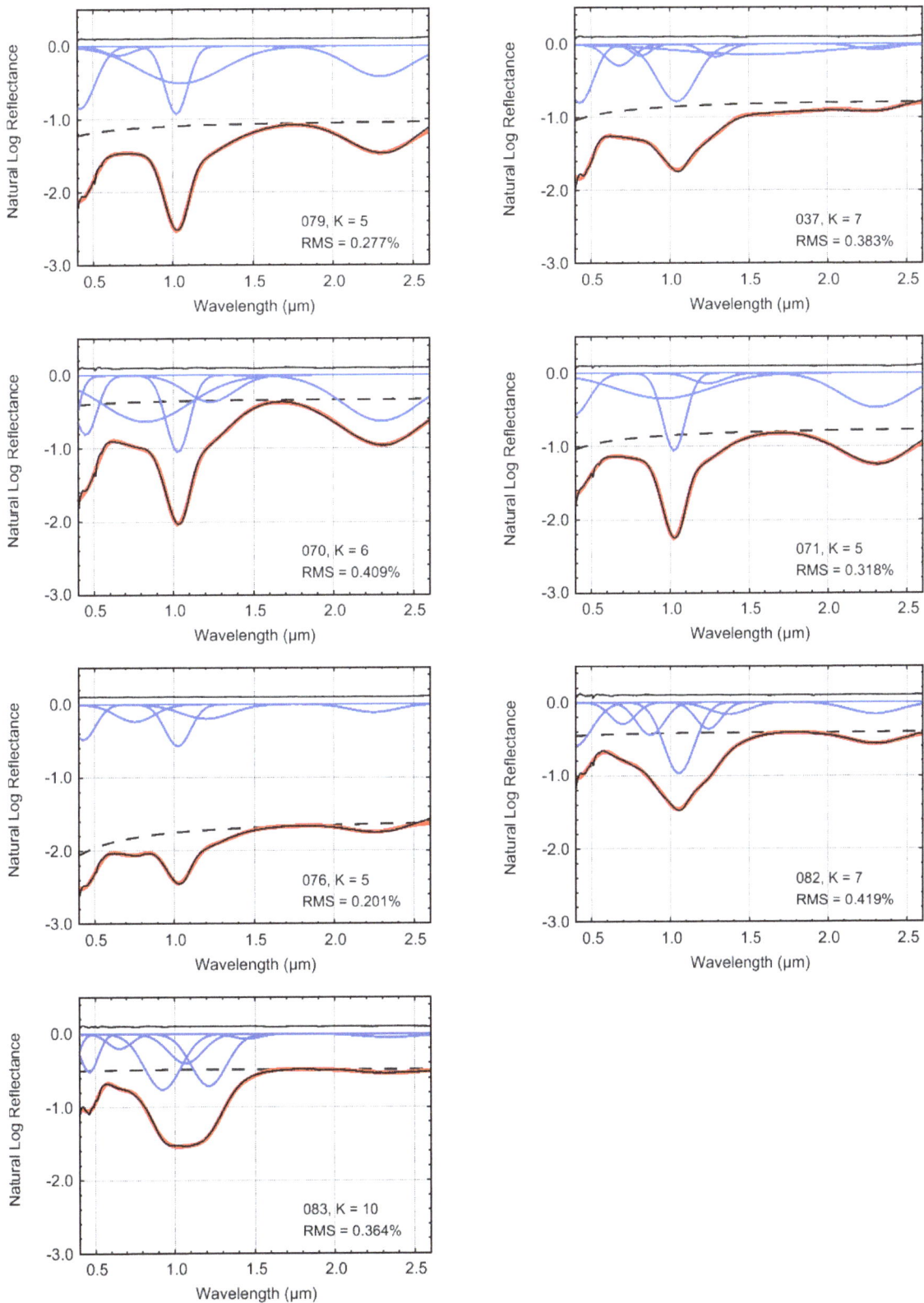

Figure 4: Spectral deconvolution results of synthetic Cpx samples using the exchange Montel Carlo method. Each Gaussian (blue solid lines) and continuum functions (black broken lines) compared to the synthetic spectra using the exchange Monte Carlo calculations (red lines) which approximate the original spectra (solid black lines). The residual errors between the modeled and the actual spectra are shown as solid black lines at top in each figure (offset +0.1 for clarity).

For example, sample 053 is fitted with 4 Gaussians with their centers locating at 0.091, 0.965, 1.258, and 2.180 μm and a nearly constant continuum. For pigeonite and augite, our results show that the spectra are mainly fitted with 3 Gaussians with their centers locating at about 1, 1.2, and 2 μm, and with one Gaussian in the wavelength region ranging from ultraviolet to visible and with a continuum being nearly constant or gradually decreasing toward shorter wavelengths (Figures 5a-5d). Each Gaussian function could be physically interpreted to represent spin-allowed crystal field transitions with 1 μm for Fe^{2+} in the M1 and M2 sites, 1.2 μm for Fe^{2+} in the M1 site and 2 μm for Fe^{2+} in the M2 site (Figures 6a-6d). A Gaussian in the wavelength region ranging from ultraviolet to visible could represent oxygen-metal charge transfers which are centered within the ultraviolet region [34] (Figure 7). Although additional small bands centered near 0.7 μm may be necessary to improve the fitting such as shown in sample 054, we find that the additional band does not significantly affect the band center for 1, 1.2, and 2 μm absorptions. Our deconvolution results for pigeonite and augite are consistent with those by Klima et al. [56].

Deconvolution analyses for type-A spectra including some of the diopside and hedenbergite are not as straightforward as type B, since the 1 μm bands are too narrow or too wide for a single Gaussian to fit, and some of the 1 and 2 μm bands are asymmetrical. The non-Gaussian shape of type-A spectra results in larger errors of band parameters when compared to type-B spectra (Table 1). Also, more Gaussian functions are needed to fit type-A spectra, reflecting the complex shape of the

spectra. For example, samples 037 and 082 were fitted with 7 Gaussians, while sample 083 with 10 Gaussians of which 4 Gaussians are centered below 0.5 μm. Comparing our fitting results of sample 082 and 083 with those reported by Klima et al. [56], we find that the exchange Monte Carlo method yields more symmetrical configurations for Gaussians in 1 μm band. We also find that our optimal deconvolution results for samples 082 and 083 do not require the very weak absorption bands which were assigned as M2 absorptions in Klima et al. [56]. These M2 absorptions in Klima et al. [56] are significantly weak compared with the strongest M1 absorption in 1 μm band, and the M2 absorption near 1 μm was mostly covered with the M2 absorptions. For type-A spectra, it is difficult to resolve the discrepancy between the results obtained by Klima et al. [56] and this study, however, since both of the modeling results can reproduce the observed spectra almost equally well. Nevertheless, the large errors of band parameters for type-A spectra indicate that caution should be taken when performing spectral deconvolution for type-A spectra using MGM. Thus, being able to estimate errors in the fitting results based on random initial parameters is also an advantage of the exchange Monte Carlo method for assessing the statistical robustness of fitting results, which has not been performed by previous MGM analyses.

Band shift as functions of Ca, Fe, and Mg Contents

The center positions of Gaussian functions corresponding to 1, 1.2, and 2 μm absorptions are summarized in Table 1. Position shifts of 1

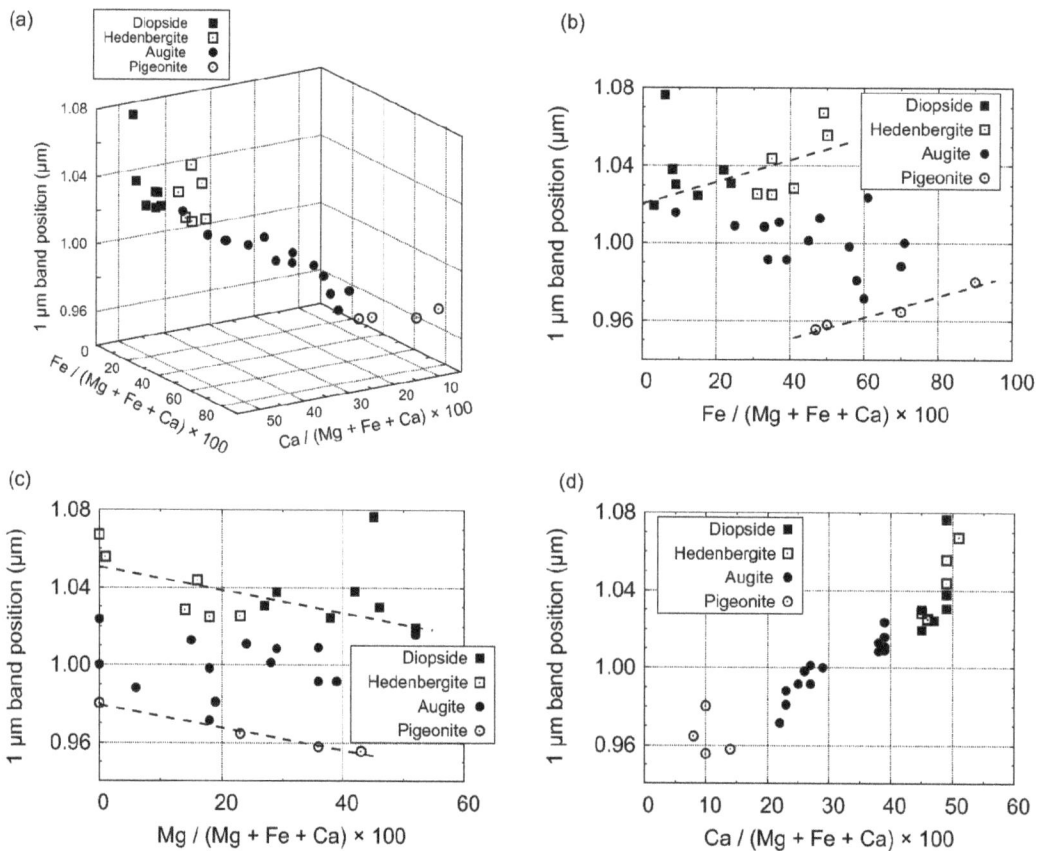

Figure 5: Center wavelengths of 1 μm band of Cpx samples as a function of the Mg, Fe and Ca contents. (a) band center as a function of the Fe and Ca contents. (b) band center as a function of the Fe content. Broken lines show linear approximations for pigeonite ($y = 0.00057x + 1.02$ where y is the band center in μm, x the Fe content) and diopside-hedenbergite ($y = 0.00054x + 0.93$). (c) band center as a function of the Mg content. Approximate lines for pigeonite ($y = -0.00059x + 1.05$) and diopside-hedenbergite ($y = -0.00058x + 0.98$) are shown. Note that sample 043 was omitted as an outlier from the linear approximation of diopside-hedenbergite in (b) and (c). (d) band center as a function of the Ca content.

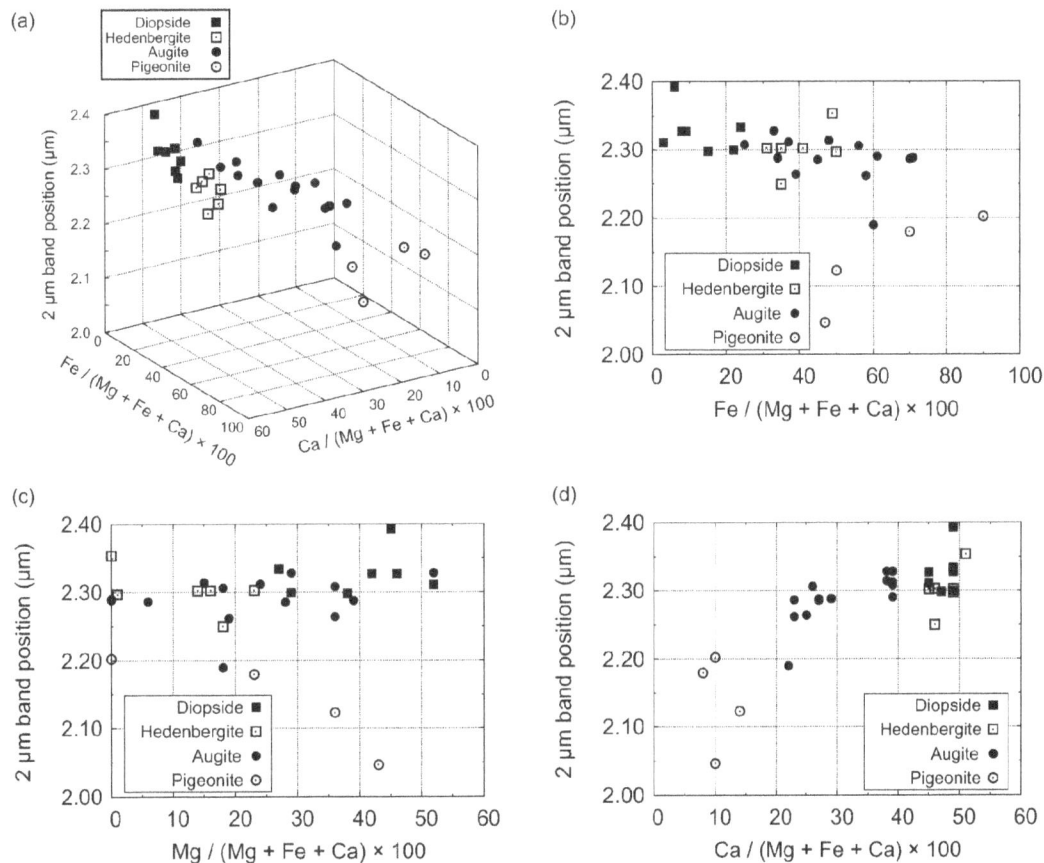

Figure 6: Center wavelengths of 2 μm band of Cpx samples as a function of the Mg, Fe and Ca contents. (a) band center as a function of the Fe and Ca contents. (b) band center as a function of the Fe content. (c) band center as a function of the Mg content. (d) band center as a function of the Ca content.

μm band as functions of Ca, Fe, and Mg components (solid solution) are shown in Figure 5, while position shifts of 2 μm band are shown in Figure 6. Our results are found to be consistent with those of Klima et al. [56], suggesting that the exchange Monte Carlo method developed in this study is able to extract results similar to those obtained by conventional MGM analysis.

The band centers of 1 μm absorption have a linear dependence on Wo (Ca/(Mg+Fe+Ca) molar%) contents, moving to longer wavelengths with increasing Wo. For diopside and hedenbergite with Wo ≈ 50, the band centers of 1 μm absorption largely scatter within the range of 1.03-1.08 μm. Pigeonite also shows a large variance as a function of Wo component, however, the band centers of 1 μm absorptions have a linear dependence on En (Mg/(Mg+Fe+Ca) molar%) or Fs (Fe/(Mg+Fe+Ca) molar%) components. The band centers of 1 μm absorption of augite scatter widely as a function of Fs content, and distinct dependence on Fs content were not observed. For diopside and hedenbergite samples, with the exception of sample 043, the band centers of 1 μm absorptions seem to have a linear dependence on Fs content, moving to longer wavelengths with increasing Fs content. En content seems to have no obvious influence on the band center of 1 μm absorption, except for pigeonite. We find that pigeonite, augite and diopside-hedenbergite can be separated with the use of 1 μm band position as a function of Fs or En content, as shown in Figure 5.

The band centers of 2 μm absorption seem to follow an approximation line on the space of Fs and Wo contents (Figure 6a). The band centers of 2 μm absorption for pigeonite and augite move to

longer wavelengths with an increase of Wo content. The average center wavelengths of 2 μm absorptions for diopside and hedenbergite remain almost constant at around 2.3 μm, but they scatter significantly around the average value, reflecting the asymmetry shape of 2 μm absorptions for some of the type-A spectra. The band centers of 2 μm absorptions of pigeonite move longer wavelengths with increasing Fs content. Overall, the band center of 2 μm absorption is separated between low-Ca clinopyroxene (pigeonite) and high-Ca pyroxene (augite, diopside, and hedenbergite) with a gap from 2.20-2.25 μm. Klima et al. [56] interpreted the gap to be a transition zone of mineral structure between $P2_1/c$ and $C2/c$. Our results generally agree with their interpretation with one exception, i.e., sample 056 is assumed to be pigeonite in Klima et al. [56].

The diagram between 1 μm and 2 μm band positions is shown in Figure 7. Generally low-Ca pyroxene locates in a shorter wavelength region, while high-Ca pyroxene locates in a longer wavelength region [19,41,52,56]. Our results are consistent with previous studies based on MGM analysis [41,56].

Discussion

In order to avoid the local minimum problem, a Bayesian spectral deconvolution method with the exchange Monte Carlo algorithm has been applied to visible to near infrared reflectance spectra of synthetic Cpx with wide ranging Mg, Fe, and Ca contents. The results obtained in this study generally agree well with conventional MGM analyses. Here, we discuss some potential interpretations for the deconvolution

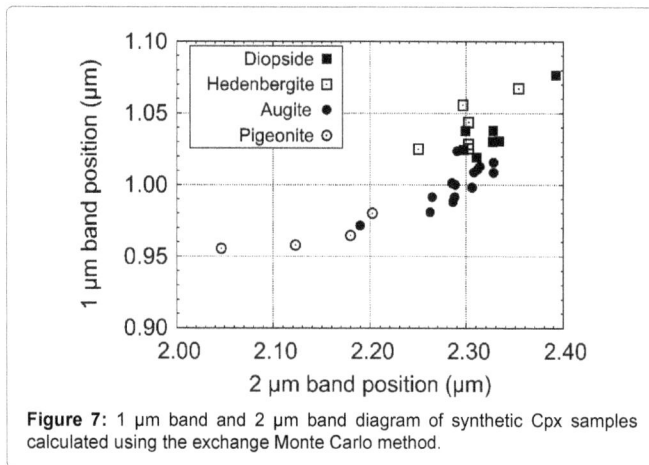

Figure 7: 1 μm band and 2 μm band diagram of synthetic Cpx samples calculated using the exchange Monte Carlo method.

results, as well as the discrepancies with previous results obtained by MGM analyses.

Fitting type-A spectra

Type-A spectra generally appear in high-Ca pyroxene such as diopside and hedenbergite, although previous studies suggest that there is no simple relationship between chemical composition of natural pyroxenes and type-A spectra [20,52,53]. The spectral data of samples 076, 037, 083, and 036 are categorized in type-A, which lack distinctive 2 μm bands. Since the absorption near 2 μm is mainly caused by spin-allowed crystal field transition of Fe^{2+} in the M2 site, the absence of 2 μm band from type-A spectra is interpreted as a result of the replacement of Fe^{2+} in M2 site by other cations such as Ca^{2+} and Fe^{3+} [52,67]. On the other hand, spin-allowed crystal field transition due to the M1 site, with the absorption center locating near 1 and 1.2 μm, appear strong, and the two bands are combined to yield single broad band around 1 μm. It is difficult to fit type-A spectra with manually provided initial parameters (e.g., center, strength of Gaussian) using a conventional MGM algorithm, since the final solutions directly depend on the initial parameters, especially for the broad 1 μm band due to its non-Gaussian shape [56]. Although it is still difficult for the exchange Monte Carlo method to delineate the optimal number of Gaussians, we find that the center wavelengths of Gaussian functions for the broad 1 μm band do not move significantly with varying K. For example, the broad 1 μm band of sample 083 was fitted with two major Gaussian functions which could correspond to crystal field absorptions due to the M1 site whose center wavelengths of absorptions locate near 1 and 1.2 μm. Although the estimated errors are large, the overall result is consistent with previous deconvolution analyses of type-A spectra [53,56], suggesting that even with manually provided initial parameters, previous analyses obtained statistically optimal results. Since manually fitting type-A spectra is more difficult than type-B spectra, the exchange Monte Carlo method can be a useful tool for future deconvolution analysis for type-A spectra, of which are likely to appear in high-Ca pyroxene [52,53]. In addition, natural Cpx incorporate many minor compositions such as Al, Ti, Mn and Cr [49,50]. Such minor elements in natural Cpx yield more complex spectra than synthetic Cpx, thus the scheme presented in this paper can be useful to deconvolve such spectra.

1 μm band position vs. Ca-Fe-Mg content

Figure 5a indicates that the Cpx seem to follow a linear relationship among the 1 μm band position, with Fs and Wo contents. As shown as broken lines in Figures 5b and 5d, linear relationships are observed for pigeonite and diopside-hedenbergite with their Fs or En content. It

should be noted that for diopside-hedenbergite, linear approximation was performed for all of the samples except sample 043. Sample 043 was omitted as an outlier because the spectrum shows higher reflectance and weaker absorptions compared with other samples, generally suggestive of the effect of glass [16,56,68]. Although it has been well documented for low-Ca pyroxene that the linear dependence of 1 μm band centers on Fs or En content [19,41,52,54,56], the relationship between the 1 μm band center and Fs or En content for high-Ca pyroxene has been poorly constrained by MGM analyses. Clénet et al. [41] analyzed only two high-Ca natural pyroxene samples in the diopside-hedenbergite region, thus no clear relationship has been derived. Although Klima et al. [56] analyzed 13 high-Ca pyroxene samples in the diopside-hedenbergite region; they observed no clear relationship between the 1 μm band center and Fs or En content. Our analyses show that the 1 μm band center of high-Ca pyroxene (i.e., diopside and hedenbergite) depends on Fs or En content with the band center moving longer wavelengths with increasing Fs content. This tendency is similar to the relationship between the 1 μm band center and the Fe/Mg ratio of Ca-free orthopyroxene [64]. For Ca-saturated synthetic Cpx in which most of the M2 site is dominated by Ca^{2+}, varying Fe/Mg ratio would affect only Fe/Mg in the M1 site because the M2 site is dominated with larger Ca^{2+} cations. Thus, the change of Fe/Mg would appear only in the 1 μm band center caused by a crystal field transition in the M1 site, but not in the 2 μm band center caused by the absorption due to the M2 site. We note, however, that large errors are included in the modeling results for type-A spectra (Table 1), thus additional data is necessary to confirm our interpretation.

Band center in three-dimensional spaces

As mentioned above, both the 1 μm and 2 μm bands seem to plot on approximate lines in the three-dimensional (3D) space of Fs-Wo content (Figures 5a and 6a). Low-Ca pyroxene samples (pigeonite) used in this study locates in the high-Fe and low-Ca regions with both the band centers being at shorter wavelengths. On the other hand, high-Ca pyroxene samples (diopside and hedenbergite) used in this study locates in the low-Fe and high-Ca regions with both the band centers being at longer wavelengths. Augite with intermediate-Wo contents distribute a in a linear trend between low-Ca and high-Ca pyroxene. Each mineral group displays three distinct clusters on these 3D spaces. The linear trend seen in both 1 μm and 2 μm bands could be understood from a crystallographic point of view. Whereas high-Fe and low-Ca Cpx samples contain more Fe^{2+} cations in both the M1 and M2 sites, leading to a decrease of the bond lengths due to the smaller size of Fe^{2+}, which results in an increase in the crystal field splitting [34,69,70], low-Fe and high-Ca Cpx samples have larger Ca^{2+} cations, which dominate the M1 and M2 sites, leading to an increase in the bond lengths and a decrease in the crystal field splitting.

Application to future remote reflectance spectroscopy

Despite requiring meticulous parameter adjustment and prior information of mineralogy, MGMs using gradient descent method have been applied to reflectance spectra not only of laboratory data but also remote sensing data of the Moon [71], Mars [4,6,72-75], and some asteroids [76,77]. By applying the exchange Monte Carlo method to spectra of synthetic Cpx samples, we have shown that it is able to yield results consistent with both conventional gradient descent methods and crystal field theory. This means that the exchange Monte Carlo method could be applicable to at least some of the previous remote sensing data which have been analyzed using conventional MGM methods. Because the application of MGM analyses has been limited due to meticulous parameter adjustment, the exchange Monte Carlo method may have a potential to expand the applicability of MGM to a variety of space/ground-based observations, especially when we cannot obtain prior information of the target body beforehand. Remote sensing data of terrestrial reflectance spectroscopy have been validated with various reference spectra observed by in situ sample analyses, increasing the confidence of interpretation of remote sensing data [12,17,78,79]. However, such an analysis is not always available due to various geological contexts. The situation is more serious for space/ground-based observations of small bodies in the solar system, because reflectance spectra are the only available compositional information for most of the small bodies such as Phobos, Deimos [80,81] and asteroid 1999 JU3 which is the target of Japanese space mission Hayabusa 2 [82]. In fact, asteroids are clustered based solely on reflectance spectra, although their detailed mineral compositions are poorly constrained [83]. Considering the large number of small bodies in the solar system, it is unlikely to send probes to each small body to conduct in situ sample analysis or sample return. Given the limited data and resource, interpreting reflectance spectra without assuming a priori information is essential not only for planetary science but also for mission planning. The exchange Monte Carlo method could be significantly useful under such circumstances, where meaningful compositional information other than reflectance spectra is not available.

Conclusions

We applied a Bayesian spectral deconvolution method with the exchange Monte Carlo algorithm to visible/near infrared reflectance spectra of synthetic clinopyroxene of diverse compositional variation, in order to avoid the local minimum problem and to remove the arbitrariness originated from initial parameters, inherent in the previous Modified Gaussian model. Our results indicate that the exchange Montel Carlo method is able to yield the consistent results obtained by conventional Modified Gaussian model. Since our model does not rely on a preliminary knowledge of the reflectance spectrum of mineral, the results suggest that previous spectrum analyses have obtained statistically optimal results. The successful application of our model to reflectance spectra of minerals indicates that this model could be applied to an automatic deconvolution analysis for a large spectral database, especially useful for space missions when preliminary knowledge of mineralogy is not available. Given the recent and future advancement of space missions, the exchange Monte Carlo method could be a useful tool for analyzing a wide range of minerals and remote sensing data of rocky bodies in the solar system.

Acknowledgment

This research utilizes spectra acquired with the NASA RELAB facility at Brown University. This work is supported in part by JSPS grant-in-aid 25120006 and TeNQ/Tokyo-dome.

References

1. Goetz AFH, Vane G, Solomon JE, Rock BN (1985) Imaging spectrometry for earth remote sensing. Science 228: 1147-1153.

2. McCord TB (1988) Reflectance Spectroscopy in Planetary Science: Review and Strategy for the Future. NASA SP-493.

3. Soderblom LA (1992) The composition and mineralogy of the martian surface from spectroscopic observations: 0.3 μm to 50 μm. In: Kieffer HH, Jakosky BM, Snyder CW, Matthews MS (eds.) Mars. The University of Arizona Press.

4. Mustard JF, Sunshine JM (1995) Seeing through the dust: Martian crustal heterogeneity and links to the SNC meteorites. Science 267: 1623-1626.

5. Bibring JP, Langevin Y, Gendrin A, Gondet B, Poulet F, et al. (2005) Mars surface diversity as revealed by the OMEGA/Mars Express Observations. Science 307: 1576-1581.

6. Mustard JF, Poulet F, Gendrin A, Bibring JP, Langevin Y, et al. (2005) Olivine and pyroxene diversity in the crust of Mars. Science 307: 1594-1597.

7. Poulet F, Gomez C, Bibring JP, Langevin Y, Gondet B, et al. (2007) Martian Surface mineralogy from the Observatory for Mineralogy, Water, Ice and the activity on board the Mars Express spacecraft (OMEGA/MEx): Global mineral maps. J Geophys Res 112: E08S02.

8. Hapke B (2012) Theory of Reflectance and Emittance Spectroscopy (2nd edn) Cambridge University Press.

9. Straub DW, Burns RG, Pratt SF (1991) Spectral signature of oxidized pyroxenes: Implications to remote sensing of terrestrial planets. J Geophys Res 96: 18819-18830.

10. Mustard JF, Hays JE (1997) Effects of hyperfine particles on reflectance spectra from 0.3 to 25 μm. Icarus 125: 145-163.

11. Pelkey SM, Mustard JF, Murchie S, Clancy RT, Wolff M (2007) CRISM multispectral summary products: parameterizing mineral diversity on Mars from reflectance. J Geophys Res 112: E08S14.

12. Piatek JL, Hardgrove C, Moersch JE, Drake DM, Wyatt MB (2007) Surface and subsurface composition of the Life in the Atacama field sites from rover data and orbital image analysis. J Geophys Res 112: G04S04.

13. Cloutis EA, Gaffey MJ, Jackowski TL, Reed KL (1986) Calibrations of phase abundance, composition, and particle size distribution for olivine-orthopyroxene mixtures from reflectance spectra. J Geophys Res 91: 11641-11653.

14. Poulet F, Erard S (2004) Nonlinear spectral mixing: Quantitative analysis of laboratory mineral mixtures. J Geophys Res 109: E02009.

15. Cheek LC, Pieters CM (2014) Reflectance spectroscopy of plagioclase-dominated mineral mixtures: Implications for characterizing lunar anorthosites remotely. Am Mineral 99: 1871-1892.

16. Horgan BHN, Cloutis EA, Mann P, Bell JF (2014) Near-infrared spectra of ferrous mineral mixtures and methods for their identification in planetary surface spectra. Icarus 234: 132-154.

17. Mustard JF, Sunshine JM (1999) Spectral analysis for Earth science: Investigations using remote sensing data. In: Rencz, AN (Ed.), Remote Sensing for the Earth Sciences: Manual of Remote Sensing (3rd edn.) John Wiley & Sons, Inc.

18. Murchie S, Arvidson R, Bedini P, Beisser K, Bibring JP (2007) Compact Reconnaissance Imaging Spectrometer for Mars (CRISM) on Mars Reconnaissance Orbiter (MRO). J Geophys Res 112: E05S03.

19. Adams JB (1974) Visible and Near-Infrared Diffuse Reflectance Spectra of Pyroxenes as Applied to Remote Sensing of Solid Objects in the solar system. J Geophys Res 79: 4829-4836.

20. Adams JB (1975) Interpretation of visible and near-infrared diffuse reflectance spectra of pyroxenes and other rock-forming minerals. In: Karr C (ed.) Infrared and Raman Spectroscopy of Lunar and Terrestrial Minerals. Academic Press, New York.

21. Gaffey MJ, Cloutis EA, Kelly MS, Reed KL (2002) Mineralogy of Asteroids. In: Bottke Jr WF, Cellino A, Paolicchi P, Binzel RP (eds.), Asteroids III, The University of Arizona Press.

22. Burbine TH, Rivkin AS, Noble SK, Mothe-Diniz T, Bottke WF (2008) Oxygen and Asteroids. Rev Mineral Geochem 68: 273-343.

23. Gietzen KM, Lacy CHS, Ostrowski DR, Sears DW (2012) IRTF observations of S complex and other asteroids: Implications for surface compositions, the presence of clinopyroxenes, and their relationship to meteorites. Meteorit Planet Sci 47: 1789-1808.

24. Reddy V, Nathues A, Le Corre L, Sierks H, Li JY (2012) Color and albedo heterogeneity of Vesta from Dawn. Science 336: 700-704.

25. Zambon F, De Sanctis MC, Schroder S, Tosi F, Longobardo A (2014) Spectral analysis of the bright materials on the asteroid Vesta. Icarus 240: 73-85.

26. Vilas F, Jarvis KS, Gaffey MJ (1994) Iron alteration minerals in the visible and near-infrared spectra of low-albedo asteroids. Icarus 109: 274-283.

27. Pieters CM, McFadden LA (1994) Meteorite and Asteroid Reflectance Spectroscopy: Clues to Early Solar System Processes. Annu Rev Earth Planet Sci 22: 457-497.

28. Hiroi T, Zolensky ME, Pieters CM, Lipschutz ME (1996) Thermal metamorphism of the C, G, B, and F asteroids seen from the 0.7 μm, 3 μm, and UV absorption

strengths in comparison with carbonaceous chondrites. Meteorit Planet Sci 31: 321-327.

29. Hiroi T, Abe M, Kitazato K, Abe S, Clark BE, et al. (2006) Developing space weathering on the asteroid 25143 Itokawa. Nature 443: 56-58.

30. Bell JF, McSween HY, Crisp JA, Morris RV, Murchie SL, et al. (2000) Mineralogic and compositional properties of Martian soil and dust: Results from Mars Pathfinder. J Geophys Res 105: 1721-1755.

31. Hunt GR, Salisbury JW (1970) Visible and near-infrared spectra of minerals and rocks: I silicate minerals. Modern Geology 1: 283-300.

32. Hunt GR (1977) Spectral signatures of particulate minerals in the visible and near infrared. Geophysics 42: 501-513.

33. Clark RN, King TVV, Klejwa M, Swayze GA (1990) High spectral resolution reflectance spectroscopy of minerals. J Geophys Res 95: 12653-12680.

34. Burns RG (1993) Mineralogical applications of crystal field theory. Cambridge University Press, Cambridge.

35. Mayne RG, Sunshine JM, McSween HY, McCoy TJ, Corrigan CM, et al. (2010) Petrologic insights from the spectra of the unbrecciated eucrites: Implications for Vesta and basaltic asteroids. Meteorit Planet Sci 45: 1074-1092.

36. Sunshine JM, Pieters CM, Pratt SF (1990) Deconvolution of Mineral Absorption Bands: An Improved Approach. J Geophys Res 95: 6955-6966.

37. Sunshine JM, Pieters CM (1993) Estimating modal abundances from the spectra of natural and laboratory pyroxene mixtures using the Modified Gaussian Model. J Geophys Res 98: 9075-9087.

38. Sunshine JM, Pieters CM, Pratt SF, McNaron-Brown KS (1999) Absorption Band Modeling in Reflectance Spectra: Availability of the Modified Gaussian Model. The 30th Lunar and Planetary Science Conference Abstract, #1306.

39. Tarantola A, Valette B (1982) Generalized nonlinear inverse problems solved using the least squares criterion. Rev Geophys Space Phys 20: 219-232.

40. Kanner LC, Mustard JF, Gendrin A (2007) Assessing the limits of the Modified Gaussian Model for remote spectroscopic studies of pyroxenes on Mars. Icarus 187: 442-456.

41. Clénet H, Pinet P, Daydou Y, Heuripeau F, Rosemberg C, et al. (2011) A new systematic approach using the Modified Gaussian Model: Insight for the characterization of chemical composition of olivines, pyroxenes and olivine-pyroxene mixtures. Icarus 213: 404-422.

42. Parente M, Makarewicz HD, Bishop JL (2011) Decomposition of mineral absorption bands using nonlinear least squares curve fitting: Application to Martian meteorites and CRISM data. Planet Space Sci 59: 423-442.

43. Makarewicz HD, Parente M, Bishop JL (2009) Deconvolution of VNIR spectra using modified Gaussian modeling (MGM) with automatic parameter initialization (API) applied to CRISM. Hyperspectral Image and Signal Processing: Evolution in Remote Sensing, WHISPERS '09 First Workshop pp: 1-5.

44. Swendsen RH, Wang JS (1986) Replica Monte Carlo Simulation of Spin-Glasses. Phys Rev Lett 57: 2607-2609.

45. Earl DJ, Deem MW (2005) Parallel tempering: Theory, applications, and new perspectives. Phys Chem Chem Phys 7: 3910-3916.

46. Nagata K, Sugita S, Okada M (2012) Bayesian spectral deconvolution with the exchange Monte Carlo method. Neural Networks 28: 82-89.

47. Hukushima K, Nemoto K (1996) Exchange Monte Carlo method and application to spin glass simulations. J Phys Soc Jpn 65: 1604-1608.

48. Kirkpatrick S (1984) Optimization by simulated annealing: quantitative studies. J Stat Phys 34: 975-986.

49. Cloutis EA (2002) Pyroxene reflectance spectra: Minor absorption bands and effects of elemental substitutions. J Geophys Res 107: 5039.

50. Morimoto N, Fabries J, Ferguson AK, Ginzburg IV, Ross M, et al. (1988) Nomenclature of pyroxenes. Mineral Mag 52: 535-550.

51. Lindsley DH (1983) Pyroxene thermometry. Am Mineral 68: 477-493.

52. Cloutis EA, Gaffey MJ (1991) Pyroxene Spectroscopy Revisited: Spectral-Compositional Correlations and Relationship to Geothermometry. J Geophys Res 96: 22809-22826.

53. Schade U, Wäsch R, Moroz L (2004) Near-infrared reflectance spectroscopy

of Ca-rich clinopyroxenes and prospects for remote spectral characterization of planetary surfaces. Icarus 168: 80-92.

54. Hazen RM, Bell PM, Mao HK (1978) Effects of compositional variation on absorption spectra of lunar pyroxenes. Proceedings of the Lunar and Planetary Science Conference 9: 2919-2934.

55. Denevi BW, Lucey PG, Hochberg EJ, Steutel D (2007) Near-infrared optical constants of pyroxene as a function of iron and calcium content. J Geophys Res 112: E05009.

56. Klima RL, Dyar MD, Pieters CM (2011) Near-infrared spectra of clinopyroxenes: Effects of calcium content and crystal structure. Meteorit Planet Sci 46: 379-395.

57. Akaike H (1980) Likelihood and the Bayes procedure. In: Bernardo JM, DeGroot MH, Lindley DV, Smith AFM (eds.), Bayesian statistics. University Press, Valencia 31: 143-166.

58. Schwarz GE (1978) Estimating the dimension of a model. The Annals of Statistics 6: 461-464.

59. Pieters CM (1983) Strength of mineral absorption features in the transmitted component of near-infrared reflected light: first results from RELAB. J Geophys Res 88: 9534-9544.

60. Pieters CM, Hiroi T (2004) RELAB (Reflectance Experiment Laboratory): A NASA multiuser spectroscopy facility. Lunar and Planetary Science Conference, XXXV: #1720.

61. Turnock AC, Lindsley DH, Grover JE (1973) Synthesis and Unit Cell Parameters of Ca-Mg-Fe Pyroxenes. Am Mineral 58: 50-59.

62. Dowty E, Lindsley DH (1973) Mössbauer spectra of synthetic hedenbergite-ferrosilite pyroxenes. Am Mineral 58: 850-868.

63. Lindsley DH, Burnham CW (1970) Pyroxferroite: Stability and X-ray Crystallography of Synthetic $Ca_{0.15}Fe_{0.85}SiO_3$ Pyroxenoid. Science 168: 364-367.

64. Klima RL, Pieters CM, Dyar MD (2007) Spectroscopy of synthetic Mg-Fe pyroxenes I: Spin-allowed and spin-forbidden crystal field bands in the visible and near-infrared. Meteorit Planet Sci 42: 235-253.

65. Goldman DS, Rossman GR (1977) The spectra of iron in orthopyroxene revisited: the splitting of the ground state. Am Mineral 62: 151-157.

66. Klima RL, Pieters CM, Dyar MD (2008) Characterization of the 1.2 μm M1 pyroxene band: Extracting cooling history from near-IR spectra of pyroxenes and pyroxene-dominated rocks. Meteorit Planet Sci 43: 1591-1604.

67. Rossman GR (1980) Pyroxene spectroscopy. Rev Mineral 7: 93-115.

68. Cloutis EA, Gaffey MJ, Smith DGW, Lambert RSTJ (1990) Reflectance spectra of glass-bearing mafic silicate mixtures and spectral deconvolution procedures. Icarus 86: 383-401.

69. Viswanathan K (1966) Unit cell dimensions and ionic substitutions in common clinopyroxenes. The American Mineralogist 51: 429-442.

70. Ohashi Y, Burnham CW, Finger LW (1975) The Effect of Ca-Fe Substitution on the Clinopyroxene Crystal Structure. Am Mineral 60: 423-434.

71. Hiroi T, Pieters C (1998) Modified Gaussian deconvolution of reflectance spectra of lunar soils. Lunar and Planetary Science conference XXIX: #1253.

72. Mustard JF, Murchie S, Erard S, Sunshine J (1997) In situ compositions of Martian volcanics: implications for the mantle. J Geophys Res 102: 25605-25615.

73. Gendrin A, Bibring JP, Mustard J, Kanner L, Mangold N, et al. (2006) Strong pyroxene absorption bands on Mars identified by OMEGA: geological counterpart. Lunar and Planetary Science conference XXXVII: #1858.

74. Baratoux D, Pinet P, Gendrin A, Kanner L, Mustard J, et al. (2007) Impact craters mineralogy from OMEGA data: implications on alteration history, ejecta emplacement, and subsurface composition. Proceedings of the Seventh International Mars Conference 3183.

75. Pinet PC, Heuripeau F, Clenet H, Chevrel S, Daydou Y, et al. (2007) Mafic mineralogy variations across Syrtis Major Shield and surroundings as inferred from visible-near-infrared spectroscopy by OMEGA Mars Express. Proceedings of the Seventh International Mars Conference 3146.

76. Binzel RP, Rivkin AS, Bus SJ, Sunshine JM, Burbine TH (2001) MUSES-C

target asteroid (25143) 1998 SF36: A reddened ordinary chondrite. Meteorit Planet Sci 36: 1167-1172.

77. Hiroi T, Vilas F, Sunshine JM (1996) Discovery and analysis of minor absorption bands in S-asteroid visible reflectance spectra. Icarus 119: 202-208.

78. Kruse FA, Lefkoff AB, Boardman JW, Heidebrecht KB, Shapiro AT, et al. (1993) The Spectral Image Processing System (SIPS) – Interactive Visualization and Analysis of Imaging Spectrometer Data. Remote Sens Environ 44: 145-163.

79. van der Meer FD, van der Werff HMA, van Ruitenbeek FJA, Hecker CA, Bakker WH, et al. (2012) Multi- and hyperspectral geologic remote sensing: A review. Int J Appl Earth Obs 14: 112-128.

80. Fraeman AA, Murchie SL, Arvidson RE, Clark RN, Morris RV, et al. (2014) Spectral absorptions on Phobos and Deimos in the visible/near infrared wavelengths and their compositional constraints. Icarus 229: 196-205.

81. Pieters CM, Murchie S, Thosmas N, Britt D (2014) Composition of Surface Materials on the Moons of Mars. Planet Space Sci 102: 144-151.

82. Tsuda Y, Yoshikawa M, Abe M, Minamino H, Nakazawa S (2013) System design of the Hayabusa 2-Asteroid sample return mission to 1999 JU3. Acta Astronaut 91: 356-362.

83. DeMeo FE, Binzel RP, Slivan SM, Bus SJ (2009) An extension of the Bus asteroid taxonomy into the near-infrared. Icarus 202: 160-180.

11

Groundwater Studies with Special Emphasis on Seasonal Variation of Groundwater Quality in a Coastal Aquifer

Kasi Viswanadh Gorthi[1]* and Mohan Babu M[2]

[1]Department of Civil Engineering, JNTUH College of Engineering Hyderabad (Autonomous), Hyderabad, India
[2]Department of Civil Engineering, Sri Venkateswara College of Engineering and Technology (Autonomous), Chittoor, Andhra Pradesh, India

Abstract

The qualitative and quantitative assessment of groundwater is obligatory pre-requisite for developing countries like India with rural based economy. It is essential to evolve a strategy for ecologically sustainable development of groundwater resources which can be done only through exploration of hydro-geological environment. In the present study, groundwater studies, with special emphasis on groundwater quality and its variation with season have been carried out for upland areas of Nellore district, situated on the East Coast of India. Groundwater quality data of samples collected during pre and post monsoon season in the year 2011 by State Groundwater Department, Nellore have been used for finding its suitability for drinking and agricultural purposes. The data pertaining to the following parameters / ions, viz., Hydrogen Concentration(pH), Total Dissolved Solids(TDS), Total Hardness(TH), Calcium(Ca^{2+}), Magnesium(Mg^{2+}), Chlorides(Cl), Sodium(Na^+), Potassium(K^+), and Sulphates(SO_4^{2+}) has been analyzed. It has been observed that in post-monsoon period most of the water samples are found not suitable for drinking. For evaluation of water quality for irrigation, US Salinity Laboratory's and Wilcox's diagrams, and percent Sodium (Na^+) are used. It is found that the majority of the groundwater samples are not good for irrigation in post-monsoon compared to that in pre-monsoon.

Keywords: Ground water quality; Seasonal variation; TDS; TH; Ca; Mg; Cl; Na; K; SO_4

Introduction

Ground water is an important natural source of water supply all over the world. Its use in irrigation, industries and domestic usage continues to increase where perennial surface water source are absent [1]. The modern civilization, over exploitation, rapid industrialization and increased population has lead to fast degradation of our environment [2]. To meet the rising demand it is imperative to recognize the fresh water resources and also to find out remedial methods for improvement of water quality [3]. The quality of water may depend on geology of particular area and also vary with depth of water table and seasonal changes and is governed by the extent and composition of the dissolved salts depending upon source of the salt and soil, subsurface environment. Good drinking water quality is essential for the well-being of all people which has affected the health and economic status of the populations. Groundwater is the major source of water for drinking, agricultural and industrial desires. Groundwater quality with respect to season has been studied by collecting data seasonally from groundwater department, Nellore. The present study aims to assess groundwater quality and its suitability for drinking, agricultural and its seasonal variation.

Study Area

Nellore district is one of the coastal districts of Andhra Pradesh. The study area lies between 13°30' – 15°05'of North Latitude, 79°05'-80°15'of East Longitude. Geographical area of the district is 13,076 sq. km and situated in the south-eastern part of the state. Nellore district is one of the coastal districts of Andhra Pradesh. The district is bounded by Bay of Bengal in the east with a coastline of 169 km and is divided into 46 revenue mandals. The Normal Rainfall of the District is 1080 mm. Agriculture is the mainstay of the population with 77.3 percent of population is being rural. The soils of the district are classified as red, black and sandy. The red soils are predominant with 40% of the area whereas black cotton soil and sandy loams occupy 23% and 34% of the area respectively. The principle crops grown in the study area are Paddy, Sugarcane and Groundnut, Bajra, Fruits & Vegetables, Chillies,

Cotton and Tobacco. The district is dependent on ground water for irrigation and domestic needs. 60% of the irrigated area is through ground water.

The objectives of the present study were to:

(1) To study seasonal variation of ground water quality and its suitability in relation to drinking water quality standards

(2) To assess suitability of groundwater for irrigation purposes.

Methods of Investigation

The methodology include, collection of monthly rainfall data, collection of periodic groundwater levels and collection of ground water samples from 20 wells in the pre-monsoon and post-monsoon period during the year 2011. The rainfall data collected for the year 2011 is shown in the Tables 1 and 2 shows the periodic groundwater levels collected (Figure 1).

Ground water analysis

The chemical quality data collected during Pre and Post-monsoon periods for 20 wells located in Anantha Sagaram, Chakali Konda, Depuru, Duthaluru, Gandipalem, Gonupalli, Jayampu, Kasumuru, Kodanda Ramapuram, Kovvurupalli, Mannarupoluru, Nandigunta, Tamidipadu, Poolathota, Rajavolu, Rapuru, Manubolu, Marlapudi, Udayagiri and Varikuntapadu villages of Nellore district for the following parameters/ ions, viz., Hydrogen ion concentration (pH),

*Corresponding author: Dr. Kasi Viswanadh Gorthi, Professor, Department of Civil Engineering, JNTU college of Hyderabad (Autonomous), Hyderabad
E-mail: gkviswanadh@jntuh.ac.in

Month	Average rainfall (mm)
January	8.34
February	28.107
March	0
April	25.718
May	11.424
June	61.578
July	57.462
August	130.108
September	60.045
October	250.916
November	329.527
December	56.621

Table 1: Average monthly rainfall distribution of Nellore (2011).

Month	Jan	Feb	Mar	Apr	May	Jun	July	Aug	Sep	Oct	Nov	Dec
Average groundwater level(DTW)(m)	3.18		3.94		5.35		6.22		6.68		3.7	

Table 2: Details of average groundwater levels.

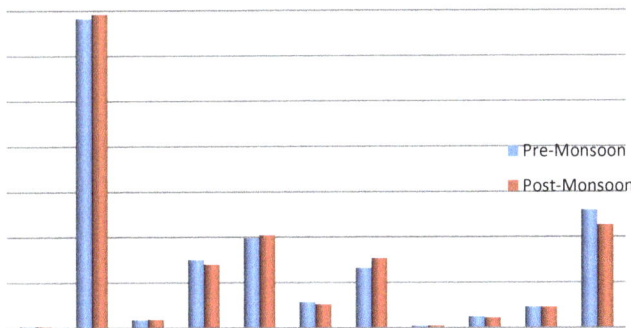

Figure 1: Comparison between Pre-monsoon and post-monsoon average chemical data.

Electrical Conductivity (EC), Total dissolved solids (TDS) were computed by multiplying the EC by a factor (0.55 to 0.75), depending on relative concentrations of ions. Total hardness (TH) as $CaCO_3$, Calcium (Ca^{2+}), Magnesium (Mg^{2+}), Sodium (Na^+), Potassium (K^+), Total Alkalinity (TA) as $CaCO_3$, Carbonate (CO_3^{2-}), Bicarbonate (HCO_3), Chlorides (Cl^-) and Sulphate (SO_4). Table 3 shows the chemical data of 20 wells for Pre-monsoon and Post-monsoon respectively.

Ground Water Quality

Quality criteria for drinking purpose

The following parameters are used to analyze water quality viz., pH, TDS, Total hardness, Calcium, Magnesium, Chlorides and Sulphates. The groundwater quality data on the following parameters/ions were presented and discussed below (Table 3).

Hydrogen ion concentration (pH): The pH of ground water in the study area ranges from 8 to 9.6. Maximum pH of more than 8 is encountered in localized pockets. According to the drinking water standards prescribed by ISI [4], WHO [5], ICMR [6] during pre-monsoon 14 samples are found to be safe i.e., under maximum permissible limits and remaining 6 are found to be above maximum permissible limits. In post monsoon all 20 samples are found to be safe.

Total dissolved solids (TDS): The simplest classification of

water is based on the total concentration of dissolved solids [7]. Total Dissolved Solids in the study area ranges from 368 mg/l to 4417 mg/l in pre monsoon and post monsoon. The variation in TDS concentration with season is shown in the following Table 4.

Total hardness (TH): Total hardness in the study area ranges from 180 mg/l – 2205 mg/l. In pre monsoon 80% of the samples are very hard, 20% are hardy category. In post monsoon 50% of samples are hard water and 35% are very hard category. The general range in the study area is above 300 mg/l. TH values for the most samples in the study area are exceeds the maximum permissible limits (ISI and WHO standards). The variation in TH concentration with season is shown in the following Table 5.

Calcium (Ca): In most natural waters, Ca is a common and wide spread element. Ca is the most abundant cation in the ground water of the study area and its values range from 5-365 mg/l. Most of the area contains Ca in the range of 20-50 mg/l. Low concentration of less than 25 mg/l is observed mostly along the upland areas of Nellore and as isolated in the other parts. Maximum range of more than 100 mg/l occurs as localized patches confining to 'kankar' zones in the soil profile and some clay zones.

Magnesium (Mg): In the study area, Mg content is less than Ca it varies from 25- 390 mg/l. Most of the area shows less than 25 mg/l. Next range of 25-50 mg/l is located on the north western and central parts of the area covering all the formations. Maximum content of above 100 mg/l is recorded in clay and in some low lying areas. Magnesium values for the most of samples are within the maximum permissible limits (ISI and WHO Standards). High concentration of Mg may cause laxative effect particularly on new users. Mg deficiency is associated

Parameter	Season	
	Pre monsoon	Post monsoon
pH	8.8	8.64
TDS (Sp*0.64)	1366	1384.3
CO_3 (mg/l)	37.002	38.999
HCO_3 (mg/l)	301.003	280.5
Chloride (mg/l)	394.99	408.998
Sulphates (SO_4)	114.15	104.65
Sodium (mg/l)	263.95	306.3
Potassium (mg/l)	9.75	10.14
Calcium (mg/l)	50.001	44.39
Magnesium (mg/l)	93.54	91.3
Total hardness ($CaCO_3$)	519.9	452.6

Table 3: Average chemical quality data both for pre and post-monsoon.

TDS (mg/l)	No. of Samples		Type of water
	Pre monsoon	Post monsoon	
<1000	12	10	Fresh water
1000 - 3000	7	9	Brackish water
>3000	1	1	Salty water

Table 4: Water quality index scale for TDS.

Total Hardness (mg/l)	No. of samples		Water class
	Pre monsoon	Post monsoon	
<75	0	0	Soft
75-150	0	0	Moderate
150-300	4	10	hard water
>300	16	7	very hard

Table 5: Water quality index scale for TH.

with structural and functional changes. It is essential as an activator of many enzyme systems (Figure 2).

Chlorides (Cl): Chlorides in study area range from 130 to 1700 mg/l. It is found that most of the samples are within the range of 150 to 400 mg/l. When compared to pre monsoon in post monsoon the Chloride content increases. The samples of Depuru, Kasumuru, Kovvurupalli, Manubolu, Nandigunta, Poolathota and Udayagiri villages are exceeding the maximum permissible limits during the both the monsoon periods (ISI and WHO standards); the maximum concentration of Chlorides with more than 500 mg/l is in isolated patches located in clay zones.

Sulphates (SO_4): Sulphur is widely distributed in reduced form in both metamorphic and sedimentary rocks as metallic sulphides though it is not a major constituent of the earth's outer crust. During weathering in contact with aerated water, the sulphides are oxidized to yield sulphate (SO_4) ions which are carried off in the water [8]. Sulphate is generally less abundant than chloride in most of the natural waters. The same is observed in the ground water of the study area also. Generally, ground water contains SO_4 at less than 100 mg/l in metamorphic or sedimentary aquifers [7]. Sulphate in the study area ranges from 25 to 420 mg/l. Most of the area shows less than 150 mg/l concentration in pre and post monsoon periods. Sulphate content in Depuru and Kasumuru villages exceeds the maximum permissible limits.

Sodium (Na) : The Na concentration in the ground water of the study area varies between 4.5 and 250 mg/l. Distribution of Na is similar to that of TDS (Figure 3). Most of the study area has the range of less than 50 mg/l. This range coincide with the low range of TDS, maximum range of more than 150 mg/l is located as small patches in the low lying areas and in the clay zones.

Potassium (K): The concentration of potassium in the groundwater of the study area varies from traces to 240 mg/l. In general, k content in the study area is much less than Na. Natural water in general, has much lower concentrations of k as compared with Na. During weathering, k ions tend to be either adsorbed or fixed in the soils or weathered products especially in the clay minerals. As a result, the concentration of k ions in natural waters is relatively less. As in the case of Na distribution, k also does not show any differential concentration.

Quality criteria for irrigation purpose

The suitability of groundwater for irrigation purposes depends upon the effect of mineral constituents of water on both plants and soils. The general criteria for assessing the irrigation water quality are: total salt concentration as measured by EC, relative proportions of Na^+ as expressed by %Na, SAR, RSC. Water quality criteria can be used as

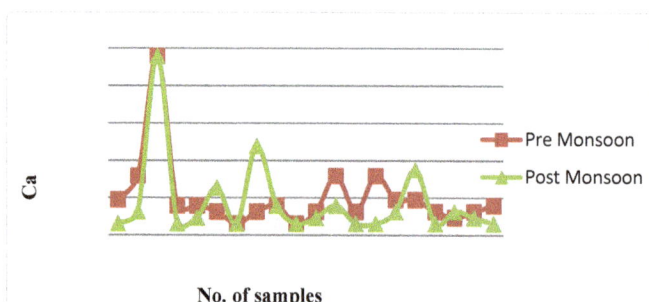

Figure 2: Distribution of Ca content in Study Area during pre and post monsoons.

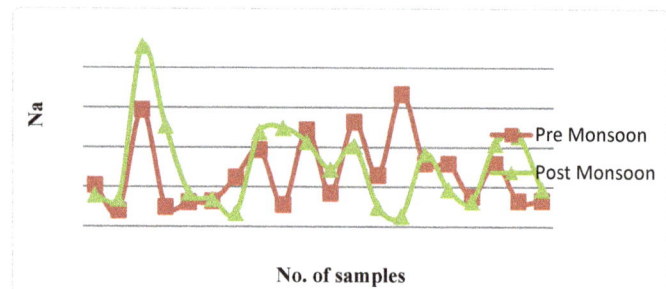

Figure 3: Sample wise Na concentration during pre and post-monsoons.

guidelines by farmers for selecting appropriate management practice to overcome potential salinity hazard, if the quality of available water would pose any problem for irrigation. The following parameters were used to assess the groundwater suitability for agricultural purpose.

Percent Sodium (% Na): The Percent sodium (%Na^+) is also widely utilized for evaluating the suitability of water quality for irrigation [9]. The %Na^+ is computed with respect to relative proportions of cat ions present in water. Various ratios of Na to (Ca+ Mg) have been used to depict the suitability of water for irrigation. Na content is usually expressed in terms of % Na defined as

$$\%Na = (Na^++K^+)*100/ (Ca^{2+}+Mg^{2+}+ Na^++K^+)$$

Where, all ionic concentrations are expressed in meq/l.

Most of the ground waters in the study area fall under good to permissible category. Very few wells come under the permissible to doubtful category. No sample fall under unsuitable ground water category according to this parameter.

Alkalinity: Alkali hazard or sodium content expressed as Sodium Adsorption Ratio (SAR) is another important chemical parameter for judging suitability of water for irrigation.

SAR is defined as

$$SAR=Na/\sqrt{(Ca+ Mg)/2}$$

Where the concentration of the constitutions are expressed in meq/l

The entire area comes under excellent class. According to salinity diagram, most of the samples of groundwater in study area are good with moderate salinity and low Na.

Residual sodium carbonate (RSC): Eaton [10] evolved a criterion on the basis of residual sodium carbonate for agricultural purpose based on alkalinity and alkaline earths.

$$RSC = (CO_3^{2-} + HCO_3^-) - (Ca^{2+} + Mg^{2+})$$

Concentrations of ions are expressed in meq/l.

According to Eaton [10], as soil alkalinity increases, plant growth diminishes. The water exceeding 2.5 meq/l of RSC, is not suitable for irrigation as it deteriorates the soil structure and air movement through soil is restricted [11]. RSC values in between 1.25 and 2.5 meq/l, are considered as marginal, RSC values less than 1.25 meq/l is said to be probably safe. Negative values of RSC are usually safe for irrigation. Ground water samples in the study area mostly possess negative values indicating predominance of alkaline earths over alkalinity. It is found that RSC values are < 1.25 meq/l in the major portion of study area and therefore it is safe.

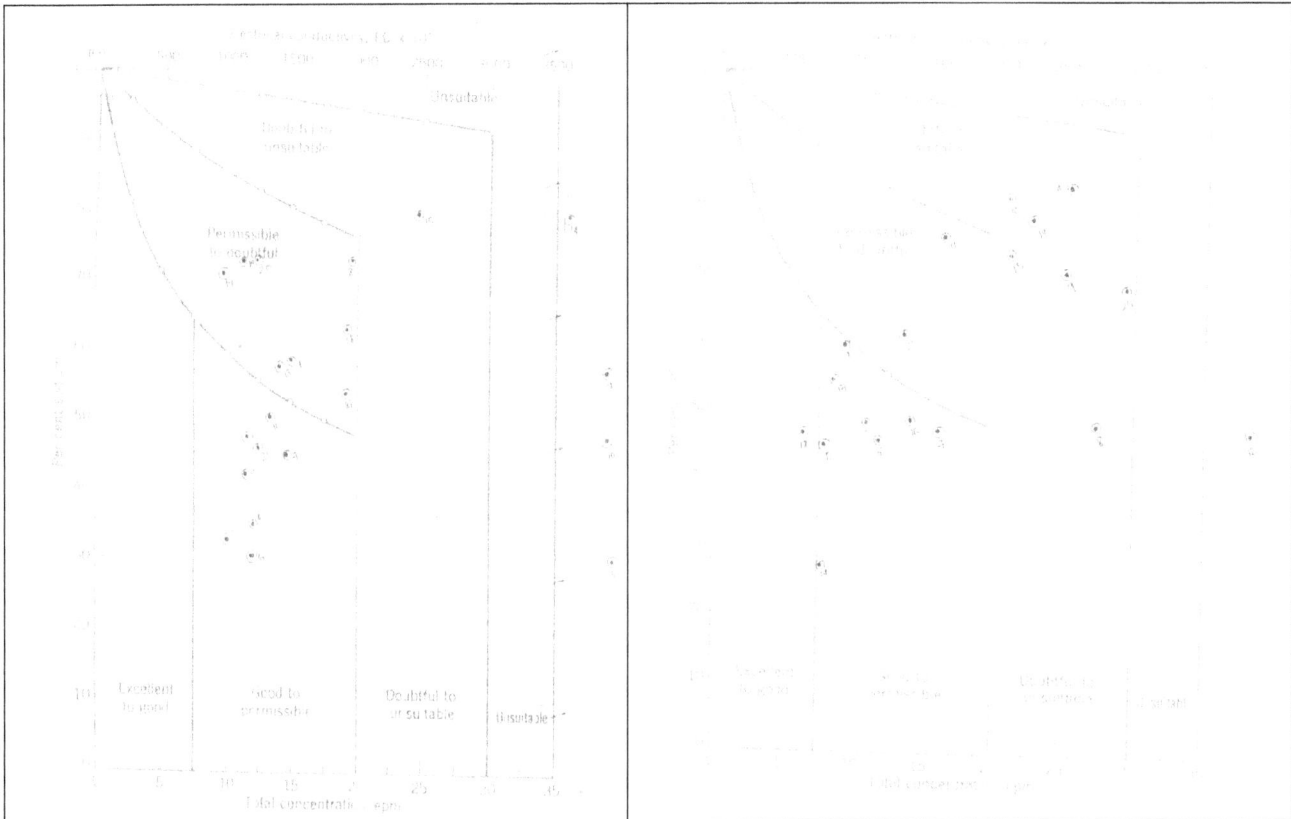

Figure 4: Classification of the ground waters based on electrical conductivity and percent of Sodium in pre & post monsoon season.

Conclusion

The dominant cation and anion in the ground water of this area are Na+ and Cl⁻, sodium chloride is thus distributed in most parts of the aquifer. TDS levels in all the ground water samples collected during pre and post monsoon periods are exceeding drinking water standards of ISI.

TDS levels in all groundwater samples in both seasons are above drinking water standards. The groundwater are exceed the maximum permissible as per ISI standards during pre- and post-monsoon seasons for the following parameters /ion, viz. TDS, TH, Na$^+$, HCO$_3^-$, Cl$^-$, SO$_4^{2-}$.

Ground water in the study area is alkaline in pre- and post-monsoon seasons non- carbonate alkali controlled by evaporation (Figure 4). The dominant cation and anion in the ground water of this area are Na+ and Cl⁻, sodium chloride is thus distributed in most parts of the aquifer.

Acknowledgements

The data provided by the State Ground Water Department & District Statistical Department, Nellore District Govt. of Andhra Pradesh, is gratefully acknowledged.

References

1. Mariappan V, Prabakaran P, Rajan MR, Ravichahandran AD (2005) A systematic study of water quality index among the physic chemical characteristic of ground water in and around Thanjavur Town. IJEP 25: 551-555.

2. Murli RD, Swasthik, Elangovan R (2011) Assessment of ground water quality in Coimbatore South Taluk. A WQI approach 10: 521-524.

3. Maruthi Devi CH, Usha Madhuri T (2011) A study on Ground water Quality in Pakistan district and its suitability for drinking nept. 10: 481- 483.

4. Indian Standard Institution (ISI) (1983). Indian Standard Specifications for Drinking Waters, IS: 10500.

5. World Health Organizations (1972) International Standards for Drinking water (3ʳᵈ edn).

6. Indian Council of Medical Research (ICMR) (1975) Manual of Standards of Quality for Drinking water Supplies. special Report Series No 44: 27.

7. Davis SN, Dewiest RJM (1966) Hydrology. John Wiley and Sons Inc, New York.

8. Hem JD (1970) Study and interpretation of the chemical characteristics of natural water. U.S. Geol. Survey, Supply paper.

9. Wilcox LV (1948) The quality of water for irrigation use. US Department of Agricultural Technical Bulletin, Washington.

10. Eaton FM (1950) Significance of carbonates in irrigation water. Soil Sci 69: 127-128.

11. Allison LE, Bernstein L, Bower CA (1954) Diagnosis and Improvement of Saline and Alkaline soils. U.S. Salinity Laboratory staff , Washington.

12

3-D Geostatistical Model and Volumetric Estimation of 'Del' Field, Niger Delta

Oluwadare OA, Osunrinde OT, Abe SJ* and Ojo BT

Department of Applied Geophysics, Federal University of Technology, Nigeria

Abstract

There is an insatiable thirst for oil and gas consumption and increased production will be made possible only through effective reservoir characterization and modeling. A suite of wire-line logs for four wells from 'DEL' oil field together with 3D seismic data were analyzed for reservoir characterization of the field. Two reservoirs were identified using the resistivity log. A synthetic seismogram was generated in order to perform seismic to well tie process as well as picking of horizons throughout the section. Time and depth structural maps were generated. Geostatistical simulation such as the sequential Gaussian stimulation and sequential indicator stimulation were carried out to provide equiprobable representations of the reservoirs, and the distribution of reservoir properties within the geological cells. The modeled reservoir properties resulted in an improved description of reservoir distribution and inter connectivity. The analysis indicated the presence of hydrocarbon in the reservoirs. There is also a fault assisted closure on the structural map which is of interest in exploration. A fluid distribution plot and map of the field were also obtained. The modeled properties gave an average porosity of 24%, average water saturation ranging from 12%-24% and moderate net-gross. The volumetric calculation of the reservoir gives a STOIIP ranging from 37.53 MMbbl-43.03 MMbbl. The result showed high hydrocarbon potential and a reservoir system whose performance is considered satisfactory for hydrocarbon production. The resulting models can also be used to predict the future performance of the reservoir.

Keywords: Porosity; Permeability; Reservoir characterization; Synthetic seismogram

Introduction

The search for hydrocarbon is becoming increasingly more difficult and expensive. Because most of the identified structural closures on the shelf and upper slope have been drilled, the search for hydrocarbon and its production requires more creativity in optimizing and integrating existing data. The reservoir is the habitat of petroleum, and therefore an enhanced understanding of the reservoir will enable a proper prediction of its characteristics and requisite inputs for modeling. A reservoir that is well understood will ultimately result in a field that is well managed. Mapping the right reservoir, understanding of reservoir characteristics most importantly porosity, permeability, water-saturation, thickness and area extent of the reservoir so as to have a maximum hydrocarbon reserves have been a major challenge in the exploration and exploitation of hydrocarbon in major hydrocarbon fields within the world today such as Niger-Delta. Steady success in the exploitation for oil and gas reserves in Niger-Delta therefore depend on having a clear understanding about the subsurface geology of the area. The ability to accurately identify the fluid present within the reservoir, predict the petro physical parameters, model and estimate accurately the amount of reserves in the reservoir will aid in a successful exploitation for hydrocarbon.

In many cases, conceptual models are essential at the scale of the reservoir unit, but their accuracy commonly remains insufficient to realistically predict the distribution of internal heterogeneities [1]. Stochastic approaches are now more frequently applied to simulate the distribution of small-scale sedimentary bodies and internal reservoir heterogeneity [2-5]. Geostatistical approaches also provide equiprobable realizations of the heterogeneity distribution. This flexibility can be used to evaluate the impact of different geological scenarios, which contribute to the optimization of a field development plan. The advances in computational technology, modern reservoir models can accommodate increasingly detailed 3D data that illustrate the spatial distribution of reservoir properties. Subsurface reservoir characterization typically incorporates well data augmented with seismic data to establish the geological model of the reservoir [6]. A successful reservoir characterization therefore involves the integration of 3D seismic data and well log data. Integrating various datasets to provide a geologically relevant subsurface image will aid interpretation and reduce uncertainty. This is with the aim of delineating the subsurface structures that are favourable for the accumulation of hydrocarbon within the 'DEL' field and also to construct a fit for purpose geologic model.

Geology and location of the study area

Niger Delta is a prolific hydrocarbon belt in the world. The formation of Niger Delta basin was initiated in the early tertiary time. The Niger Delta is situated in the Gulf of Guinea and extends throughout the Niger Delta province [7]. From the Eocene to the present, the Delta has prograded Southwest ward, forming depobelts that represent the most active portion of the Delta at each stage of its development [8]. These depobelts form one of the largest regressive deltas in the world with an area of some 300,000 km^2 a sediment volume of 500,000 km^3 and a sediment thickness of over 10 km in the basin depocenter [9,10].

The Niger Delta province contains only one identified petroleum system [9,11]. This system is referred to here as the tertiary Niger Delta (Akata-Agbada) Petroleum System. Deposition of the three formations occurred in each of the five off lapping siliciclastic sedimentation cycles that comprise the Niger Delta. These cycles (depobelts) are 30-60

*Corresponding author: Abe SJ, Department of Applied Geophysics, Federal University of Technology, Nigeria, E-mail: jsabe@futa.edu.ng

kilometres wide, prograde south-westward 250 kilometres over oceanic coast into the gulf of guinea and are defined by synsedimentary faulting that occurred in response to variable rates of subsidence and sediment supply [8,12].

The Delta formed at the site of a rift triple junction related to the opening of the Southern Atlantic starting in the late Jurassic from interbedded marine shale of the lower most Agbada formation and continuing into the cretaceous. The Delta proper began developing in the Eocene, accumulating sediments that now are over 10 km thick. The primary source rock is the upper Akata formation, the marine-shale facies of the Delta, with possibly contribution from interbedded marine shale of the lowermost Agbada formation. Oil is produced from sandstone facies within the Agbada formation, however, turbidite sand in the upper Akata Formation is a potential target in deep water offshore and possibly beneath currently producing intervals onshore. The intervals, however, rarely reach thickness sufficient to produce a world class oil province and are immature in various parts of the delta [12]. The Akata shale is present in large volumes beneath the Agbada Formation and is at least volumetrically sufficient to generate enough oil. Based on organic-matter content and its types, Evamy et al. proposed that both the marine shale (Akata Formation) and the shale interbedded with paralic sandstone (lower Agbada Formation) were the source rocks for the Niger Delta oils [13,14]. Petroleum occurs throughout the Agbada Formation of the Niger Delta, however, several directional trends form an "oil-rich belt" having the largest field and lowest gas:oil ratio [8,15]. The belt extends from the northwest off-shore area to the southeast offshore and along a number of north-south trends in the area of Port Harcourt. It roughly corresponds to the transition between the Continental and Oceanic crust and is within the axis of maximum sedimentary thickness.

This hydrocarbon distribution was originally attributed to the timing of trap formation relative to petroleum migration (earlier landward structures trapped earlier migrating oil), however, showed that in many rollovers, movement on the structure building fault and resulting growth continued and was relayed progressively southward into the younger part of the section by successive crestal faults [13]. He also concluded that there was no relation between growth along a fault and distribution of petroleum.

The study area ('DEL' FIELD) falls within the western margin of off-shore depobelt of Niger Delta (Figure 1). The fault pattern is NW-SE and the traps involved in this field are mainly structural in nature. The study area is within the Parasequence set of Agbada formation. The field covers approximately 720 sq km.

Methodology

Reservoir geology

The lithology was delineated by first setting a range for the gamma ray log. The gamma ray log ranges from 0 API to 150 API. The shale formations have high radioactive contents, thus deflecting to the right of the baseline while the sand formations will deflect to the left of the baseline since they have low radioactive content. Two reservoirs were correlated across the four (4) wells using both the gamma ray log and the resistivity log. The fluid types within these reservoirs were identified using both the neutron and density logs.

Seismic interpretation

Prominent geologic structures such as faults were identified across the seismic section. The check shot data from "DEL 4" was

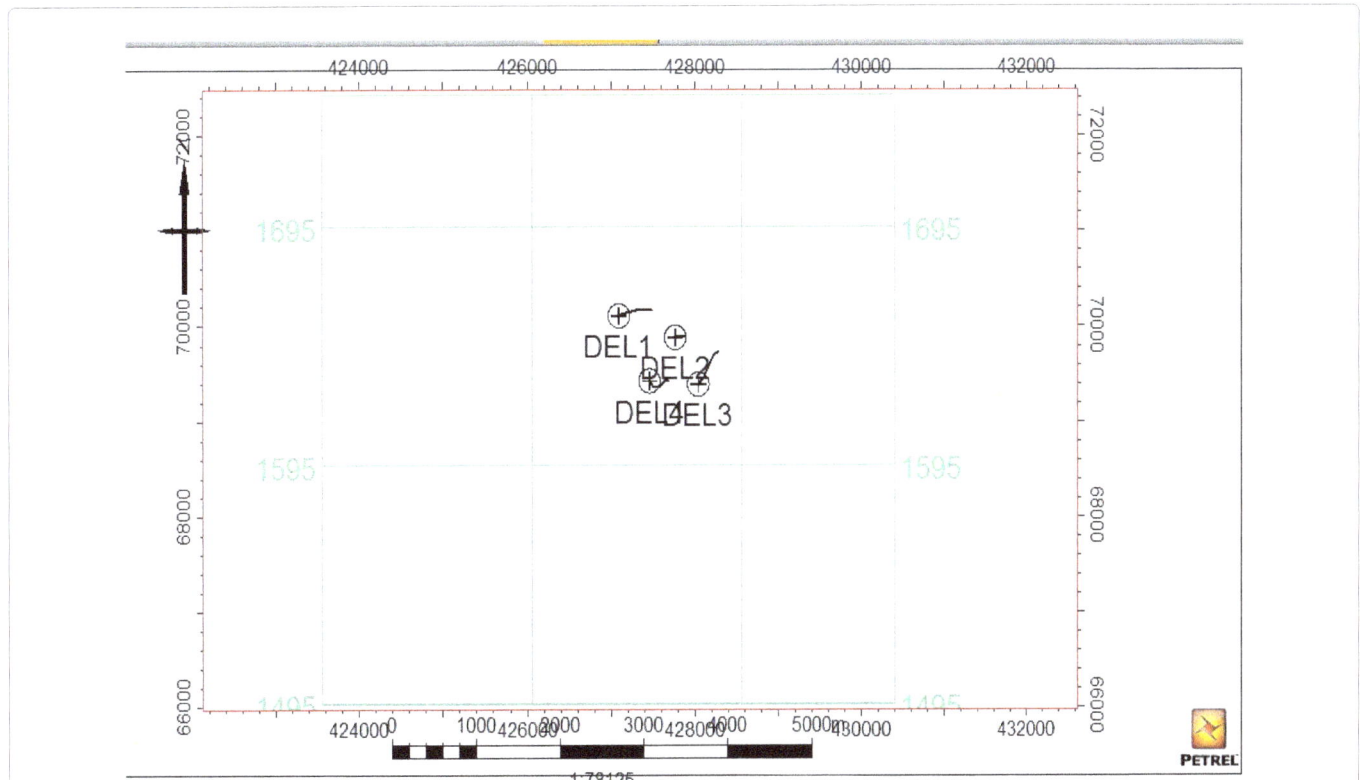

Figure 1: Base map of 'DEL' field showing the well locations and seismic lines orientation.

used to generate synthetic seismogram which was later used to tie the information from the well to our seismic data. The seismic events that correspond to the two reservoirs sand via the synthetic seismogram were mapped across the field. The mapped horizons were then contoured by the software to generate a structural time map. The check shot data shows a relationship between the two-way time (TWT) and True Vertical Depth (TVD) plotted against each other (Figure 2). A polynomial equation of the second order was derived from the plot. The equation was then used to build a velocity model which was used to convert the time structural map to depth structural map.

Petrophysical evaluation

Volume of shale: The volume of shale within the reservoir was determined from the gamma ray log. This was achieved by first calculating the gamma ray index using the equation below:

$$I_{GR} = \frac{GR_{LOGM} - GR_{IN}}{GR_{MAXM} - GR_{IN}} \tag{1}$$

where: IGR=gamma ray index, GRLOG=gamma ray reading of the formation, GRMIN=minimum gamma ray (clean sand); GRMAX=maximum gamma ray (shale). The gamma ray index was then used to calculate the volume of shale using the Steiber equation [16].

$$V_{SH} = \frac{0.5 \times I_{GR}}{1.5 - I_{GR}} \tag{2}$$

Total porosity: The total porosity gives the ratio of pore volume to the total volume of the reservoir. It was evaluated using the Wylllie equation [17].

$$\Phi = \frac{\rho_{ma} - \rho_b}{\rho_{ma} - \rho_f} \tag{3}$$

where: ρma=matrix density=2.684, ρb=formation bulk density, ρf=fluid density (1.1 salt mud, 1.0 fresh mud, 0.8 oil and 0.6 gas) and Φ=porosity

Effective porosity: The effective porosity gives the volume of the interconnected pore spaces within the reservoir i.e. it gives the volume of pore spaces that is contributing to the production of fluid within the reservoir. The effective porosity was evaluated using the equation below;

$$\Phi_{eff} = \Phi \times \left(1 - V_{SH}\right) \tag{4}$$

Permeability: This gives the rate at which the fluid can move within the interconnected pore spaces [18].

$$k = \left(250\left(\frac{\Phi^3}{SW_{irr}}\right)\right)^2 \tag{5}$$

where: K=permeability, Φ=porosity and SWirr=irreducible water saturation.

Water Saturation: This was estimated using the Archie's equation [19].

$$S_W = \sqrt{R_W / \left(ILD \times \Phi^{1.74}\right)} \tag{6}$$

where SW=water saturation, RW=water resistivity and ILD=true resistivity

Static model

The first step in building a 3D geologic model is to construct a skeletal framework where both the discreet and continuous properties will be distributed into the geologic cells. This was done through a process called "pillar gridding". Up scaling which is the process where values are assigned to the cells penetrated by the well logs in the 3D grid was also carried out. Since each cell can only hold one value, the well logs must be averaged, i.e., the lithology, resistivity, porosity, permeability, water saturation and EOD logs are up scaled into the 3D grid using the arithmetic mean method. The quality of the up scaled logs is checked by inspecting their histogram produced by the software. This is used for comparing the raw logs with the up scaled logs. If there is no much disparity between them, the up scaled logs

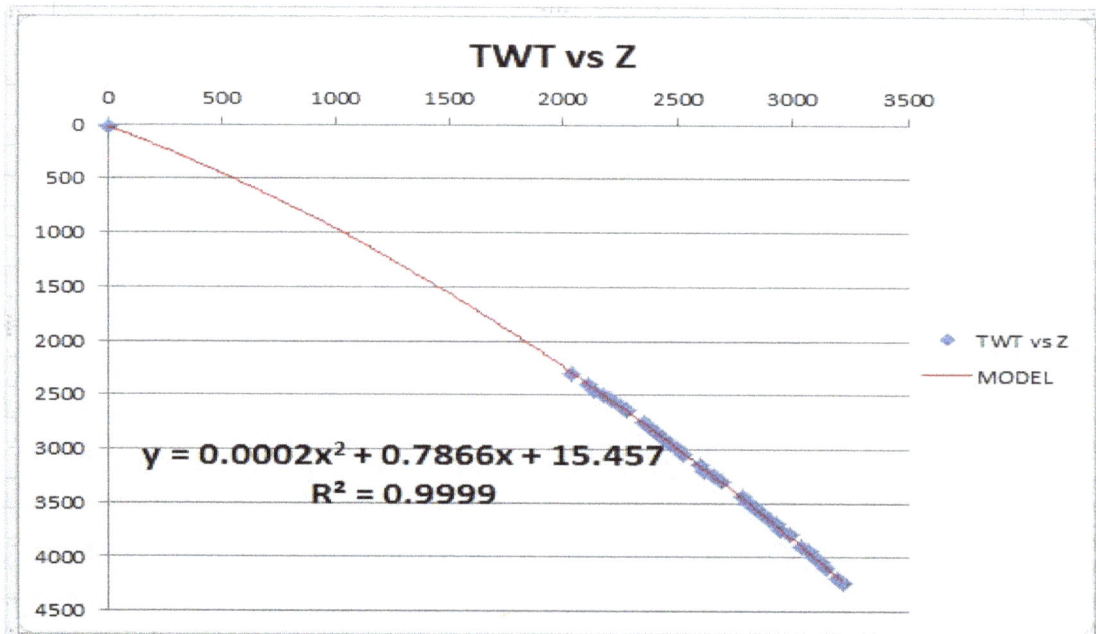

Figure 2: TWT-Z curve used for depth conversion using polynomial method.

are acceptable. Variogram analysis which is a function describing the degree of spatial dependence of a spatial random field or stochastic process was carried out. This variogram gives a measure of how data taking from the field varies in percentage depending on the distance between the data. Samples taking far apart will vary more than sample taking close to each other. The larger the separation distances between two points, the larger the variability. A variogram must be specified when a discrete property is "populated" (that is extrapolated to densely spaced points) using a stochastic simulation algorithm. The variogram analysis was carried out for all zones along the vertical direction. Cross plots which is also a statistical tools/ methods that provides the relationship between two variables was carried out. It was used to determine the relationship between some reservoir properties such as water saturation, permeability and effective porosity. These cross plots were later used as a secondary variable in the co-simulation technique.

The next step involves facies and property modeling. Facies and Property modeling is the distribution of reservoir rock properties into the 3D geocellular models using geostatistical principles. 3D property modeling without adequate data analysis will result into models that have little or no relation to geology. Therefore, a good knowledge about the geology of the area is needed to get meaningful property modeling. This is where the variogram analysis comes into play. Stochastic simulation which is a method of generating multiple equally probable realizations of reservoir properties was employed [20]. Sequential Indicator Stimulation (SIS) and Sequential Gaussian simulation (SGS), one of the dominant forms of stochastic simulation for reservoir modeling applications, was utilized. The algorithm was used to generate geologic models that honor the local conditioning (well) data, the global histogram, areal and vertical geological trends of the data and patterns of spatial correlation. The relationship between the water saturation and porosity also known as transformation plot was used to build the water-saturation model. The transformation plot was used to calculate for uncertainties within the reservoirs when the

volumetric is been computed. The upper-case, base-case and low-case relationship between porosity and water-saturation was determined.

Result and Discusson

Reservoir geology

The well correlation of "Del" field was carried out along the strike. Four wells namely; DEL1, DEL2, DEL3, and DEL4 were correlated together along the strike direction (Figure 3). One of the major reasons for carrying out correlation exercise within the given wells is to have an idea of the occurrences of horizontal sand packages from one well to the other that were deposited at the same time and space within the field. It was observed from the correlation panel that the sand packages are thinning towards the N-E direction. Two reservoirs were identified within this field. They include; Reservoir B and Reservoir E.

The fluid types (oil and gas) within these reservoirs were identified using both the neutron log and density log respectively. It was observed that the wells within this field were saturated mostly with oil (Figures 4a and 4b).

Structural interpretation

One major fault was identified across the seismic section (Figure 5). Other faults identified within this field are synthetic and antithetic to the major fault. It was also observed from the seismic-well tie that the synthetic trace with the well tops also tie very well on the seismic section (Figure 6). From the structural time and depth maps of both reservoirs (Figures 7 and 8), it was observed that the maps are having an anticlinal structure (four-way dip closure) which is of importance to the petroleum geologist, geophysicists and engineers since hydrocarbon do accumulate within an anticlinal structure. The direction of the major fault is along the NW-SE. The other faults within this field are either synthetic or antithetic to the major faults. It was also observed from the elevation legend that the orange color indicates the shallowest part of the field while the green color indicates the deepest

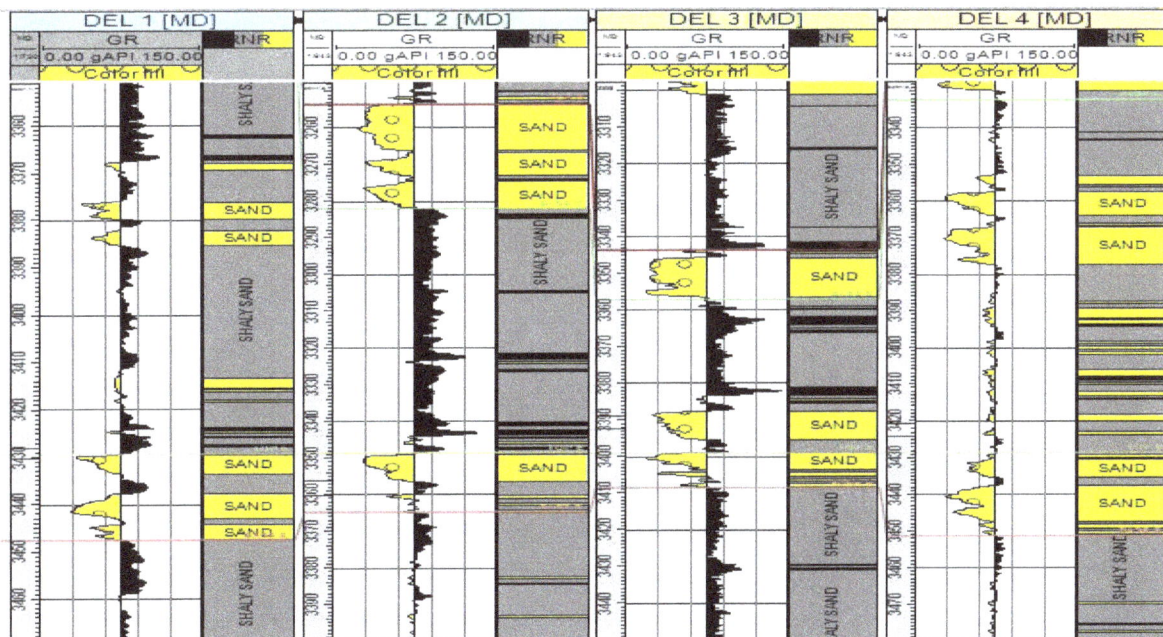

Figure 3: Lithology delineation and well correlation.

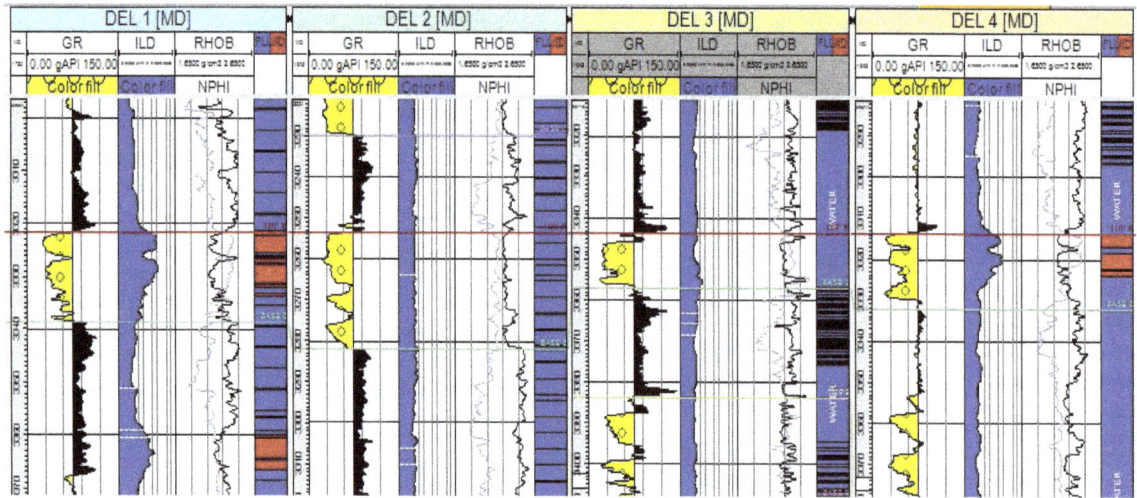

Figure 4a: Reservoir identification in reservoir B.

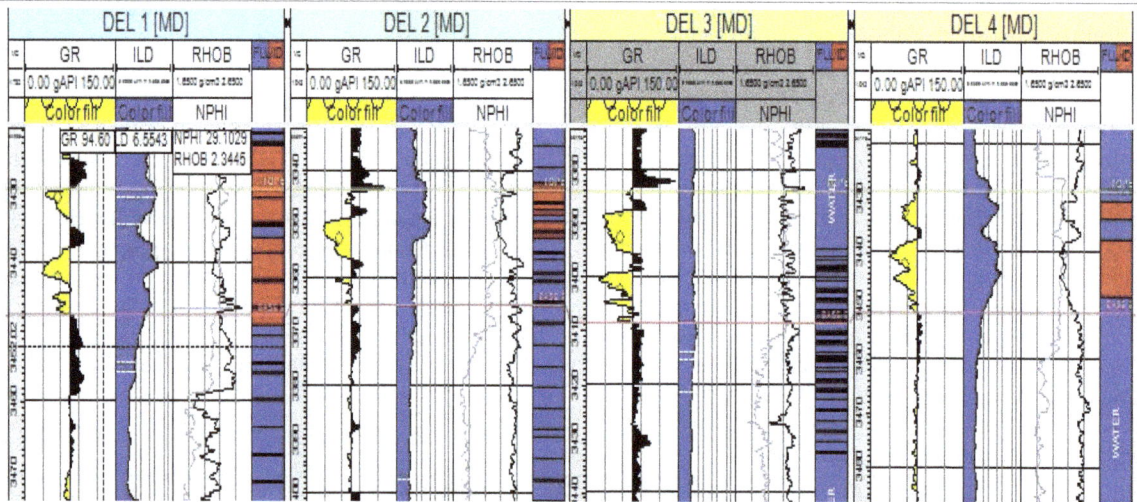

Figure 4b: Reservoir identification in reservoir E.

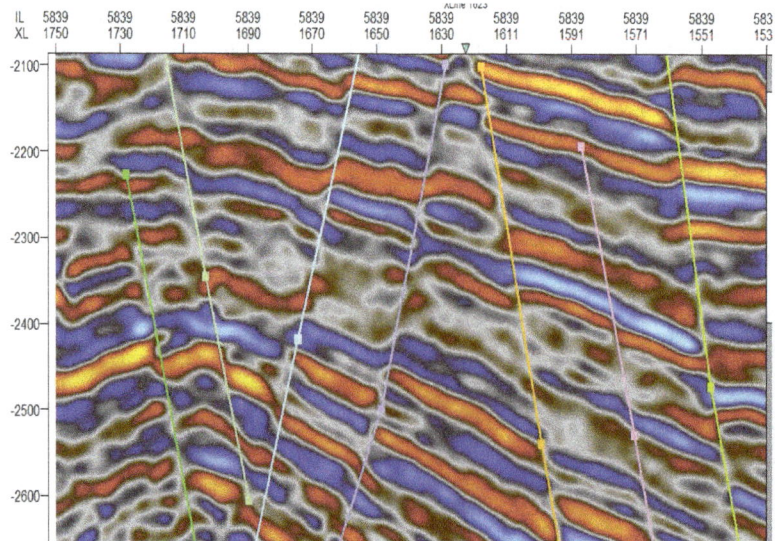

Figure 5: Fault mapped on the seismic section.

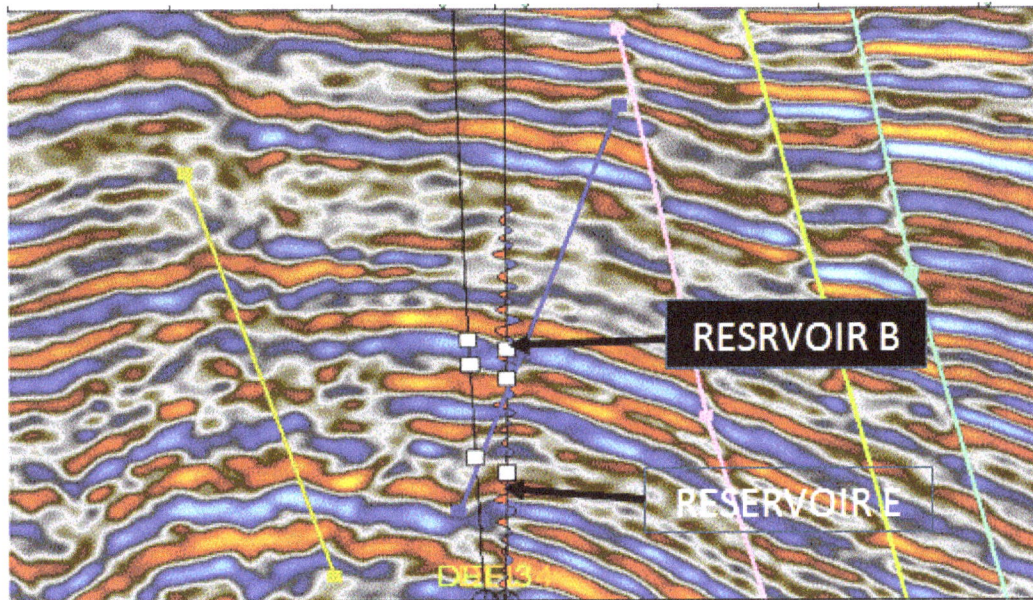

Figure 6: Seismic to well tie.

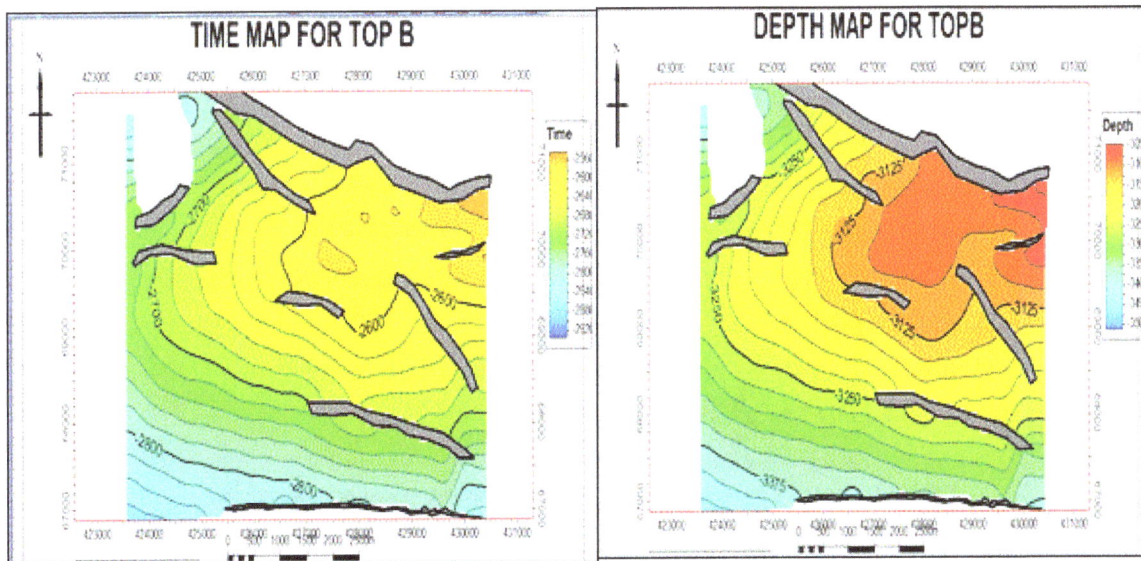

Figure 7: Structural time map and depth map for reservoir.

part of the field. The closure within this field is a fault assisted closure which serves as a seal that prevented further migration of hydrocarbon.

Static model

From the result, it was observed that the top, mid, and base skeleton (Figure 9) serve as the structural framework for reservoir modelling. The skeleton frame work of the reservoir has total grid cells of 323,850. This implies that the geological heterogeneities will be captured with grid resolution for the construction of a fit for purpose geological model. From the petrophysical parameters evaluated for both reservoirs, it was observed that much hydrocarbon can be economically

exploited from both reservoirs due to their high porosity and low water saturation. Based on this, 3D geologic model was constructed for reservoir E so as to know how these properties are being distributed within the subsurface. The total porosity and effective porosity which is a continuous properties of the reservoir was distributed properly within the geologic cells of the reservoir E ad modeled. The total porosity gives the ratio of pore volume to the total volume of the reservoir. The total porosity model shows that reservoir E has a minimum porosity of 0.16 and a maximum porosity of 0.30 (Figure 10). The model shows that the reservoir has an average total porosity of 24%. (Figure10). These indicate that the reservoir is very porous and well completed.

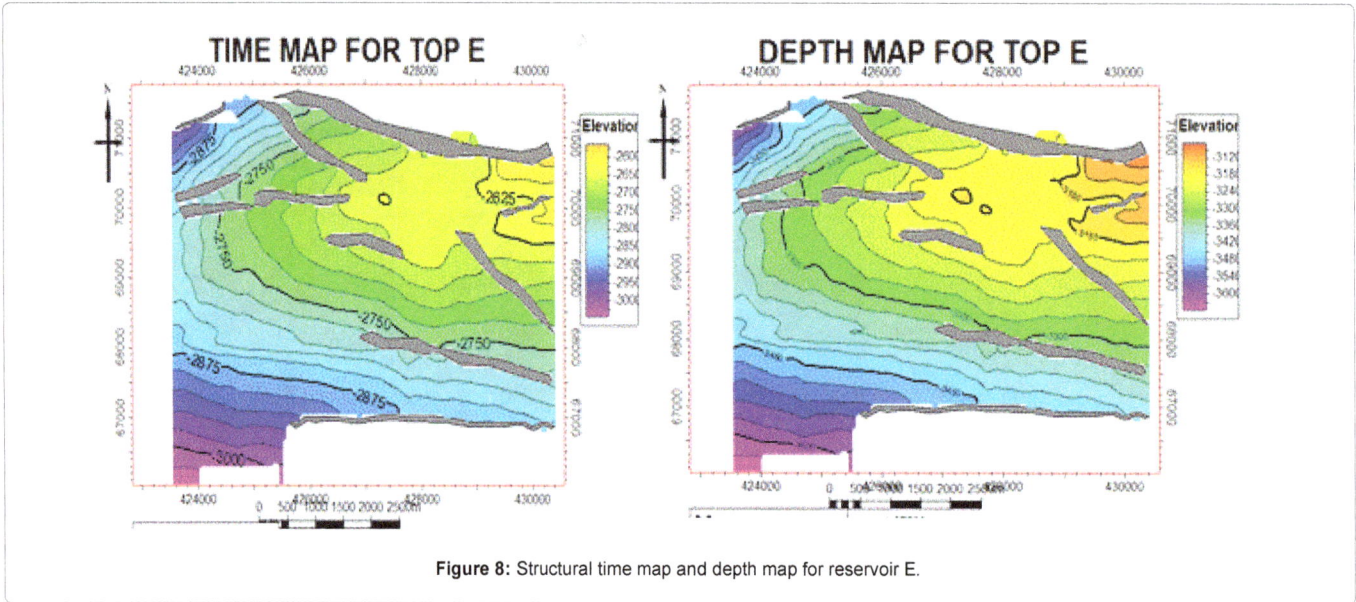

Figure 8: Structural time map and depth map for reservoir E.

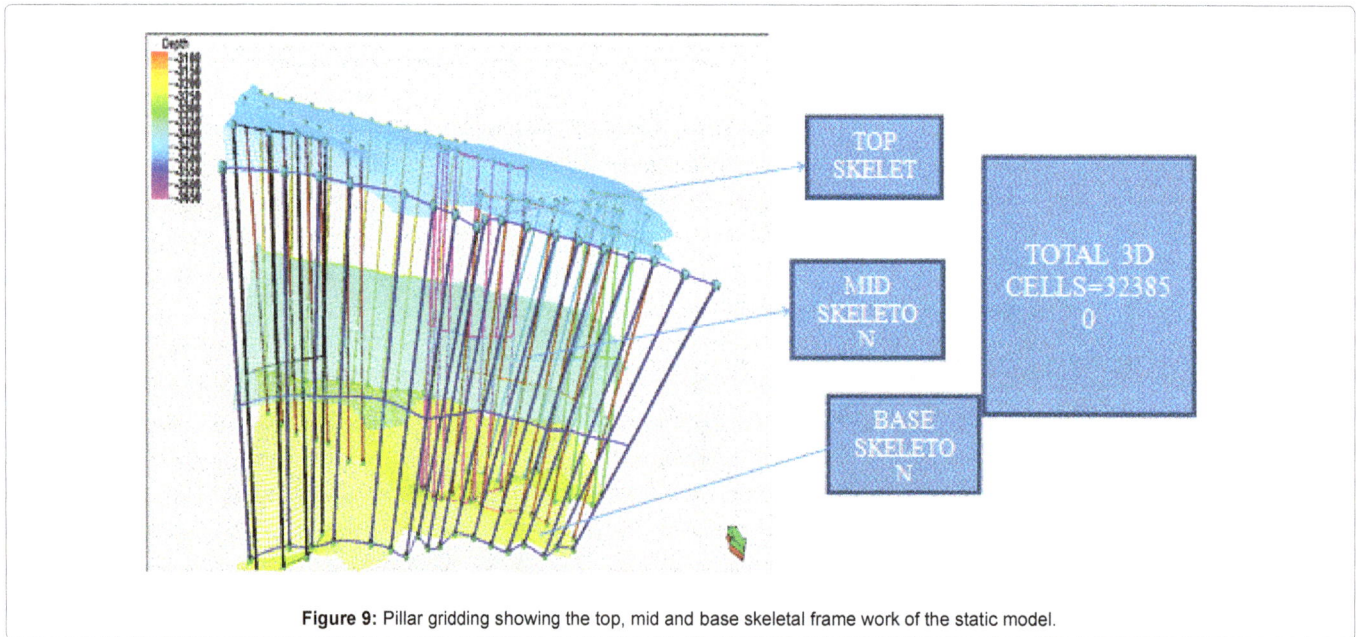

Figure 9: Pillar gridding showing the top, mid and base skeletal frame work of the static model.

From the colour legend, areas with very low porosity are indicated by the purple colour while areas with high porosity are indicated with orange colour. The effective porosity gives the volume of the interconnected pore spaces within the reservoir i.e. it gives the volume of pore spaces that is contributing to the production of fluid within the reservoir. The effective porosity model shows that reservoir E has a minimum effective porosity of 0.07 and a maximum effective porosity of 0.28. The average effective porosity within this reservoir is 0.22. These indicate that the pore spaces within the reservoir are well connected. The permeability model (Figures 11a and b) indicate that the rate of flow within the reservoir is high.

From the transformation plot (Figure 12a); three different water saturation cases were identified. These include the base case water saturation, low case water saturation, and high case water saturation.

The low case water saturation model gives an average water saturation of 0.23, while the high case water saturation model gives an average water saturation of 0.12. This implies that a lot of hydrocarbon can be exploited at the high case than the low case water saturation. The water saturation model for the base case shows that reservoir has a minimum water saturation of 0.06 and a maximum water saturation of 0.38 (Figure 12b). The average water saturation within this reservoir is 0.23. These indicate that the reservoir is 0.77 saturated with hydrocarbon. Areas with blue colour indicate part of the reservoir that is completely saturated with water.

Volumetric Analysis

$STOIIP = GRV \times POROSITY \times NTG \times (1-S_w) / B_o$

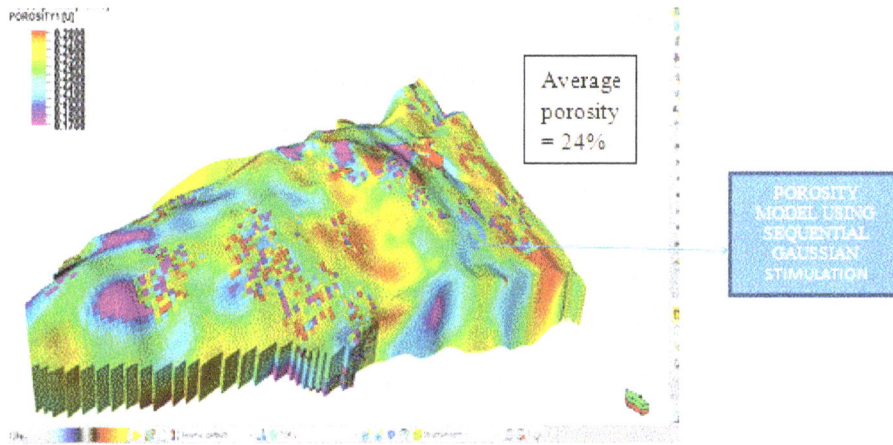

Figure 10: Porosity model for reservoir E.

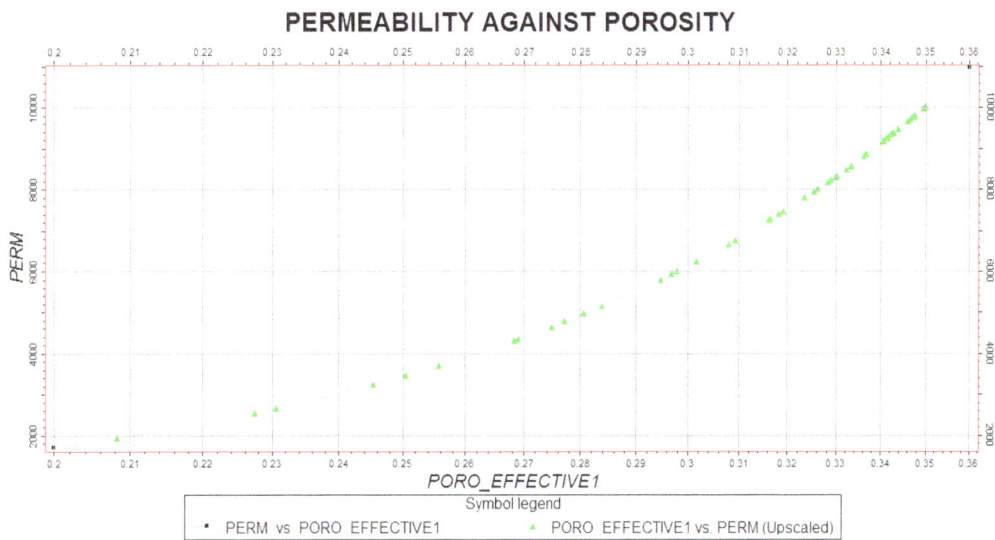

Figure 11a: Cross plot of porosity against permeability.

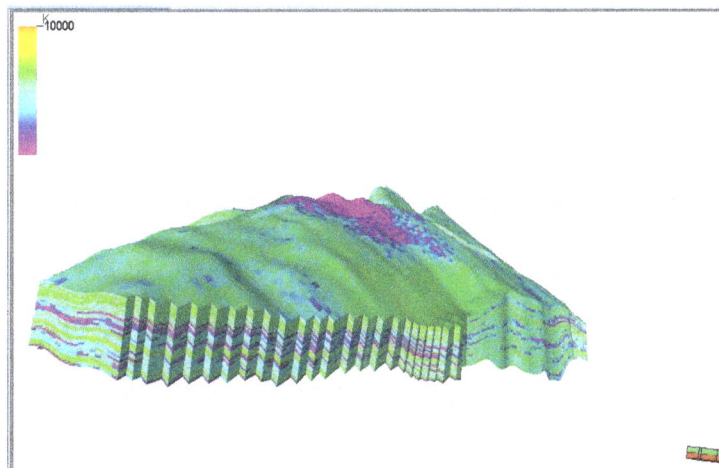

Figure 11b: Permeability model for reservoir E.

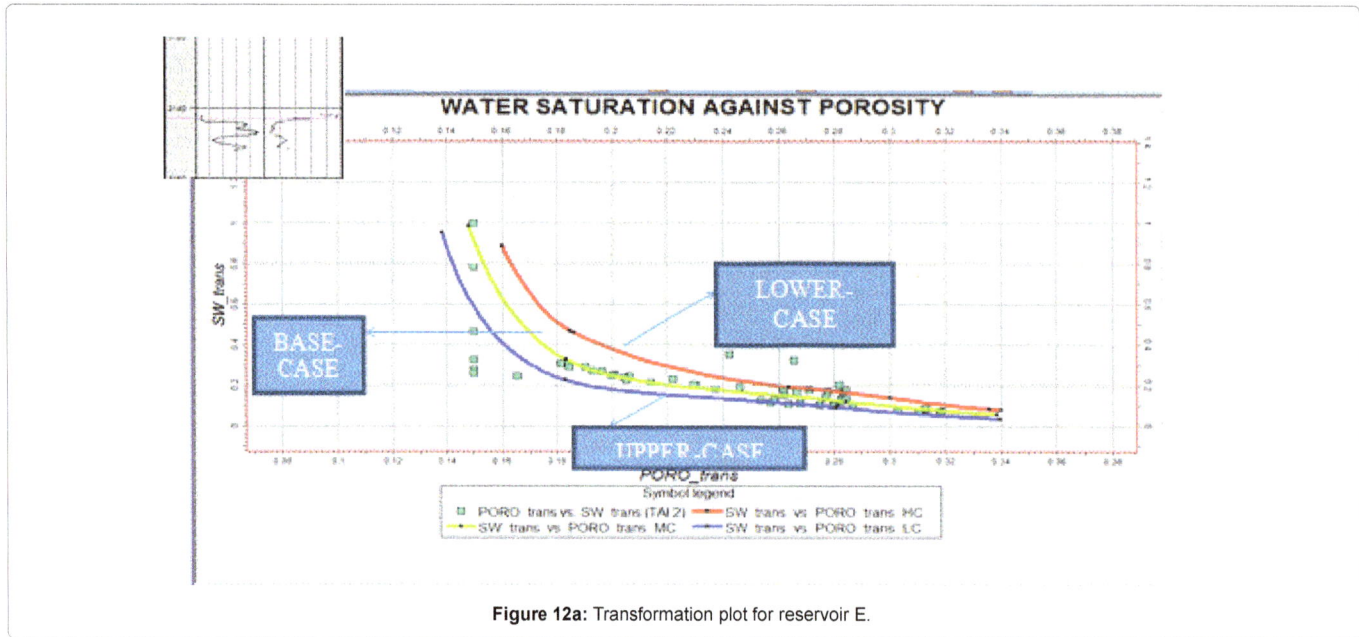

Figure 12a: Transformation plot for reservoir E.

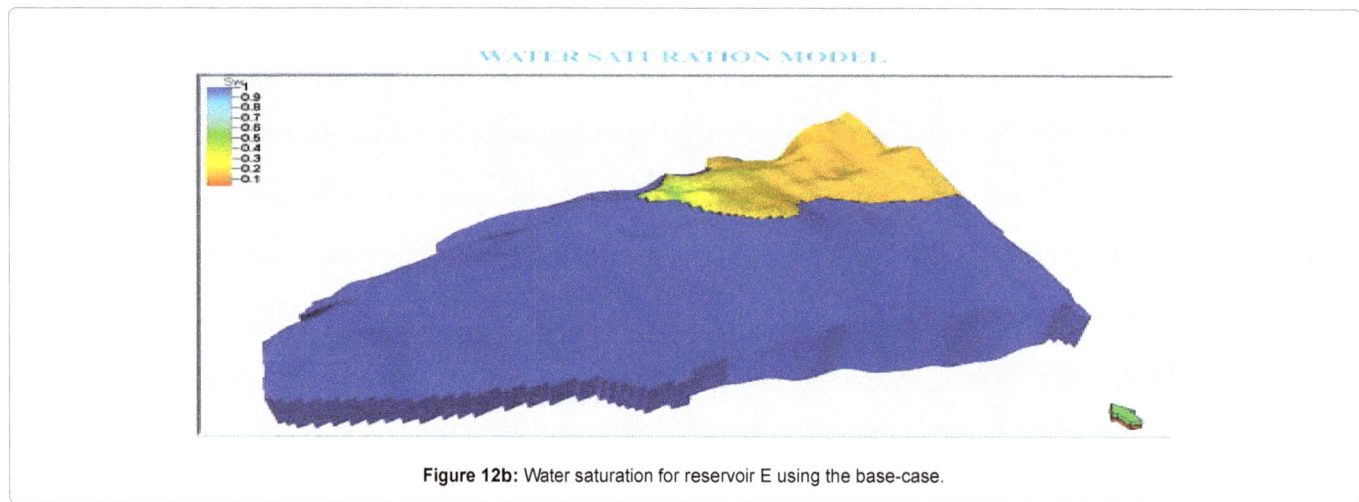

Figure 12b: Water saturation for reservoir E using the base-case.

Conclusion

This research work shows the versatility of integrating 3D seismic reflection, and well logs data for interpretation, petro physical analysis and construction of a 3D geologic model. The accumulation and trapping of hydrocarbon in this field is fault assisted. Reservoirs B &E are very promising because of its very good porosity values, low water saturation, high hydrocarbon saturation (SH), good permeability and moderate net to gross. The static model of reservoir E indicate that the reservoir has an average total porosity of 24% (Table 1), effective porosity of 22%, water saturation ranging from 12%-23% (Table 1) and permeability of 4476 mD. The volumetric calculation indicates that Reservoir E has a STOIIP ranging from 37.53-43.03 × 10^6 STB (Table 1) respectively.

Recommendations

Further interpretation such as Seismic attribute analysis, AVO analysis, Seismic inversion, Rock physics, etc. should be carried out on the field so as to affirm the existence of hydrocarbon and the tapping mechanism in this field. Uncertainty analysis such as Monte Carlo stimulation, material balance analysis, etc. should also be used to evaluate and affirm the STOIIP value. This will help to affirm if the field is economically fit for exploitation. More wells should be drilled along the west and east direction of the field. This will help in the proper optimization of the field.

Acknowledgement

Our appreciation goes to Integrated Data Services Limited, Benin-City, Nigeria for releasing the data used to carry out this research work.

GRV (FT)	POROSITY	NTG	SW	HS	STOIIP (MMBL)	
950356	0.24	0.67	0.17	0.83	40.84	BASE-CASE
950356	0.24	0.67	0.23	0.77	37.53	LOW-CASE
950356	0.24	0.67	0.12	0.88	43.03	HIGH-CASE

Table 1: Summary of volumetric analysis for reservoir E from 3-D model.

References

1. Haldorsen HH, Damsleth E (1990) Stochastic modeling. JPT 42: 404-412.

2. Alabert FG, Corre B (1991) Heterogeneity in a complex turbiditic reservoir: Impact on field and Antarctica. Berlin, Gebr Aijder Borntraeger, pp: 143-172.

3. MacDonald AC, Hayye TH, Lowry P, Jacobsen T, Aasen JO, et al. (1992) Stochastic flow unit modelling of a North Sea coastal-deltaic reservoir. First Break 10: 124-133.

4. Massonnat G, Alabert F, Guidicelli C (1993) Anguille Marine, a deep sea-fan reservoir offshore Gabon: from geology to stochastic modelling. SEG Technical Program Expanded Abstracts 1993: 345-345.

5. Shanor GG, Bahman S, Bagherpour H, Karakas M, Buck S, et al. (1993) An integrated reservoir characterization study of a giant middle east oil field: Part 1-geological modelling: Proceedings of 8th Middle East Oil Show and Conference 2: 491-504.

6. Patrick DD, Gerilyn SS, John PC (2002) Outcrop-base reservoir characterization: A composite phyllo- id-algal mound, Western Orogrande Basin (New Mexico). AAPG Bulletin 86: 780.

7. Klett TR, Ahlbrandt TS, Schmoker JW, Dolton JL (1997) Ranking of the world's oil and gas provinces by known petroleum volumes: U.S. geological survey open-file report.

8. Doust H, Omatsola E (1990) Niger Delta, Divergent/passive margin basins. American Association of Petroleum Geologists 239-248.

9. Kulke H (1995) Regional Petroleum Geology of the World. Part II: Africa, America, Australia and Antarctica. Gebrüder Borntraeger, Berlin pp: 143-172.

10. Hospers J (1965) Gravity field and structure of the Niger Delta, Nigeria, West Africa. Geol Society Am Bull 76; 407-422.

11. Ekweozor CM, Daukoru EM (1994) Northern delta depobelt portion of the Akata-Agbada petroleum system, Niger Delta, Nigeria. Am Assoc Petroleum Geol pp: 599-614.

12. Stacher P (1995) Present understanding of the Niger Delta hydrocarbon habitat. Geology of Deltas: Rotterdam, Balkema AA, pp: 257-267.

13. Evamy BD, Haremboure J, Kamerling P, Knaap WA, Molly FF et al. (1978) Hydrocarbon habitat of the tertiary Niger Delta. Am Assoc Pet Geol Bull 62; 125-142.

14. Michele LW, Tuttle, Charpentier RR, Brownfield ME (1999) The Niger Delta petroleum system: Niger Delta province, Nigeria, Cameroon and Equatorial Guinea, Africa. Denver, Colorado. US Department of the Interior and Geological Survey, Colorado.

15. Ejedawe JE (1981) Patterns of incidence of oil reserves in Niger Delta basin. Am Assoc Pet Geol 65; 1574-1585.

16. Stieber SJ (1970) Pulsed neutron capture log evaluation-Louisiana Gulf Coast: Society of petroleum engineers annual fall meeting proceedings.

17. Wyllie MRJ, Gregory AR, Gardner LW (1956) Elastic wave velocities in heterogeneous and porous media. Geophys 21: 41-70.

18. Tixier MP (1949) Evaluation of permeability from electric-log resistivity gradients. Oil and Gas J.

19. Archie GE (1942) The electrical resistivity log as an aid in determining some reservoir characteristics. Ins Pet Tech.

20. Esfahani NM, Asghari O (2013) Fault detection in 3D by sequential gaussian simulation of rock quality designation (RQD) case study: Gazestan phosphate ore deposit, Central Iran. Arab J Geosci 6: 3737-3747.

13

Geotechnical Considerations in Shoreline Protection and Land Reclamation in Kula, Eastern Niger Delta

Nwankwoala HO[1]* and Orji MO[2]

[1]Department of Geology, University of Port Harcourt, Port Harcourt, Nigeria
[2]Department of Petroleum Engineering and Geosciences, Petroleum Training Institute, Effurun, Warri, Delta State, Nigeria

Abstract

The investigation is to determine the suitability of the study site for the design and construction of a shoreline protection and also carry out reclamation exercise at the adjoining lands. Nine (9) number boreholes were drilled to a maximum depth of 20.0 m below the existing ground level using a cable percussion rig and nine (9) numbers Cone Penetrometer Testing using 2.5 tonne CPT equipment. The lithology reveals intercalations of clay and sand in thin layers to a depth of 2.0 m below the existing ground level. Only borehole 3 revealed the clay layer to a depth of 5.0 m. Underlying this clay is a stratum of loose to medium dense sand and dense sand. The sand is well sorted grading from fine to medium as the borehole advances. The laboratory analysis showed that the silty clay has undrained shear strength of 48 kPa. The loose sand has a maximum SPT (N) value of 12 while the medium dense sand has maximum SPT (N) value of 28. Considering the nature of the intended structure, the anticipated load and the moderate compressibility of this near surface silty clay and the underlying loose silty sand, it is suggested that the cellar slab be supported by means of raft foundation founded within the clay layer. Where the proposed project precludes the use of raft foundation, pile foundation should be employed to transmit the anticipated load from the cellar slab to the underlying sand stratum and that such piles should be closed-ended, straight-shaft steel pipe piles driven into the sand stratum and all driven piles should undergo pile load test to confirm their working load and consequent estimated settlement.

Keywords: Engineering geology; Foundation; Subsoil; Boreholes; Stratigraphy; Niger delta

Introduction

Increased population in various cities of the Niger Delta Region of Nigeria and the consequent demand for increased residential space have necessitated the need for reclamation of coastal marginal lands which comprise mainly swampy soils [1] and the protection of the shoreline. A marginal land is a one which is unsuitable for development in its original condition [2]. The low and flat nature and the dense criss-cross network of rivers in the area render extensive portions of the land mass seasonally flooded. Some studies have been carried out on geotechnical properties of the subsoils generally [3-6].

The study area is situated in Kula Community (Figure 1). Kula Community is located in Akuku-Toru Local Government Area of Rivers State in the Niger Delta Area of Nigeria. The local geology of the location is composed of sediments which are characteristic of several depositional environments. Deposits are geologically young, ranging from the Eocene to the recent Pliocene. They include river mouth bar, delta front platform, delta slope and open shelf sediments. The river mouth bar sediments generally consist of coarse grained sands which extend out in shallow water depths before merging with the sands and clays of the sub-horizontal delta front platform. The increased pressure on land in Kula Town, Eastern Niger Delta has led to the use of marginal lands for development and more seriously, the fact that most of the rivers are actively eroding, scouring and cutting their banks, thus exposing more landmass to excessive flooding calls for proper definition of engineering solutions for construction purposes. Against this background, this study provides a detailed assessment of the suitability of the soils of the area, the sub-soil conditions and suggests relevant soil improvements where necessary as well as recommend appropriate foundation type and design parameters (Figures 2 and 3).

Study Techniques

Sampling/borehole drilling

The investigation comprised mainly exploring nine (9) geotechnical boreholes with soil sampling and measurement of water table and the execution of nine (9) cone penetration testing. The boreholes were drilled by the shell and auger cable percussive drilling method, using a hand rig. The hand rig is fitted with a free fall auger. The auger is lifted to a height of about 1.0 m above ground level, using gloved hands, and allowed to free-fall under gravity to advance the boring. As the auger falls it cuts through the soil such that the cut soil material is retained inside it by means of a clerk (Figures 4 and 5). The auger is then brought to the surface where the soil retained in it is emptied out. To prevent collapse of the borehole wall, the hole is lined with casings or shell corresponding to the size of the auger being used for the drilling. As the drilling continues, the auger drops into the open hole until the time sample is to be taken (Tables 1 and 2).

Representative undisturbed and disturbed samples were taken at regular intervals of 1.0 m depth, and also when a change in soil type was observed. The samples were used for a detailed and systematic description of the soil in each stratum in terms of its visual and haptic properties and for laboratory analysis. The borehole log obtained is presented in Figure 5. In the cohesive soils, six undisturbed samples were taken for examination and laboratory analysis. The laboratory test results are shown in Tables 3-6, respectively.

Standard Penetration Tests (SPT) was carried out at regular intervals of depth in the granular sediments in order to assess the *in situ* densities. In this test, the number of blows required to drive the

*Corresponding author: Nwankwoala HO, Department of Geology, University of Port Harcourt, Port Harcourt, Nigeria
E-mail: nwankwoala_ho@yahoo.com

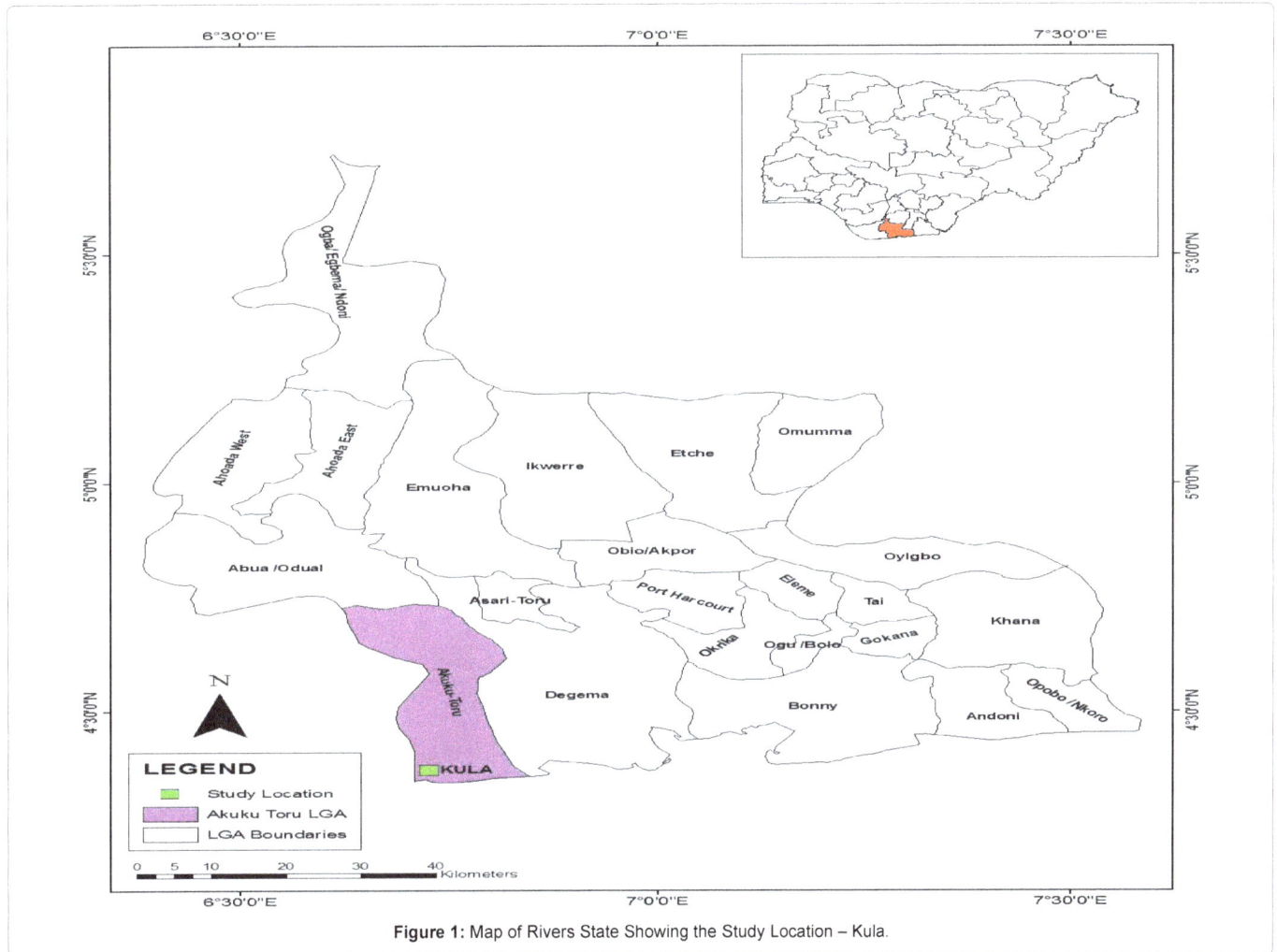

Figure 1: Map of Rivers State Showing the Study Location – Kula.

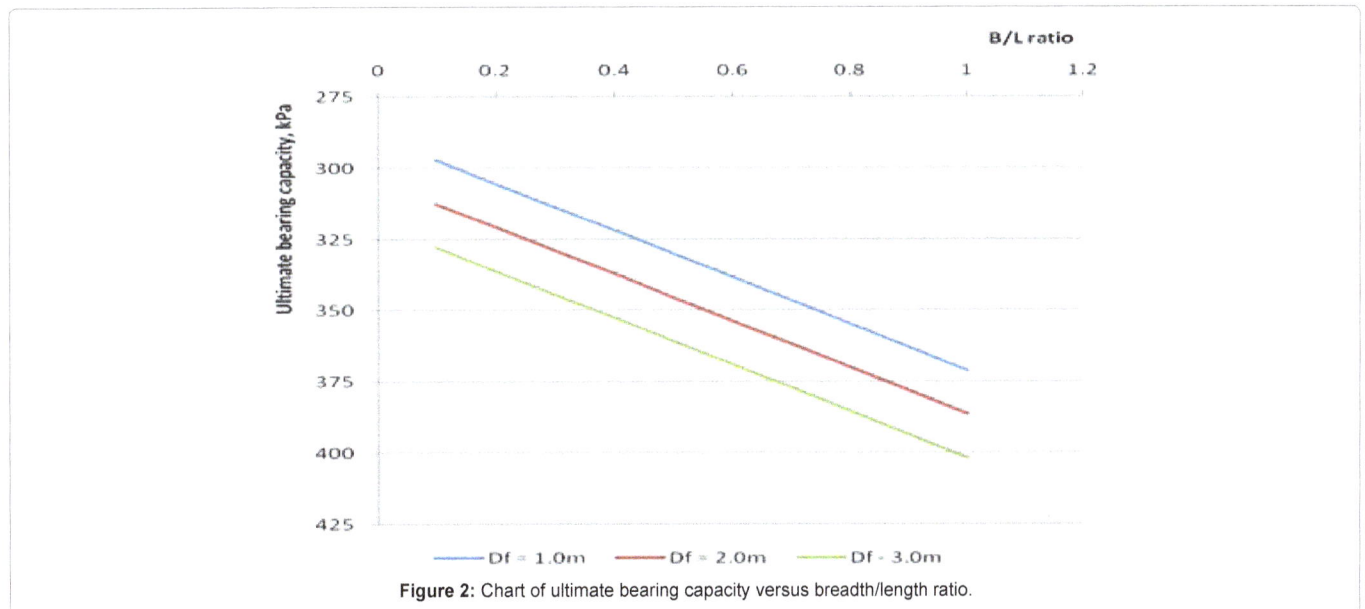

Figure 2: Chart of ultimate bearing capacity versus breadth/length ratio.

standard sampling spoon 300 m penetration after the initial sitting drive was recorded as the SPT (N) value. Six (6) numbers Cone Penetration tests were carried out to refusal. Readings of cone tip resistance and sleeve friction were taken at every 0.2 m interval of depth. A CPT rig and cones were used for the tests. Field measurements ground water showed that the ground water levels stood at between 2.50 and 3.0 m

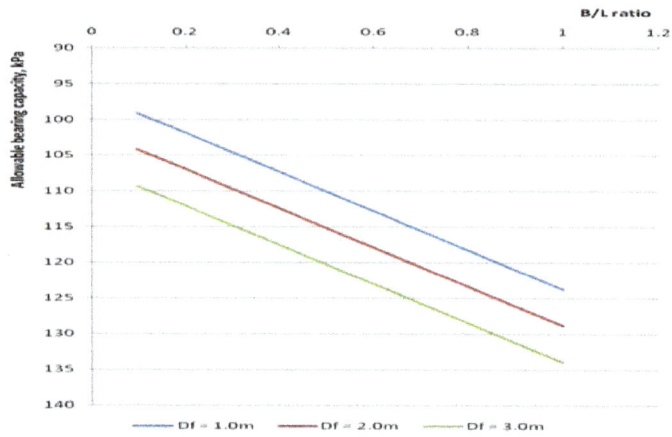

Figure 3: Chart of allowable bearing capacity versus breadth/length ratio.

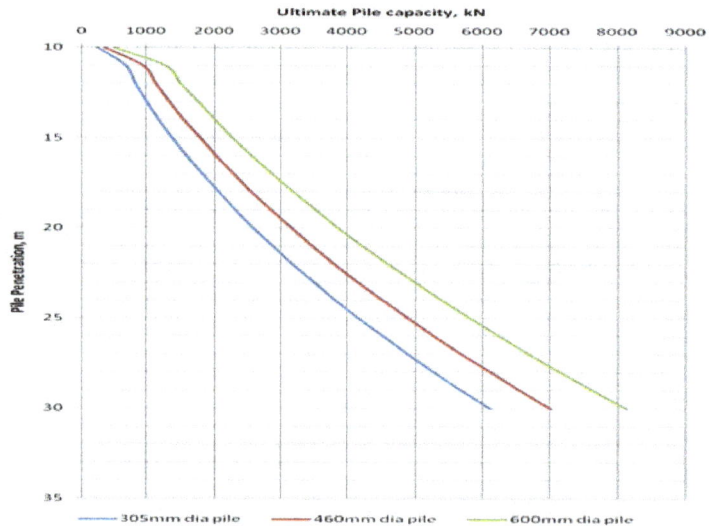

Figure 4: Chart of Ultimate Pile Capacity.

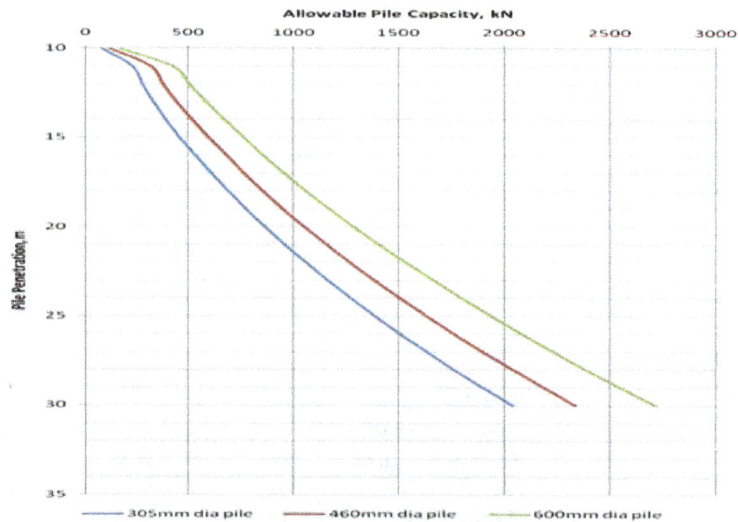

Figure 5: Chart of allowable pile capacity.

Stratum No.	Description	Average depth range (m)
1	Clay, and sand intercalations, silty, medium mottled brown and grey	0-2
2	Fine sand, silty, loose to dense, grey In borehole 3 (clay to 5.0 m)	2-30

Table 1: Soil Stratigraphy to the depth of the borings.

	Min	Max	Mean
Natural Moisture Content (%)	18	39	26
Liquid Limit (%)	35	51	43
Plastic Limit (%)	17	33	25
Plasticity Index (%)	13	22	17
Liquidity Index	0.06	0.71	0.32
Consistency Index	0.94	0.29	0.68
Bulk Unit Weight(kN/m³)	18.76	18.76	18.76
Dry Unit Weight(kN/m³)	15.34	15.34	15.34
Final Void Ratio	0.67	0.67	0.67
Final Porosity (%)	47.48	47.48	47.48
Undrained Strength (kPa)	48	48	48
Coefficient of consolidation, m²/yr	3.66	5.12	4.22
Coeff. of Compressibility, mv,m²/MN	0.33	0.44	0.38

Table 2: Range of variations in the Index and Engineering Properties of near surface soils to the Depth of 5.0 m.

	Min	Max	Ave
Effective Particle Size, d_{10} (mm)	0.075	0.6	0.15
Mean Particle Size, d_{30} (mm)	0.16	0.45	0.32
Particle Size, d_{60} (mm)	0.16	0.45	0.32
Coefficient of Uniformity, $Cu = d_{60}/d_{10}$	2.82	9.83	4.34
Coefficient of Curvature, $Cc = d_{30}^2/d_{10}.d_{60}$	0.35	1.39	1.02

Table 3: Range of Variations of the Geotechnical Parameters to a Depth of 25.0 m.

Foundation depth, D_f (m)	Allowable, Q_{all} (kPa) and Safe Bearing Capacity, Q_s (kPa), for Various Width, B(m) of Raft Foundation					
	2 m		5 m		10 m	
	Q_{all}	Q_s	Q_{all}	Q_s	Q_{all}	Q_s
1	98	60	102	52	110	35
2	104	95	107	75	115	55
3	109	105	113	110	120	75

Table 4: Allowable and safe bearing capacities for various foundation widths at different foundation depths.

BH No	Depth, m	Angle of internal friction, φ	Pile-wall friction, δ
1	5	34	26
	17	40	30
2	8	38	29
	27	32	24
3	8	41	31
	28	35	26

Table 5: Design parameters for cohesionless soil, using the angle of internal friction as obtained from the laboratory shear box test.

below the existing ground surface in all boreholes explored at the time of the field work.

Laboratory tests

Detailed laboratory investigations were carried out on representative undisturbed and disturbed samples obtained from the boreholes for the classification tests and other tests. All tests were

carried out in accordance with BS 1377 (1990). Atterberg consistency limit tests were carried out on the cohesive samples. The results show that the samples are low to medium plasticity silty clay. The particle size distributions of a number of representative samples of the cohesionless soils were determined by sieve analysis. The results disclosed that the samples are predominantly, fine, fine to medium and medium sands.

Unconsolidated Undrained triaxial (UU) tests were performed on relatively undisturbed samples obtained from the boreholes. Test results show the average unconsolidated undrained shear strength parameters for the clays encountered in the study area. Laboratory consolidated tests were carried out on relatively undisturbed samples with the objectives of determining the compressibility properties of the soils. The plot of void ratio (e) against effective pressure (P) for the samples tested and calculated values of the coefficients of consolidation (Cv) and of the coefficients of compressibility (Mv).

Results and Discussion

Test results showed that the samples were of moderately high compressibility and predominantly exhibiting negligible swelling potentials (Table 2).

Soil stratigraphy

The soil stratigraphy encountered on the study site as obtained from the explored boreholes are as presented in Table 1. The lithology revealed intercalations of clay and sand in thin layers to a depth of 2.0 m below the existing ground level as presented in boreholes 1 and 2. Below this depth, the formation presents a stratum of medium dense fine sand that increases in gradation and density with depth to becoming dense fine to medium sand at about 7.0 m below ground level. Borehole 3, however, revealed a 5.0 m thick near surface clay layer overlying the medium dense fine sand. The sand increases in density to become very dense at about 10 m. Below this depth, at about 25.0 m below the existing ground level medium dense sand is encountered again. This medium dense layer is observed to the final depth of the boring.

Engineering soil properties

The near surface soil encountered during the investigation is firm clay extending from the ground level to a depth of 2.0 below the ground surface and extending to 5.0 m in Borehole 3. This firm clay is characterized by moderate compressibility, low moisture content and low undrained strength. The range of variations in the index and engineering parameters of this near surface soil are shown in Table 2.

Loose to medium dense sand was encountered immediately beneath the near surface silty clay soil. This loose to medium dense sand increases in density to becoming dense to very dense sand from about 7.0 m below the existing ground level. Deeper down in the boring from about 25.0 m, it is observed that the sand loosens to becoming medium dense. This medium dense sand continues to the final depth

Pile depth (m)	Allowable Pile Capacity, kPa, for Various Pile diameter		
	305mm (12") Pile	460 mm (18") Pile	600 mm (24") Pile
10	93	140	198
15	553	713	914
20	1036	1263	1547
25	1668	1961	2329
30	2448	2808	3259

Table 6: Allowable bearing capacity and specific pile diameter for specific depths.

of the investigation. The ranges of variations of the geotechnical parameters are shown in Table 3.

Soil foundation design parameters

The study has revealed the relevant soil parameters for the design of the foundation of the cellar slab. The near surface soil is 2.0 m thick intercalations of clay and sand in Boreholes 1 and 2 and a 5.0 m thick clay layer in Borehole 3. Underlying this near surface clay is a formation of loose sand becoming medium dense and dense sand with depth.

The lithology revealed a stratum of graded bed. The upper sand stratum being loose and fine sand immediately beneath the clay layer and grading to become medium dense sand with depth. This gradation continues as the borehole advances and deeper down some loosening of the sand is observed. This loosening is observed to the final depth of the borehole. From consideration of the nature of the intended structure, the anticipated load, the moderate compressibility of the near surface clay and the underlying loose silty sand, it is suggested that the cellar slab be supported on raft foundations founded in the upper clay. However, where the requirements preclude the use of raft foundation, pile foundation should be employed to transmit the load to the underlying soil stratum. Using a safety factor of 3 on the ultimate bearing capacity, the chart for the allowable bearing capacity is as presented in Figure 2 while also using a safety factor of 3 on the ultimate pile capacity, the chart for the allowable pile capacity is as presented in Figure 5.

Bearing capacity calculations

The bearing capacity for the foundation was determined using the Terzaghi [7] bearing capacity formulae as stated below:

$$qd = 5.7c\left\{1 + 0.3x\frac{B}{L}\right\} + \gamma D_f \text{ for } \Phi = 0 \tag{1}$$

where qd = Ultimate bearing capacity

c = Undrained cohesion of the soil

B = Width of footing

L = Length of footing

γ = Unit weight of soil

D_f = Depth of footing

Φ = Angle of friction taking as zero for undrained condition of the soil.

Settlement calculations

The settlement for the footing is determined using the Terzaghi [7], Skempton and MacDonald [8] formulae and the Bousinesq's chart.

$$S_t = S_i + S_c \tag{2}$$

where S_t = total settlement

S_i = immediate settlement for

S_c = consolidation settlement at depth of footing

Calculation of immediate settlement, S_i

$$S_t = qB(1-\mu^2) I_p/E \tag{3}$$

Q = imposed load

B = footing width, m

M = Poisson's ratio for undrained shear strength

I_p = Influence factor for a rectangular footing

E = Stiffness modulus for the firm sandy clay

Calculation of consolidation settlement, S_c

$$S_c = 0.7 \times S_{oed} \tag{4}$$

$$S_{oed} = m_v \times \sigma \times H$$

m_v = coefficient of volume compressibility

σ = Applied pressure at point under consideration = qxI_f

H = Thickness of strata under consideration = 2B

I_f = Influence factor from Bousinesq's chart

0.7 = geological coefficient that relates oedometer results to actual field estimates

Calculation of total settlement, S_t

$$S_t = S_i + S_c \tag{5}$$

Safe bearing capacity, Qs

Table 4 shows the allowable, Q_{all} (kPa) and safe Q_s (kPa) bearing capacities for various foundation width, B(m) of raft footing at different foundation depth, D_p, m. The safe bearing capacity for the raft foundation is limited by a maximum settlement value of 50 mm.

Bearing capacity calculations – pile foundation

The ultimate bearing capacity, Q_u, of driven piles is determined by the equation below:

$$Q_u = Q_p + Q_f \tag{6}$$

where Q_p = q x A_p = total end bearing, kN

Q_f = f x A_s = skin friction resistance, kN

And, q = unit end bearing capacity = kPa

f = unit skin friction = kPa

A_p = gross end area of pile, m^2

A_s = side surface area of pile, m^2

End bearing and skin friction in cohesive soils

For piles in cohesive soils,

The unit skin friction, f = $\alpha.S_u$ (7)

The unit end bearing, q = 9. S_u (8)

Where $\alpha = 0.5\psi^{-0.50}$ for $\psi \leq 1.0$

$\alpha = 0.5\psi^{-0.25}$ for $\psi > 1.0$

and $\alpha = S_u/P_o$

S_u = undrained shear strength of the soil at the point, kPa

P_o = effective overburden pressure of the soil at the point, kPa

End bearing and skin friction in cohesionless soils

For piles in cohesionless soils,

The unit skin friction, f = K P_o tanδ (9)

The unit end bearing, q = $P_o N_q$ (10)

Where

K = coefficient of lateral earth pressure

δ = friction angle between the soil and pile wall

N_q = bearing capacity factor

Conclusion

This study has revealed a near surface stratigraphy of silty clay to a depth of 5 m underlain by loose silty sand to a depth of 9.0 m below the existing ground level. Underlying this layer of loose sand is a 1.0 m thick layer of plastic clay. Considering the nature of the intended structure, the anticipated load and the moderate compressibility of the near surface silty clay and the underlying loose silty sand, it is suggested that the cellar slab be supported by means of raft foundation founded within the upper clay layer where it is uneconomical to take it deeper. The plastic clay beneath the cellar slab, however, will undergo consolidation along with the compression and creep that will result from loading the loose sand beneath it. Adequate consideration should be taken of this settlement during the design and construction of the cellar slab.

References

1. Abam TKS, Okogbue CO (1993) Utilization of marginal lands for construction in the Niger Delta. Bulletin of the International Association of Engineering Geology 48: 5-14.

2. British Standard Methods of Test for soils for Civil Engineering Purposes. B.S 1377: Part 2 (1990). Published by the British Standards Institution pp: 8-200.

3. Nwankwoala HO, Amadi AN (2013) Geotechnical Investigation of Sub-soil and Rock Characteristics in parts of Shiroro-Muya-Chanchaga Area of Niger State, Nigeria. International Journal of Earth Sciences and Engineering 6: 8-17.

4. Youdeowei PO, Nwankwoala HO (2013) Suitability of soils as bearing media at a freshwater swamp terrain in the Niger Delta. Journal of Geology and Mining Research 5: 58-64.

5. Oke SA, Amadi AN (2008) An Assessment of the Geotechnical Properties of the Sub-soil of parts of Federal University of Technology, Minna, Gidan Kwano Campus, for Foundation Design and Construction. Journal of Science, Education and Technology pp: 87-102.

6. Oke SA, Okeke OE, Amadi AN, Onoduku US (2009) Geotechnical Properties of the Sub-soil for Designing Shallow Foundation in some selected parts of Chanchaga Area, Minna, Nigeria.

7. Terzaghi K (1943) Theoretical Soil Mechanics. Wiley & Sons, New York.

8. Skempton AW, MacDonald DH (1956) The Allowable Settlement of Buildings, Proc. Inst. of Civil Engineers Part pp: 3727-784.

14

Exploration of Lead-Zinc (Pb-Zn) Mineralization Using Very Low Frequency Electromanetic (VLF-EM) in Ishiagu, Ebonyi State

Mbah Victor O*, Onwuemesi AG, Aniwetalu and Emmanuel U

Geophysical Science Department, Nnamdi Azikiwe University, Awka, Nigeria

Abstract

Very low frequency Electromagnetic (VLF-EM) exploration over the Ishiagu area of Abakaliki Basin, Lower Benue Trough, Nigeria was carried out to determine the Pb-Zn mineralization in the sedimentary bedrock. The conductivity contrast between the conductive mineralized veins and the host rock as generated by induction mechanism was used to delineate the potential zones of Pb-Zn mineralization. Results of high in-phase and quadrature readings due to strong EM induction were detected in the survey area and on the average, the deeper sources response range from 4.7% to about 7.6%, while the shallower sources response range from 8.8% to about 17.1% and these probably indicate the presence of Pb-Zn deposits with thick overburden in the northern part. Current density maps show the Pb-Zn mineralized veins trends in NW-SE direction with their subordinates in N-S direction. The central part of the study area displays very sharp VLF tipper responses indicating shallow sources, while the northern part displays broad VLF tipper responses indicating deeper sources. These readings correlate to the depth values over the Pb-Zn mineralized veins which range from 10 to 13 m in the central part and 16 to 22 m in the northern part of the mapped area. The high VLF anomalies delineated from Ishiagu area, in the SW part of Abakaliki basin as a rule yielded a high conductivity contrast result well-suited with the Pb-Zn mineralization and geologic information of the area.

Keywords: VLF-EM; Pb-Zn mineralization; Tipper response; In-phase and quadrature anomalies; Current density maps; Ellipse polarization; Wave superposition; KHF filtering techniques

Introduction

The Pb-Zn mineralization in Ishiagu has instigated repeated studies of the area using different geological and geophysical techniques, though less has been done with electromagnetic techniques. This paper presents the remarkable results obtained with VLF-EM method to detect the conductive ore-bearing mineralized veins within the ultrabasic rocks in Ishiagu, Ebonyi State, Nigeria. The well-established very low frequency electromagnetic (VLF-EM) method is a rapid, wide coverage and cost effective technique for locating both hidden ores and the structures associated with the mineralization, in use for over 30 years [1-4]. The proficiency of VLF method for high-grading mineralized area in preparation for competent mine development is of significant contribution to an integrated geophysical investigation effort. The readings of VLF technique, just like every other EM geophysical prospecting is based on variations in subsurface electrical conductivity which is the inverse of resistivity. This method therefore, provides a quick and powerful tool for the study of 2-D geological structures to a maximum skin depth of about 100 m, though variation in the skin depth is based on changes in subsurface conductivity.

Careful study reveals the Pb-Zn mineralization in Ishiagu to be structurally controlled such that both ore and gangue minerals occur in successive and symmetrical layers along vertically and/ or steeply dipping fractures which often have parallel or matched walls, thus indicating its fissure filling mode of occurrence which is so pinpointing to VLF-EM prospecting [1,5]. The ground VLF-EM method is a quick and powerful geophysical technique for the study of shallow 2-D geological structures most especially in respect of mineral exploration. In addition, it has been used to high echelon of success to map weathered basement layer and detection of water-filled fractures or shallow faults. This method is very useful and pertinent because of how quickly data can be collected, and large survey area can also be covered quickly with the portable instruments. The surveyed area was selected in view of the uncertainty in the extent of concealed Pb-Zn lodes, overburden thickness and the geological context of the Ishiagu areas.

Location of the Study Area

Ishiagu Field is located in Ivo LGA of Ebonyi State between latitude 5°54' – 5°59' N and longitudes 7°30' – 7°35' E. The area (about 25 sq.km), is situated in the SW tip of the Abakaliki Basin on the Lower-Benue Trough geologic complex, SE Nigeria and is composed of a low-lying sedimentary terrain with some intrusions of different episodes. The Ishiagu area of the Abakaliki Basin is delineated by geology of the Abakaliki Basin as shown in Figure 1. Evolution of this generally low-lying to gently undulating shaly terrain is correlated to basement fragmentation, block faulting, subsidence and rifting of the Lower Benue Trough during the early Cretaceous separation of Africa and America [6]. The Pb-Zn deposits in Ishiagu area appear to be the southern limit of mineralization in the Benue Trough and the Pb-Zn mineralized zone extends over a distance of 500 km in a narrow belt from Ishiagu in the lower Benue Trough to Zurak in the upper Benue Trough, likewise the extent of igneous intrusions in the Benue Trough. Majority of the geologic and topographic features of the area align in the NW-SE direction, and conform to orientation of the folds from the Santonian orogenic deformation.

Geologic model of Pb-Zn mineralization in ishiagu

The Ishiagu Pb-Zn deposits represent an integral part of Benue Trough sedimentary basin evolution and strategies for exploring them must take into account the pertinent geologic model. The Ishiagu Pb-Zn mineralization is attributed to a sedimentary geological model based on the geotectonic setting, the mode of occurrence and fluid-inclusion characteristics. This model is noted to connate brines set

---

*Corresponding author: Mbah Victor O, Geophysical Science Department, Nnamdi Azikiwe University, Awka, Nigeria, E-mail: vctrmbah@yahoo.com

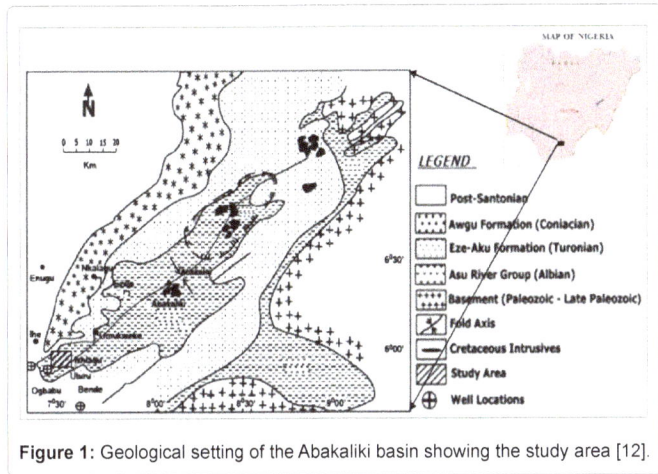

Figure 1: Geological setting of the Abakaliki basin showing the study area [12].

into motion by a high geothermal gradient accompanying continental rifting. The model (Figure 2) interprets the Pb-Zn deposits in Ishiagu area as distinct epigenetic ore bodies generated by convectional flow of hot brines that leached the metals from arkosic sediments and later precipitated them as metallic sulphides [7]. Fractures generated by the Cenomanian tectonic events in the Asu River Group (ARG) is marked the key control of the precipitated base metals. This model thus, suggests the primary ore target to be in the ARG sediments of the Abakaliki Basin in which the metal ore bodies are strata-bound.

The Pb-Zn mineralized zone extends over a distance of 500 km in a narrow belt from Ishiagu in the Lower Benue Trough to Zurak in the Upper Benue Trough, likewise the extent of igneous intrusions in the Benue Trough.

Methodology

The field data for this research was acquired from ten measurement profiles, 300 m each in the survey area. The well pegged geophysical grid in Figure 3 was established from an east-west trending baseline 300 m long at a bearing of 90°, while the profiles, parallel to each other and to the baseline with 100 m spacing were approximately perpendicular to the transmitter and to the strike direction. The field data was then acquired by systematically traversing along these profiles at a 10 m interval with an ABEM Wadi VLF receiver, Model-9133001869, operating on the VLF principle of recording tipper responses at every measurement station using radio waves from transmitters at remote distance. The ABEM WADI receiver requires no physical contact with the transmitter and the ground during VLF survey as it operates on induction mechanism. The optimal configuration of VLF survey as in Figure 4 is to have the geologic strike oriented parallel to the transmitter direction so that a vertical magnetic component is generated for any conductivity contrast by the propagating horizontal and concentric magnetic and orthogonal electrical fields due to induction [8]. Thus, the DMB transmitter located in Germany that is oriented to the north from the site and of 26.9 kHz VLF frequency was chosen for this survey in consideration to the two prevailing fracture sets in the Abakaliki Basin trending northwest and northeast respectively, with subordinates trending north directly. The survey area entirely covering Amata Village is in the vicinity of already existing mines in Ihietutu.

The VLF-EM prospecting is fundamentally based on the primary EM wave impedance over 2-D structures and this wave impedance depends upon the orientation of the EM field components with respect to the geologic strike of the 2-D geologic structures. The EM wave

impedance is therefore, evaluated with the "E-parallel" polarization mode being the most apposite for detecting the associated anomalous secondary magnetic field. In VLF-EM prospecting, the superposition of the primary magnetic field (H_y) from the transmitter and the secondary magnetic field (H_z) from the subsurface gives an important diagnostic parameter, Tipper, (B) that feasibly reflects the conductivity contrast of every 2-D structures in the subsurface.

$$B = H_z / H_y \qquad (1)$$

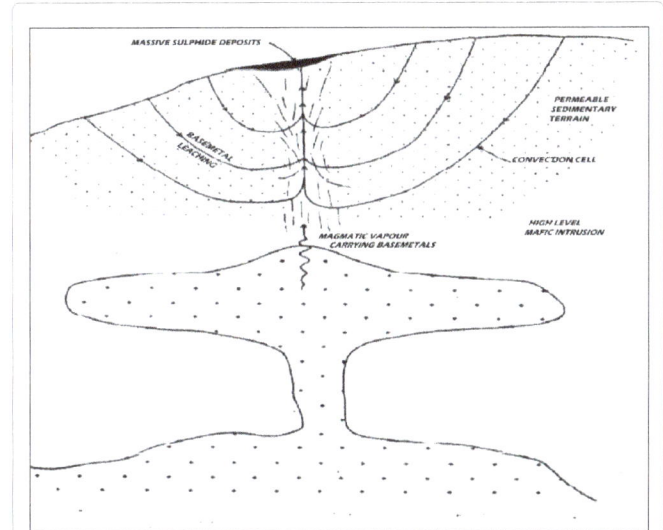

Figure 2: Genetic evolution for an active hydrothermal system precipitating massive sulfide deposits [7].

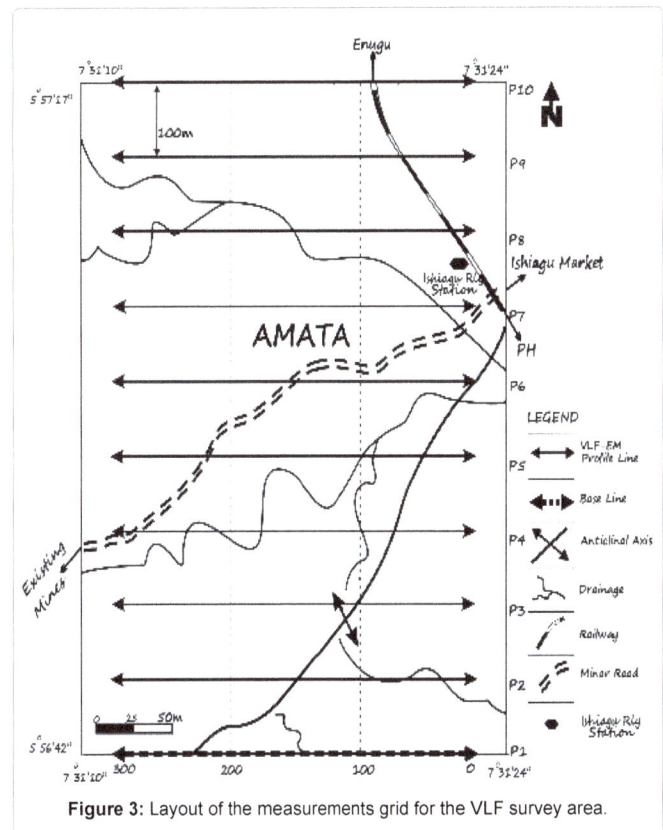

Figure 3: Layout of the measurements grid for the VLF survey area.

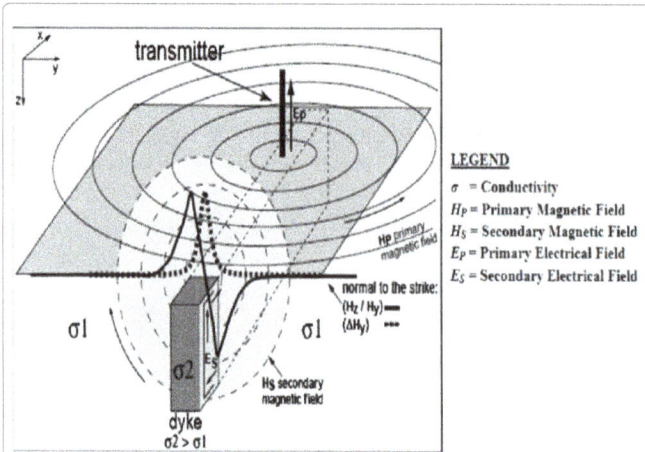

Figure 4: Field components of VLF Field from transmitter at remote distance [8].

Tipper is derived by the phase lag between (H_y) and (H_z) magnetic fields due to induction thus, allowing this method to become widely used for geophysical prospecting. Recorded in the field by the equipment are the in-phase and quadrature components of this complex quantity, Tipper, obtainable mathematically by evaluating the shape of the polarized ellipse. This measurement initiates the detection of structures and subsurface features of interest even in areas concealed with overburden containing relatively thin horizons of anomalously high conductivity. Figure 5a and 5b is the typical raw VLF-EM readings attained from Ishiagu field presented as nomograms.

Anomalies of the raw VLF-EM profiles are usually complicated and this calls for data processing so as to improve the resolution of local anomalies and limit the possible interpretational error. This was achieved in this work using the prominent KHF-filtering techniques as outlined by Fraser [9], Karous and Hjelt [10]. The Fraser filtering technique transforms the zero-crossing on a raw in-phase reading to maximum peak and the corresponding quadrature component transformed to minimum peak for anomalously conductive structures. The analogous apparent current density pseudo-section with depth as gotten by Karous-Hjelt filtering offer the possibility to generate a pictorial image of the profile indicating the geometry of the 2-D structure that instigated the anomaly. This pseudo-section is shown as colour codes with red colour indicating high conductivity (i.e. positive) and blue colour indicating low conductivity (i.e. negative). Figure 6a and 6b is a representation of various anomalies of varying degree of conductivity trending in different directions as delineated from the honed profile sections of Ishiagu field.

Generally, VLF analysis is on the bases that the higher the in-phase values of the anomaly, the greater the conductivity of the underlying structures in relation to the surrounding rock [11]. Succinctly, the extent of the Tipper responses in Ishiagu field is highly controlled by the conductivity contrast of some geologic features and these anomalies vary greatly; some of the anomaly peaks are sharp and of high intensity while others are broad and of lower intensity. Suspected mineralized veins, $(F_1–F_{16})$ were delineated on the gridded survey area using characteristic coincidence of positive inflections on filtered in-phase anomaly and were further interpreted on current density maps [12-14]. With good geologic information of the area, it is therefore logical to interpret the VLF-EM anomalies caused by the mineralized veins.

Results and Interpretation

Finally, the in-phase readings on stacked profiles (Figure 7) and the current density maps (Figures 8 and 9) were used for the interpretation of the geologic structures in terms of conductive mineralized veins and this provided more detailed information on the extension, thickness and depth distribution of the Pb-Zn mineralized veins in Ishiagu field [15-19]. Generally, good 2-D structures with less overburden were detected by induction, hence, the Pb-Zn mineralized veins lying in the area of survey with less overburden (in this case closer to the Abakaliki anticlinal axis) were utterly inducted; while those concealed by very thick overburden were not easily inducted, but accomplished through current gathering and are thus, of moderate responses [20-22].

The VLF readings of P_{Anom-1} in the southern part of the surveyed area is attributed to less mineralization as the conductive anomaly obtained directly from the shallow and outcropped Pb-Zn host rock (Asu River Group) is still of moderate intensity. But the strong P_{Anom-3} and P_{Anom-4} in the central part of the surveyed area are attributed to the shallow Pb-Zn mineralized veins, though deeper in their northern section as thick overburden (Eze-Aku Shale) in that zone reduces the skin depth of VLF signal, causing moderate VLF readings in the zone [23]. Evaluation of the positive and very close to zero quadrature components in this northern outskirt also substantiates the thick overburden over the attributed mineralized veins in the zone.

The collocation of the shallow part of the core positive in-phase anomalies to the anticlinal axis and to the corresponding negative in-phase anomalies create a prominent criterion of using VLF readings to analyze the composition and structural control of the characterized

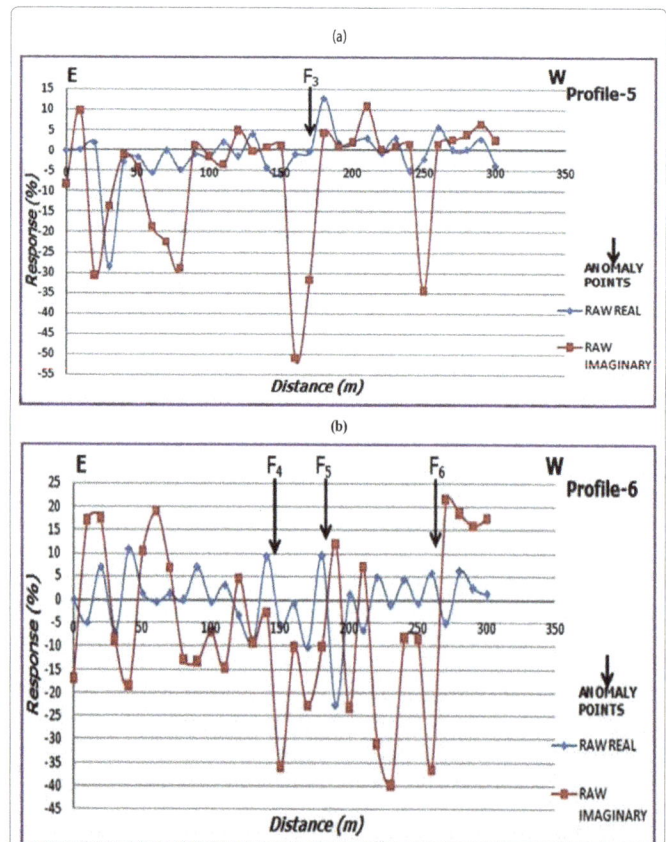

Figure 5: Typical Raw In-phase and quadrature components of tipper (%) Obtained from the Ishiagu Field (Trend: E –W); (a) Profile-5; (b) Profile-6.

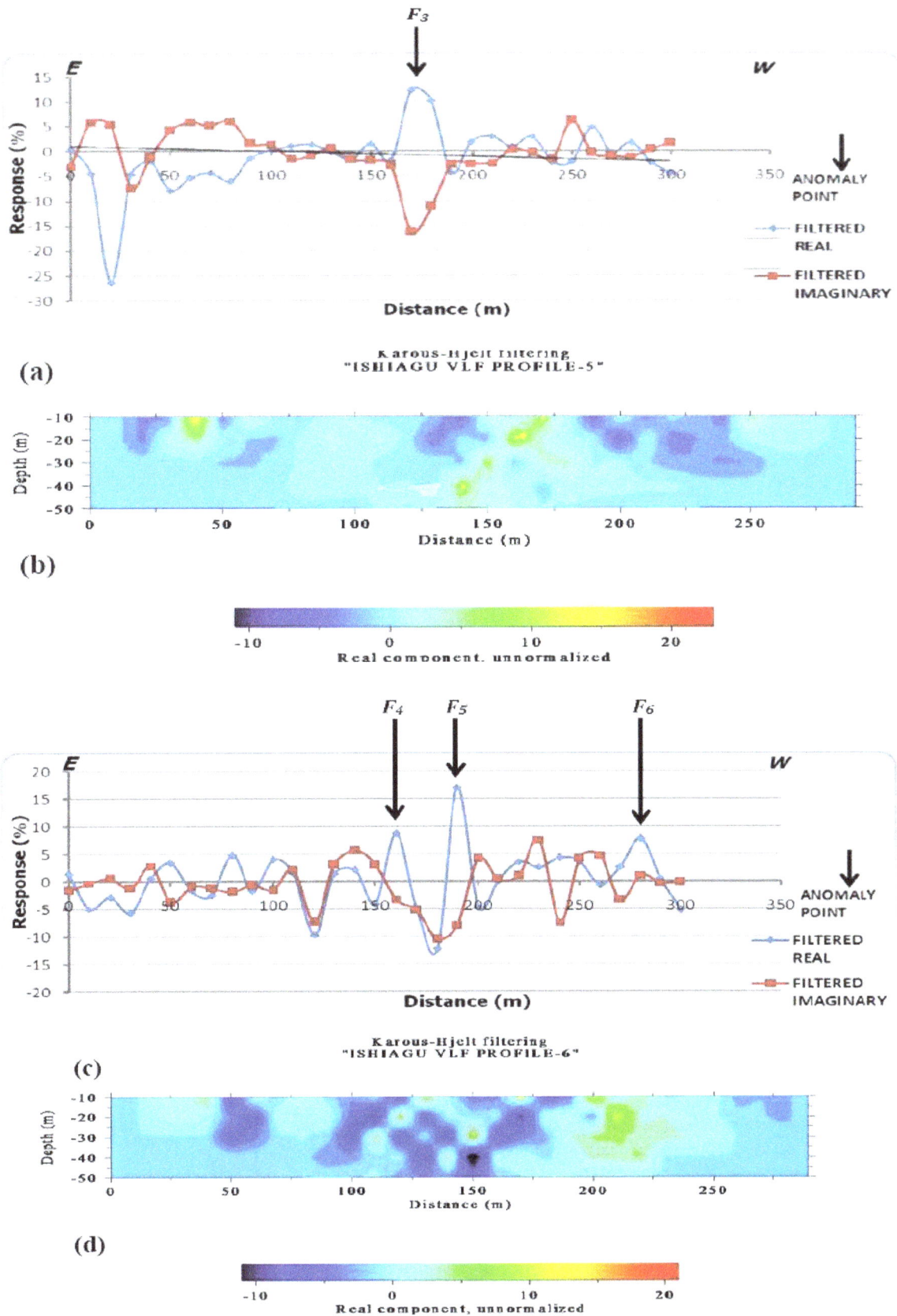

Figure 6: Filtered VLF-EM In-phase and quadrature components of tipper (%) and corresponding current density pseudo-sections of; (a-b) profile-5 (*Trend: E –W*) and (c-d) Profile-6 (*Trend: E –W*).

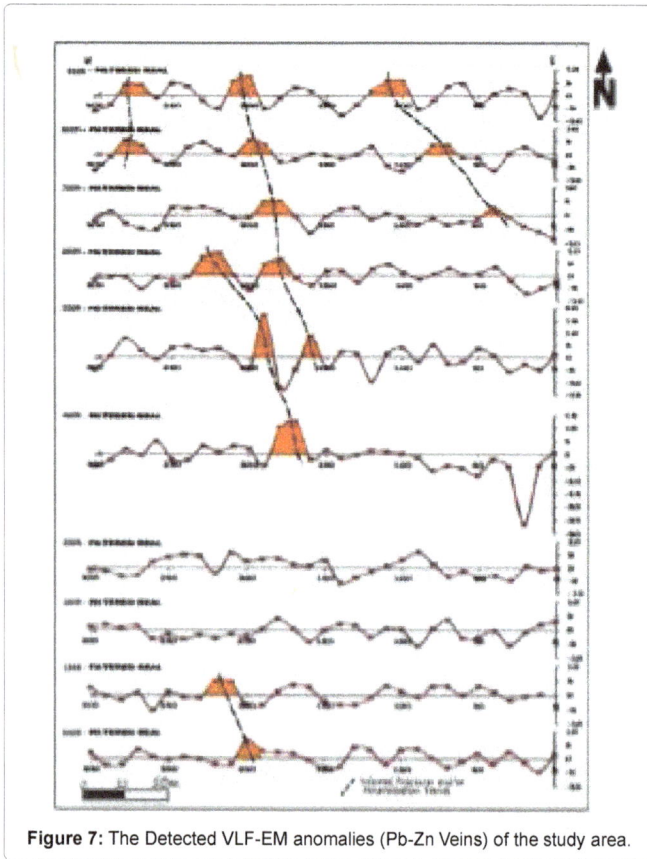

Figure 7: The Detected VLF-EM anomalies (Pb-Zn Veins) of the study area.

mineralized veins in Ishiagu field [24-26]. The sharpness of the in-phase anomalies of P_{Anom-3} and P_{Anom-4} in the central zone due to their shallow depths and their proximity to the anticlinal axis elucidates the characterized mineralized veins in Ishiagu field to be structurally controlled by the NW–SE trending fractures produced by the tensional tectonic deformation of Cenomanian episode [27]. The northern parts of these anomalies along with P_{Anom-5} outlying from the anticlinal axis (Figure 8) are of broad and moderate VLF responses indicating their occurrence at a profound depth with enormous overburden of Eze-Aku Shale. The broad and very close to zero values of the quadrature components of the anomalies at this outlying distance from the anticlinal axis also anticipate the northern parts of the detected mineralized veins in Ishiagu to be concealed by thick overburden [28].

As the igneous intrusions of the Abakaliki Basin are structurally controlled due to their occurrence in steeply dipping fractures, the negative in-phase anomalies are thus, interpreted to be for the cretaceous intrusions in Ishiagu area. The N_{Anom-1} anomaly is on the same trend with the prospective mineralized veins. Its parallel and proximity to the P_{Anom-3} and P_{Anom-4} as emerged on profiles P-5, P-6 and P-7 (Figure 9) is thus, a lucid indication of the occurrence of both ore and gangue minerals in successive layers along mineralized fractures, thus signifying their fissure filling mode of occurrence. The NW and NE trends of N_{Anom-1} and N_{Anom-2} respectively also suggest these intrusions to be of Cenomanian and Santonian episodes respectively [29-32].

The depth to these anomalous bodies was also obtained in this work from the shape of the raw VLF-EM readings. The peak-to-peak horizontal distance on a raw in-phase anomaly is about the same as the depth of the conductive body known to be positioned at the crossover point of the anomaly [33]. This simple "rule of thumb" was applied

in this work to evaluate the depth of these prospective mineralized veins as outlined in Table 1. The computed depth values obtained from the Ishiagu field was plotted and contoured using surfer-10 software (Figure 10). This depth to the prospective mineralized veins is shallow in the entire southern and central parts of the study area contiguous to the NE trending anticlinal axis, but deeper in the northern part [34].

Consequently, the overall trace of the VLF anomalies most proximate to the anticlinal axis likely indicates the near surface trace

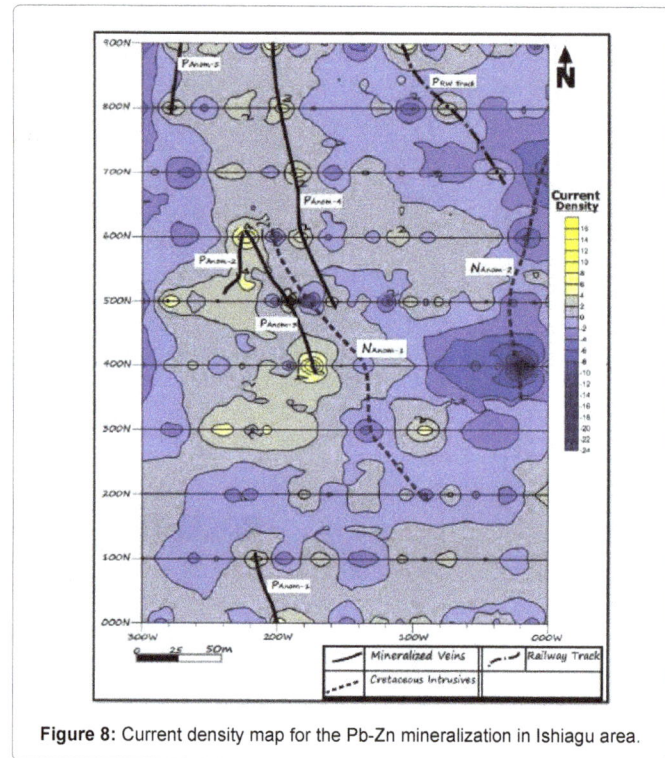

Figure 8: Current density map for the Pb-Zn mineralization in Ishiagu area.

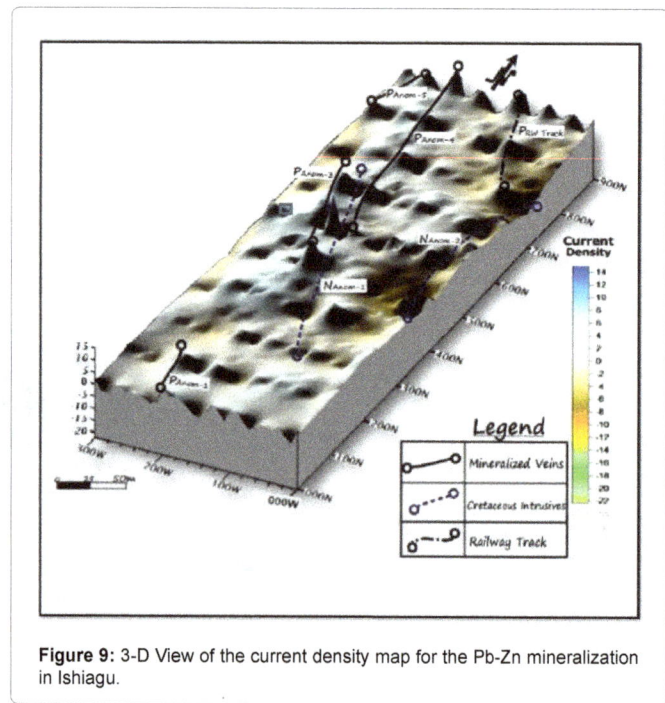

Figure 9: 3-D View of the current density map for the Pb-Zn mineralization in Ishiagu.

Profiles	Anomaly Points	Measured Horizontal Distance (m)	Vein Numbers	Vein Trend	Vein Length (m)	Depths (m)
P1	F1	200	Anom1	NNW-SSE	100	10
P2	F2	212	Anom1	NNW-SSE	100	16
P3		80	Intru1	NW-SE	400	10
P4		132	Intru1	NW-SE	400	12
		20	Intru2	N-S	350	22
P5		135	Intru1	NW-SE	400	10
	F3	171	Anom3	NW-SE	200	12
		30	Intru2	N-S	350	12
	F4	160	Anom4	NNW-SSE	400	10
P6		175	Intru1	NW-SE	400	7
	F5	190	Anom3	NW-SE	200	11
	F6	278	Anom2	N-S	80	12
		18	Intru2	N-S	350	9
	F7	182	Anom4	NNW-SSE	400	17
P7		180	Intru1	NW-SE	400	10
	F8	223	Anom2	N-S	80	19
			Anom3	NW-SE	200	
		3	Intru2	N-S	350	14
P8	F9	35	RW Track	NW-SE	-	-
	F10	175	Anom4	NNW-SSE	400	19
	F11	71	RW Track	NW-SE	-	-
P9	F12	194	Anom4	NNW-SSE	400	19
	F13	276	Anom5	NE-SW	100	13
	F14	100	RW Track	NE-SW	-	-
P10	F15	202	Anom4	NNW-SSE	400	22
	F16	244	Anom5	NE-SW	100	12

Table 1: Summary of the Obtained VLF-EM Parameters of Ishiagu Area.

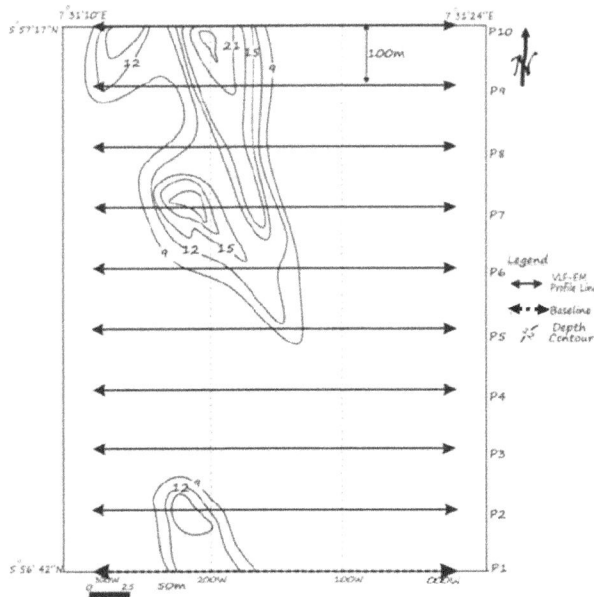

Figure 10: Depth map of the prospective Pb-Zn mineralized veins (*contour interval~3 m*).

of significant Pb-Zn mineralized veins while the parts outlying from the anticlinal axis, though with moderate VLF anomalies due to thick overburden still indicates significant mineralized veins. Previous studies indicate that the Ishiagu mineralized fractures normally fade out farther away from the Pb-Zn lodes, thereby alluding to fracture extension as a clear clue to extension of Pb-Zn mineralization in Ishiagu area [35]. The NW trending of these outlying fractures in correlation to the NE trending Abakaliki anticlinal axis is also an apparent evidence

of Pb-Zn mineralization potentials of the ascertained VLF anomalies in the surveyed area of Ishiagu. To this upshot, concerted efforts involving VLF-EM geophysical method and detailed geologic investigation has given a better definition of the parameters of these interesting structures that revealed several Pb-Zn mineralized veins in Ishiagu, thus calling for trenching across these prospective mineralized veins for further authentication.

Conclusion

The high VLF anomalies delineated from Ishiagu area, in the SW part of Abakaliki basin as a rule yielded a high conductivity contrast result well-suited for Pb-Zn mineralization and the geologic information of the area. KHF filtering techniques were integrated and applied in the VLF data interpretation of the Ishiagu area and results of strong EM induction were detected, probably indicating the presence of Pb-Zn deposits. Current density maps also revealed the inferred mineralized veins' trends in NW–SE direction with their subordinates in N–S direction and this conforms to the tensional tectonic deformations of Cenomanian episode in the Abakaliki Basin. The broad tipper responses in the northern part of the survey area indicate deeper sources while the sharp tipper responses observed in the central part indicate shallow sources in correspondence to the NE trending anticlinal axis.

All these deductions in queue to the geologic information of the area definitely confirm the efficacy of VLF technique to enhance geological mapping over well concealed mineralized areas. In all cases where there is VLF data available on a given profile, the VLF responses correlate with changes in the apparent resistivity. The outcome of this scrutinized VLF technique in terms of its intrinsic characteristics, rather than through the obtained results obviously revealed the mineralization potential of the concealed fractures in the Ishiagu area, though at greater depth in the northern outskirt. Recommended for full exploitation of Pb-Zn deposits in this surveyed area of Ishiagu is the trenching across the southern parts of the inferred Pb-Zn mineralized veins so as to establish the suitable ground plans for effective mine development. Some highly negative in-phase anomalies attributed to some intrusive bodies as detected in this work, also calls for the wariness of the locations and trends of these bodies in the course of efficient and cost effective mine development in Ishiagu area.

Because of the economic efficiency of this method, speed of field survey and cost effectiveness of the operation, the VLF method is recommended for rapid geophysical exploration of Pb-Zn mineralized zones.

References

1. Fischer G, Le Quang BV, Muller I (1983) VLF Ground Surveys, a Powerful Tool for the Study of Shallow 2-D Structures. Geophys Prosp 31: 977-991.

2. McNeill JD, Labson VF (1991) Geological Mapping using VLF Radio Fields in Nabighian MN Electromagnetic Methods in Applied Geophysics II. Soc Expl Geophys 521-640.

3. Ogilvy RD, Lee AC (1991) Interpretation of VLF-EM In-phase Data using Current Density Pseudo-sections. Geophys Prospect 39: 567-580.

4. Bayrak M (2002) Exploration of Chrome Ore in Southwestern Turkey by VLF-EM. Jour of the Balkan Geophy Society 5: 35-46.

5. Ezepue MC (1984) The Geologic Setting of Lead-Zinc Deposits at Ishiagu, Southeastern Nigeria. Jour of Afri Earth Sci 2: 97-101.

6. Grant FK (1971) South Atlantic, Benue Trough and Gulf of Guinea Cretaceous Tripple Junction. Geol Soc Amer Bull 82: 2295-2298.

7. Akande SO (2003) Minerals and Fossil Fuels Discovery: The Adventure of Exploration. 67th Inaugural Lecture, Univ. of Ilorin, Nigeria.

8. Bosch FP, Muller I (2005) Improved Karst Exploration by VLF-EM Gradient Survey: Comparison with other Geophysical Methods. Near Surface Geophysics 3: 299- 310.

9. Fraser DC (1969) Contouring of VLF-EM data. Geophysics 34: 958-967.

10. Karous M, Hjelt SE (1983) Linear Filtering of VLF Dip-Angle Measurements. Geophy Prosp 31: 782-794.

11. Smith BD, Ward SH (1974) On the Computation of Polarization Ellipse Parameters. Geophysics 39: 867-869.

12. Akande SO, Egenhoff SO, Obaje NG, Erdtmann BD (2011) Cretaceous Sediments in the Lower and Middle Benue Trough, Nigeria: Insights from New Source Rock Facies Evaluation. Petro Tech Dev Jour 1: 1-34.

13. Becken M, Pedersen LB (2003) Transformation of VLF Anomaly Maps into Apparent Resistivity and Phase. Geophysics 68: 497-505.

14. Benkhelil J (1989) The Origin and Evolution of the Cretaceous Benue Trough, Nigeria. Jour Afr Earth Sci 8: 251-282.

15. Bruce PJ (2005) The History of Electromagnetic Theory. University of Aberdeen publ, USA.

16. Etuk EE, Ukpabi N, Ukaegbu VU, Akpabio IO (2008) Structural Evolution, Magmatism and Effects of Hydrocarbon Maturation in Lower Benue Trough, Nigeria: A Case Study of Lokpaukwu, Uturu and Ishiagu. The Pacific Jour of Sci & Tech 9: 526-532.

17. Grant FS,West GF (1965) Interpretation Theory in Applied Geophysics: McGraw-Hill Book Co.

18. Gurer A, Bayrak M, Gurer OF (2009) A VLF Survey using Current Gathering Phenomena for Tracing Buried Faults of Fethiye-Burdur Fault Zone, Turkey. Jour of Applied Geoph 68: 437-447.

19. Hutchinson PJ, Barta L (2002) VLF Surveying to Delineate Long-wall Mine-Induced Fractures: The Leading Edge 21: 491-498.

20. Khalil MA, Santos FM (2011) Comparative Study between Filtering and Inversion of VLF-EM Profile Data. Arab Jour of Geosci 4: 309-317.

21. Livelybrooks D, Mareschal M, Blais E, Smith JT (1996) Magnetotelluric Delineation of the Trillabelle Massive Sulfide Body in Sudbury, Ontario. Geophysics 61: 971-986.

22. Nwachukwu SO (1972) The Tectonic Evolution of the Southern Portion of the Benue Trough, Nigeria. Geol Mag 109: 411-419.

23. Ofoegbu CO, Odigi MI (1990) Basement Structures and Ore Mineralization in the Benue Trough. In: The Benue Trough Structures and Evolution. Ofoegbu.

24. Olade MA, Morton RD (1980) Temperature of Ore Formation and Origin of the Ishiagu Lead-Zinc Deposit, Southern Benue Trough, Nigeria. Jour Min Geol 17: 19-127.

25. Olorunfemi MO, Fatoba JO, Ademilua LO (2005) Integrated VLF-Electromagnetic and Electrical Resistivity Survey for Groundwater in a Crystalline Basement Complex Terrain of Southwestern Nigeria. Global Journal of Geological Science 3: 71-80.

26. Olubambi PA, Ndlovu S, Potgieter JH, Borode JO (2008) Mineralogical Characterization of Ishiagu (Nigeria) Complex Sulphide Ore. Int J Miner Process 87: 83-89.

27. Pedersen LB, Becken M (2005) Equivalent Image Derived from Very-Low-Frequency (VLF-EM) Profile Data. Geophysics 70: 43-50.

28. Schlumbeger (1985) Well Evaluation Conference, Nigeria. Schlumbeger: Paris 290: 122-131.

29. Sinha AK (1990) Interpretation of Ground VLF-EM Data in Terms of Inclined Sheet-Like Conductor Models. PAGEOPH Journ 132: 733-756.

30. Srigutomo W, Sutarno D, Harja A, Kagiyama T (2005) VLF Data Analysis through Transformation into Resistivity Value: Application to Synthetic and Field Data. Indonesian Journal of Physics 16: 127-136.

31. Telford WM, King WF, Becker A (1977) "VLF Mapping of Geological Structure". Geological Survey of Canada 76: 25.

32. Ugwu SA, Eze CL (2009) Mapping Bedrock Topography using Electromagnetic Profiling. Jour Applied Sci Environ Manage 13: 43-46.

33. Umeji AC (2000) Evolution of the Abakaliki and Anambra Sedimentary Basins Southeastern Nigeria. A report submitted to Shell Petrol Dev Co Nig Ltd 147: 43-49.

34. Vargemezis G (2007) Interpretation of VLF Measurements Related to Hydrogeological Surveys. Bulletin of the Geological Society of Greece 40: 593-604.

35. Wright JB (1976) Fracture Systems in Nigeria and Initiation of Fracture Zones in the South Atlantic. Tectonophysics 34: 43-47.

15

The Catchment Area of Kadey in East-Cameroon: Assessment of Arsenic Contamination in Deep Groundwater Resources

Kouassy Kalédjé PS[1,2]*, Ndam Noupayou JR[1], Djomou Djomga PN[3] and Mvondo Ondoua J[1]

[1]Laboratory of Engineering Geology and Alterology, Department of Earth Sciences, Faculty of Science, University of Yaounde 1, PO Box: 812, Yaounde, Cameroon
[2]Department of Mining and Geological, Sub-Regional Bilingual University of Mining, Sciences, Technology, Management and Professional Training, PO Box: 863, Yaounde, Cameroon
[3]Laboratory of Material and Inorganic Industrial Chemistry, Department of Applied Chemistry, National School of Agro-Industrial Sciences (ENSAI), University of Ngaoundéré, PO Box: 455 Ngaoundere, Cameroon

Abstract

This present study, in general, was carried out to assess arsenic in deep groundwater resources in the catchment area of Kadey, East-Cameroon and to predict arsenic mobilization process in relation to copper (indicatives aspects), iron, manganese, pH and ORP. Seventy-two (72) deep groundwater samples were collected during twenty four months between January 2014 and December 2015. The depths of the wells were ranged from 7 to 34 m. In year 2014 and year 2015, arsenic concentration in 17% and 26% of examined groundwater wells, respectively exceeded permissible World Health Organization (WHO) guideline value of 0.010 mg/L for drinking water. The concentrations of arsenic were in the range between <0.003 to 0.137 mg/L. The study demonstrated elevated concentrations of iron and manganese in the groundwater. Arsenic is highly correlated with iron and manganese in the first time and in the second time, arsenic is medium correlated with copper. The strong negative correlation between arsenic and ORP indicates that arsenic mobilization occurs under reducing condition. These distinct relationships indicate that arsenic release is considered to be affected by the reductive dissolution of Fe/Mn oxides in the groundwater. Arsenic has very weak negative correlation with pH suggesting less effect of pH on arsenic mobilization. Arsenic is not significantly correlated with the season which infers similar distribution of arsenic in both seasons. Arsenic varies spatially in groundwater of the valley showing high concentrations in central groundwater district.

Keywords: Deep groundwater; Reductive dissolution; Arsenic; Catchment area of Kadey; WHO

Introduction

Today, heavy metal contamination of groundwater is one of the major problems in the world. The heavy metals that occur as natural contaminants of groundwater and are potentially bio-hazardous include manganese (Mn), lead (Pb), cadmium (Cd), mercury (Hg), cyanide (Cn) and arsenic (As) [1,2]. Like cyanide an mercury, arsenic is recognized as a toxic element and has been classified as a human carcinogen affecting skin and lungs [3]. Arsenic has strong toxicity at even low concentrations and can accumulate in body tissues over long periods of time and is nonessential for human health [4,5]. So, elevated levels of arsenic constitute problems for water supplies around the world [6]. In recent past years, the occurrence of high concentrations of arsenic has been detected in groundwater from a number of regions across the world. The problem has increased greatly in recent years in several regions of Africa. In those regions, countries affected with arsenic in groundwater include South Africa, several states of Nigeria (Kano, Adamawa, Taraba, Oyo, Delta Imo Lagos and Edo), Chad, Central Africa Republic, Gabon and Cameroon (East and South regions) [7,8].

The demand for water is increasing due to rapid growth of urban population and industrial activities (mining exploration; mining extraction and treatment of precious stones) in the Kadey catchment. There is a high decadal population growth rate found in Kadey district i.e., 57.19% [9]. As a result there is an immense pressure on groundwater resources in the catchment. Groundwater was first exploited for water supply in 1990 in the catchment. Mechanized extraction of groundwater resources began in 2005 with the implantation of Non-Governmental Organizations (NGOs). Groundwater is an important water resource in the Kadey catchment. It contributes about 57% of the total water supply in the catchment [10]. In dry season, 67% to 74% of the water supply is met by ground water [11].

There have not been several studies of assessment of arsenic in groundwater of Cameroon in general and specially in groundwater of the Kadey catchment. Groundwater survey of the catchment reported the presence of arsenic and the concentrations were below World Health Organization (1993) guideline values [12]. Similar study revealed that the ground water resource in the Delta state of Nigeria is contaminated with arsenic in deep aquifers [13]. Likewise, other studies have reported elevated levels of arsenic in groundwater of Bouar and Mbaïki (Central Africa Republic) [14]. With the mercury and cyanide, arsenic in groundwater wells in the Kadey catchment is one of the major environmental issues due to its negative health impact and more than 55% of water supply in the area is derived from groundwater resources. So, this paper presents an overall study on the occurrence, mobilization and distribution of arsenic in deep groundwater resources in the catchment. The study attempted to visualize spatial distribution pattern of arsenic in Kadey catchment groundwater in GIS environment. The study also aimed to demonstrate temporal (seasonal) variation of arsenic. The possible relationship between arsenic and depth of the groundwater was also examined.

*Corresponding author: Kouassy Kalédjé PS, Laboratory of Engineering Geology and Alterology, Department of Earth Sciences, Faculty of Science, University of Yaounde 1, PO Box: 812 Yaounde, Cameroon, Department of Mining and Geological, Sub-Regional Bilingual University of Mining, Sciences, Technology, Management and Professional Training, PO Box: 863 Yaounde, Cameroon
E-mail: kkaledje@yahoo.fr

Materials and Methods

Study area

The Kadey catchment area (2 647 km^2) is located in Eastern Cameroon, between longitudes 14°29'E and 14°45'E and the latitudes 4°12'N and 4°36'N (Figure 1). The basin is located in the equatorial climate transition upstream from the part of the basin located south of the Sanaga. As for the entire Congo Basin, annual average rainfall is 1 428.9 mm between 1960s and 2015s. The rainfall pattern has four seasons marking the equatorial influence. The distribution of seasons varies following stations (Bertoua; Abong-Mbang; Batouri and Meiganga). However, the aspect "camel's back" characterizes all stations of the Sangha Basin, with two maxima and two minima well marked, the second maxima always been the most powerful. The highest temperatures are observed between March and April and are respectively 25.4°C Batouri, 25.9°C Dem and 26°C to Bertoua [15]. In terms of the lowest temperatures, they correspond to the months of July and August, with 23°C in Batouri, Bertoua 23.4C and 23.9°C to Dem (Figure 2). The hydrographic network of the basin narrows upstream downstream. The relief consists of upstream basin plateaus, plains in the middle and downstream lowlands whose altitude varies between 600 m and average 880 m. However, there are isolated ancient massifs whose peaks reach around 927 m (Figure 1). The geology of the catchment area consists of base for the most part [15]. The population is predominantly rural and relatively dense. The main economic activity remains artisanal gold mining although agriculture remains the main activity in the whole of Congo.

Water sampling and analysis

The study was carried out in 72 deep groundwater wells, six points in the Kadey river and two sources during the years 2014 and 2015 (Figure 3). The study covered the groundwater wells of 7 to 34 m depth. The geo-positions of groundwater sampling locations were determined using global positioning system (GPS). Random sampling technique was used to collect groundwater samples. The locations of all the samples were recorded by handheld Garmin-E GPS and referenced to WGS 84 coordinate system. The high density polyethylene (HDPE) sampling bottles were treated with 5% HNO and then rinsed with double distilled water. Samples from wells were collected by using 500-mL Nalgene (UK) HDPE bottles. A set of samples were collected in sampling bottles after pumping water for five minutes to get the representative samples of groundwater wells. The bottles were labeled with the sample code number. These samples were preserved as per APHA-AWWA-WEF [13] and then brought to the laboratory for the analysis. The samples were kept at 4°C prior to analysis. Oxidation reduction potential (ORP), electrical conductivity (EC), pH and temperature were measured *in situ* at each sampling locations. The ORP and pH were measured by Hanna HI 8314 pH/ORP meter (Italy). The EC was measured by Jenway 4200 conductivity meter (UK).

The analysis of total arsenic, total iron, total manganese, total copper and total lead were carried out in Tanzania standard (NS) certified-CEMAT Water Laboratory by using Varian AA 240 atomic absorption spectrometer (AAS). The analysis of arsenic was carried out using Varian AA 240 atomic absorption spectrometer (Nigeria) with vapor generation accessory VGA-77 (Nigeria). The standard solutions produced by Merck, Germany traceable to standard reference material (SRM) of NIST (National Institute of Standards and Technology, Gaithersburg, MD, USA) were used to prepare calibration standards.

The samples for the analysis of arsenic, iron and manganese were digested with high purity HNO_3 (Merck) within a week of sample collection as per APHA-AWWA-WEF (2005). Sample digestion with the HNO_3 allows total extraction of the metals from the samples. Three replications of each analysis were performed and mean values were used for calculations. Analytical reagent blanks were prepared for each batch of the digestion set and then analyzed for the same elements as the samples. Analytical precision was in good agreement, generally better than 5% RSD. SPSS version 18 was used for all statistical analyses.

Today, Cameroon adopted universal transverse mercator (UTM) projection for the base mapping of the country with some modifications suited to its shape. This is named as modified universal transverse mercator projection. So, all the spatial data layers were maintained in a standard Cameroonian coordinate system of modified universal transverse mercator, central meridian 84° longitude (i.e., MUTM 84). The software used for mapping and spatial analysis was ArcGIS version 9.3.

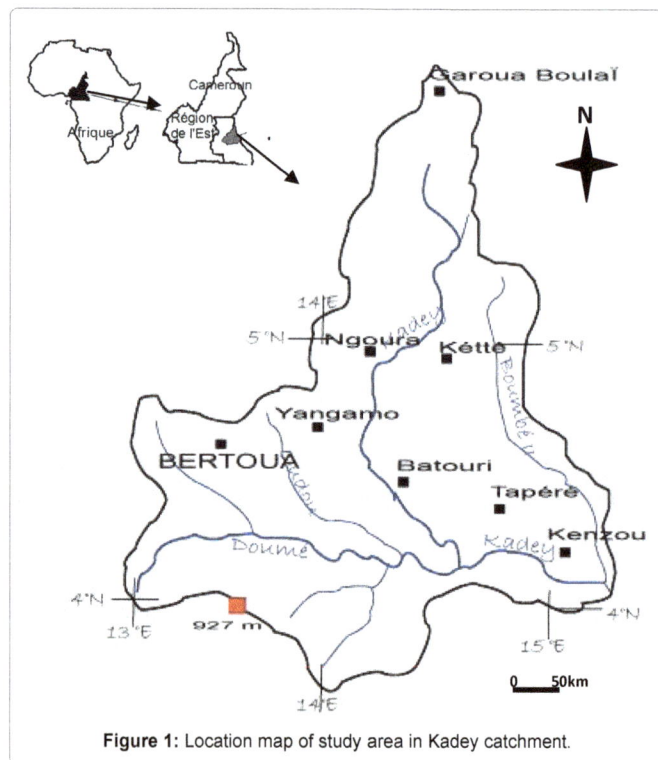

Figure 1: Location map of study area in Kadey catchment.

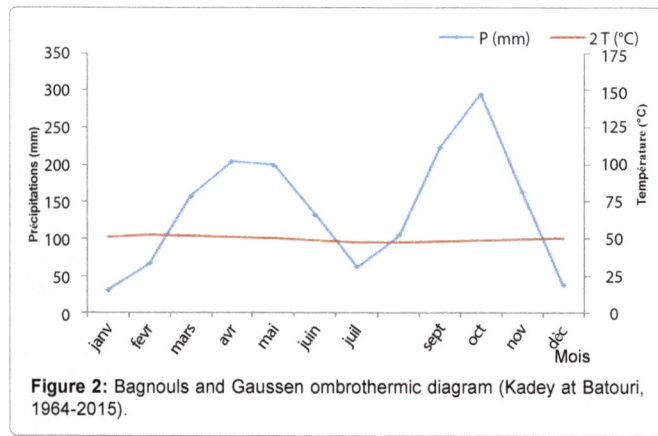

Figure 2: Bagnouls and Gaussen ombrothermic diagram (Kadey at Batouri, 1964-2015).

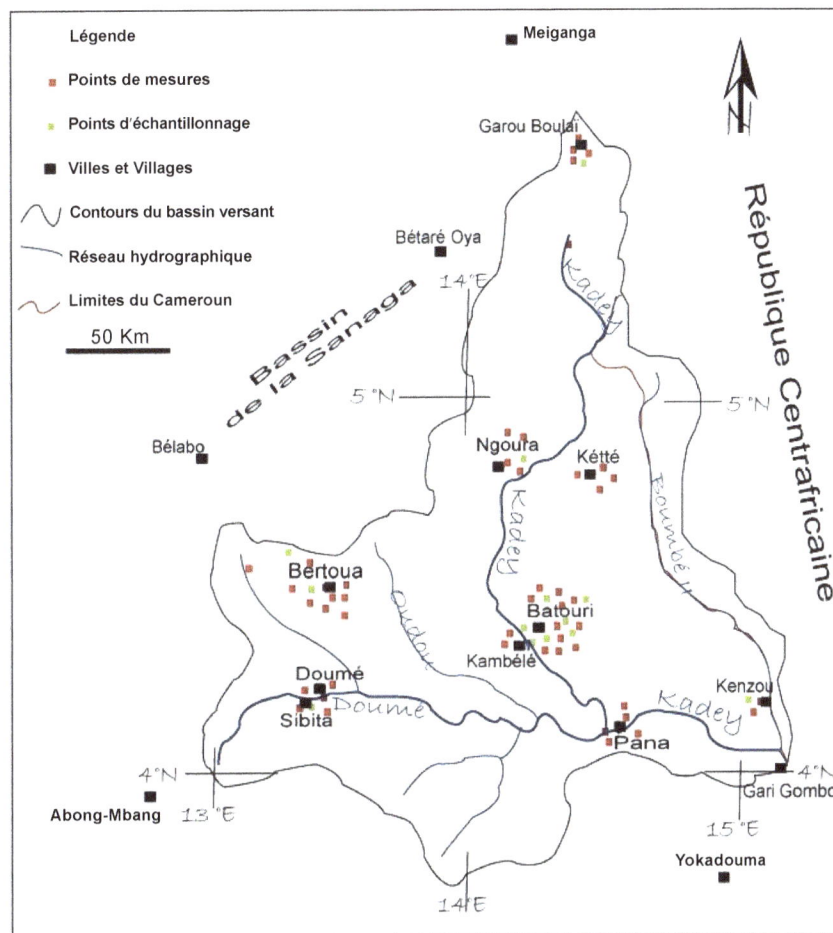

Figure 3: Sampling locations map.

Variable	Unit	Year 2014					Year 2015				
		Min.	Max.	Mean	Med.	SD	Min.	Max.	Mean	Med.	SD
PH		5.4	6.4	5.7	5.6	0.5	5.2	6.4	5.6	5.4	0.5
EC	µS/cm	92	1715	567	452	381	93	1665	567	456	379
ORP	mV	-161	134	-59	-76	68	-120	135	-49	-56	55
Depth	m	7	34	13	17	57	7	34	13	17	57
As	mg/L	BDL	0.137	0.011	0.003	0.024	BDL	0.131	0.013	0.005	0.023
Fe	mg/L	BDL	11.09	3.61	2.88	2.88	0.1	12.99	3.76	2.79	2.87
Mn	mg/L	BDL	1.55	0.44	0.37	0.35	BDL	1.75	0.42	0.31	0.4

Min. = Minimum; Max. = Maximum; Med. = Median; SD = Standard Déviation; BDL = Below Detection Limit.

Table 1: Summary of statistical data for physicochemical parameters and metals.

Results and Discussion

Physicochemical parameters and metals

In the Table 1, we have presented the synthesis of values of pH, EC, ORP and the concentrations of metals in the groundwater wells. All the pH was acid and ranged from 5.4 to 6.4. The EC ranged from 92 to 1715 µS/cm (mean = 567 µS/cm), 93 to 1666 µS/cm (mean = 565 µS/cm) in pre monsoon and post monsoon, respectively. The elevated EC value is mainly due to geological sources since groundwater contamination is less in deep aquifers due to presence of widespread thick lacustrine clay beds that significantly restrict downward percolation [16] and leaching of ions contributing conductivity would be limited.

The mean values for ORP in the groundwater wells were -59 mV in pre monsoon and -49 mV in post monsoon. The negative ORP value was up to -161 mV indicates that the groundwater wells are under reducing condition. A study reported negative ORP value up to -195 mV in deep groundwater [14]. Likewise, another study indicated low ORP value in deep groundwater with mean value of -82 mV in pre monsoon and -56 mV in monsoon [17]. The groundwater wells in the study area are relatively anoxic, as indicated by low ORP values. The dominance of thick lacustrine clay probably restricts the downward diffusion of oxidants such as oxygen in deep groundwater [16].

The mean iron concentration in deep groundwater wells were 3.61 mg/L in pre monsoon and 3.76 mg/L in post monsoon with the highest concentration of 12.99 mg/L in post monsoon. Iron oxides dissolve under strongly acidic and reducing environment. If present in water in excessive amounts, it forms red Fe oxyhydroxide. Reductive dissolution of Fe (III) oxides accounts for the high Fe (II) content in anaerobic water [17,18]. The chemical composition of the major elements of the sediments, i.e., Fe_2O_3 ranged from 1.48 to 9.55 wt% [12] could be the source of iron in groundwater of the valley. A similar study indicated the Fe_2O_3 contents of the sediments are generally high (ranges < 0.5 to 15 wt%), and are uniformly higher in the fine sediments of the central basin (average 7 wt%) [8]. The overall bulk elemental concentrations in the central basin are greater in comparison with the northern part. The total iron content in the black sticky clay is high (>7 wt%) [19]. The variation in Fe_2O_3% and redox state of groundwater might have affected the levels of iron concentration in the groundwater.

Manganese concentration ranged from <0.02 to 1.55 mg/L (mean = 0.44 mg/L) in pre monsoon, whereas <0.02 to 1.75 mg/L (mean = 0.42 mg/L) in post monsoon. The chemical composition of the major element of the sediments, i.e., MnO from 0.01 to 0.18 wt% [20] is probable source of manganese in the groundwater of the valley. In the study area, arsenic concentration in the groundwater wells ranged from < 0.003 to 0.137 mg/L (mean = 0.013 mg/L) in pre monsoon and <0.003 to 0.131 mg/L (mean = 0.011 mg/L) in post monsoon. Arsenic concentration in 17% in pre monsoon and 26% in post monsoon exceeded provisional World Health Organization guideline value for drinking water of 0.010 mg/L [17]. A similar study reported arsenic concentration up to 0.265 mg/L [13]. Iron oxyhydroxides are the common host matter for arsenic, either adsorbed into the surface or co-precipitated [14] and reductive process is responsible for arsenic mobilization by dissolution or desorption in the groundwater [15,19], so it is possible that the higher concentration of arsenic in the groundwater is due to the more reducing environment as indicated by lower ORP value. Likewise, under the aerobic and acidic to near neutral conditions typical of many natural environments, arsenic is strongly adsorbed by oxide minerals as the arsenate ion and the concentrations in solution are low [20].

The groundwater wells observed elevated concentrations of iron and manganese. It is observed that the concentrations of iron and manganese are pH dependent and higher aggressiveness of iron and manganese in low pH [11]. Likewise, it is suggested that presence of a reductant (such as organic carbon) is a dominant factor controlling iron and manganese concentrations in groundwater. The oxidation of the reductant would leads to the reduction and solubilisation of iron and manganese [20] and the lacustrine clay in the Kadey catchment area is rich is organic matter [13] would contribute to the reducing environment. The organic matter may play an important role in the mobilization of arsenic, as reported by many studies from West Bengal (India) many states in Nigeria and Central Africa Republic [5,14,15]. It has been indicated that the fluvio-lacustrine sediments in the Catchment are rich in organic matter, and this organic matter may contribute in mobilization of arsenic.

Correlation between physicochemical parameters and arsenic

All the relationships of pH, EC, ORP and metals in the groundwater were examined by Spearman's rank correlation coefficient (Table 1).

The pH has strong negative correlations with iron and manganese, and weak negative correlation with arsenic (r = -0.534, $p < 0.01$; r = -0.402, $p < 0.01$; r = -0.169, $p < 0.129$) which can be explained by the higher aggressiveness of acidic media towards soil and host rocks that increase the concentrations of the rest of the ions [11]. Though arsenic

has negative correlation, p-value suggests that there is an insignificant negative correlation between arsenic concentration and pH in the groundwater.

The EC shows strong positive correlations (at $p < 0.01$) with iron, manganese and arsenic. Arsenic has strong positive correlations with iron and manganese (r = 0.384, $p < 0.01$; r = 0.447, $p < 0.01$), which is attributed to common geogenic origin of these metals.

The ORP has strong negative correlation with arsenic (r = -0.492, $p < 0.01$), which can be explained by reductive arsenic mobilization mechanisms in the groundwater. Likewise, ORP also shows negative correlations with iron and manganese. Reducing environment is responsible for the release of iron as well as manganese through the reduction of Mn (III, IV) hydroxides to soluble Mn (II) and of Fe (III) hydroxides to soluble Fe (II), respectively.

Correlation between arsenic and depth of groundwater

The depth of the deep groundwater wells tested arsenic ranged from 84 to 304 m. The mean and standard deviation (SD) of depth were 222.0 m and 63.4 m respectively (Table 2). The study showed weak positive correlations between arsenic and depth of groundwater in pre monsoon and post monsoon (r = 0.206, $p = 0.196$; r = 0.178, $p = 0.266$), respectively. Though, it showed positive correlations in both seasons, p-values suggest that there are insignificant positive correlations between arsenic concentration and depth of groundwater. However, it contradicts with the results shown by the some previous studies [13,15].

Temporal (seasonal) variation of arsenic

The temporal (seasonal) variation of the physicochemical parameters and metals were evaluated through season-parameter Spearman's correlation matrix in the groundwater. The measured parameters are not significantly ($p > 0.05$) correlated with the season except for pH (r = -0.238, $p < 0.05$), which infers similar distribution of arsenic in both seasons. Arsenic concentrations were insignificantly varied between seasonal groundwater. A study also reported very similar distributions of arsenic for pre monsoon and monsoon [15].

The lack of temporal (seasonal) variation is attributed to less dilution effect of monsoon rainfall in the groundwater. Additionally, contribution of anthropogenic metal contaminations is reluctant in the studied time-series in the groundwater wells. These findings are consistent with the results of previous study which pointed out seasonal variability has no significant effect on deep groundwater quality [17]. The similar studies also indicated no seasonal variability of arsenic in the groundwater [15,16]. The reports on the temporal variation of arsenic concentration in other parts of the world are inconsistent. Significant variation in arsenic concentration among the seasons was observed in a study [20]. A study spotted no significant monsoonal effects on arsenic distribution [17]. The seasonal variability has little effect in the groundwater [18]. Likewise, limited temporal variability observed in arsenic concentrations in groundwater [21-23].

Parameters	As	Mn	Fe	ORP	EC	PH
As	1					
Mn	0.447*	1				
Fe	0.384*	0.656*	1			
ORP	-0.492*	-0.447*	-0.570*	1		
EC	0.463*	0.654*	0.591*	-0.664*	1	
PH	-0.169	-0.402	-0.534*	0.052	-0.332*	1

Table 2: Spearman's rank correlation coefficients of physicochemical parameters and metals.

Spatial distribution of arsenic

The concentration of arsenic, iron, manganese and ORP value vary significantly in central, northern and southern groundwater districts in the valley. The spatial distribution pattern reveals higher values of arsenic, iron and manganese in central groundwater district. Northern and southern groundwater districts have lower arsenic concentrations in most of the groundwater wells. The concentration of the arsenic in groundwater of the study area increases from northern to southern and showing highest towards central groundwater district. Arsenic concentration was relatively lower in southern groundwater district in comparison with central groundwater district. The spatial distribution of ORP in the groundwater clearly shows most of the groundwater wells have lower ORP value.

The variations of elemental concentration are mainly clay controlled in both the margin and central parts. There is progressive increase in the finer particles and trace elements towards the central part of the sediments from the northern part in the valley [21] which is attributed to decrease in grain size and the concentration of metals in sediments tend to increase in fine grained sediments [2,11,12]. The larger particles in sediments have less surface area available for metal hydroxide coatings to form and adsorb arsenic and less adsorbed arsenic contributes a smaller amount of aqueous arsenic in equilibrium with adsorbed arsenic. Therefore, there is in less potential for release of arsenic through reductive mobilization mechanisms [19]. The variation in grain size has role in mobilization of metals in groundwater. Therefore, higher metal concentration is associated with the fine grained sediments in the central groundwater district. Furthermore, higher concentrations of metals in central groundwater district might be due to the fact that the central groundwater district is considered as poorly recharging due to the presence of a thick black clay layer [22].

Groundwater quality depends on the composition of recharging water, the mineralogy and reactivity of the geological formations in aquifers, anthropogenic activities and environmental conditions that may affect the geochemical mobility of certain constituents [23-25]. Arsenic concentrations in groundwater of the Kadey catchment area show a wide range and some of groundwater sources investigated were found to be in elevated levels in some parts of the basin could be due to the nature of the sediments there. The high degree of spatial variability in groundwater quality over short distances indicates that groundwater movement has been limited and is poorly mixed [24].

Arsenic concentrations in the sediments of the Kadey catchment area averaged 8 mg/kg (ranging 3 to 25 mg/kg) similar to the general level seen in modern unconsolidated sediments, typically 5 to 10 mg/kg [26,27]. The widespread lacustrine clay of the Kadey catchment area could have greater potentiality for arsenic release [28]. Arsenic mobilization is high in the reducing conditions [17,26,29]. The higher concentration of arsenic under reduced groundwater environment may be due to Fe/Mn oxides and direct reduction of As (V) into As (III). After an initial increase, arsenic concentration often decreases again as a function of time below water table due to sulfide precipitation, whereas it increases with increasing sulfate concentrations above water table [30]. Under moderately reduced environment (0 to 100 mV), arsenic solubility seemed to be controlled by the dissolution of Fe oxyhydroxides. But at highly reduced condition, e.g., at -250 mV, arsenic chemistry is dominated by the formation of insoluble sulfides $FeAsS$, AsS, As_2S_3 [22] attenuating concentration of arsenic in the groundwater.

Conclusion

Firstly, this study has confirmed the presence of higher levels of arsenic in deep groundwater of the Kadey catchment area. Arsenic showed wide spatial variation ranged from <0.003 to 0.137 mg/L. In pre monsoon and post monsoon, 17% and 26% of groundwater wells, respectively exceeded arsenic concentration of permissible WHO guideline value of 0.010 mg/L for drinking water. The arsenic varies spatially with high concentration towards central groundwater district. Strong negative correlation between arsenic and ORP demonstrated reductive arsenic mobilization mechanisms in deep groundwater. Arsenic showed strong correlations with iron, manganese and EC. Iron and manganese are presumably the main factors in regulating release of As in groundwater through reductive mobilization mechanism. Secondly, the study revealed weak positive correlation between arsenic concentration and depth of deep groundwater. Arsenic is not significantly correlated with season suggesting similar distribution of arsenic in both seasons. The high concentration of arsenic in deep groundwater of some parts of study area particularly in central groundwater district is attributed to the groundwater geochemistry of the study area. Release of arsenic into the groundwater is considered to be due to the natural source under the reductive process.

Acknowledgements

The authors would highly acknowledge Sub-Regional Bilingual University of Mining, Sciences, Technology, Yaounde-Cameroon for the support of this research work. We would like to thank First Aid – International Medical Assistance (PU-AMI).

References

1. Chapagain SK, Shrestha S, Nakamura T, Pandey VP, Kazama F (2009) Arsenic Occurrence in Ground water of Kathmandu Valley, Nepal. Desalination and Water Treatment 4: 248-254.

2. Garbarino JR, Hayes HC, Roth DA, Antweiler RC, Brinton TI, et al. (1995) Heavy Metals in the Mississippi River. US Geological Survey Circular 1133, Virginia.

3. IARC (2004) IARC Monographs on the Evaluation of Carcinogenic Risks to Humans. Some Drinking-water Disinfectants and Contaminants, Including Arsenic. International Agency for Research on Cancer, Lyons 84.

4. Sigha-Nkamdjou L (1994) Hydrochemical functioning of forest ecosystems in Central Africa: Ngoko to Moloundou (Southeast of Cameroon) 380.

5. Marcovecchio JE, Botte SE, Freije RH (2007) In: Nollet ML (ed.) Heavy Metals, Major Metals, Trace Elements. Handbook of Water Analysis. 2nd Edition, CRC Press, London pp: 275-311.

6. Appelo CAJ, Postma D (2005) In: Balkema AA (ed.) Geochemistry, Groundwater and Pollution. 2nd Edition, Publishers, Amsterdam.

7. Mitchell E, Frisbie S, Sarkar B (2011) Exposure to Multiple Metals from Groundwater - A Global Crisis: Geology, Climate Change, Health Effects, Testing and Mitigation. Metallomics 3: 874-908.

8. Hossain MF (2006) Arsenic Contamination in Bangladesh - An Overview. Agriculture, Ecosystems and Environment 113: 1-16.

9. Mukherjee A, Sengupta MK, Hossain MA, Ahamed S, Das B, et al. (2006) Arsenic Contamination in Groundwater: A Global Perspective with Emphasis on the Asian Scenario. J Health, Popul Nutr 24: 142-163.

10. Acres International (2004) Optimizing Water Use in Kathmandu Valley (ADB-TA) Project. Final Report. Acres International in Association with Arcadis Euroconsult Land and Water Product Management Group, East Consult (P) Ltd. and Water Asia (P) Ltd.

11. ICIMOD (2007) Kathmandu Valley Environment Outlook. International Centre for Integrated Mountain Development, Kathmandu.

12. CBS (2012) National Population and Housing Census 2011. National Report. NPHC 2011. Nepal Bureau of Statistics, Kathmandu 1.

13. JICA/ENPHO (2005) Arsenic Vulnerability in Groundwater Resources in Nigeria Delta. National Final Report II. Japan International Cooperation Agency / Environment and Public Health Organization, Nigeria.

14. Khatiwada NR, Takizawa S, Tran TVN, Inoue M (2002) Groundwater Contamination Assessment for Sustainable Water Supply in Kathmandu Valley, Nepal. Water Science Technology 46: 147-154.

15. Kalédjé PSK, Ngoupayou JRN, Kpoumié A, Takounjou AF, Ondoua JM (2016) Analysis of climate variability and its influence on the hydrological response of the catchment area of Kadey (east Cameroon). International Journal of Geosciences 47: 127-138.

16. Jha MG, Khadka MS, Shrestha MP, Regmi S, Bauld J, et al. (1997) The Assessment of Groundwater Pollution in Kathmandu, Nepal. Report on Joint Nepal-Australia Project, 1995-96. Australian Geological Survey Organization.

17. CBS (2013) Environment Statistics of Nigeria. Nigeria Bureau of Statistics, Lagos.

18. Shrestha SM, Rijal K, Pokhrel MR (2013) Arsenic Contamination in the Deep and Shallow Groundwater of Kathmandu Valley, Nepal. Scientific World 10: 25-31.

19. Shrestha SM, Rijal K, Pokhrel MR (2014) Spatial Distribution and Seasonal Variation of Arsenic in Groundwater of Kathmandu Valley, Nepal. Journal of Institute of Science and Technology 19: 7-13.

20. APHA-AWWA-WEF (2005) Standard Methods for the Examination of Water and Wastewater. 21st Edition, American Public Health Association, American Water Works Association, Water Environment Federation, Washington DC.

21. Yoshida M, Igarashi Y (1984) Neogene to Quaternary Lacustrine Sediments in the Kathmandu Valley, Nepal. Journal of Nepal Geological Society 4: 73-100.

22. Shrestha SD, Karmacharya R, Rao GK (1996) Estimation of Groundwater Resources in Kathmandu Valley, Nepal. Journal of Groundwater Hydrology 38: 29-40.

23. JICA (1990) Groundwater Management Project in the Kathmandu Valley. Final Report to Nepal Water Supply Corporation. Japan International Cooperation Agency, Kathmandu.

24. Chapagain SK, Pandey VP, Shrestha S, Nakamura T, Kazama F (2010) Assessment of Deep GroundwaterQuality in Kathmandu Valley Using Multivariate Statistical Techniques. Water Air Soil Pollution 210: 277-288.

25. Smedley PL, Kinniburgh DG (2013) In: Selinus O (ed.) Arsenic in Groundwater and the Environment. Essential of Medical Geology pp: 279-310.

26. Sakai T, Gajurel AP, Tabata H, Uprety BN (2001) Small Amplitude Lake-Level Fluctuations Recorded in Aggrading Deltaic of the Upper Pleistocene Thimi and Gokarna Formations, Kathmandu Valley, Nepal. Journal of Nepal Geological Society 25: 43-52.

27. Kalédjé PSK, Ngoupayou JRN, Kpoumié A, Takounjou AF, Ondoua JM (2016) Hydrogeology of the watershed of the river Kadey: approach of the water exchange surface area/underground continuous medium. Journal of Water Resources and Protection.

28. Gurung JK (2007) Geochemical Studies of Sediments and Water, and Implications for Mobilization of Arsenic into Groundwater in Nepal and Japan. Ph.D. Thesis, Shimane University, Japan.

29. Hem JD (1985) Study and Interpretation of Chemical Characteristics of Natural Water. 3rd Edition, US Geological Survey Water-Supply Paper 2254.

30. Dixit A, Upadhya M (2005) Augmenting Groundwater in Kathmandu Valley: Challenges and Possibilities. Report to Nepal Water Conservation Foundation, Kathmandu, Nepal.

16

A Preliminary Estimate of the Reserve of the Marble Deposit in Itobe Area, Central Nigeria

Onimisi M[1]*, Abaa SI[2], Obaje NG[3] and Sule VI[4]

[1]Department of Earth Sciences, Kogi State University, Anyigba, Nigeria
[2]Department of Geology and Mining, Nasarawa State University, Keffi, Nigeria
[3]Department of Geography and Geology, Ibrahim Babangida University, Lapai, Nigeria
[4]Department of Physics, Kogi State University, Anyigba, Nigeria

Abstract

A preliminary estimate of the reserve of the marble deposit in Itobe area, central Nigeria was determined from both outcrop geological mapping and geophysical resistivity surveying. Two outcrops (designated as mass I and mass II) of the marble deposit occurring about 850 m apart along a NE–SW axis were identified. The areal extent, thickness and density of the marble have been used in computing the reserve of the marble deposit. The reserve of mass I is estimated at 1418.4 tons and that of mass II has been estimated at 142643.2 tons. The total reserve of the marble deposit is 144061.6 tons.

Keywords: Nigeria; Marble; Deposit; Mass

Introduction

Marble, a major raw material for industries, is a crystalline, non-foliated metamorphic rock formed from the metamorphism of limestone (a carbonate sedimentary rock formed at the bottom of lakes and seas). The Itobe marble deposit is located about 1 km from Itobe town along the Ajaokuta-Anyigba road, Kogi State, central Nigeria. The study area lies between longitudes 6°40' E and 6°48'E and latitudes 7°22'N and 7°30' N (Figure 1).

The industrial uses of a marble deposit are determined largely by its chemical composition and the reserve of the deposit. Geological and geophysical investigation methods have been employed in the exploration of marble deposits. The electrical resistivity method of geophysical prospecting in particular has been employed to delineate the areal extent and thickness of marble deposits. The choice of the resistivity method was informed by the local resistivity contrast characteristics of marble deposits (10^2-2.5×10^8 Ωm) compared to the immediate host rocks of schists (20-10^4 Ωm) and quartzites (10-2×10^8 Ωm) [1-4]. This study aims at estimating the reserve of the marble deposit in Itobe from both outcrop geological studies and geophysical resistivity investigations.

Field Geologic Occurrence

The marble deposit in Itobe is associated with crystalline rocks of the Precambrian Basement Complex. The marble deposit occurs within a host rock of quartz-mica schist and feldspathic quartzite. Two outcrops (designated as mass I and mass II) of the marble deposit have been identified in the study area (Figure 2). Mass I with a surface area of 0.84 m², trends NE-SW with a strike azimuth of 033°. It is poorly exposed near Alo village, along the Anyigba-Itobe road. It is light grey in color and fine grained in texture. Mass II, which is about 800 meters from mass I on a NE-SW axis, is a much larger deposit, and outcrops on the Ayanka hill as massive boulders. It is medium grained in texture, dark grey in color and has an elevation of 132 m at the bottom and 138 m at the top. It has a sharp contact with the quartz mica schist to the east and the Feld spathic quartzite to the west.

Method of Study

The lateral extent and thickness of the marble deposit was obtained both from outcrop geological and geophysical resistivity surveys. The Etrex Global Positioning System was used to measure the elevations at the bottom and top of the outcropping marble. The Vertical Electrical Sounding (VES) method using the Schlumberger configuration was employed to determine the thickness of the marble deposit. The horizontal profiling method using the Wenner configuration was employed to determine the lateral extent of the marble deposit (Figures 3 and 4).

The density of the marble was determined by weighing a fresh and clean sample of the marble rock using a Top Weighing Balance, and the mass was noted. A well calibrated beaker was filled with water to a certain mark and the mark noted. The marble sample was lowered into the beaker containing water using a thin thread. The change in the volume of the water in the beaker, which is equal to the volume of the marble sample, was noted. The density of the marble sample was obtained by dividing the mass of the marble sample by the change in the volume of the water on immersion of the marble sample.

The estimated total reserve of the marble deposit was determined by adding the reserves of the individual masses (masses I and II). The reserve of each mass of the marble deposit has been computed using the block method [5]. The surface extent of mass I and mass II (Figures 5 and 6) were subdivided into blocks of regular geometries (rectangular block of 0.072 cm × 0.096 cm) and then summing up the area of the entire blocks. The reserve of each mass of the marble deposit is obtained as follows:

Area of the marble outcrop (m²) × thickness of the marble outcrop (m) × the density of the marble (kg/m³)

Results and Discussions

The apparent resistivity data for the seven horizontal profiles

*Corresponding author: Onimisi M, Department of Earth Sciences, Kogi State University, Anyigba, Nigeria, E-mail: matflo65@yahoo.com

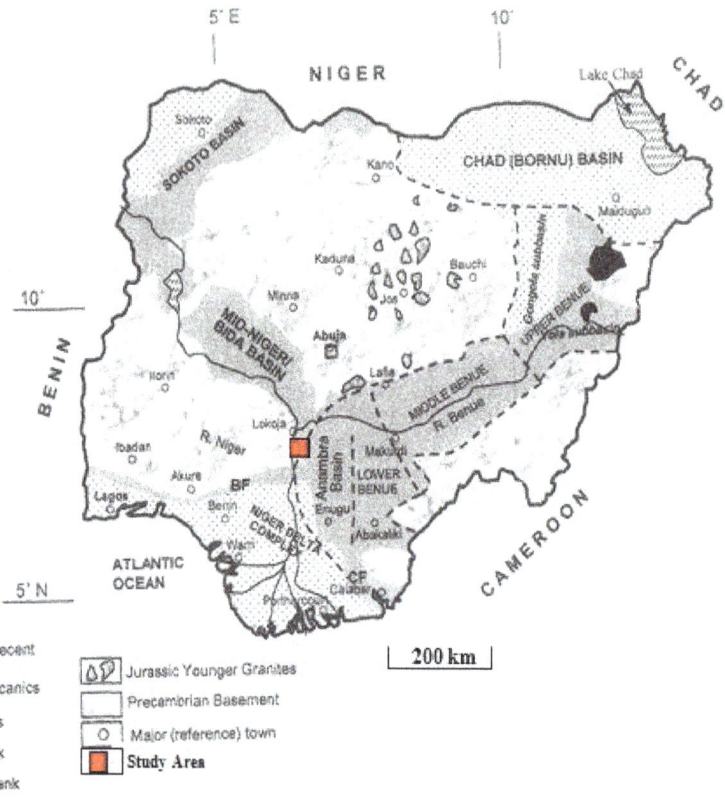

Figure 1: Geological sketch map of Nigeria showing the location of the study area (After Obaje, 2009).

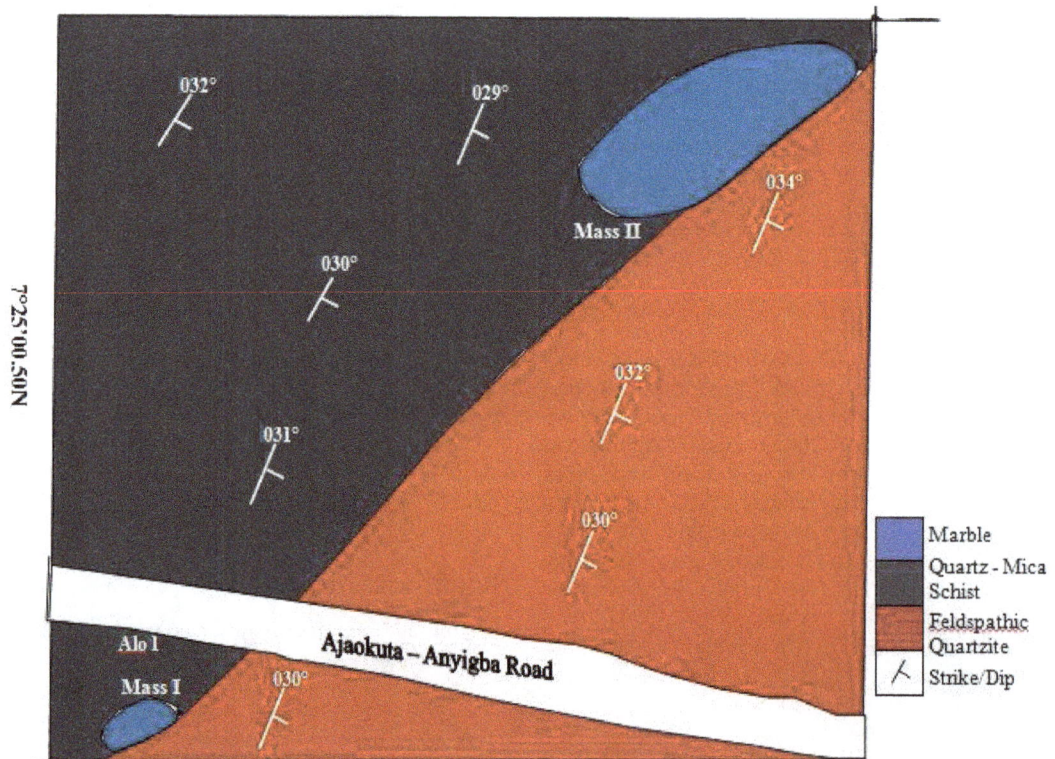

Figure 2: Geologic map of the study area showing the two masses of the marble deposit.

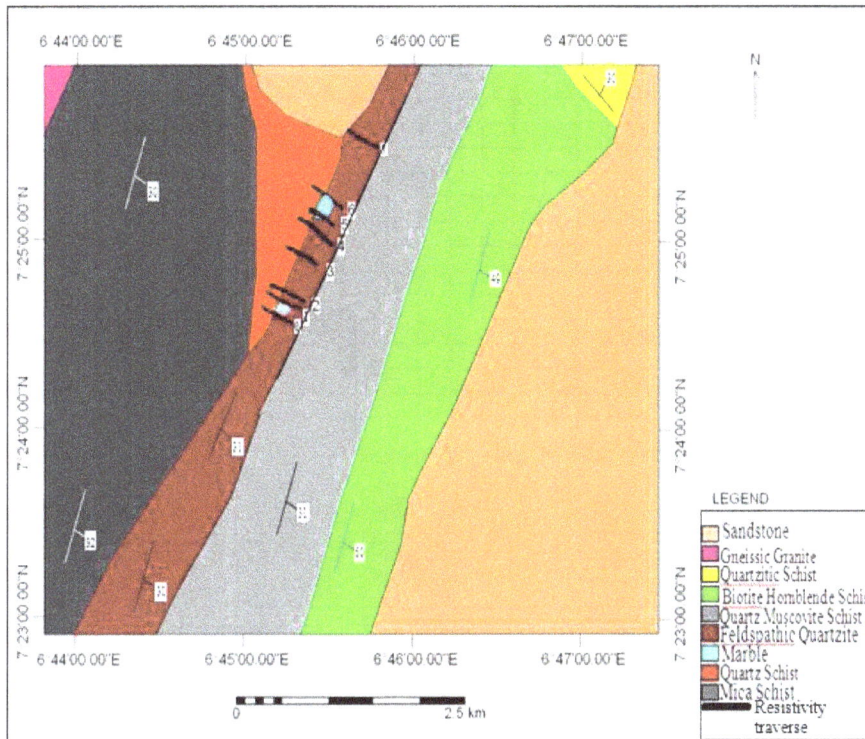

Figure 3: Geologic map showing the resistivity profiling traverses across the marble outcrops.

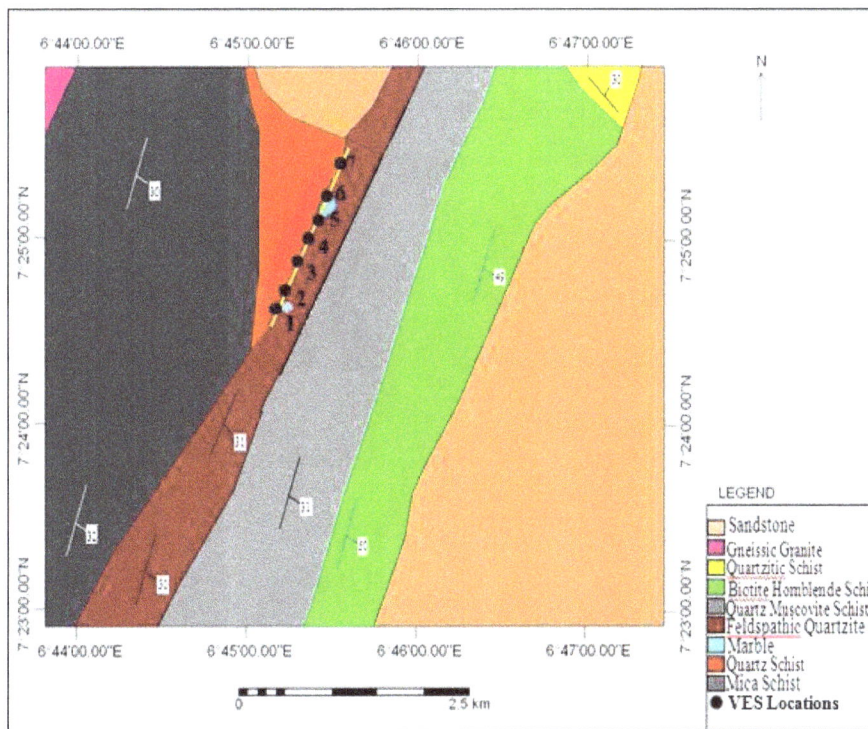

Figure 4: Geologic map showing the VES locations along the marble outcrops.

carried out across the marble deposit in Itobe are presented in Table 1 below. The interpretation of the resistivity profiles is qualitative, and it involves identifying signatures characteristic of the marble. Figure 7 shows a correlation of the resistivity profiles along the seven traverses. The outcropping marble along traverses 5 and 6 is characterized by a relatively high resistivity (500-600 Ωm) in a background of low resistivity

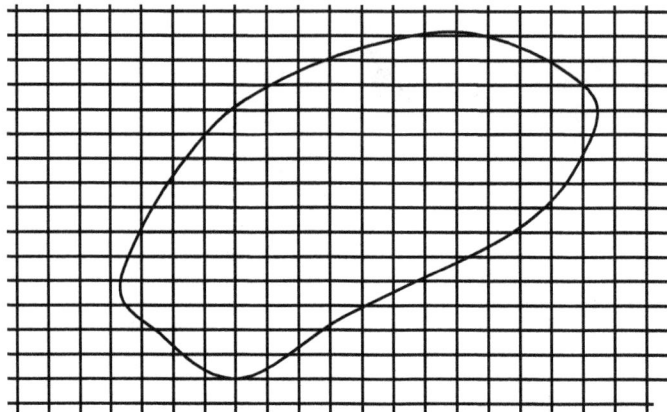

Figure 5: Rectangular grids on the surface area of mass I (× 7 of Figure 2).

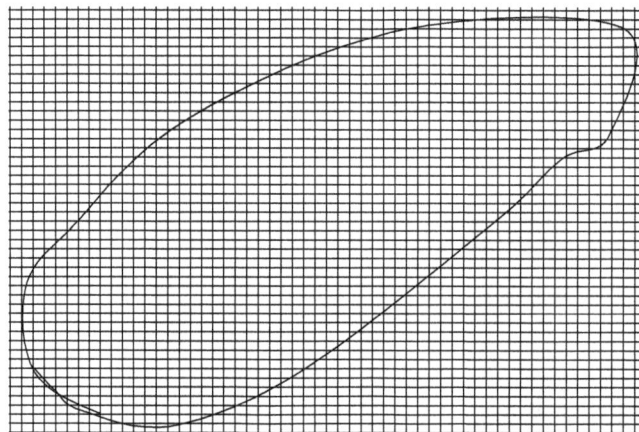

Figure 6: Rectangular grids on the surface area of mass II (× 7 of Figure 2).

Station Number (Si)	Apparent Resistivity, (Ωm)Profile 1	Apparent Resistivity (Ωm)Profile 2	Apparent Resistivity (Ωm)Profile 3	Apparent Resistivity (Ωm)Profile 4	Apparent Resistivity (Ωm)Profile 5	Apparent Resistivity (Ωm)Profile 6	Apparent Resistivity (Ωm)Profile 7
S1	46.70	7.61	14.61	6.91	7.23	6.60	49.47
S2	39.10	13.51	5.94	5.66	7.23	5.66	41.49
S3	84.92	16.69	8.01	3.46	8.17	3.46	64.18
S4	98.44	15.65	17.63	2.51	13.83	7.23	58.08
S5	55.76	16.22	18.26	6.91	25.14	7.54	65.37
S6	55.19	21.66	21.69	4.71	25.46	503.0	13.39
S7	22.88	27.85	25.14	4.71	506.0	524.0	17.92
S8	35.20	28.73	44.00	4.71	518.6	530.0	17.85
S9	60.22	22.32	19.80	25.14	512.3	20.12	33.32
S10	60.22	23.57	24.83	25.14	24.83	23.26	23.70
S11	85.43	24.20	24.20	25.14	25.14	23.89	24.70
S12	94.42	30.49	24.83	25.46	25.14	25.41	20.24
S13	96.18	30.80	30.49	25.77	25.46	30.17	98.06
S14	37.72	30.49	31.12	29.23	30.80	30.49	38.34
S15	38.34	30.80	32.37	30.49	31.12	30.49	40.23
S16	40.23	30.49	31.12	30.80	31.12	30.49	23.26
S17	22.63	30.17	31.74	30.80	31.43	30.80	22.00
S18	16.34	43.69	30.80	31.74	30.80	30.49	19.49
S19	23.26	43.37	30.49	31.12	31.12	30.49	23.89
S20	20.12	47.46	44.32	30.80	30.49	30.80	20.12

Table 1: Horizontal resistivity profiling data in the study area.

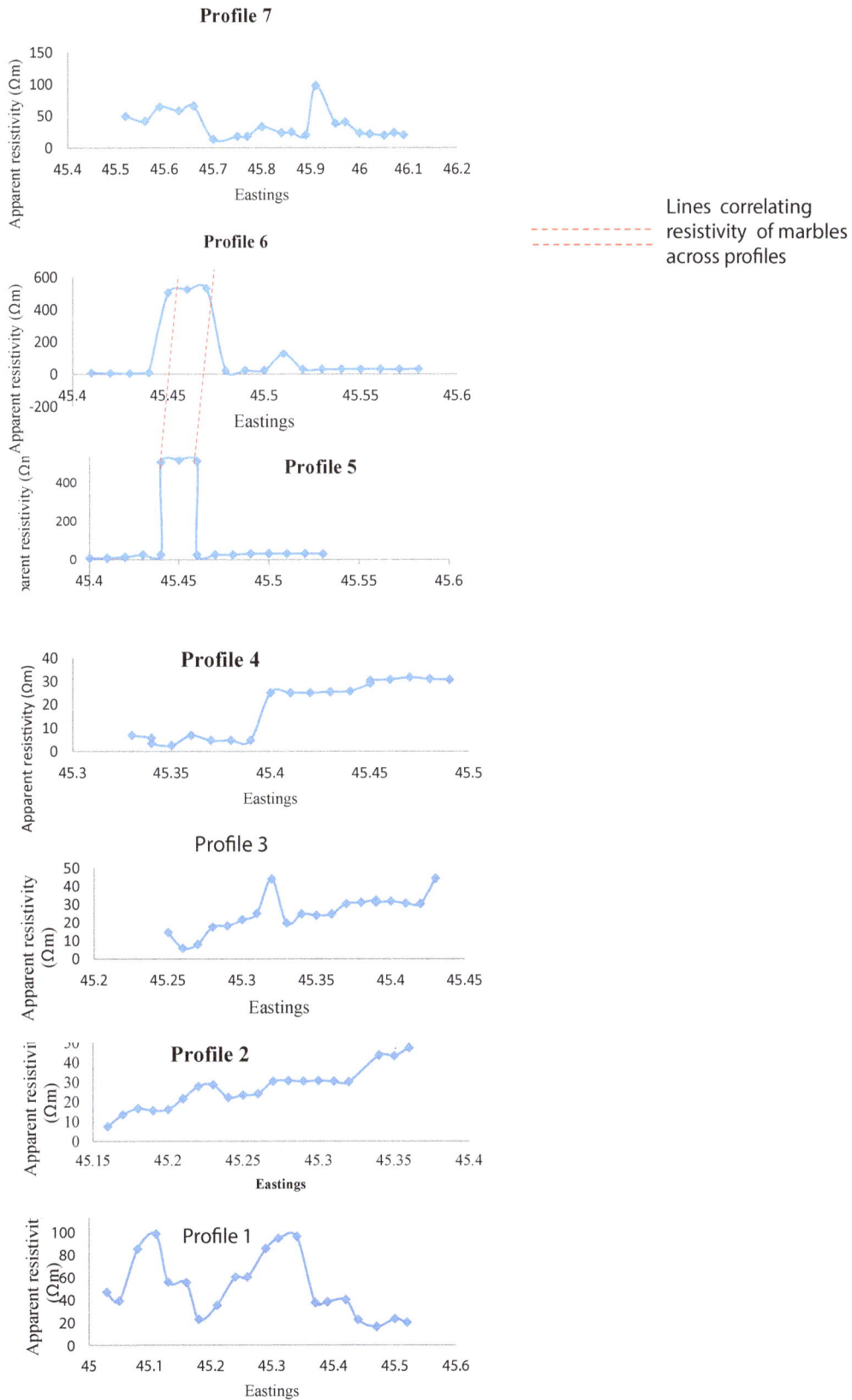

Figure 7: Correlation of the resistivity profiles across mass I and mass II.

(6-250 Ωm) characteristic of the host rocks (mica schist and quartz mica schist). Similar geophysical patterns were observed over marble deposits in Igarra area of southwestern Nigeria [2,3]. The interpretation of the resistivity profiles across the marble deposit in Itobe using the above characteristic shows that there is no continuity between masses I and II of the marble deposit. The marble deposit in Itobe shows a NE-SW trend.

The resistivity sounding data obtained for eight vertical electrical sounding surveys carried out along the trend of the marble deposit are presented in Table 2, and the corresponding resistivity sounding curves are shown in Figures 8-15. The interpretation of the VES data involves the determination of the geoelectric layer parameters (resistivity and thickness) of the marble and the surrounding rocks (Table 3).

AB/2 (m)	Apparent resistivity (Ωm)					AB/2 (m)	Apparent resistivity (Ωm)		
	VES 1 Elevation =116 m	VES 2 Elevaton=114m	VES 3 Elevation=113 m	VES 4 Elevation =112 m	VES 5 Elevation =117 m		VES 6 Elevation = 118 m	VES 7 Elevation =114 m	VES 8 Elevation = 112 m
1.0	1185	380	340	1100	1776	1.1	61	110	105
1.3	1000	300	284	985	1544	1.6	69	95	93
1.8	642	160	147	778	1154	2.3	80	78	75
2.4	398	130	95	700	822	3.4	99	70	60
3.2	177	96	80	589	570	5.0	105	80	50
4.2	100	81	70	486	389	7.3	119	05	60
5.6	77	43	61	403	324	10.7	122	142	69
7.5	67	70	61	274	263	15.8	130	180	85
10.0	74	75	69	184	260	3.2	150	250	120
13.3	84	85	75	132	285	4.1	190	320	170
18.0	81		90	120	300	0.0	255	400	240
24.0	97		98	130	330	3.5	340	00	340
32.0	110			150	340				

Table 2: Vertical electrical sounding data in the study area.

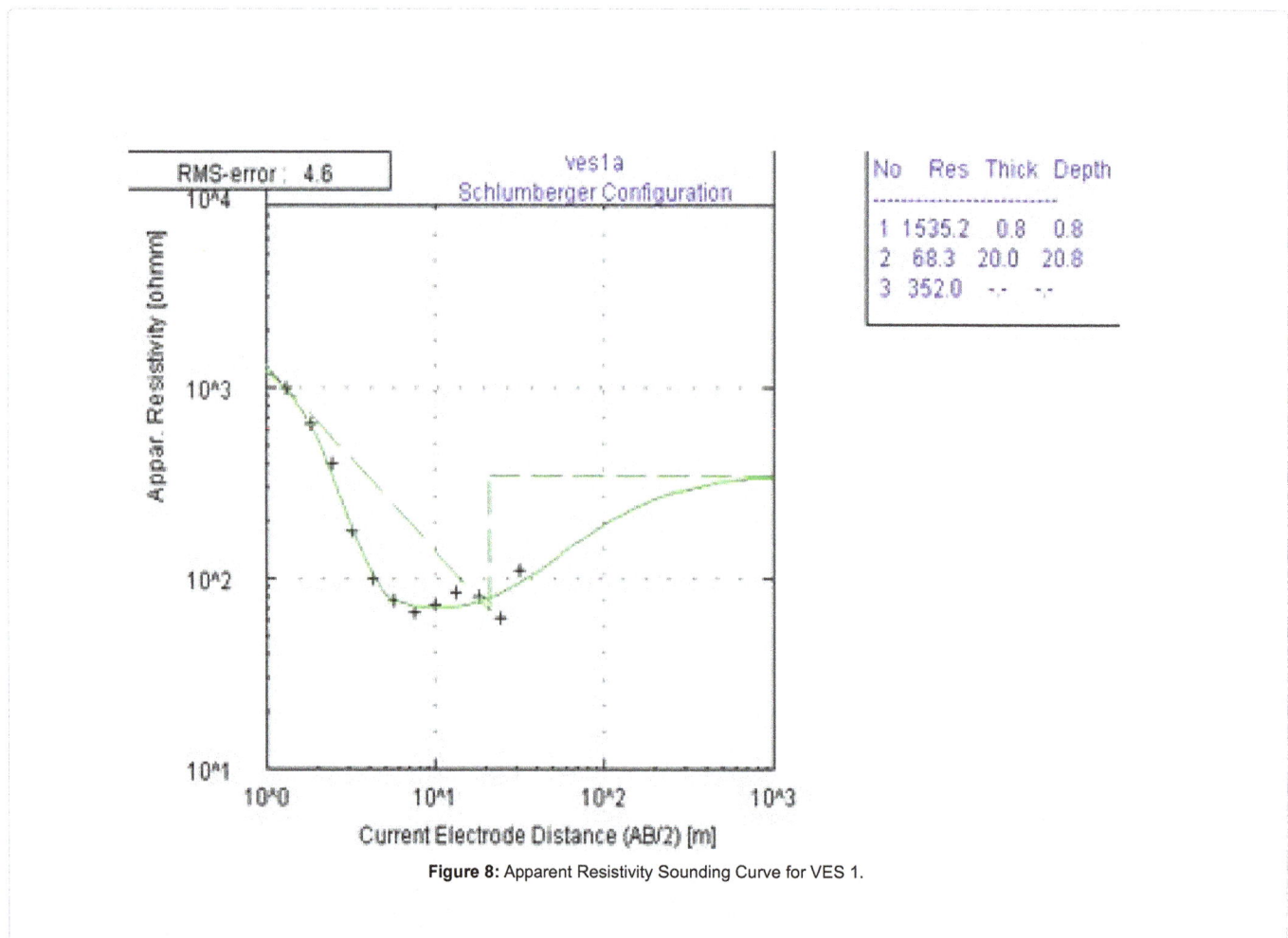

Figure 8: Apparent Resistivity Sounding Curve for VES 1.

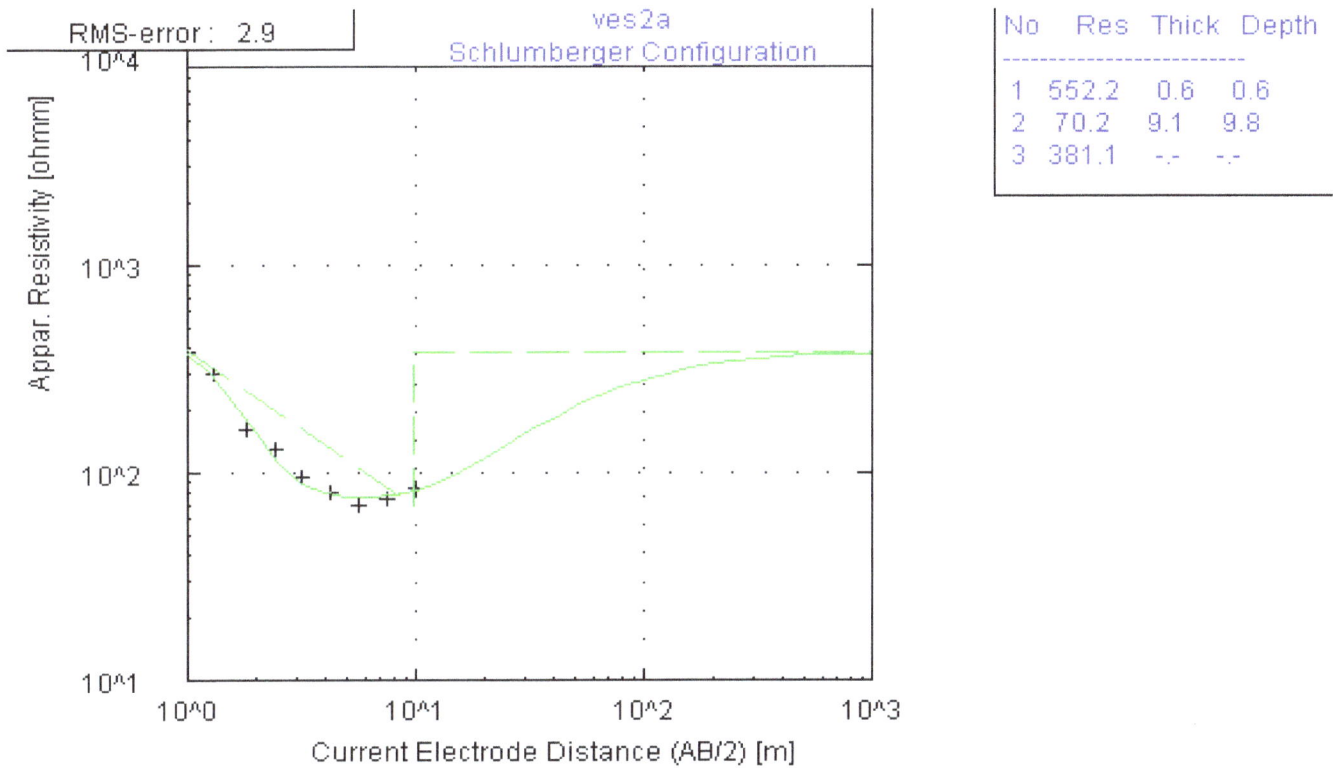

Figure 9: Apparent Resistivity Sounding Curve for VES 2.

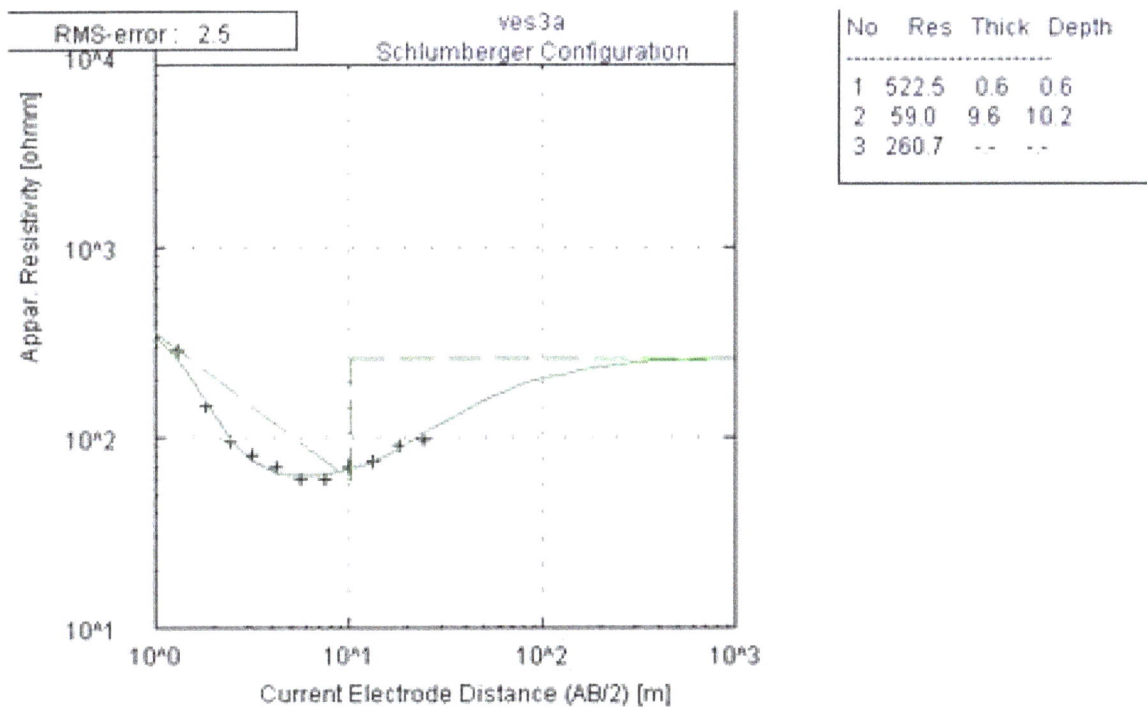

Figure 10: Apparent Resistivity Sounding Curve for VES 3.

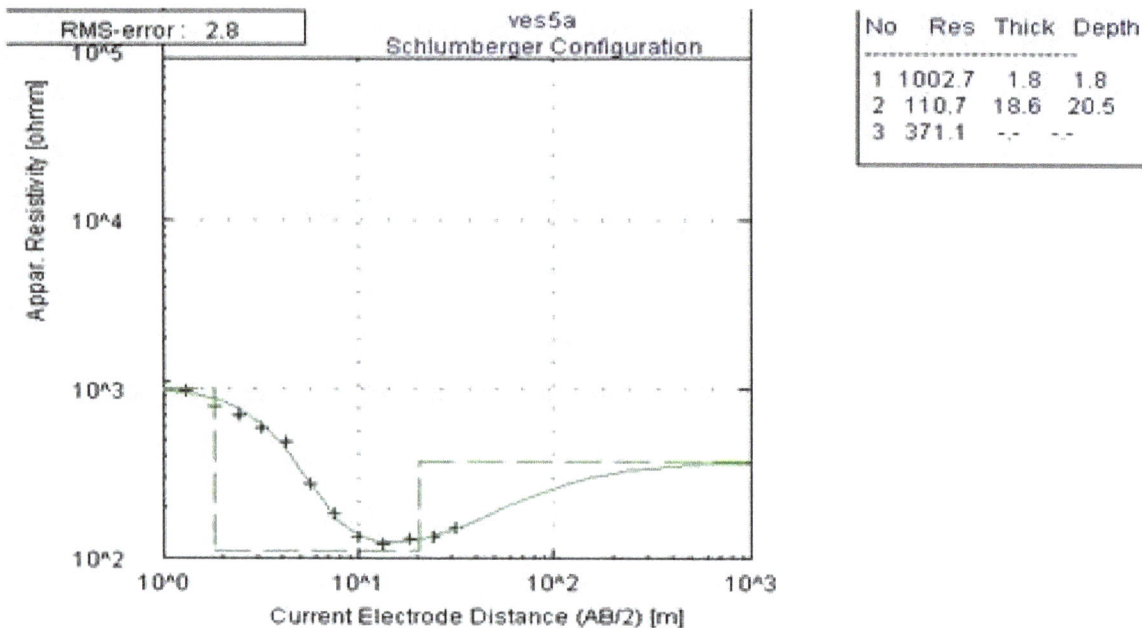

Figure 11: Apparent Resistivity Sounding Curve for VES 4.

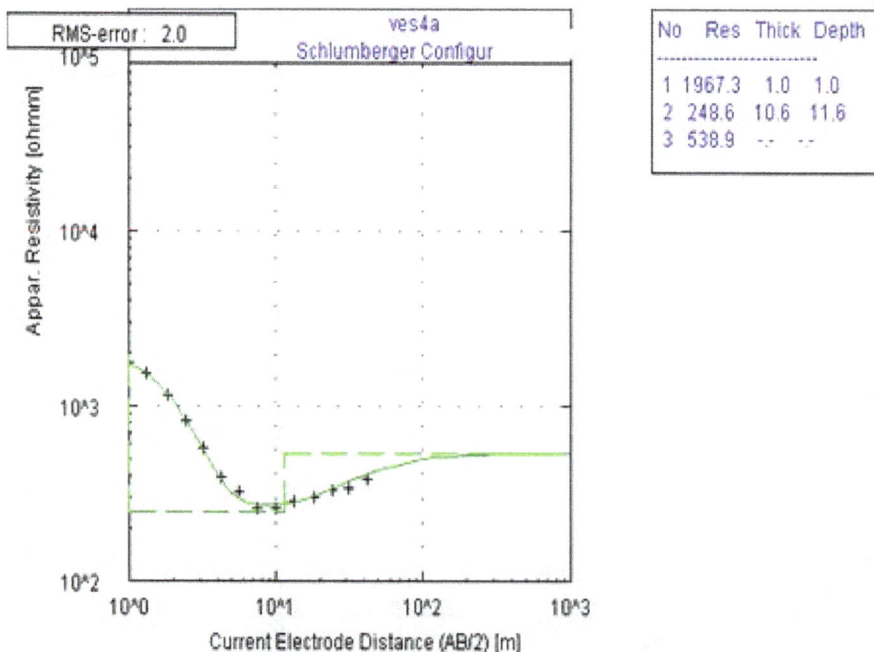

Figure 12: Apparent Resistivity Sounding Curve for VES 5.

Fig.11:

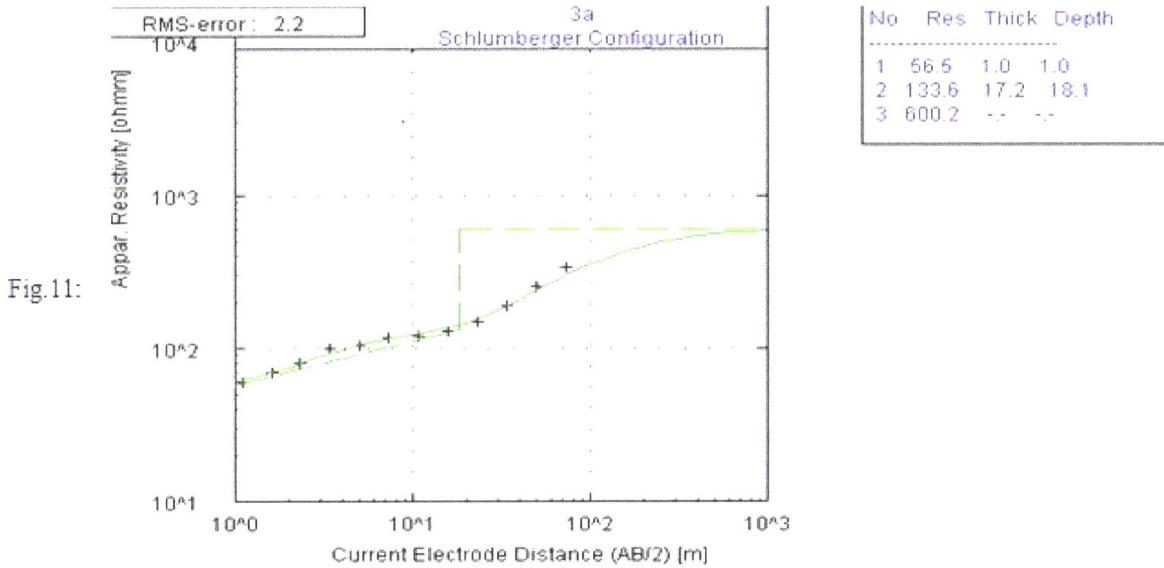

Figure 13: Apparent resistivity sounding curve for VES 6.

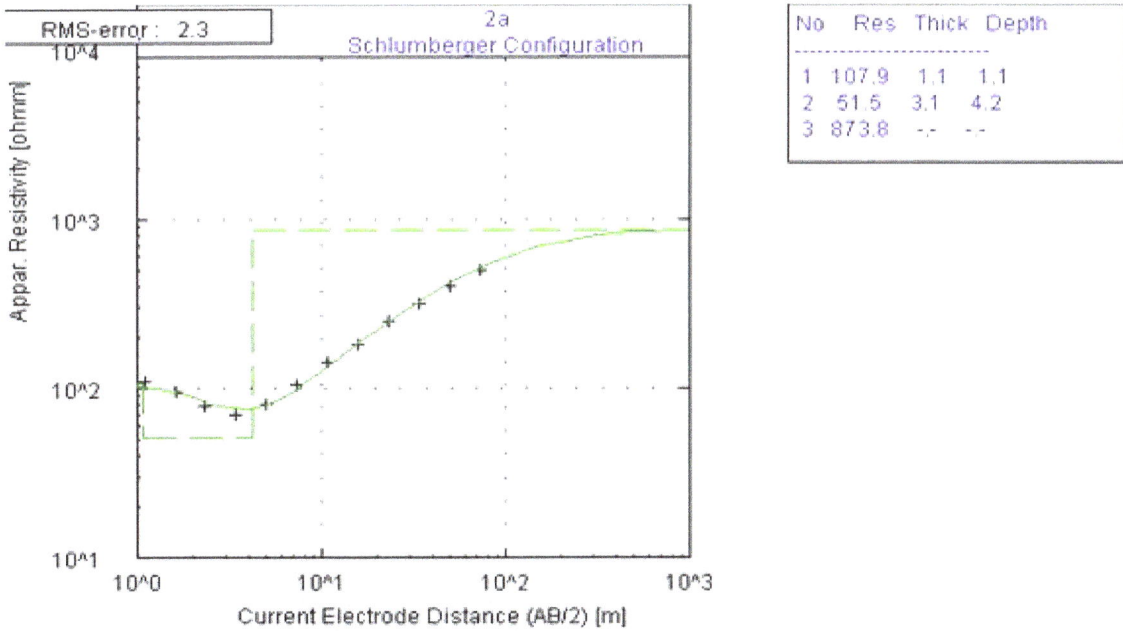

Figure 14: Apparent Resistivity Sounding Curve for VES 7.

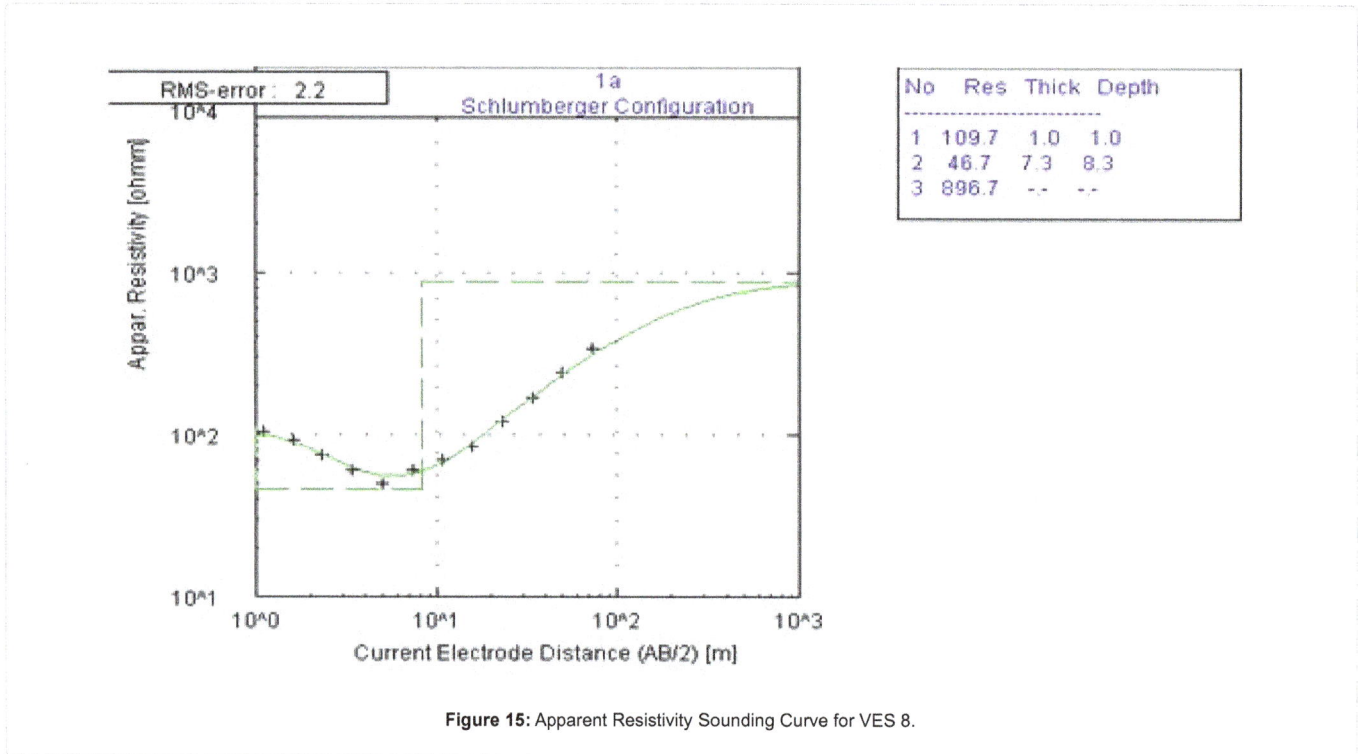

Figure 15: Apparent Resistivity Sounding Curve for VES 8.

VES Location	RMS Errors	Layer Resistivity (Ωm)			Layer Thickness (m)			Depth to layer (m)		
		1	2	3	1	2	3	1	2	3
VES 1	4.6	1535.2	68.3	352.0	0.8	20.0		0.8	20.8	
VES 2	2.9	552.2	70.2	381.1	0.6	9.1		0.6	9.8	
VES 3	2.5	522.5	59.0	260.7	0.6	9.6		0.6	10.2	
VES 4	2.8	1002.7	110.7	371.1	1.8	18.6		1.8	20.5	
VES 5	2.0	1967.3	248.6	538.9	1.0	10.6		1.0	11.6	
VES 6	2.2	56.5	133.6	600	1.0	17.2	-	1.0	18.1	
VES 7	2.3	107.9	51.5	873.8	1.1	3.1		1.1	4.2	-
VES 8	2.2	109.7	46.7	896.7	1.0	7.3		1.0	8.3	

Table 3: Geoelectric layer parameters (resistivities, thicknesses and depth) in the study area.

Sample Number	Mass I (kg/m³)	Mass II (kg/m³)
1	2610	2430
2	2620	2470
3	2640	2550
4	2660	2580
5	2660	2450
6	2670	2440
7	2665	2560
8	2670	2490
9	2650	2570
10	2658	2575
Mean	2650.3	2511.5
Standard Deviation	20.8	61.2

Table 4: Density of the marble samples.

The resistivity value for marble in the study area (obtained from the outcropping marble at VES 2 location), varies from 520 to 555 Ωm, depending on water content and porosity.

Three geoelectric layers have been recognized within the subsurface of the VES sounding locations. The geoelectric parameters of the resistivity sounding survey (Table 3) shows a thickness of 0.6 m for the outcropping marble (mass I) at VES 2 location. This marble unit is suspected to extend to VES 3 as shown by the resistivity values of the first geoelectric layer (552-523 Ωm) between VES 2 and VES 3 locations. This marble unit is underlain by a second geoelectric layer interpreted as mica schist with a resistivity ranging from 7050 Ωm, and a thickness ranging from 9.2-9.6 m. The third geoelectric layert, with a resistivity value ranging from 381-261 Ωm, is a highly weathered Basement rock. VES locations 5 and 6, in close contact with the outcropping marble in mass II is a lateritic layer with a resistivity of 1003 1967 Ωm and a thickness ranging from 1.0-1.8 m. It is underlain by a quartz-mica schist layer with resistivity values in the range (249-134 Ωm) and a thickness ranging from 10.617.2 m. Underlying the quartz-mica schist is the weathered Basement rock with a layer resistivity ranging from 539-630 Ωm. This is a more competent layer than that below mass I.

The average density value for the two marble outcrops in the study area is presented in Table 4.

(i) Calculation of reserve of marble in mass I:

Area of mass I=Area of triangle + Area of rectangle

= 0.089856 m² + 0.801792 m²

=0.891648 m²

Thickness of mass I=0.6 m

Density of the marble=2650.3 kg/m³

Reserve of mass I=0.892 m² × 0.6 m × 2650.3 kg/m³=1418.4 tons

(ii) Calculation of reserve of marble in mass II:

Area of mass I=Area of triangle + Area of rectangle

=0.390528 m² + 9.075456 m²

=9.465984 m²

Area of mass II=9.466 m²

Thickness of mass II=6.0 m

Density of the marble=2511.5 kg/m³

Reserve of ass II=9.466 m² × 6 m × 2511.5 kg/m³=142643.2 tons

(iii) Total reserve of the marble in Itobe=reserve of mass I + reserve of mass II

=1418.4 tons + 142643.2 tons

=144061.6 tons

Conclusion

Field geological and geophysical investigations of the marble deposit in Itobe area, Kogi State, central Nigeria reveals the occurrence of two separate marble outcrops (mass I and mass II) occurring 800 m apart along a NE–SW trend. The estimated reserve of the marble in mass I is 1418.4 tons and the estimated reserve of marble in mass II is 142643.2 tons. The total reserve of the marble deposit in Itobe is 144061.6 tons.

References

1. Folami SL, Ojo JS (1991) Gravity and magnetic investigations over marble deposits in the Igarra area, Bendel State. *J Mining and Geology* 27: 49-59.

2. Ojo JS, Olorunfemi MO (1992) Geophysical investigation of Ogurabe marble deposit. Geowork Ltd.

3. Aina AO, Olorunfemi MO (1996) Comparative field results for electrical resistivity, magnetic, gravity and VLF methods over a marble lens. Exploration Geophysics 27: 217-222.

4. Odeyemi IB, Oloruniwo MA, Folami SL (1997) Geological and geophysical characteristics of Ikpeshi marble deposit, Igarra area, southwestern Nigeria. Journal of Mining and Geology 33: 63-79.

5. Po Poff CC (1966) Computing reserves of mineral deposits: principles and conventional methods. Information Circular 8283, USBM, Washington.

17

Petrophysical and Mechanical Evaluation of the Moghra Sandstones, Qattara Depression, North Western Desert, Egypt

Mohamed K Salah[1]*, Ahmed Abd El-Al[2,3] and Abdel-Hameed AT[4]

[1]Department of Geology, American University of Beirut, Lebanon
[2]Civil Engineering Department, Najran University, Saudi Arabia
[3]Geology Department, Al Azhar University, Egypt
[4]Geology Department, Tanta University, Egypt

Abstract

In the present study, we conduct a number of petrophysical and geomechanical investigations on a large number of sandstone core samples collected from the Lower Miocene Moghra Formation exposed at Qattara Depression, North Western Desert, Egypt to determine their reservoir characteristics and to investigate the effect of the provenance and digenetic processes on their petrophysical and geomechanical characteristics. Results of petrographical, scanning electron microscope (SEM) and the X-ray diffraction (XRD) analyses show that the studied sandstones are composed mainly of quartzarenites with little limestone and shale interbeddings and can be categorized into three main sedimentary microfacies: fossiliferous dolomitic quartzarenites, ferruginous quartzarenites and calcareous quartz arenite. The mainly-recognized diagenetic processes that prevailed during the post-depositional history of the Moghra sandstones are compaction, cementation, and dissolution. These processes impacted the porosity and influenced the petrophysical and geomechanical parameters of the studied sandstones. The Moghra sandstones possess average values of 14.74%, 2.21 g/cc, 2.76 g/cc, 23.02 mD, 49.44%, and 3341.17 m/s, for porosity ($\emptyset$), bulk (ρ_b) and grain (ρ_g) densities, permeability (K), irreducible water saturation (S_{Wirr}), and the P wave velocity (Vp), respectively. In addition, these rocks have average values of 81.92 MPa, 5.84 MPa, and 58 for the unconfined compressive strength (UCS_{dry}), the point load strength index (IS50) and the Schmidt hammer number (SHV) respectively. Significant relationships, with high correlation coefficients, between the investigated parameters have been obtained for the studied sandstones. The results indicate that both porosity and bulk density are the major parameters which control other petrophysical and geomechanical parameters

Keywords: Moghra formation; Petrophysical and mechanical properties; Diagenetic impacts; Qattara depression

Introduction

Assessment of both the petrophysical and the mechanical characteristics of rock types are among the most important tasks of rock mechanics [1]. These properties affect the rocks' response to applied stresses and are therefore fundamental elements of rock classifications [2]. They also give important information regarding construction plans (building upon or within the bedrock), as well as slope instability. Different methods have been employed to determine these important parameters. In the oil industry, a full understanding of the petrophysical and geomechanical properties of hydrocarbon-bearing rocks plays an essential role in exploration, production, and the development of petroleum reservoirs. They are fundamental in assessing the pore fluid pressure, adjusting the drilling fluid pressure to successfully drill and complete the wells. Moreover, investigating the physical and mechanical properties of rocks underlying foundations of new urbanizations is a key parameter in the structural and financial viability of buildings, and is also an important step dealing with the problem of construction instabilities. The load of any engineering structure placed on soil or bedrock must be safely maintained without causing a shear failure.

The physical properties of the intact rocks are mainly controlled by rock microstructures such as cleavage in minerals, boundaries of grain and microfractures [3]. There is a correspondence between the strength of the rocks and their mineral content and weathering processes [4,5]. Sandstones, as typical sedimentary rocks with a very wide geographic distribution, possess very complicated strength characteristics and are very essential for many engineering projects and other industrial purposes [6]. They also constitute an important category of reservoir rocks where they, alone, represent almost 50% of known hydrocarbon reservoirs. The geomechanical properties of sandstones are not only affected by external factors, such as stress, the geological environment and the depositional and sedimentary conditions, but are also strongly dependent on the internal components, structures, and fabric of the rocks [7]. In addition, diagenetic processes have significant effects on their mechanical and petrophysical properties and their overall reservoir quality. Being influenced by both depositional facies and diagenetic processes; reservoir quality is one of the critical aspects in understanding the basic elements of the play in sedimentary basins [8-10]. The sandstone diagenesis is controlled by a variety of interrelated parameters, including composition of framework grains, pore-water chemistry, tectonic setting of the basin and burial-thermal history of the succession [11,12].

Although the sedimentary facies and general geology of the Qattara Depression have been investigated by many researchers very few studies on their petrophysical and mechanical characteristics are available [13-17]. The relationship between sediment characteristics and the mechanical properties of sandstone rocks is very complicated and needs to be more thoroughly addressed [7]. The present study is

*Corresponding author: Mohamed K Salah, Department of Geology, American University of Beirut, Lebanon; E-mail: ms264@aub.edu.lb

concerned specifically with the petrophysical and mechanical aspects of a large number of rock samples collected from the Moghra sandstones compared to those considered in the study of Salah et al. [17]. The petrophysical characteristics comprise porosity ($\varnothing$), bulk (ρ_b) and grain (ρ_g) densities, permeability (K), irreducible water saturation (S_{Wirr}), P wave velocity (Vp), whereas the mechanical properties include the unconfined compressive strength (UCS_{dry}, MPa), point load strength index (IS50, MPa) and Schmidt hammer number (SHV). The study also relates these measured parameters with the diagenetic processes (e.g., compaction, dissolution, cementation) affecting the Moghra Formation siliciclastics. Diagenesis includes all the physical, chemical and biological changes that sediments are subjected to after the grains are deposited, but before they are metamorphosed [18]. Some of these changes occur at the water-sediment interface, but the bulk of diagenetic activity such as compaction and lithification takes place after burial. The processes that affect the sediments after deposition include pedogenic activity and bioturbation, the dissolution and reprecipitation of detrital components, the precipitation of pore-filling and cementing materials from solution. In the present discussion, diagenesis will be considered in the loose sense of Blatt et al. as the sum of all the processes by which an original sedimentary assemblage attempts to reach equilibrium with its environment [19].

Location and geologic setting

Large parts of Egypt, especially in the Western Desert, are dominated by clastic sedimentary rocks ranging in age from Lower Cretaceous to the Quaternary [20]. These rocks essentially comprise sandstone, shale and mudstone. Being famous with their vertebrate fossil content, the thick sandstones of the Lower Miocene Moghra Formation in the northern cliffs of the Qattara Depression are ideal outcrops for sedimentological, petrophysical, and geotechnical studies [20].

The Qattara Depression, with its triangular shape and vertex at about 67 km distance from the Mediterranean Sea, is considered as the largest, natural, closed, land depression (19,605 km^2) of the Eastern Sahara [21]. It represents the most important geomorphologic feature in the northern part of the Egyptian Western Desert (Figure 1a). The depression was first discovered and mapped by Ball [22,23]. It is surrounded from the north and west by steep escarpments; the top of which averages about 200 m above sea level. Several investigators described the tectonic framework of the Qattara Depression [24-28].

Geologically, the origin of the Qattara Depression remains a complicated problem and a matter of controversy. Many hypotheses were proposed since the beginning of the twentieth century. The first hypothesis was proposed by Ball, who suggested that the Qattara Depression originated from the deflation of the wind to a base-level controlled by the groundwater table [23]. Other explanations propose solution, mass-wasting followed by wind deflation or assume that the depression was originally excavated as a stream valley which was

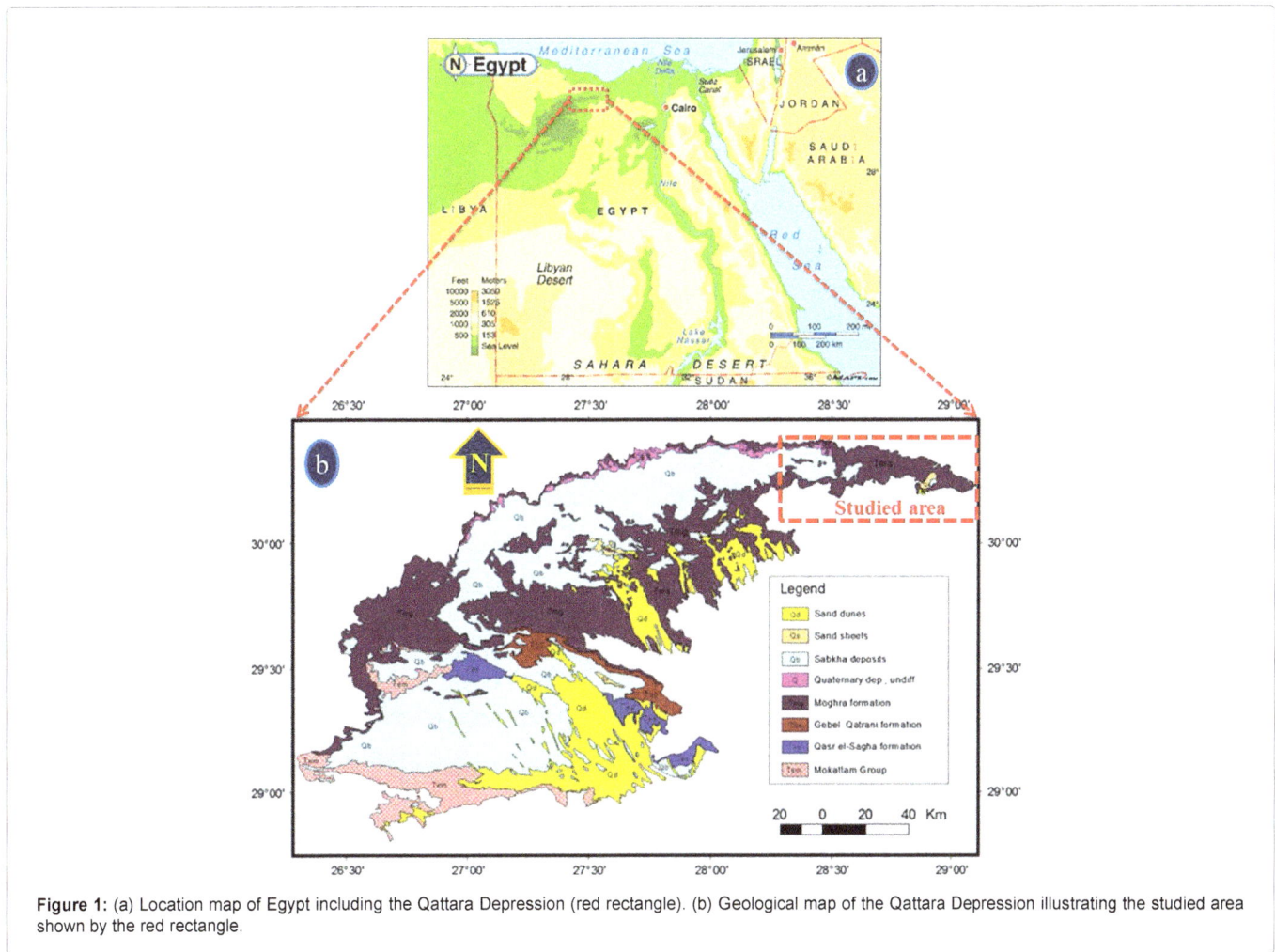

Figure 1: (a) Location map of Egypt including the Qattara Depression (red rectangle). (b) Geological map of the Qattara Depression illustrating the studied area shown by the red rectangle.

subsequently dismembered by karst development processes, and was further deepened and extended by mass-wasting and fluviatile processes [13,29]. It has also been suggested that the depression is mainly of structural origin [30]. Recently, Aref et al. proposed that the high salinity of the near-surface groundwater, the sodium chloride nature and the high rate of evaporation caused the disintegration of the bedrock [31].

The area under investigation in the northeastern tip of the Qattara Depression is situated between latitudes 30° 10′ and 30° 30′ N and longitudes 28° 15′ and 29° 10′ E (Figure 1b). Said divided the Miocene sediments of the Egyptian Western Desert into two distinct rock units from base to top: the Moghra Formation and the Marmarica Limestone Formation [32,33]. The Moghra Formation is mainly a fluviomarine clastic unit and comprises all the facies exhibited by the Lower Miocene sediments of the North Western Desert. Misak subdivided the Moghra Formation into the following three members from base to top: 1) The basal sand member (El-Raml member), 2) The clay member (Deir El-Tarfaya member), and 3) The upper sand member (Qaret El–Rikab member) [34].

From a sedimentological point of view, the Qattara Depression is cut into nearly horizontal beds of Miocene to Eocene age [32]. Sand and clay-rich units of the Early Miocene Moghra Formation form the bottom and the surroundings of the northeastern part of the depression, while calcareous sands and clay-rich sediments of middle-late Eocene and Oligocene age (Mokattam, Qasr El Sagha and Gebel Qatrani formations, respectively) form the southern and western boundaries of the depression. At its type section, the Moghra Formation is ~ 230 m thick sequence of sandstone, siltstone, and calcareous shale, with vertebrate remains and petrified tree trunks (Figures 2a and 2b). The northern border of the Qattara Depression is marked by a steep escarpment (250 m a.s.l.) of white limestone of the Middle Miocene Marmarica Formation. Over large areas of the depression's floor, the bedrock is covered with surficial deposits, including sand dunes, sabkha deposits, and Quaternary evaporites (Figures 2c and 2d). The area in the vicinity of the Moghra Lake is mostly covered by Quaternary deposits with occasional outcrops of Moghra Formation as small mesas and residual hills within the Quaternary deposits. The floor of the Qattara Depression reaches 50 to 80 m below sea level [32].

The studied Moghra clastics consist mainly of fining-upward cycles of laminated, massive, and cross-bedded sandstones. Occasionally, the sandstones are thinly-bedded and have different colors varying from white, yellow, orange to brown (Figures 2b and 2c). Some beds exhibit vertical variations in the primary sedimentary structures, with trough cross-bedded, pebbly, coarse-grained sandstones at the base, followed upwards by planar cross-bedding with subordinate convolute bedding, and terminated by flat-bedded to laminated, fine-grained sandstones with gypsum veinlets (Figure 2d). Yellowish, finely-laminated, grey and green shale are also recorded where they constitute about 28% of the section (Figure 2b). Thin beds of yellowish brown sandy dolomitic limestone (3.5 m thick), about 2%, are also present in the studied section.

Sampling and Methodology

A total of 60 rock samples were collected from the Moghra Formation exposed at the northeastern part of the Qattara Depression. Core samples of 6.0 cm in diameter and 12.0 cm in length for sandstone samples were prepared to fit the different petrophysical and mechanical tests and all specimens were first oven-dried (105 °C) to prevent dehydration of the contained clay minerals and to keep other soft evaporite minerals

like gypsum and halite. Moreover, the presence of moisture in tested samples may also influence the geotechanical characteristics of the rocks. For example, it has been shown that moisture decreases the UCS of not only the weaker sandstones but also the strong sandstones as well [35,36]. From the collected samples, we have carefully selected 28 for petrographic analysis and the investigation of the different diagenetic imprints of the studied rocks. The petrographical study was carried out using the polarizing microscope to identify the various rock components, the mineral composition, the different types of grains, the textural and fabric relationships and to describe the different types of pores. The sandstone samples were classified according to Pettijohn [37]. The scanning electron microscope (SEM) technique yields a three-dimensional (3D) image useful for understanding the sedimentological fabric and structure of the studied samples. It provides a wide range of information about the morphology, mineral composition, distribution of authigenic clay minerals, mechanically infiltrated clays, transformational clays and pedogenic mud aggregates. The technique is also a useful tool in understanding the clay mineralogy and their effect on porosity, permeability and other reservoir characteristics [38-40]. Therefore, 18 fresh samples were studied using the SEM technique for a visual examination of the internal pore spaces between the very fine cement materials. Some textural parameters such as fabric, sorting, grain contacts and different types of porosity were also considered. Additionally, X-ray diffraction (XRD) analysis was applied to 25 rock samples using a Philips X-ray diffractometer with Ni-filtered copper radiation at 40 KV and 30 MA and scanning speed of 0.02 /s [41]. The analysis was applied according to their main d-spacing and the main results are shown in Table 1.

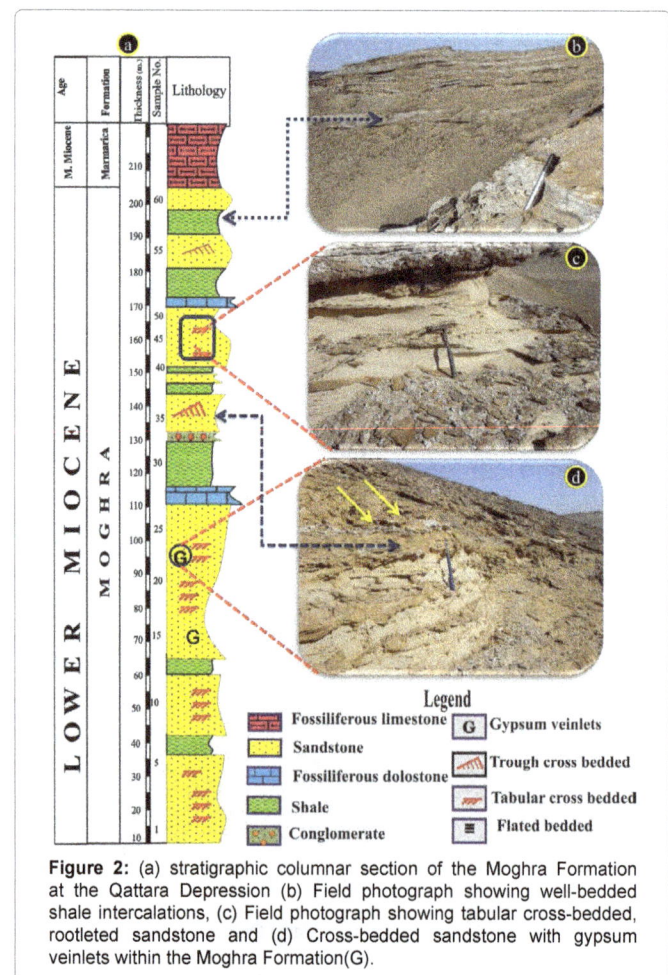

Figure 2: (a) stratigraphic columnar section of the Moghra Formation at the Qattara Depression (b) Field photograph showing well-bedded shale intercalations, (c) Field photograph showing tabular cross-bedded, rootleted sandstone and (d) Cross-bedded sandstone with gypsum veinlets within the Moghra Formation(G).

The plug samples are used for measuring the effective porosity ($\varnothing$), bulk (ρ_b), and grain (ρ_g) densities, permeability (K, in mD), as well as other mechanical characteristics. Porosity is the percentage of pore spaces in the total volume of the rock.

$$\varnothing = V_p/V_b \qquad (1)$$

where V_p is the pore volume in a rock sample, and V_b is the sample bulk volume.

To determine porosity, only two of three volumes are necessary; namely the bulk volume, interconnected pore volume, and the grain volume (V_g), where:

$$V_b = V_p + V_g \qquad (2)$$

The matrix-cup helium porosimeter (Heise Gauge type) is used for grain volume estimation; then the grain density (ρ_g) was determined using the following equation:

$$\rho_g = W_d/V_g \qquad (3)$$

where W_d is the dry weight of the core sample.

The bulk density (ρ_b) is defined as the mass per unit volume of a rock in its natural state. It is determined as the ratio of the dry weight of the sample to its bulk volume according to the following equation:

$$\rho_b = W_d/V_b \qquad (4)$$

Permeability is a measure of the ease with which a fluid of a certain viscosity can flow through a rock sample under a pressure gradient [42,43]. It is controlled by many factors such as rock pore geometry, cement, texture, grain size, grain shape, sphericity and roundness. Permeability increases with increasing effective porosity and grain size (especially in unconsolidated sediments from shale/clay to gravel), and decreases with compaction and cementation (due to decrease of porosity and pore-throat radii). Permeability was measured in this study by using the CoreLab Permeameter and is expressed in units of millidarcies (mD). One darcy defines the ability of the porous rock to transmit a fluid of one centipoise viscosity (μ) at a flow rate (Q) of 1 cm^3/s through a cross sectional area (A) of 1 cm^2, when the pressure gradient ($\Delta P/L$) is 1 atm/cm. The permeability (K) is estimated using the following equation:

$$K = Q\mu L/A\Delta P \qquad (5)$$

In addition, the irreducible water saturation 'S_{Wirr}' is estimated as the percentage of the volume of the pore spaces not filled by mercury during injection [43,44]. Increasing the pore volume mostly contributes to increasing percentage of the macropore spaces and the pore connectivity; therefore it reasonably decreases the percentage of the micropores, thus decreasing the irreducible water saturation.

The P-wave velocity (Vp) measurements can be applied both *in situ* and in the laboratory. As these investigations are highly significant, non-destructive and relatively easy to apply, they are increasingly used in many geological and engineering applications. They are used specifically in civil, geotechnical and mining projects such as underground opening, quarrying, blasting, ripping, and mining [45,46]. The velocity of ultrasonic pulses travelling in a solid material depends on mineral compositions and textures, density, porosity, pore water, confining pressure, temperature, weathering and alteration, bedding planes, joint properties (roughness, filling material, water, attitude, etc.), and anisotropy [47]. In the present study, the P wave velocity was determined using a Portable Ultrasonic Nondestructive Digital Indicating Tester. The Vp (in m/s) is calculated from the sample length (L) in meters divided by the pulse travel time (T) in seconds, as described in detail by Rao et al. [48].

$$Vp = \frac{L}{T} \qquad (6)$$

The geomechanical tests conducted in this study include the unconfined compressive strength at dry conditions (UCS_{dry}) and the point load strength index (IS50) tests which are measured according to ASTM D2938-95 [49]. Uniaxial compression tests were carried out on smooth core samples which have a diameter of 54 mm and a length-to-diameter ratio of 2. The stress rate was applied within the limits of 0.5-1.0 MPa/s. Cohesion and friction angle values were calculated using the formulae suggested by ISRM [3]. The diametral point load test was carried out on cores having a diameter of 38 mm. The core specimens had a length-to-diameter ratio of 1.2. While testing, a specimen was positioned between the platens, taking care that the minimum distance to either end was 0.5 times the core diameter. The load on the sample was then increased to failure with the maximum load being recorded. The calculated point load strength was corrected to a specimen diameter of 50 mm according to the method proposed by ISRM [3].

The Schmidt hammer number is a non-destructive testing of concrete hardness, where it has long been used to estimate the strength of different rock types [50,51]. It is routinely used to estimate the strength and the quality of a rock, where it provides a quick and an inexpensive measure of surface hardness which is widely used to estimate the mechanical properties of a rock material [52]. The test may be used in various configurations (vertical or inclined), depending on the type of material and the purpose of the test. During our measurements, the hammer was moved and 30 impacts were made on each sample, the highest and lowest five (i.e., a total of 10) values were excluded and the average of the remaining twenty impacts was taken as the sample hardness [3]. The tests were performed using an N-type hammer having impact energy of 2.207 Nm. All tests were conducted with the hammer held vertically downwards and at a right angle to the horizontal rock surface. Results of the petrophysical and mechanical parameters measured for the 60 rock samples are listed in Table 2.

Results

Petrographic and microfacies analysis

The petrographic investigations of these rocks have been carried out to emphasize the different types of lithofacies in each lithostratigraphic unit based on the type and amount of allochemical constituents (skeletal and non-skeletal grains), textures, diagenetic processes and the cement. The studied sandstones are mainly quartzarenites formed from

Sample No.	Qz%	Clay Minerals			Feldspar		Dolo%	Cal%	Ha%	He%
		S%	I%	K%	Pla%	Micr%				
Maximum	89.64	12.4	5.75	12.28	23.26	14.92	9.43	10.37	7.91	7.15
Minimum	51.46	2.25	0	1.2	0	0	0	0	0	0
Average	63.36	6.44	1.49	4.89	7.86	5.27	2.66	4.84	1.74	1.45

Table 1: Mineralogical composition of the Moghra sandstones based on the semi-quantitative X-ray diffraction (XRD) analysis. [Qz: Quartz; S: Smectite; I: Illite; K: Kaolinite; Pla: Plagioclase; Micr=Microcline; Dolo: Dolomite; Cal: Calcite; Ha: Halite; He: Hematite].

Sample No.	Bulk density (g/cc)	Porosity (%)	Grain Density (g/cc)	Permeability (mD)	S_{Wirr} %	Vp (m/s)	UCSdry (MPa)	IS50 (MPa)	SHV
60	2.16	13.23	2.71	27.55	66.00	3967.4	89.184	6.21	60
59	2.22	13.56	2.76	26.87	67.00	3825.4	88.134	6.30	63
58	2.12	15.67	2.74	34.33	43.00	3601.0	66.025	4.72	47
57	2.12	14.21	2.81	29.65	55.00	3319.1	70.477	5.03	50
56	2.19	15.56	2.64	27.20	45.00	3151.9	78.797	5.63	56
55	2.22	14.78	2.77	30.65	62.00	3677.4	76.934	5.50	56
54	2.23	12.54	2.87	15.32	74.00	3909.4	91.987	6.57	66
53	2.13	16.60	2.70	38.65	34.55	2774.6	61.865	4.42	44
52	2.18	17.05	2.65	37.43	23.44	2345.0	56.543	4.04	40
51	2.03	17.89	2.64	30.54	24.33	2764.0	52.459	3.75	37
50	2.21	16.28	2.61	35.32	34.65	2881.9	75.540	5.40	54
49	2.19	15.78	2.74	33.32	43.12	3142.2	78.100	5.58	56
48	2.12	18.65	2.64	40.32	22.12	2233.0	65.826	4.70	47
47	2.11	19.53	2.61	33.45	11.23	2086.1	77.152	5.51	54
46	2.14	15.33	2.76	32.43	42.45	3134.3	71.358	5.10	51
45	2.16	17.54	2.63	39.00	26.56	2687.7	77.192	5.51	55
44	2.02	19.54	2.65	44.56	13.55	2444.0	45.178	3.23	32
43	2.06	17.56	2.65	38.44	27.54	2341.3	55.033	3.93	39
42	2.01	18.54	2.69	40.55	25.67	2321.2	50.529	3.61	55
41	2.01	19.67	2.61	39.54	21.43	2123.3	47.082	3.36	34
40	2.12	16.43	2.70	32.30	36.55	3108.5	77.713	5.55	56
39	2.03	17.67	2.70	33.50	25.43	2726.1	78.152	5.58	56
38	2.15	15.65	2.77	34.20	43.26	2957.6	73.941	5.28	53
37	2.11	16.64	2.68	32.40	35.68	2914.2	72.855	5.20	52
36	2.32	13.65	2.84	24.50	64.55	2951.9	73.798	5.27	53
35	2.01	20.43	2.61	43.20	7.54	2080.3	44.507	3.18	32
34	2.02	18.78	2.63	36.40	23.54	2372.0	74.300	5.31	54
33	2.19	16.57	2.68	34.60	32.38	2963.1	74.078	5.29	53
32	2.12	17.67	2.69	36.50	24.54	3295.6	82.390	5.88	56
31	2.32	12.43	2.89	25.70	43.22	3701.1	82.527	5.89	59
30	2.31	14.54	2.79	23.20	30.43	3680.9	79.524	5.68	52
29	2.31	15.32	2.76	25.40	31.36	3229.0	70.724	5.05	51
28	2.33	12.43	2.84	23.40	78.50	3996.3	102.408	7.31	73
27	2.32	11.10	2.87	22.43	75.89	3969.5	114.237	8.16	82
26	2.31	11.43	2.86	14.60	72.44	3923.0	111.075	7.93	79
25	2.32	11.67	2.87	24.30	77.54	3890.2	109.754	7.84	78
24	2.12	15.21	2.73	30.54	34.54	3948.4	81.211	5.80	58
23	2.21	14.17	2.79	12.10	44.32	3270.3	81.757	5.84	58
22	2.11	17.54	2.65	10.07	22.33	2681.0	66.525	4.75	48
21	2.13	15.65	2.68	10.33	33.54	3693.1	69.826	4.99	50
20	2.15	15.65	2.77	7.13	36.50	3267.9	61.697	4.41	44
19	2.05	17.43	2.66	44.33	26.50	3415.1	45.377	3.24	32
18	2.26	14.32	2.77	4.56	32.45	3367.3	74.182	5.30	53
17	2.32	13.33	2.85	0.52	61.43	3868.8	96.719	6.91	69
16	2.24	14.23	2.80	0.92	54.32	2445.4	76.359	5.45	55
15	2.23	16.43	2.66	37.55	41.32	3123.1	53.076	3.79	38
14	2.39	11.21	2.88	2.44	76.43	4167.0	104.174	7.44	74
13	2.35	11.44	2.88	2.68	81.33	4173.0	104.325	7.45	75
12	2.32	13.23	2.88	22.00	66.32	3864.5	106.613	7.61	76
11	2.22	15.03	2.68	29.00	35.44	3705.0	72.624	5.19	52
10	2.37	11.72	2.83	0.22	92.30	4030.3	100.756	7.20	72
9	2.36	11.43	2.90	0.20	92.30	4168.8	104.220	7.44	74
8	2.33	11.54	2.84	0.22	97.30	4082.7	102.068	7.29	73
7	2.35	11.48	2.88	0.22	76.43	4094.8	102.369	7.31	73
6	2.39	9.09	2.90	0.17	87.43	4247.9	126.196	9.01	90
5	2.39	8.91	2.90	0.28	90.33	4255.3	136.382	9.74	66
4	2.35	8.71	2.88	0.35	92.32	4287.6	127.191	7.34	67
3	2.39	9.41	2.89	0.18	92.95	4300.5	127.514	7.43	66
2	2.36	10.32	2.89	0.19	83.23	4182.0	114.549	8.18	82
1	2.34	11.22	2.89	0.23	72.36	4021.0	122.000	8.71	87

Min.	2.01	8.71	2.61	0.17	7.54	2080.27	44.510	3.18	32
Max.	2.39	20.43	2.90	44.56	97.30	4300.54	136.380	9.74	90
Ava.	2.21	14.74	2.76	23.02	49.44	3341.17	81.920	5.84	58

Table 2: Petrophysical and mechanical properties of the selected 60 core samples representing the Moghra sandstones.

detrital quartz grains, bounded together with cementing material of calcite, dolomite or iron oxides. Description of the different microfacies observed petrographically is presented as follows.

Fossiliferous dolomitic quartzarenite

In thin section, the rock consists of quartz graines (51.0-65.0%), dolomite rhombs, and some fine bioclastic fragments (F) embedded in a dolomitic matrix (Figure 3a). Bioclastic particles (Figure 3b) are mainly represented by bryozoa and foraminiferal tests (*Globigerinoides sp.*), that are scattered with no definite orientation. Most of these components are susceptible to dolomitization and the binding material is mainly the dolomitic cement. The majority of the dolomite rhombs possess xenotopic and hypidiotopic texture of equigranular fabric of dolomite that is stained with iron oxide. Between the dolomite rhombs, there are clear blocky and massive calcite crystals that show a gradational relation to the adjacent dolomite rhombs (Figure 3c). The subangular to subrounded monocrystalline quartz grains have an average size of about 210 μm; although few polycrystalline types also exist. The bryozoa are filled with dolomitic cement (Figure 3d).

Calcareous quartzarenite

The calcareous quartzarenite microfacies are composed of about 77% quartz grains. The grains are attached with silica and calcite. The calcite cement is noticed as a diagenetic mineral filling in between the quartz grains and replaces some parts of the quartz overgrowth at the surface of the quartz grains (Figure 4a). The calcite content as a filling material is shown in Figure 4b.

Ferruginous quartzarenite

This microfacies constitutes the base of the cycle (10-17 m thick) and is topped by fossiliferous dolomitic quartzarenites. Rocks belonging to this lithofacies are usually hard, massive, fine to medium-grained, well-sorted, and yellow in color in some parts. Petrographically, the rock is essentially made up of quartz grains (84%) and iron oxide (mainly hematite) cement along with other components (Figures 5a and 5b). The quartz grains are subrounded to rounded with straight undulose extinction and vacuole inclusions. They range in size from 0.1 to 0.24 mm. Frequently; large grains of up to 0.75 mm size are present and mostly cracked (Figure 5c). Most of the quartz grains are highly corroded due to their replacement by the iron oxide cement.

The X-ray diffraction analysis

Results of the mineralogical composition of the studied samples as estimated by the XRD analysis are shown in Table 1. The data of bulk sandstone analysis show that quartz is the main constituent besides other minors such as feldspar, hematite, clay minerals, dolomite and halite. Quartz (3.34 A°) ranges from 51.46 to 89.64% with an average of 63.36% in the studied samples. Clay minerals are represented mainly by smectite, kaolinite, and illite. Smectite varies between 2.25 and 12.40% with an average of 6.44%; illite attains a maximum of 5.75% with a lower average of 1.49; whereas kaolinite varies between 1.2% and 12.28% with an average of 4.89%. Feldspars are represented by plagioclase and microcline where they have average values of 7.86% and 5.27%, respectively. Calcite, dolomite, halite, and hematite are represented by average values of 4.84, 2.66, 1.74, and 1.45%, respectively.

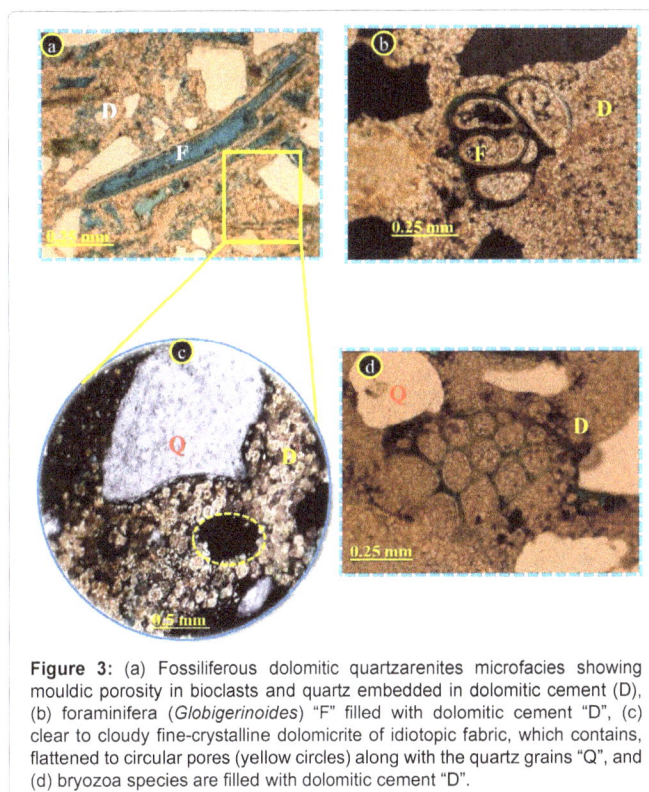

Figure 3: (a) Fossiliferous dolomitic quartzarenites microfacies showing mouldic porosity in bioclasts and quartz embedded in dolomitic cement (D), (b) foraminifera (*Globigerinoides*) "F" filled with dolomitic cement "D", (c) clear to cloudy fine-crystalline dolomicrite of idiotopic fabric, which contains, flattened to circular pores (yellow circles) along with the quartz grains "Q", and (d) bryozoa species are filled with dolomitic cement "D".

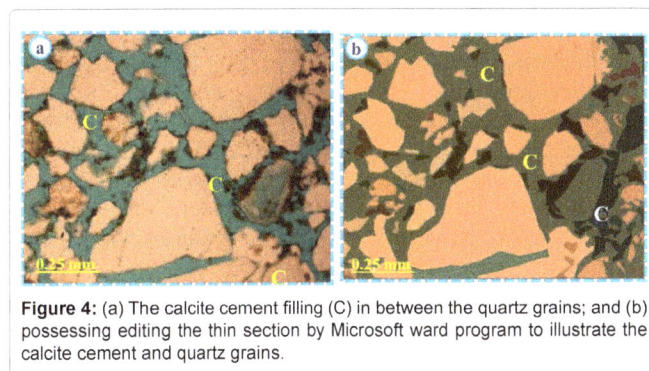

Figure 4: (a) The calcite cement filling (C) in between the quartz grains; and (b) possessing editing the thin section by Microsoft ward program to illustrate the calcite cement and quartz grains.

Petrophysical and mechanical parameters

Porosity, density and permeability data for the 60 plug samples are listed in Table 2 along with the minimum, maximum and average values. The petrophysical data seem to be highly heterogeneous which could be attributed to differences in the rock types, heterogeneity in the particle size, clay amount and distribution and the complexity of the pore spaces [53,54]. Density and porosity of sandstone are influenced by the amount of cement and/or matrix material filling the pore spaces. The porosity of sandstones is developed during sedimentary deposition with the packing density of the grains as one of the main parameters [55].

Porosity varies from 8.71 to 20.43%, with an average of 14.74%, indicating that the Moghra sandstones are characterized by moderate to high porosity [56]. The grain density, which depends mainly on

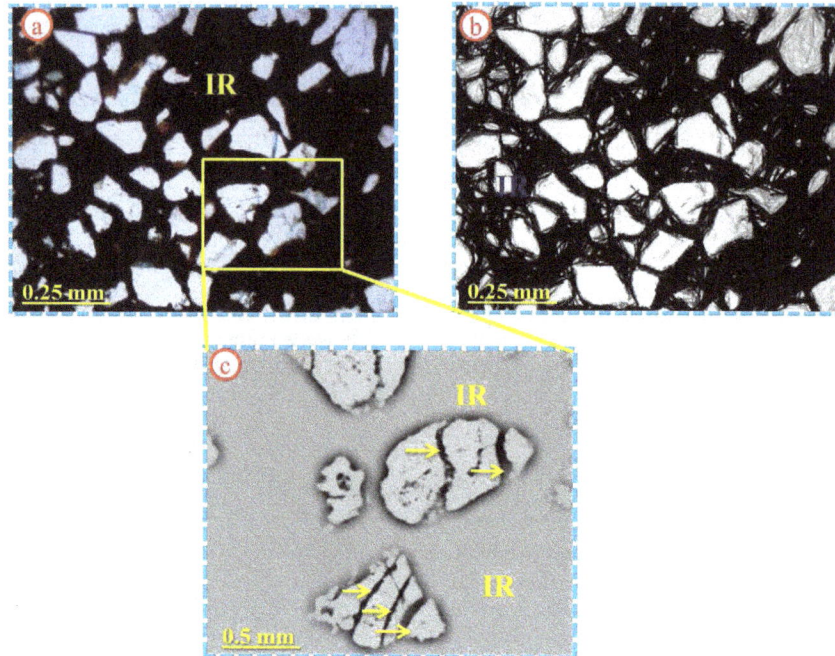

Figure 5: (a) Optical micrograph of primary and secondary pores filled with authigenic hematite "IR"; (b) Possessing the section to show the cement and quartz grains; and (c) Monocrystalline quartz with fracture-healing iron material in the Moghra Formation.

composition, varies from 2.61 to 2.89 g/cm³ with an average of 2.76 g/cm³. Bulk density varies from 2.01 to 2.39 g/cm³, with an average of 2.21 g/cm³. The bulk density of a rock is mainly affected by the grain size, grain shape, the volume of pores, and the grain sorting which is a result of the energy of the depositional environment.

The permeability values obtained for the core samples range from 0. 17 to 44.56 mD with an average of 23.02 mD. Both porosity and permeability decrease downward in the Moghra Formation (Table 2) possibly due to increased compaction and higher clay amount both of which have a high impact on the porosity, permeability, bulk density and mechanical characteristics of sandstones [57]. The irreducible water saturation varies widely between 7.54% and 97.30%, with an average of 49.44%. The percentage of irreducible water saturation depends on the mineralogy of the samples, their overall composition and the degree of alteration, pore size, throat radii, grain size and shape [58].

The velocity of acoustic waves through a material is a good indicator of other properties such as strength, cohesion and elastic moduli. The average value of Vp for the Moghra sandstones is 3341.17 m/s with the minimum and maximum values at 2080.27 and 4300.55 m/s, respectively. According to the classification of Anon [56], the studied sandstone samples are classified as of very low to moderately high seismic wave velocities. The low P wave velocities are commonly associated with the highly porous and permeable samples. Of the measured P wave velocities, more than 25% of the samples have low P-wave velocities with equivalent high porosities.

Strength properties such as the tensile strength and compressive strength are important rock parameters which are influenced by the burial diagenetic processes of the rocks. The minimum and maximum values of the unconfined compressive strength obtained for the Moghra Formation sandstones are 44.51 and 136.38 MPa, respectively, with an average of 81.92 MPa (Table 2). According to the classification of rock material based on unconfined compressive strength, the strength of the studied samples can be described as moderate to very high [3,59,60]. Results of the point load test vary from 3.18 to 9.74 MPa, with an average of 5.84 MPa. This indicates that the strength of the samples under investigation is medium to high (Table 2) [61]. The results of the SHV for the Moghra Formation sandstones vary from 32 to 90, with an average of 57.8, which can be considered as moderate to strong [56]. The minimum, maximum and average values of the bulk density (ρ_b, g/cc), porosity ($\varnothing$, %), grain density (ρ_g, g/cc), permeability (K, mD), irreducible water saturation (S_{Wirr}, %), P wave velocity (Vp, m/s), unconfined compressive strength (UCS_{dry}, MPa), point load strength index (IS50, MPa) and Schmidt hammer number are listed in Table 2.

Data analysis and empirical relationships

To check the impact of the diagenetic processes on the petrophysical and mechanical characteristics of the studied sandstones, a number of relationships were constructed between the measured parameters and are presented in the following paragraphs. These relationships introduce a number of reliable empirical equations with high correlation coefficients which can be used to predict some petrophysical parameters from other easily-measurable ones. Nonetheless, the given empirical equations should be used, with great care, for only similar rocks at more or less analogous depositional settings.

Porosity ($\varnothing$) *vs.* bulk density (ρ_b)

Porosity and bulk density are standard petrophysical parameters that have been determined for all specimens. The bulk density reflects variations in pore space volume, detrital grains, and cement amounts and distribution. The variation in detrital grains has the minimal effect and we can consider that bulk density is affected mainly by the variation in pore space and cement distribution [62]. Usually, the density of sandstone tends to increase with increasing depth below the surface, while the porosity decreases due the burial compaction [63].

A very reliable, inverse, linear relationship between the bulk density and porosity for the sandstone samples of Moghra Formation has been obtained. This relationship (Figure 6a) shows that the bulk density decreases with the increase of porosity. The observed scatter in this inverse relationship is attributed to the cement distribution, heterogeneous mineralogical composition, grain density and heterogeneous porosity types in the samples. The porosity-bulk density relationship has the highest correlation coefficient (R^2=0.85) and is controlled by the following equation:

$$\varnothing = -28.135\rho_b + 92.292$$

Porosity ($\varnothing$) vs. permeability (K)

Porosity and permeability exhibit a direct proportional relationship (Figure 6b). Generally, porosity alone is not only the factor affecting the permeability of a rock. Other factors that may have a significant effect on permeability include grain size, shape, packing, pore throat size, etc. Moderate coefficient of correlation is obtained between porosity and permeability of the tested sandstone rocks (R^2=0.65). This may be attributed to the presence of clays as pore lining, which have a large effect on pore throat radii rather than the pore volume. The low permeability values in the Moghra sandstones correspond principally to the abundance of smectite, kaolinite and feldspars, and some evaporites. Sandstones with low amount of kaolinite and feldspars are highly permeable, while those with abundant kaolinite and feldspars are less permeable [64]. Porosity and permeability relationship for the studied sandstone samples is represented by the following equation:

$$\varnothing = 0.1596K + 11.065$$

Porosity ($\varnothing$) vs. irreducible water saturation 'S$_{Wirr}$'

Plotting the porosity as a function of irreducible water saturation "S_{Wirr}" displays an inverse relationship (Figure 6c) with a relatively high correlation coefficient (R^2=0.89). In the present study, the existence of clay minerals, represented by smectite and kaolinite, increases the irreducible water saturation and therefore decreases the permeability. High porosity contributes to higher permeability and is observed as a decrease in the irreducible water saturation (Figure 6c). The empirical relation between porosity and irreducible water saturation is given as:

$$\varnothing = -0.1116S_{Wirr} + 20.236$$

Porosity ($\varnothing$) vs. dry unconfined compressive strength 'UCS$_{dry}$'

Porosity has a strong effect on the strength properties of rocks. As porosity increases, the compressive and tensile strengths decrease, i.e., the increasing void space reduces the integrity of the material. The results of the present analysis (Figure 6d) indicate that porosity has a significant effect on the mechanical properties of the studied sandstone samples. The UCS$_{dry}$ is inversely related to porosity with R^2=0.81 and the UCS$_{dry}$ can be calculated from the following equation:

$$UCS_{dry} = -6.8945\varnothing + 183.78$$

Porosity ($\varnothing$) vs. ultrasonic P-wave velocity (Vp)

The relationship between rock velocity and other petrophysical parameters, such as porosity and density is essential for identifying many rock characteristics and the vertical changes in the physical properties of rocks including the origin of reflectivity on seismic lines [65]. The P-wave velocity is inversely proportional to porosity (Figure 6e). The following empirical equation, with a high correlation coefficient (R^2=0.81) relates porosity to the P-wave velocity.

$$\varnothing = -0.004Vp + 28.008$$

Porosity ($\varnothing$) vs. acoustic impedance (Z)

Acoustic impedance (Z) is the product of bulk density and the P-wave velocity and represents the opposition of a medium to a longitudinal wave motion. It characterizes the relationship between the acting sound pressure and the resulting particle velocity, and therefore reflects the medium properties. When the bulk density is expressed in kg/m^3 and the P wave velocity in m/s, the acoustic impedance will be expressed in kgm^{-2}s^{-1}. Previous research results [65] indicated that, rather than rock composition, porosity has the strongest control on acoustic impedance. The acoustic impedance is inversely proportional to porosity (Figure 6f). The following empirical equation, with a high correlation coefficient (R^2=0.88), relates porosity to the acoustic impedance of the studied rock samples.

$$Z = -570247\varnothing + 2E+07$$

Bulk density (ρ_b) vs. the P-wave velocity (Vp)

The compressional wave velocity is directly proportional to bulk density (Figure 6g). An increase in the rock density is accompanied by higher efficiency to transmit seismic waves [66,67]. The samples show some scatter which may be induced by the orientation of grains, their size range, and the voids in the rock sample. The following empirical equation, with a high correlation coefficient (R^2=0.63) relates porosity to the acoustic impedance of the studied rock samples.

$$Vp = 4500.4\rho_b - 6599.1$$

Discussions

Depositional Environments

The sandstones of Moghra Formation are mainly composed of quartz, rock fragments and feldspar; accordingly, these sandstones can be classified as or quartzaranites based on Pettijohn's definition [68]. The clastic succession of the Moghra Formation (Figure 2a) suggests a fluvial environment, where the Western Desert was emerged and a regressive stage commenced during the Early Miocene times. This fluvial deposition was interrupted by two minor marine invasions during the accumulation of the Moghra clastics which are represented by small pittes grooves (Figure 7). The occurrence of the fossiliferous carbonate units within the Moghra clastics furnishes an evidence for a shallow marine transgression. The fluvial environment of the clastic facies, on the other hand, is evidenced from the frequent presence of tabular cross-bedding (Figure 2b), which is known to be very common in the deposits of fluvial channels [69,70]. The abundance of drifted silicified tree trunks and the absence of detrital ferrous-iron minerals such as siderite and chlorite. Investigation of clay minerals has demonstrated that smectite and kaolinite are the most predominant clay minerals in the Moghra Formation; suggesting a continental or non-marine type of environment. The high degree of crystallinity of kaolinite also supports this view. The detrital clastics were provided to the environment by fluvial water from the provenance; whereas the carbonate facies were deposited during the marine invasions accompanying the accumulation of the Moghra clastics [71].

Diagenetic processes and their impact on the Moghra sandstones

Porosity and permeability of a reservoir rock are controlled by several

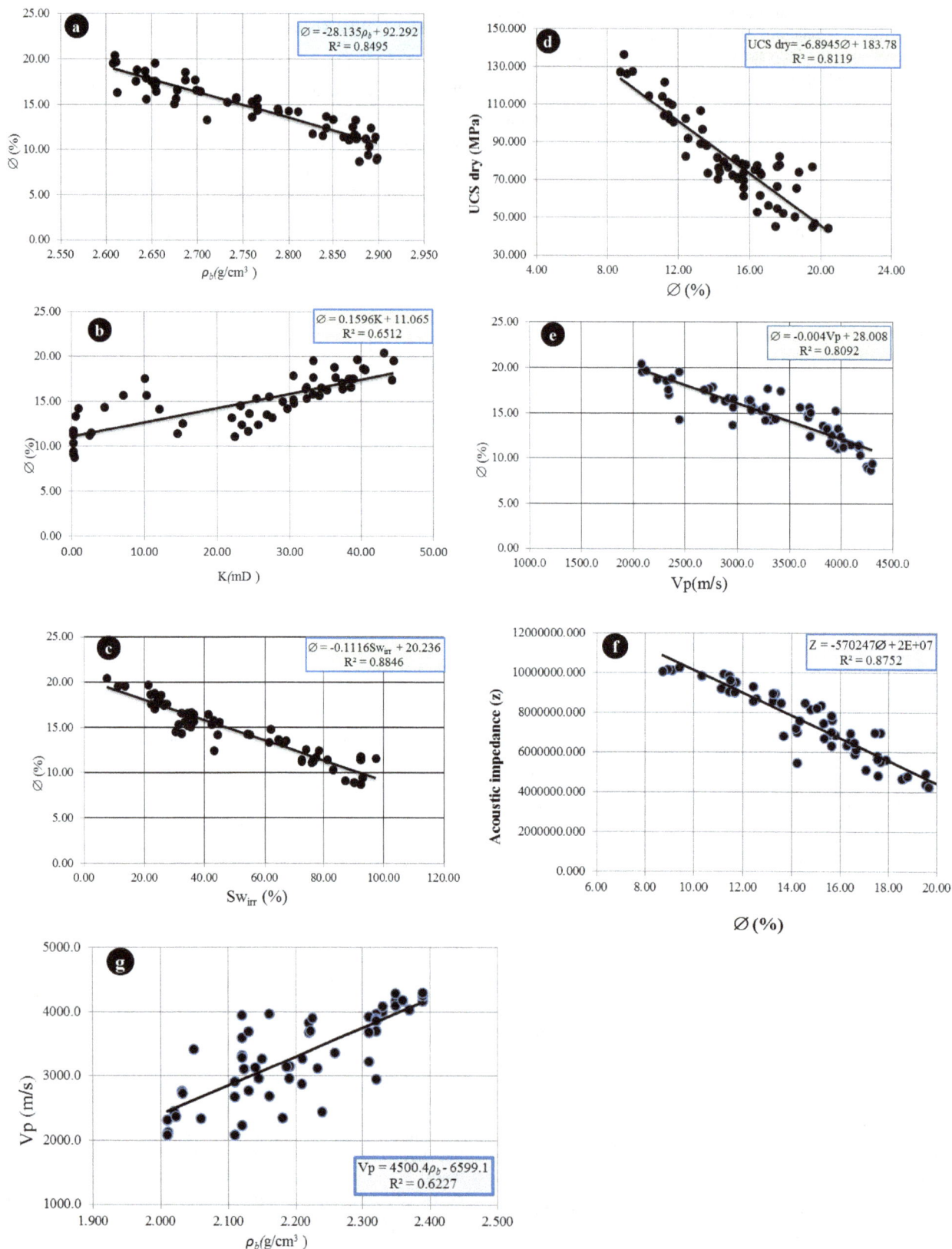

Figure 6: Cross plots showing the relationships between some petrophysical and mechanical parameters: (a) porosity *vs.* bulk density, (b) porosity *vs.* permeability, (c) porosity *vs.* irreducible water saturation, (d) porosity *vs.* dry unconfined compressive strength, (e) porosity *vs.* the ultrasonic compressional wave velocity, (f) porosity *vs.* acoustic impedance, and (g) bulk density *vs.* the ultrasonic compressional wave velocity.

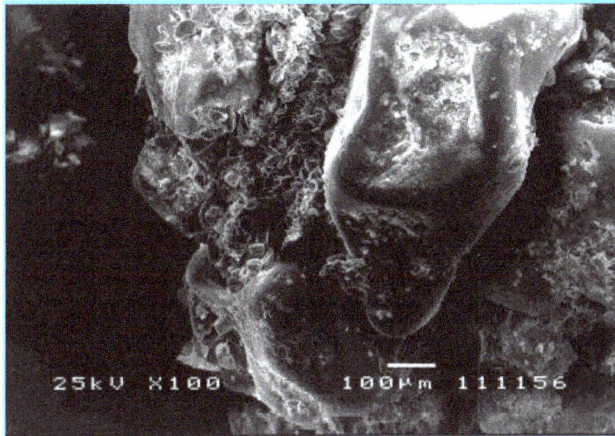

Figure 7: Scanning electron micrograph showing curved and pittes grooves; characteristic of high energy fluvial environments, Moghra Formation.

geological and diagenetic (enhancing or reducing) factors [40,72]. The presence of good porosity does not mean that good permeability exists as in the case of the pumice stone, clays, and shales. In addition, a high permeability value may be obtained from a low porosity rock such as the case of the highly-fractured carbonates or crystalline rocks [73]. To be permeable, the pores in a rock must be interconnected, and the pore-throats should be large enough to allow fluids to flow through. The diagenetic processes are very important due to their effect upon the composition, texture, physical and mechanical aspects of the rocks and due to their reducing or enhancing effect on the pore spaces of the affected rocks. Although diagenetic processes are marked mostly by porosity-permeability reduction, porosity enhancement, such as through dissolution is observed in some cases. In addition, diagenetic processes induce chemical and mineralogical changes through alteration, dissolution mechanical compaction, chemical compaction, cementation and precipitation. Diagenetic changes are further complicated and controlled by the prevailing environment and the duration of exposure to that environment. During later diagenetic processes, the primary porosity will change through compaction and water loss, through pressure solution at grain boundaries, and through cementation.

Diagenetic processes enhancing porosity

In studying sandstone rocks, dissolution is expected to be the main diagenetic process responsible for enhancing porosity and permeability. This could be achieved by replacing the cement by a more highly porous one. However, due to the high intercrystalline porosity of most samples and the absence of allochems and the scarcity of quartz grains, it is not easy to assign an indication for the dissolution even though it was present. Due to the dissolution affecting the Moghra sandstones, about 2.2% oversized pores were produced (Figure 8a and 8b). Such oversized pores are due to the complete dissolution of the detrital grains; possibly feldspars [74]. Based on the data of McBride [75] and Milliken et al. [76], three-fourths of these pores could be feldspars and the other part was rock fragments. However, a part of the oversized vuggy and mouldic pores could form due to the dissolution of the early intergranular carbonate cement; thus enhancing the porosity of many arenites [72]. In general, the dissolution takes place preferentially along the cleavage planes of the grains. Dissolution of the quartz and K-feldspar is also observed (Figure 8a). As the quartzarenite facies are sometimes fossiliferous, significant mouldic porosity is found due to the complete dissolution of fossils (Figures 8c and 8d).

Dissolution of the authigenic minerals, particularly dolomite, is a common feature in the present sandstones especially in the fossiliferous dolomitic quartzarenite microfacies. This could be detected either by loose packing or dolomite relics within pores (Figure 8e). The irregular oversized pores larger than the surrounding corroded quartz grains [74] might be originally filled by carbonate [77-79] that dissolved later on.

Diagenetic processes reducing porosity

A large number of diagenetic processes which collectively reduce porosity exist in the studied sandstones. Examples of these include the following:

a.   **Infiltrated Clays**

The infiltrated clays are one of the most important diagenetic processes in the present sandstones. Clays may occur as coating around the framework components similar to the soil cutans identified by Brewer [80]. Infiltrated clays are formed during the near-surface diagenesis of continental sediments under arid conditions by episodic floods [81,82]. The variable arid to humid climatic conditions prevailing during the deposition of the Moghra Formation enhanced the formation of infiltrated clays. Clays were formed in the sediments slightly after deposition as a result of the mechanical infiltration of water containing suspended clay particles [83]. The infiltrated clays occur around the detrital grains and in places forming meniscus-shaped pore bridges

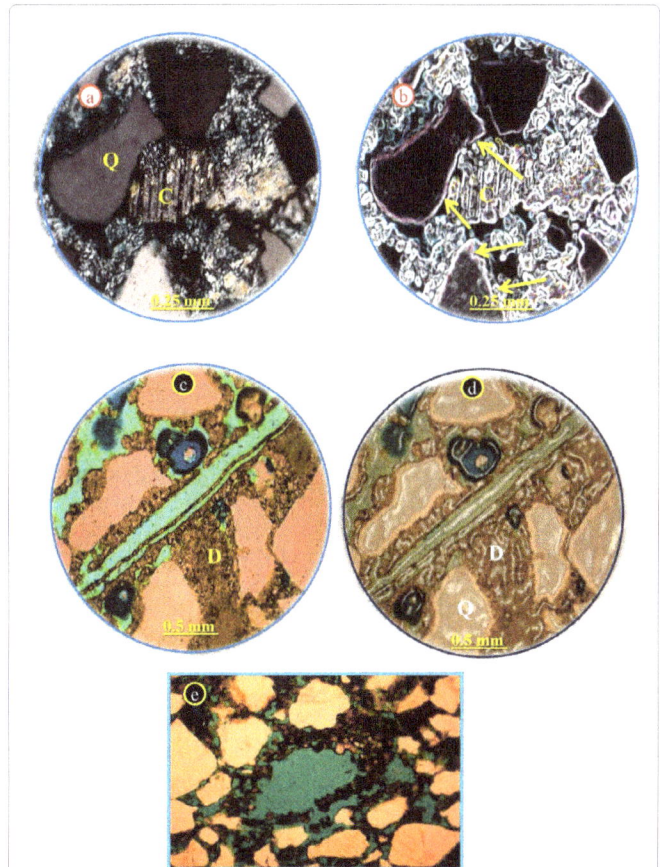

Figure 8: (a) optical micrograph showing partial dissolution and replacement of k-feldspar and quartz "Q" by calcite "C", (b) possessing the section to show the dissolution from the margins to the center of grains; (c) and (d) optical micrographs of mouldic porosity formed due to selective dissolution "D", and (e) dissolution of authigenic dolomite that is detected by loose packing and dolomite relics within pores.

(Figures 9a and 9b). The mechanically-infiltrated clays are thought to have proceeded before or after the quartz overgrowth (Figure 9c). Infiltrated clays did not significantly reduce the porosity, but it may have inhibited the precipitation of quartz cement in most samples in the manner described by Heald and Larese [84].

b. Cementation

Cementation is the occlusion of intergranular volume by the precipitation of authigenic minerals, with no direct relation to the reduction of bulk volume. Cementation results always in the reduction of intergranular porosity. Compaction and early cementation, and accordingly the resulting total porosity, are fundamental controls of the reservoir quality, carbonate rock strength and compressibility, as well as other parameters like elastic moduli [85]. A variety of authigenic minerals were observed through the thin section and SEM analyses including iron oxides, quartz, calcite and kaolinite and all reduced porosity. Silica cement is also observed as quartz overgrowths on some detrital quartz grains. However, significant amount of secondary porosity was created chiefly by dissolution of feldspar and calcite as a response to progressive burial.

c. Clay minerals

Diagenetic clay minerals are found in a wide variety of morphologies detected by the SEM analysis. The presence of such clay minerals is also confirmed by the XRD mineralogy. The authigenic minerals in the studied rocks are mainly represented by high amounts of clay minerals, mostly smectite (Figures 10a and 10b) and lesser amounts of kaolinite and illite. Smectite occurs as coating materials and range in amount from 2.25 to 12.40% (avg. 6.44%) of the total rock volume. Smectite may be formed from the mica and other silicate minerals under alkaline conditions [86]. Illite is recorded in some pore spaces of the studied samples and has a lower average of 1.49%. Kaolinite is the second abundant clay mineral in the Moghra sandstones, and is thought to have formed by full dissolution of feldspars. It occupies oversized, irregular or elongated pores (Figure 10c). Kaolinite in the Moghra sandstone ranges in amount from 1.20 to 12.28% (avg. 4.89%) of the total rock volume. The favorable environment for the formation of kaolinite is the dilute acidic water [87]. Petrologic data indicate that kaolinite occurs in the form of face-to-face stacking of microporous pseudo-hexagonal book-shaped plates.

d. Other cementing materials

The iron oxide cement is predominant in the ferruginous quartzarenite lithofacies in the Moghra Formation and occupies the pore spaces between the quartz grains in the form of minute crystals (Figures 11a and 11b). The iron oxide cement is represented mainly by hematite which shows a preferential distribution in the form of irregular patches concentrated within areas of high matrix and/or cement. Iron oxides are identified in the studied sandstones through the petrographic study and SEM observation as well as the XRD analysis. Hematite content ranges from 0.0 to 7.15% (av 1.45%) and occurs as grain coatings, commonly intergrown with the infiltrated clays in both intergranular primary and mouldic porosities (Figures 11c and 11d). Though the iron oxides are present through the entire section, their percentage is low (not more than 7.15%; Table 1).

Two main carbonate cements are recognized, dolomite and non-ferroan calcite. The carbonate cement occurs as pore-filling material and as a complete or partial replacing mineral (Figures 12a and 12b). Carbonate cement ranges from 0.0 to 10.37% (average 7.5%) of the whole rock volume in the Moghra sandstones. Carbonate cement fills both the intergranular primary pores (Figure 12a) and the oversized pores. Dolomite cement occurs as coatings with an average of 2.66%

(Table 1). However, the possibility of dissolution of dolomite cement in these sandstones has to be considered. The common occurrence of micritic dolomite as a few scattered patches in thin sections is a criterion commonly cited as evidence of carbonate dissolution [74]. Halite cement, with an average of 1.74%, occurs also as coatings, or may fill intergranular primary pores (Figures 12c and 12d).

Interdependence of the petrophysical characteristics and the mechanical properties

The mineralogical composition of sandstone has a great influence on the strength parameters. It has been found that the strength of some sandstones increases with increasing the quartz content [88]. In contrast, it has also been reported that the textural characteristics are more effective than mineral composition on the mechanical behavior of sandstone [89]. Porosity, in particular, has a strong influence on the strength of sandstone [90].

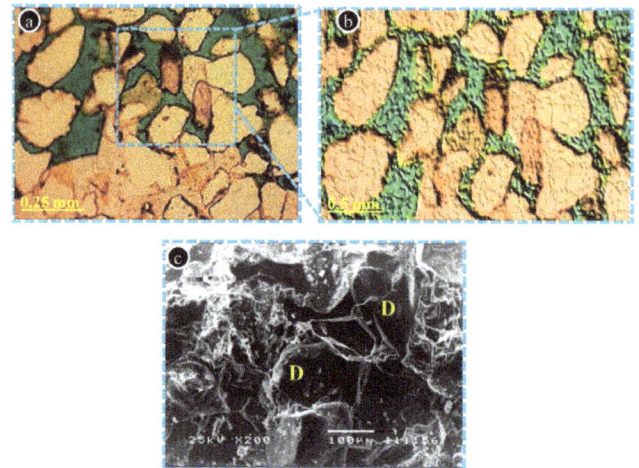

Figure 9: (a) and (b) optical micrographs of mechanically infiltrated clays filling interstitial voids. (c) Scanning electron micrograph showing mechanically infiltrated clay coating detrital quartz (D) oriented parallel to grain surface.

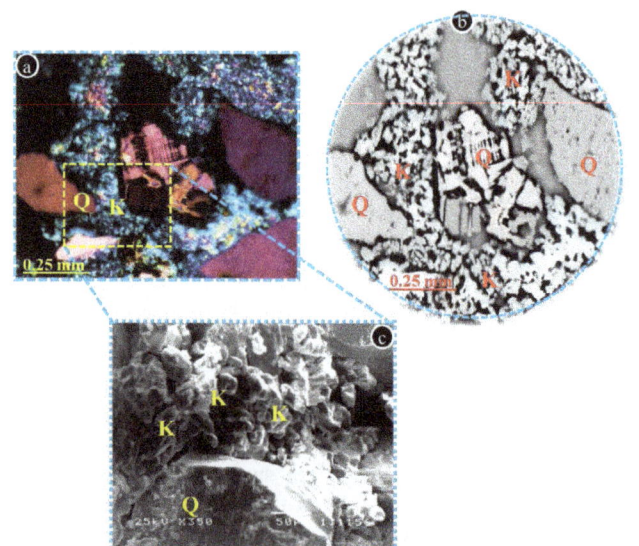

Figure 10: (a) and (b) Optical micrographs showing pore-filling kaolinite "K" intergrowths between quartz grains "Q". (c) Scanning electron micrograph of authigenic kaolinite cement "K" and quartz grains (Q).

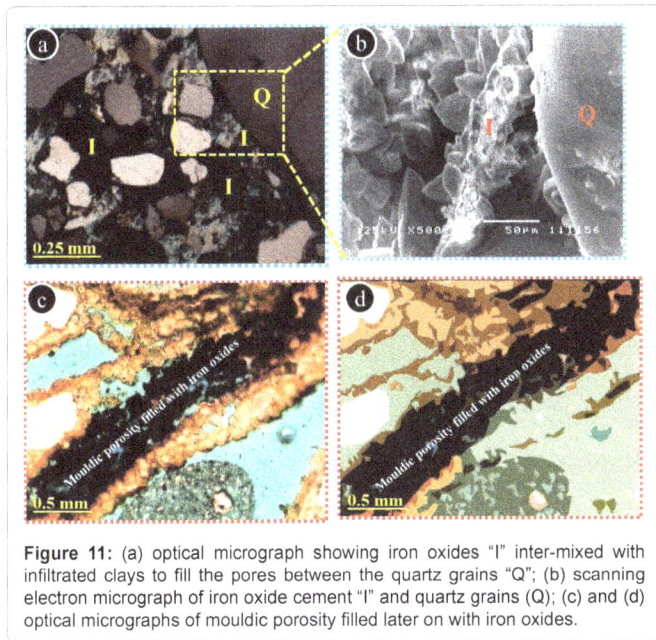

Figure 11: (a) optical micrograph showing iron oxides "I" inter-mixed with infiltrated clays to fill the pores between the quartz grains "Q"; (b) scanning electron micrograph of iron oxide cement "I" and quartz grains (Q); (c) and (d) optical micrographs of mouldic porosity filled later on with iron oxides.

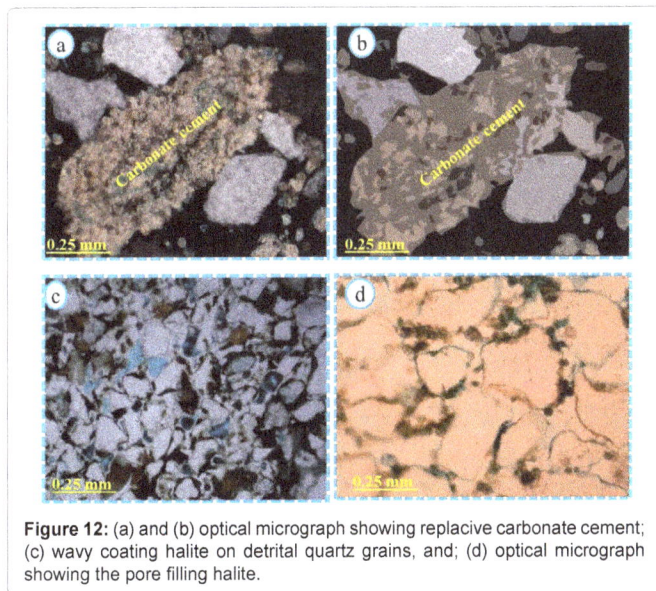

Figure 12: (a) and (b) optical micrograph showing replacive carbonate cement; (c) wavy coating halite on detrital quartz grains, and; (d) optical micrograph showing the pore filling halite.

The diagenetic processes have many influences on the petrophysical and mechanical properties of the Moghra sandstones and caused major changes in these properties. These changes are detected by the petrographic investigation of the thin sections of rock samples as well as the petrophysical and mechanical measurements. Many researchers [88,89,91-96] have also noticed these various effects. Since the mechanical properties of sandstones are affected by several parameters, the influence of each of them should be discussed separately. The investigated parameters also influence each other; for instance, grain density may result in greater porosity and bulk density [94]. Therefore, assessing the influence of the petrophysical parameters on the mechanical aspects (UCS$_{dry}$, IS50 and SHN), was inter-correlated and analyzed.

The results illustrated in Table 3 show that porosity ($\varnothing$, %) is correlated with the grain density (g/cc), permeability (mD), irreducible

Correlation Coefficient	ρ_b (g/cc)	$\varnothing$ (%)	ρ_g (g/cc)	K (mD)	S_{Wirr} %	V_p (m/s)	UCS$_{dry}$ (MPa)	IS50 (MPa)	SHV
ρ_b (g/cc)	1								
$\varnothing$ (%)	-0.91	1							
ρ_g (g/cc)	0.85	-0.92	1						
K (mD)	-0.77	0.81	-0.76	1					
S_{Wirr} %	0.84	-0.94	0.87	-0.76	1				
V_p (m/s)	0.79	-0.9	0.82	-0.7	0.84	1			
UCS$_{dry}$ (MPa)	0.83	-0.9	0.82	-0.74	0.87	0.81	1		
IS50 (MPa)	0.83	-0.88	0.82	-0.73	0.85	0.8	0.98	1	
SHV	0.81	-0.89	0.82	-0.74	0.87	0.79	0.99	0.97	1

Table 3: Correlations between the variously measured parameters of the studied sandstone samples from the Moghra formation shown as the 'r' values. [ρ_b (g/cc): Bulk Density; $\varnothing$ (%): Porosity; ρ_g (g/cc): Grain Density; K (mD): Permeability; S_{Wirr} (%): Irreducible Water Saturation; V_p (m/s): P-wave Velocity; UCS$_{dry}$ (MPa): Unconfined Compressive Strength; IS50 (MPa): Point Load Strength Index and (SHV): Schmidt Hammer Number].

water saturation (%), P-wave velocity, unconfined compressive strength, point load strength index (IS50), and Schmidt hammer number with 'r' values of -0.92, 0.81, -0.94, -0.9, -0.9, -0.88, and -0.89, respectively. Bulk density has a strong correlation (direct or inverse) with the petrophysical and geomechanical aspects of the studied samples including porosity, grain density, permeability, irreducible water saturation, P-wave velocity, unconfined compressive strength, point load strength index, and the Schmidt hammer number with 'r' values of -0.91, 0.85, -0.77, 0.84, 0.79, 0.83, 0.83 and 0.81, respectively.

The unconfined compressive strength is correlated significantly with grain density, permeability, irreducible water saturation, and the P-wave velocity (Table 3). The corresponding r values are 0.82, -0.74, 0.87 and 0.81, respectively. Similarly, the point load strength index is correlated significantly with grain density, permeability, irreducible water saturation, and the P-wave velocity, with r values of 0.82, -0.73, 0.85 and 0.8, respectively. Finally the Schmidt hammer number is also correlated with petrophysical and mechanical parameters. As far as the influence of one physical or mechanical property on the other is concerned, the Schmidt Hammer number increases with increasing density and decrease with increasing porosity and permeability. The relationships between Schmidt Hammer number and permeability, irreducible water saturation, P-wave velocity, unconfined compressive strength, and the point load strength index have the following r values -0.74, 0.87, 0.87, 0.79, 0.99 and 0.97, respectively. From all previous data and correlations among the petrophysical and mechanical aspects, it is evident that both porosity and bulk density are the major parameters influencing other physical and mechanical properties of the studied sandstone samples of the Moghra Formation.

Conclusions

The present study deals with the evaluation of the petrophysical and geomechanical aspects of the Lower Miocene Moghra Formation sandstones exposed at the Qattara Depression, north Western Desert, Egypt. For this task, 60 plug samples were collected from the studied rocks. The rocks belonging to the Moghra Formation are about 2.5 m thick and are overlained by the carbonates of the Marmarica Formation. Field investigations showed that the Moghra Formation consists mainly of laminated, massive, and cross-bedded sandstone. Moreover, the sandstones are thinly bedded and multicolored (white, yellow, orange and brown). Some primary sedimentary structures such with trough cross-bedding and pebbly coarse-grained sandstones at the base are also recorded. We also investigated the lithofacies, mineralogy and the various diagenetic processes which characterize the investigated sandstones. The petrophysical analysis focuses mainly on the evaluation of porosity, permeability and density.

Petrographically, the sandstones of Moghra Formation comprise three main microfacies: 1) fossiliferous dolomitic quartzarenites; 2) calcareous quartzarenites; and 3) ferruginous quartzarenites. The burial history of the Lower Miocene sandstones involved several diagenetic processes. The most common of these are dissolution, compaction and cementation. The results show that cementation resulted in porosity and permeability reduction. The cementing materials include iron oxides which occur as dark brown hematite filling the pore spaces between the detrital quartz grains. Calcite cement occurs in the form of microcrystalline calcite between the detrital quartz grains of the calcareous quartzarenites microfacies. The calcareous cement is an important factor limiting further diagenetic changes. The third type of cement is the clay minerals (kaolinite and smectite) are found in a wide variety of morphologies as revealed by the SEM and XRD observations. Under the microscope, kaolinite content scales down porosity and permeability by filling and lining the pore spaces and pore throats. The different cementing materials, along with compaction as well as the formation of secondary pores by feldspar, carbonate, and fossil dissolution control the overall reservoir quality of these sandstones. Dissolution processes potentially decrease the mechanical strength of the rock, whereas cementation has healed and filled some of the pore spaces of the rocks increasing their strength and stiffness.

The Moghra Formation was deposited in a fluvial environment. However, the occurrence of the fossiliferous carbonate beds within the Moghra clastic units provides an evidence for shallow marine conditions. The tabular cross-bedding, the abundance of drifted silicified tree trunks and the absence of detrital ferrous-iron minerals are strong evidences for the fluvial environment of the Moghra Formation.

The measured bulk density varies from 2.01 to 2.39 g/cm³; thus the sandstones of the Moghra Formation are classified as dense sandstones. Porosity ranges from 8.71 to 20.43% which means that the studied rocks are medium to highly porous. The measured grain density and permeability vary from 2.61 to 2.90 g/cm³, and 0.17 to 44.0 mD, respectively. Irreducible water saturation varies widely from 7.54 to 97.30%; while P-wave velocity ranges from 2080.27 to 4300.54 m/s. Both the bulk density and the P-wave velocity are inversely proportional to porosity. Permeability is directly proportional to porosity, but inversely proportional to the irreducible water saturation.

The dry uniaxial unconfined compressive strength measured for the studied sandstones varies from 44.51 to 136.38 MPa, which indicates that their compressive strength is moderate to high. The minimum and maximum values of the point load strength index are 3.18 and 9.74 MPa, respectively. This means the shear strength ranges from medium to high. The Schmidt hammer number fluctuates from 32 to 90. All of the measured mechanical parameters are closely related to each other.

The statistical correlation between various petrophysical properties showed that there are high to very high correlation between the bulk density and porosity in one hand and the other petrophysical and geomechanical parameters, on the other, which clearly indicated that both porosity and bulk density had been considered as the major parameters influencing other physical and mechanical properties of the studied sandstone samples of Moghra Formation. The linear relationship between the bulk density and porosity indicates that the rock samples have similar mineralogical composition, grain shape, packing and fabric; therefore, the pore framework is expected to be uniform and homogeneous.

References

1. Özbek A, Murat G (2015) The geotechnical evaluation of sandstone claystone alternations based on geological strength index. Arab J Geosci 8: 5257-5268.

2. Şen Z (2014) Rock quality designation-fracture intensity index method for geomechanical classification. Arab J Geosci 7: 2915-2922.

3. ISRM (1981) In: E.T.B (edn.) Rock characterization testing and monitoring-ISRM suggested methods. New York, Pergamon p: 211.

4. Wuerker RJ, McWilliams JR (1969) Microstructural techniques in the study of physical properties of rock. Int J Rock Mech Min Sci 6: 1-12.

5. Tugrul A, Zarif IH (2000) Engineering aspects of limestone weathering in Istanbul, Turkey. Bull Eng Geol Environ 58: 191-206.

6. Fujii Y, Ishijima Y (2004) Consideration of fracture growth from an inclined slit and inclined initial fracture at the surface of rock and mortar in compression. Int J Rock Mech Min Sci 41: 1035-1041.

7. Meng Z, Zhang J, Peng S (2006) Influence of sedimentary environments on mechanical properties of clastic rocks. Environ Geol 51: 113-120.

8. Stringfield VT, LaMoreaux PE, LaGrand HE (1974) Karst and paleohydrology of carbonate rock terranes in semiarid regions, with a comparison to humid karst of Alabama. Geological Survey of Alabama Bulletin p: 106.

9. Meshref WM, Rafai EM, Sadek HS, Abdel-Baki SH, El-Sirafe AMH, et al. (1980) Structural geophysical interpretation of basement rocks of the north western desert of Egypt. Ann Geol Surv Egypt 10: 923-987.

10. Said R (1981) The geological evolution of the river Nile. New York, Springer-Verlag p: 151.

11. Albritton CC, Brooks JE, Issawi B, Swedan A (1990) Origin of the Qattara depression, Egypt. Geol Soc Am Bull 102: 952-960.

12. Gindy AR (1991) Origin of Qattara depression-discussion. Geol Soc Am Bull 103: 1374-1375.

13. Aref MAM, El-Khoriby E, Hamdan MA (2002) The role of salt weathering in the origin of the Qattara depression, western desert, Egypt. Geomorphol 45: 181-195.

14. Said R (1962a) The geology of Egypt. Elsevier Publishing Company, Amsterdam p: 370.

15. Said R (1962b) Das Miozän in der westlichen wuste agyptens. Geologisches Jahrbuch 80: 349-366.

16. Misak RF (1979) Geology of the area between the Moghra oasis and the Mediterranean sea, western desert, Egypt. Ain Shams University, Cairo.

17. Dyke CG, Dobereiner L (1991) Evaluating the strength and deformability of sandstones. Quarterly J Eng Geol 24: 123-134.

18. Hawkins AB, McConnell BJ (1992) Sensitivity of sandstone strength and deformability to changes in moisture content. Quarterly J Eng Geol 25: 115-130.

19. Pettijohn FJ (1975) Sedimentary rocks (3rd edn), Harper and Row, New York p: 628.

20. Jiang S (2012) Clay minerals from the perspective of oil and gas exploration. Clay Minerals in Nature.

21. Nabawy BS (2013) Impacts of dolomitization on the petrophysical properties of the cenomanian El-Halal formation, north Sinai, Egypt. Arab J Geosci 6: 359-373.

22. Nabway BS, Kassab MA (2014) Porosity-reducing and porosity-enhancing diagenetic factors for some carbonate microfacies: A guide for petrophysical facies discrimination. Arab J Geosci 7: 4523-4539.

23. Cullity BD (1978) Elements of X-Ray diffraction. (2nd edn) Addison Wesley Publishing Company, Massachusetts.

24. Lynch EJ (1962) Formation evaluation. Harper and Row Publishers, New York p: 422.

25. Serra O (1984) Fundamentals of well-log interpretation. Amsterdam Oxford New York Tokyo p: 250.

26. El Naggar OM (2009) Geological application of capillary pressure and permeability data using reservoir anisotropy concept, Western Desert, Egypt.

27. Nabawy BS, Rochette P, Geraud Y (2009) Petrophysical and magnetic pore network anisotropy of some cretaceous sandstone from tushka basin, Egypt. Geophys J Int 177: 43-61.

28. Kahraman S, Yeken T (2008) Determination of physical properties of carbonate rocks from P-wave velocity. Bull Eng Geol Environ 67: 277-281.

29. Singh TN, Kanchan R, Saigal K, Verma AK (2004) Prediction of p-wave velocity and anisotropic property of rock using artificial neural network technique. J Sci Indust Res 63: 32-38.

30. Sharma PK, Singh TN (2008) A correlation between P-wave velocity, impact strength index, slake durability index and uniaxial compressive strength. Bull Eng Geol Environ 67: 17-22.

31. Rao MVMS, Sarma LP, Prasanna Lakshmi KJ (2002) Ultrasonic pulse broadening and attenuation in volcanic rock-a case study. Ind J Pure Appl Phys 40: 396-401.

32. ASTM International Designation D2938-95 (2002) Standard test method for unconfined compressive strength of intact rock core specimens. USA.

33. Schmidt E (1951) Quality control of concrete by rebound hammers testing. Schweizer Archiv fur angewandte Wissenschaft und Technik 17: 139-143.

34. Cargill, JS, Shakoor A (1990) Evaluation of empirical methods for measuring the uniaxial compressive strength of rock. Int J Rock Mech Min Sci Geomech Abstr 27: 495-503.

35. Kahraman S (2001) Evaluation of simple methods for assessing the uniaxial compressive strength of rock. Int J Rock Mech Mining Sci 38: 981-994.

36. Nabawy BS (2011) Impacts of dolomitization on the petrophysical properties of El-Halal formation, north Sinai, Egypt. Arab J Geosci 6: 359-373.

37. Kassab MA, Abdou AA, El Gendy NH, Shehata MG, Abuhagaza AA (2013) Mutual relations between petrographical and petrophysical properties of cretaceous rock samples for some wells in the North Western Desert, Egypt. Egypt J Petroleum 22: 73-90.

38. Tucker ME (2003) Sedimentary rocks in the Field. (3rd edn.) The Geological Field Guide Series. Wiley, Chichester, p: 250.

39. Anon (1979) Classification of rocks and soils for engineering geological mapping, part 1-Rock and soil materials. Report Comm Eng Geol Mapp Bull Int Assoc Eng Geol 19: 364-371.

40. Kassab MA, Abu Hashish MF, Nabawy BS, Elnaggar OM (2017) Effect of kaolinite as a key factor controlling the petrophysical properties of the Nubia sandstone in central Eastern Desert, Egypt. J Afr Earth Sci 125: 103-117.

41. Yıldız Ş, Özgökçe MS, Özgökçe F, Karaca İ, Polat E (2010) Zooplankton composition of Van Lake Coastline in Turkey. Afr J Biotech 9: 8248-8252.

42. Deere D, Miller U (1966) Engineering classification and index properties for intact rock. Kirtland Air Base, New Mexico.

43. Piteau DR (1970) Engineering geology contribution to the study of stability in rock with particular reference to De Beer's mine.

44. ISRM (1985) Suggested methods for determining point load strength: International Journal of Rock Mechanics and Mineral Sciences Geomech 22: 53-60.

45. Salah A (2008) The Impact of Rock Pore Throat Distribution on Different Reservoir Parameters. Egypt.

46. Bell FG (2007) Engineering Geology. (2nd edn) Butterworth- Heinemann Publications, Burlington, MA p: 581.

47. Kassab MA, Hassanain IM, Salem AM (2014) Petrography, diagenesis and reservoir characteristics of the Pre-Cenomanian sandstone, Sheikh Attia area, East Central Sinai, Egypt. J Afr Earth Sci 96: 122-138.

48. Zampetti V, Sattler U, Braaksma H (2005) Well log and seismic character of Liuhua 11-1 Field, South China Sea; relationship between diagenesis and seismic reflections. Sediment Geol 175: 217-236.

49. Gardner GHF, Gardner LW, Gregory AR (1974) Formation velocity and density-the diagnostic basis for stratigraphic traps. Geophysics 39: 770-780.

50. Anselmetti FS, Eberli GP (1997) Sonic velocity in carbonate sediments and rocks. Geophysical Development Series 6: 53-74.

51. Pettijohn FJ, Potter PE, Siever R (1987) Sand and sandstone. Springer-Verlag, Berlin p: 306.

52. Collinson JD (1970) Bedforms of the Tana River. Geogr Annal Norway 52: 31-56.

53. Jackson RG (1976) Depositional model of point bars in the lower Wabash River. J Sed Petrol 46: 579-594.

54. Farouk S, Khalifa MA (2010) Facies tracts and sequence development of the Middle Eocene-Middle Miocene successions of the southwestern Qattara Depression, northern Western Desert, Egypt. Paläontologie, Stratigraphie 536: 195-215.

55. Zaid SM (2013) Provenance, diagenesis, tectonic setting and reservoir quality of the sandstones of the Kareem Formation, Gulf of Suez, Egypt. J Afr Earth Sci 85: 31-52.

56. Tiab D, Donaldson EC (2004) Petrophysics: Theory and practice of measuring reservoir rock and fluid transport properties. Gulf Professional Publishing, USA p: 880.

57. Schmidt V, McDonald DA (1979) The role of secondary porosity in the course of sandstone diagenesis. Aspects of diagenesis. Spec Pub pp: 209-225.

58. McBride EF (1985) Diagenetic processes that affect provenance determinations in sandstone. Provenance of arenites, Boston pp: 95-113.

59. Milliken KL, McBride EF, Land LS (1989) Numerical assessment of dissolution vs. replacement in the subsurface destruction of detrital feldspar, Oligocene Frio Formation, south Texas. J Sed Perol 59: 740-757.

60. Pittman ED (1979) Porosity, diagenesis and productive capability of sandstone reservoirs. Spec Pub 26: 159-173.

61. Abdel-Wahab AA (1988) Lithofacies and diagenesis of the Nubia Formation at Central Eastern Desert, Egypt. 9th EGPC Expl Conf Cairo.

62. Abdel-Wahab AA (1999) Petrography, fabric analysis and diagenetic history of Upper Cretaceous sandstone, Kharga Oasis, Western Desert, Egypt. Sediment Egypt 7: 99-117.

63. Brewer R (1964) Fabric and mineral analysis of soils. Wiley, New York p: 470.

64. Keller WD (1970) Environment aspects of clay minerals. J Sed Petrol 40: 788-813.

65. Walker TR, Waugh B, Crone AJ (1978) Diagenesis in first-cycle desert alluvium of Cenozoic age, south western United States and north western Mexico. Geol Soc Am Bull 89: 19-32.

66. Wilson MD, Pittman ED (1977) Authigenic clays in sandstone: recognition and influence on reservoir properties and paleoenvironmental analsysis. J Sed Petrol 47: 3-31.

67. Heald MT, Larese RF (1974) Influence of coatings on quartz cementation. J Sed Petrol 44: 1269-1274.

68. Croizé D, Ehrenberg SN, Bjørlykke K, Renard F, Jahren J (2010) Petrophysical properties of bioclastic platform carbonates: implications for porosity controls during burial. Marine Petrol Geol 27: 1765-1774.

69. Dunoyer D, Segonzac G (1970) The transformation of clay minerals during diagenesis and low-grade metamorphism: A review. Sedimentol 15: 281-346.

70. Krauskopf KB (1979) Introduction to geochemistry. McGrow-Hill, Kogakusha Ltd. New York p: 617.

71. Bell FG, Lindsay P (1999) The petrographic and geomechanical properties of sandstones from the newspaper member of the Natal Group near Durban, South Africa. Eng Geol 53: 57-81.

72. Ulusay R, Tureli K, Ider MH (1994) Prediction of engineering properties of selected litharenite sandstone from its petrographic characteristics using correlation and multivariate statistical techniques. Eng Geol 37: 135-157.

73. Plachik V (1999) Influence of porosity and elastic modulus on uniaxial compressive strength in soft brittle porous sandstones. Rock Mec Rock Eng 32: 303-309.

74. Bell FG (1978) The physical and mechanical properties of the fell sandstones, Northumberland, England. Eng Geol 12: 1-29.

75. Shakoor A, Bonelli RE (1991) Relationship between petrographic characteristics, engineering index properties, and mechanical properties of selected sandstones.

76. Bell FG, Culshaw MG (1993) A survey of the geotechnical properties of some relatively weak Triassic sandstones. Bull Assoc Eng Geol XXVIII-1: 55-71.

77. Jeng FS, Weng MC, Lin ML, Huang TH (2004) Influence of petrographic parameters on geotechnical properties of tertiary sandstones from Taiwan. Eng Geol 73: 71-91.

78. Abu Seif AS (2016) Evaluation of geotechnical properties of cretaceous sandstone, Western Desert, Egypt. Arab J Geosci 9: 299.

18

Variability in the Highway Geotechnical Properties of Two Residual Lateritic Soils from Central Nigeria

Owoyemi OO[1]* and Adeyemi GO[2]

[1]*Department of Geology, Kwara State University, Malete, Nigeria*
[2]*Department of Geology, University of Ibadan, Nigeria*

Abstract

Sixty-four bulk samples of two residual lateritic soils forming the subgrade of the failed sections flexible highway pavement linking Ilorin to Mokwa in central Nigeria were investigated. This was with a view to determining the level of variation in the geotechnical properties of soil samples taken systematically within restricted area in two locations underlain by different bed rocks. One set was developed over sandstone formation of the Southern Bida Basin while the other set was developed over migmatite-gneiss. Consistency limits, grain size distribution, specific gravity, compaction, California Bearing Ratio (CBR), permeability and compressibility characteristics of these soils were determined using the British standard procedures 1377. Coefficient of variation was used to measure the degree of variation in the determined properties. The coefficients of variations for the sandstone derived soil (1.68% and 56.86%) are higher than that of the migmatite-gneiss derived soil (1.28%-54.40%). Permeability, linear shrinkage, and coefficient of volume compressibility possess the highest variability. Atterberg limits and derived indices, amount of fines, soaked and unsoaked CBR possess moderate variability, while moisture density parameters (MDD and OMC), natural moisture content and specific gravity exhibits the least variability. In order to prevent design errors, field sampling should be very thorough involving collection of several samples. This approach will eliminate wrong inferences often associated with results of testing of few samples

Keywords: Highway geotechnics; Coefficient of variation; Sampling; Subgrade; Parent rock

Introduction

The Ilorin-Mokwa road is an economically important highway linking the Nigerian Southwest with the Northwest. This highway is the major transport route for agricultural goods, services and petroleum products between these regions. It is characterized by all manner of structural failures ranging from waviness to large potholes and completely failed sections. A number of reasons ranging from faulty designs, lack of drainage, thin wearing course coverings, negligible quality control, inadequate maintenance funding, geological and pedogenic factors to geotechnical factors have been adduced to the general poor conditions of Nigerian roads [1-3]. Although, there are many probable reasons for the failure of this important highway, certain observations were made while carrying out preliminary studies on it. Figure 1 shows that the thickness of the highway flexible pavement and its foundation are not good enough and do not meet up with recommended specification for upper layers of highly trafficked flexible highway pavements. It was also noticed that the highway lacks adequate crown and possesses little slope that give room for efficient drainage of rainwater away from the pavement surface. Water penetrates the pavement from the surface and infiltrates from the sides of the road because of holes in the pavement and the fact that most sections do not have shoulders bordered by ditches as shown by Figure 2. The road therefore sometimes serves as drainage path for rainwater because the pavement is not well elevated. Incidentally, this federal highway has the highest average daily traffic and percentage of heavy vehicles in Nigeria [3]. This road therefore, requires more skillful design and careful considerations of all possible factors that might affect its service life.

Researchers have reported the variability in geotechnical properties of Nigerian soils [4-7]. However the study area has not been covered in such research. Fundamentally there are many sources of variabilities and uncertainties associated with site characterization. This includes measurement errors and statistical uncertainty. While the actual variability involves only the variability of soil properties, the total variability includes other additional sources of uncertainties such as measurement errors and statistical uncertainties [8]. This research however assumes that the other sources of variabilities aside the actual

Figure 1: A deplorable section of the Ilorin-Mokwa Road at Onipako near Jebba.

*Corresponding author: Owoyemi OO, Department of Geology, Kwara State University, Malete, Nigeria, E-mail: bunmmydot@yahoo.com

variability have been cancelled out by comparing two genetically different residual soils and observing the trend of variabilities in both soils.

It is often assumed that soil properties are the same throughout

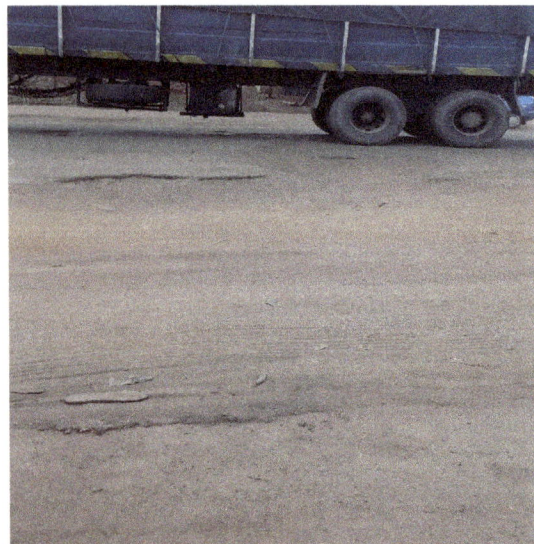

Figure 2: A failed section of the road exposing the roads foundation and relatively thin asphalt riding surface.

a rather wide area when sampling for geotechnical tests. However, undetailed sampling can lead to conclusions that significantly differ from true soil behavior. Therefore, there is a need to quantify the amount of uncertainties attached to highway subgrade sampling so as to minimize design errors attached to undetailed sampling. The aim of this paper was to establish the degree of variation of some highway geotechnical parameters within restricted area underlain by different bedrocks. Phoon and Kulhawy recommend a statistical analysis including the coefficient of variation and the scale of fluctuation for this purpose [9]. However scale of fluctuation for evaluating spatial variability of parameters that can be obtained from *in situ* tests because they provide continuous record of ground properties [10].

Location of Study Area

A study location was located in Shao near Ilorin on latitude N08°33.268' and longitude E004°30.730', while the second location was located near Mokwa and it lies on latitude 09°12.484' and longitude E 004°52.459'. These two locations belong to the same climatic belt and but underlain by different bedrocks. Ilorin is underlain by Precambrian Basement Complex rocks, while Mokwa is underlain by the Cretaceous Sandstone Formation of the Bida Basin. The Bida Basin is a NW-SE trending intracratonic sedimentary basin extending from Kontagora in Niger State of Nigeria to areas slightly beyond Lokoja in the south. It is delimited in the Northeast and Southwest by the Precambrian Basement Complex [11]. Figure 3 shows the location and geology of the study areas. Nigeria is located within the tropics and therefore experiences high temperatures throughout the year. The mean temperature for

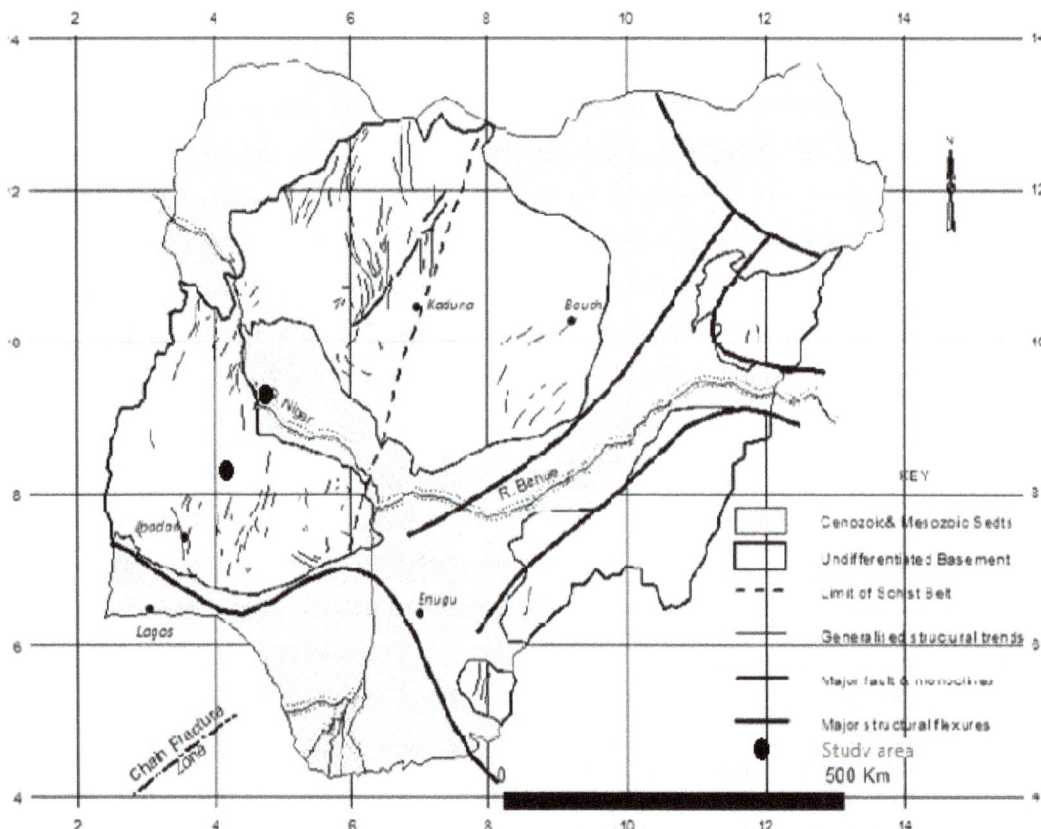

Figure 3: Generalized geology of Nigeria showing the study area [12].

the country is 27 °C. Average maximum temperatures vary from 32°C along the coast to 41°C in the far north, while minimum figures range from 21°C in the coast to below 13°C in the north. The climate of the country varies from a very wet coastal area with annual rainfall greater than 3,500 mm to the Sahel region in the north eastern parts with annual rainfall less than 600 mm. Generally, there is a distinct wet and the dry season within a year. The length of the rainy season decreases from 9-12 months in the south to only 3-4 months in the extreme Northeast. Average rainfall in the northern limit of the belt is about 254 mm annually. Mean monthly relative humidity is about 29%. The study area falls within the zone that receives 140-160 mm of rain per annum.

Method

Two locations 15 m away from the highway with different exposed bedrocks (Migmatite-Gneiss and Sandstones) were selected. In each location, disturbed samples were taken at 5 m sampling interval within gridded as shown in Figure 4. Index and engineering tests relevant in highway geotechnics were carried out on the samples. All tests were carried out in accordance with the British standard method of testing soil 1377, modifications where necessary were however made [13]. Determined parameters include consistency limits, grain size distribution, specific gravity, compaction, soaked and unsoaked California bearing ratio (CBR) permeability and compressibility. The variability in the values of measured parameters was presented using contour plots while coefficient of variation was used to measure the degree of variation of these properties. The contour plots were made using MATLAB curve fitting method. The higher the coefficient of variation, the greater the dispersion in a set of variables and values up to 10% is believed to show significant variability within any set of data.

Results and Discussions

Linear shrinkage

The linear shrinkage of the sandstone derived soil ranges from 1.5%

to 7.3% while that of the soil developed over migmatite ranges from 2.2% to 10.1%. Although the linear shrinkage of the sandstone soil is averagely lower than that of migmatite soil, it has higher standard deviation, variance and coefficient of variation. Figure 5 compares the contour plots of the linear shrinkage values of the migmatite derived (MG) soil samples and the sandstone derived (SS) soil samples. It can be seen from this figure, that the contours for SS are more closely spaced with highly contrasting colours than MG. This indicates higher variability in the linear shrinkage values of the SS soil samples.

Atterberg limits

Table 1 shows the values obtained from the statistical treatment of data obtained from liquid limit, plastic limit and plasticity index of the studied soils. The degree of variability in Atterberg limits values for the sandstone soil is generally higher than that of the migmatite soil. This can be observed in the consistently higher values of the coefficient of variation characteristic of the sandstone soils. The coefficient of variation in the Atterberg limit for the migmatite derived soil ranges from 17.00% to 21.64% while that of the sandstone soil ranges from 11.26% to 17.32%. Figures 6-8 show the contour plots of the Atterberg limit of the studied soils. It can be noticed from these figures that the Atterberg limit values of the SS soil samples vary more within the gridded sampling area than the MG soil samples.

Grain size distribution parameters

Table 2 shows coefficient of variation of the grain size distribution parameters of the studied soils. Variation up to 150% was recorded in the grainsize distribution parameters of the studied soils. The high coefficient of variation values associated with the grain size distribution characteristics of these soils implies that their grain size distribution characteristics vary significantly within restricted area Except for coefficient of curvature, the sandstone derived soil has higher coefficient of variation than the MG soil. This implies that the sandstone derived soil has higher level of heterogeneity than the SS one. Figure 9 compares

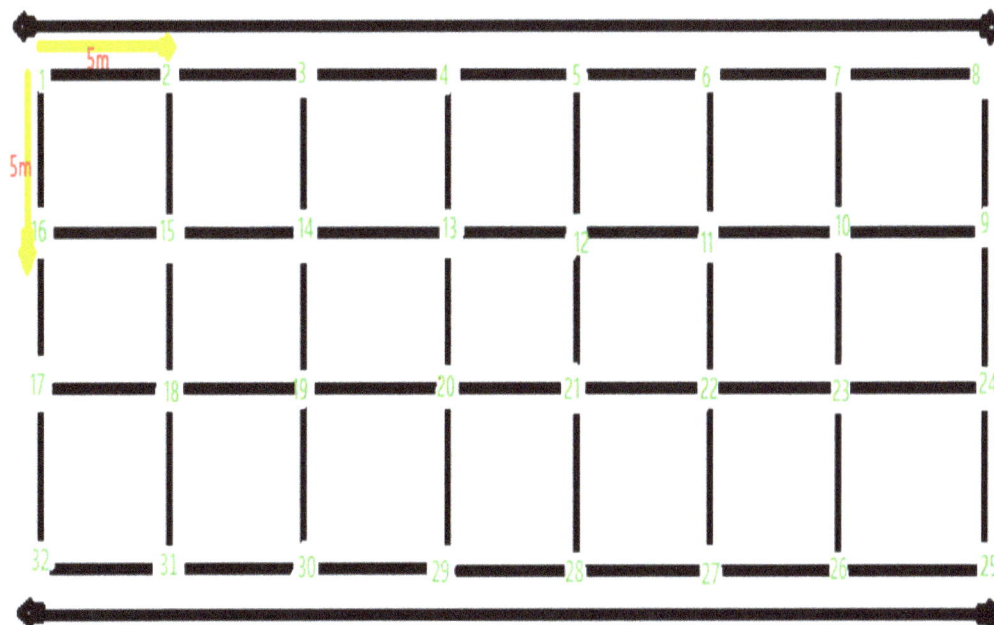

Figure 4: A sketch of the sampling area showing the spatial distribution of the sampling points.

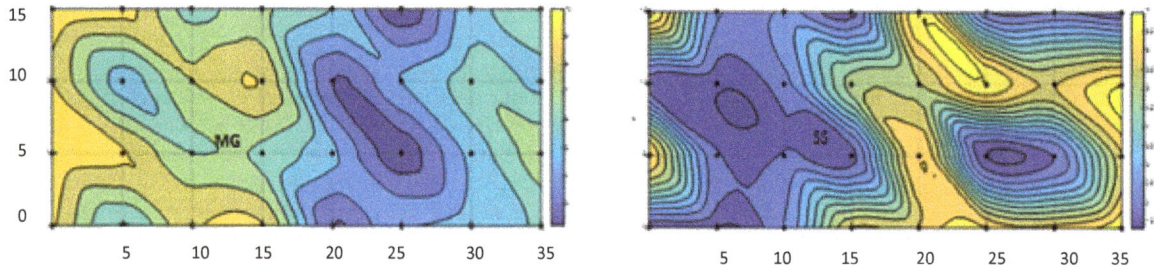

Figure 5: Contour plots of the linear shrinkage values of the studied soil samples.

Soil type	Statistical parameter	Liquid limit (%)	Plastic limit (%)	Plasticity index (%)
Migmatite derived soil	Range	28.4-49.1	16.0-30.8	11.03-27.06
	Variance	21.05	14.26	8.87
	Standard deviation	4.59	3.78	2.98
	Coefficient of variation (%)	11.29	16.35	17.32
Sandstone derived soil	Range	20.8-44.2	10.9-27.7	8.34-17 .04
	Variance	32.07	18.34	8.2
	Standard deviation	5.66	4.28	2.86
	Coefficient of variation (%)	17	21.34	21.64

Table 1: Statistical analysis of Atterberg Limits data.

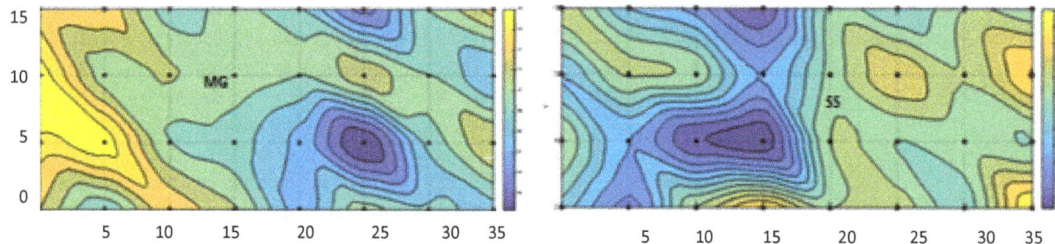

Figure 6: Contour plots of the liquid limit values of the studied soil samples.

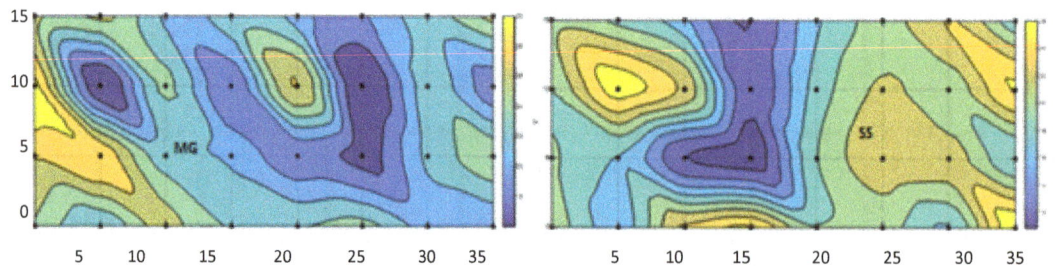

Figure 7: Contour plots of the Plastic limit values of the studied soil samples.

the contour plots of the amount of fines present in the SS soil with the amount present in the MG soil.

Specific gravity, moisture content and derived atterberg indices

Table 3 shows the summary of the statistical evaluation of spatial variability in the specific gravity, moisture content derived indices of both soils. The coefficient of variation in specific gravity values for

the MG soil is similar to that of the sandstone ones. There is also little spatial variation in specific gravity values across the gridded area. Since specific gravity is a measure of degree of weathering, it implies that the two set of soil have similar degree of weathering and are uniformly weathered [14]. There is little variation in the spatial distribution of moisture content for both set of soils. The degree of variations in the values of parameters derived from index properties for the MG soil is lower than those recorded for the SS soil. This trend can also be

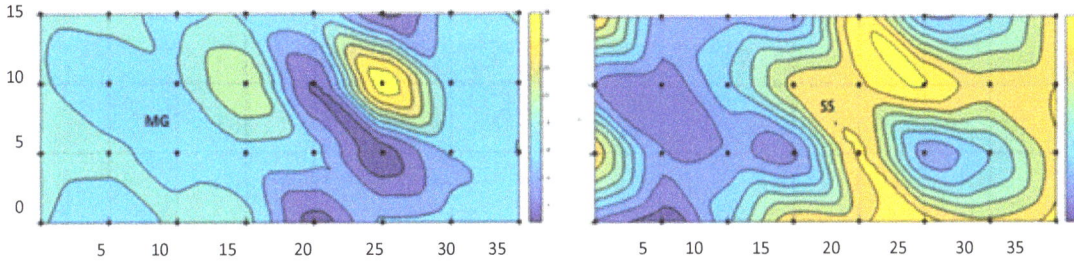

Figure 8: Contour plots of the plasticity index values of the studied soil samples.

Grain size distribution parameter	Coefficient of variation (%)	
	Migmatite derived soil	Sandstone derived soil
Amount of gravel sized particles	70.22	145.37
Amount of Sand sized particles	23.34	22
Amount of silt sized particles	33.58	53.92
Amount of clay sized particles	68.94	95.86
Amount of fine particles	32.47	45.22
Coefficient of Uniformity	115.54	130.63
Coefficient of curvature	136.75	150.9

Table 2: Statistical analysis of grain size parameters.

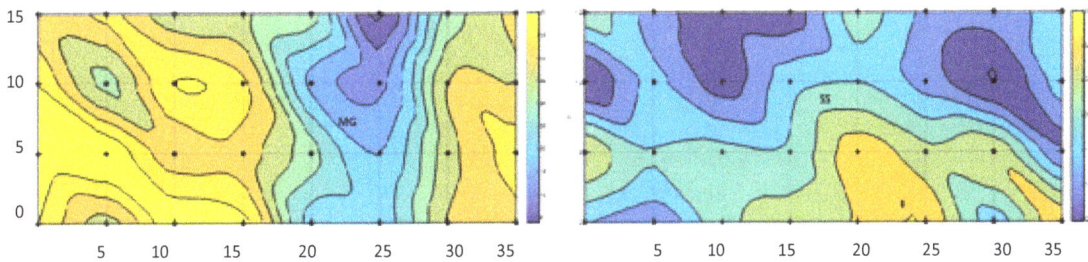

Figure 9: Contour plots of the amount of fines present in the studied soil samples.

Soil	Statistical parameter	Specific gravity	Moisture content (%)	Derived atterberg indices	
				Flow index	Toughness index
Migmatite derived soil	Range	2.6 –2.7	19.2 -25.66	15.30-32.4	0.45-1.22
	Variance	0.001	3.81	12.05	0.04
	Standard deviation	0.03	1.95	3.47	0.19
	Coefficient of variation (%)	1.28	8.59	17.14	21.67
Sandstone derived soil	Range	2.55 –2.7	13.98 -1.52	7.36-4.90	0.47-1.60
	Variance	0.002	3.27	19.86	0.08
	Standard deviation	0.04	1.81	4.46	0.29
	Coefficient of variation (%)	1.68	10.51	26.45	34.55

Table 3: Statistical analysis of data for specific gravity, moisture content and derived units.

observed in the contour plots of the values of natural moisture content and specific gravity of the studied soil shown in Figures 10 and 11.

Moisture density relationship and California bearing ratio

CBR is a measure of road subgrade strength and important parameter used in highway design. Table 4 presents the summary of the statistical treatment of the results of CBR and compaction test carried out on both soils. While the coefficient of variation of the unsoaked CBR and maximum dry density for the MG soil is higher than that of the SS soil, the soaked CBR is higher for the sandstone soil. The coefficient of variation for optimum moisture content for the SS soil is higher than that of the MG soil. The differences in the coefficient of variation for CBR and OMC of the studied soils are marginal. This is

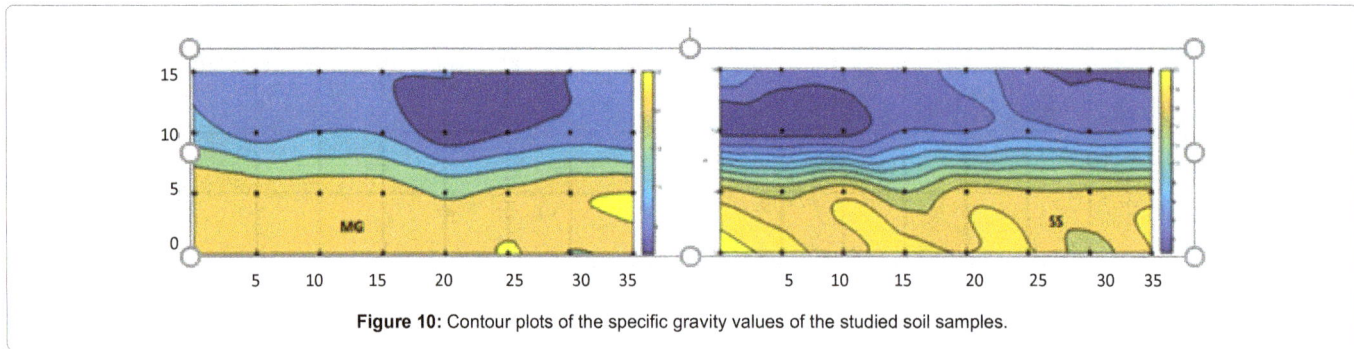

Figure 10: Contour plots of the specific gravity values of the studied soil samples.

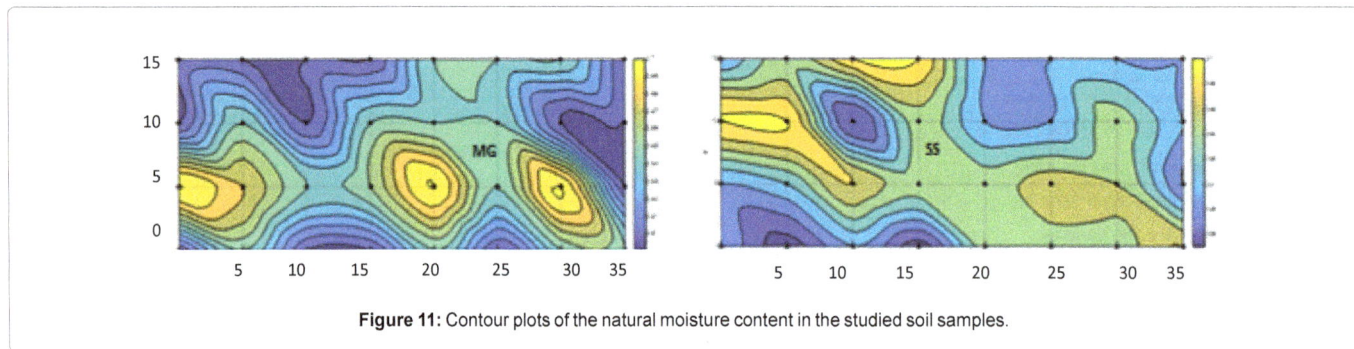

Figure 11: Contour plots of the natural moisture content in the studied soil samples.

Soil	Statistical parameter	MDD(Kg/m³)	OMC (%)	Unsoaked CBR (%)	Soaked CBR (%)
Migmatite derived soil	Range	1830.05-1970.16	16.0-19.20	22.0-88.7	12.53-36.25
	Variance	41996.37	0.72	274.48	37.98
	Standard deviation	204.93	0.85	16.57	6.16
	Coefficient of variation (%)	11.3	4.86	29.07	27.39
Sandstone derived soil	Range	1070.02-1960.04	13.0-16.0	43.1-126.9	16.53-45.32
	Variance	1811.84	0.77	608.66	84.31
	Standard deviation	42.57	0.88	24.67	9.18
	Coefficient of variation (%)	2.23	6.03	27.51	28.31

Table 4: Summary of the statistical treatment of the CBR and moisture dry density values of the studied soil.

also evident from Figures 12 and 13 which shows the contour plots for OMC and CBR values respectively.

Permeability and compressibility

The coefficient of variation of the coefficient of permeability for the MG soil is higher than that of the SS soil. Table 5 shows the summary of the determined statistical parameters for permeability and coefficient of (volume) compressibility. The coefficient of variation of the coefficient of (volume) compressibility for the MG soil is higher than that of the SS soil.

Figure 14 shows the contour plot of the coefficient of Permeability values of the studied soil, while Figure 15 shows the contour plot of the coefficient of (volume) compressibility values. These figures also show that the variability in the coefficient of permeability values of the MG soil samples is more than that of the SS ones.

Degree of variation and sampling

Comparing Figures 16 and 17, it can be observed that the degree of variability of the laboratory determined highway geotechnical parameters for both set of soil vary similarly. The variation associated

with the properties of the studied soils appears to be higher for some parameters than it is for others. On this basis, for this work, the observed coefficient of variations of the studied soils has been grouped into three. Category one consist of parameters with relatively high coefficient of variation, these include permeability, linear shrinkage and coefficient of volume compressibility. Category two consist of parameters with relatively moderate coefficient of variation, which include atterberg limits, amount of fines, soaked and unsoaked CBR. Category three consists of parameters with relatively low coefficient of variation, which include moisture density parameters (MDD and OMC), natural moisture content and specific gravity. Therefore, sampling for the determination of properties belonging to category one should be most detailed and more samples should be tested to minimise design error due to using values that do not correctly represent the soil mass being investigated.

Conclusions

Highway geotechnical parameters of two genetically different residual lateritic soils were treated statistically to determine the degree of variation associated with them within a restricted area. The coefficient of variation for the migmatite derived soil samples range

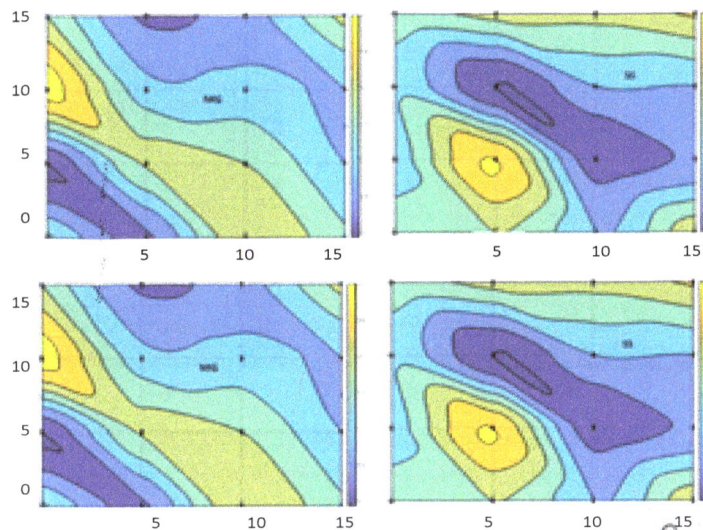

Figure 12: Contour plots of the OMC values of the studied soil.

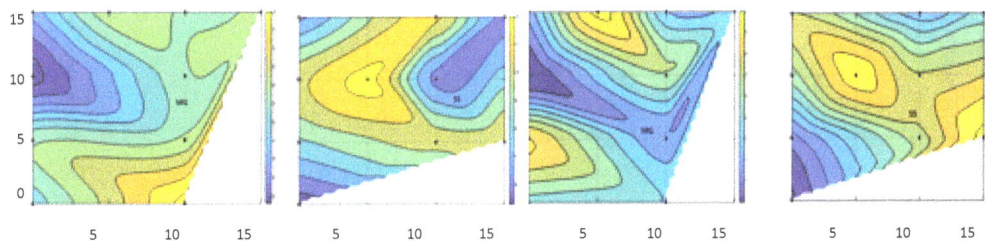

Figure 13: Contour plots of the soaked and unsoaked CBR values of the studied soil.

Soil	Statistical Parameter	Permeability coefficient (cm/sec)	Coefficient of Volume Compressibility (m^2/KN)
Migmatite soil	Range	9.2×10^{-7}-3.5×10^{-6}	4.4×10^{-5}-6.1×10^{-4}
	Variance	9.91×10^{-13}	3.2×10^{-8}
	Standard deviation	9.96×10^{-7}	1.8×10^{-4}
	Coefficient of variation (%)	54.35	43.38
Sandstone soil	Range	1.0×10^{-6}-3.1×10^{6}	2.4×10^{-4}-7.1×10^{-4}
	Variance	8.4×10^{-13}	3.3×10^{-8}
	Standard deviation	9.17×10^{-7}	1.8×10^{-4}
	Coefficient of variation (%)	46	54.38

Table 5: Summary of the statistical treatment of the permeability and consolidation data.

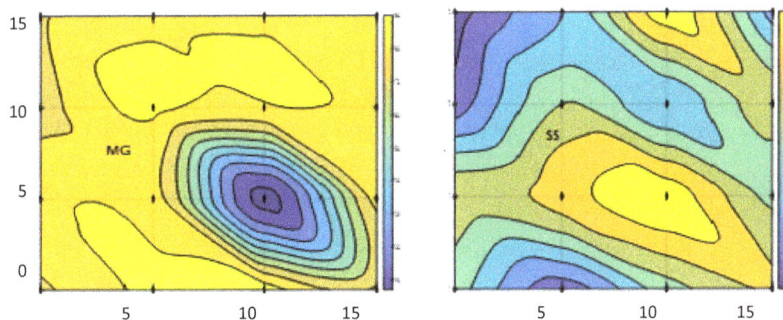

Figure 14: Contour plots of the coefficient of Permeability values of the studied soil.

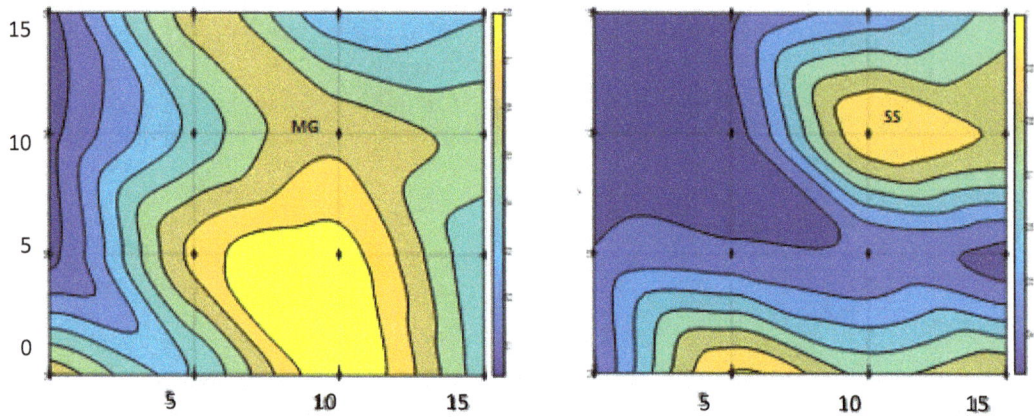

Figure 15: Contour plots of the coefficient of volume compressibility values of the studied soil.

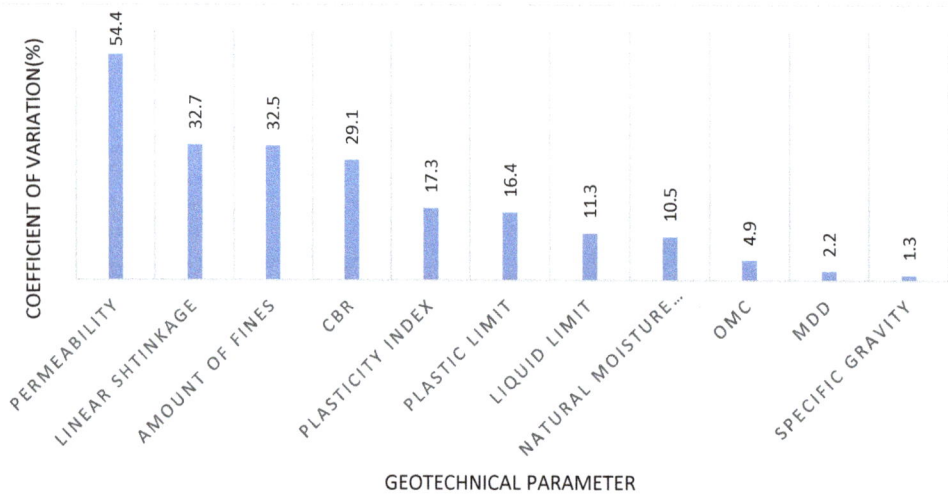

Figure 16: Coefficient of variation characteristic of the highway geotechnical properties of the sandstone derived soil.

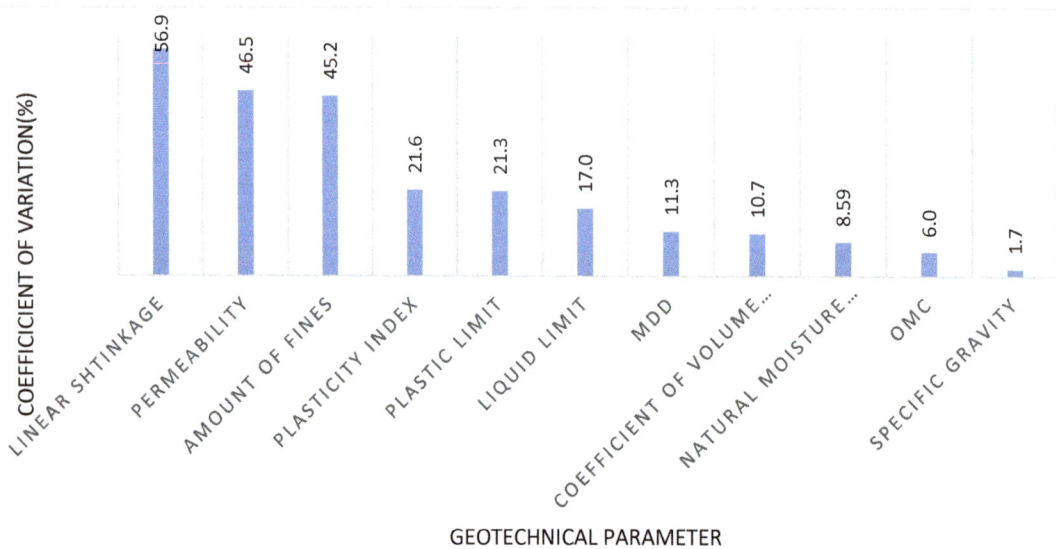

Figure 17: Coefficient of variation characteristic of the highway geotechnical properties of the migmatite derived soil.

between 1.28% and 54.40% while those of the sandstone derived soil range between 1.68% and 56.86%. Except for specific gravity, significant variability exists in the values of all determined highway geotechnical parameters of the studied soil samples within restricted area. Except for Permeability coefficient and unsoaked CBR, the Mokwa sandstone derived lateritic soil exhibits more heterogeneity than the Shao migmatite-gneiss derived one. Permeability coefficient, linear shrinkage and coefficient of volume compressibility possesses relatively high variability. Atterberg limits, amount of fines, soaked and unsoaked CBR have relatively moderate coefficient of variation while moisture density parameters (MDD and OMC), natural moisture content and specific gravity have relatively low variability. Therefore, detailed sampling, statistical analysis and geological considerations should be the basis for determination of parameters often utilised for foundation design for flexible highway pavement.

References

1. Adeyemi GO, Oyeyemi F (1998) Geotechncial basis for failure of sections of the Lagos-Ibadan expressway, Southwestern Nigeria. Bull Eng Geol Environ 39; 39-45.

2. Abam TKS, Ofoegbu CO, Osadebe CC, Gobo AE (2000) Impact of hydrology on the Port-Harcourt-Patani-Warri Road. Environ Geol 40: 153-162.

3. Campbell AE (2009) Federal road management for sub-Saharan African nations: A Nigerian case study. Ontario, Canada, pp: 125.

4. Adeyemi GO, Wahab KA (2008) Variability in the geotechnical properties of a lateritic soil from southwestern Nigeria. J Int Ass Eng Geo Environ 67: 579-584.

5. Mustapha M (2008) Physical properties of residual profile found in Minna. AU JT 11: 91-98.

6. Adebisi NO, Adeyemi GO, Oluwafemi OS, Songca SP (2013) Important properties of clay content of lateritic soils for engineering project. J Geograp Geo.

7. Eze EO, Adeyemi GO, Fasanmade PA (2014) Variability in some geotechnical properties of three lateritic sub-base soils along Ibadan oyo road. J Applied Geol Geophy.

8. Wang Yu, Zijun Cao, Dianqing Li (2016) Bayesian perspective on geotechnical variability and site characterization. Eng Geo 203: 117-125.

9. Phoon KK, Kulhawy FH (1999) Evaluation of geotechnical variability. Canadian Geotech J 36: 625-639.

10. Bronco LP, Gomes AT, Cardoso AS, Pereira CS (2014) Natural variability of shear strength in a granite residual soil from Porto. Geotech Geol Eng 32: 911-922.

11. Adeleye DR (1974) Sedimentology of the fluvial bida sandstones (cretaceous) Nigeria. Sediment Geo 12: 1-24.

12. Oyawoye M (1972) The basement complex of Nigeria. Ibadan. Afri Geol pp: 66-102.

13. British Standard Institution (1990) Laboratory testing. London.

14. Tuncer ER, Lohnes RA (1977) An engineering classification for basalt-derived lateritic soils. Eng Geo 4: 319-339.

19

Petrography, geochemistry and Alteration Studies of Kanawa Uranium Occurrences, Wuyo-Gubrunde Horst, Northeastern Nigeria

Saleh Ibrahim Bute*

Department of Geology, Gombe State University, Nigeria

Abstract

The Kanawa uranium occurrence is situated along a northerly fault zone at Gubrunde horst, underlain by migmatites-gneiss, syntectonic S-type granites and minor volcanic rock. The sheared zoned is highly altered. The alteration products are sericite, chlorite, and hematite. The uranium occurrences are epigenic derived by remobilization from the host rock to the sheared zones, probably through metasomatic process where feldspars has been replaced by U-Fe-Mg in an oxidized conditions, the mineralizing fluids may be sourced from the numerous volcanic bodies around the area. The host rock showed enrichment of uranium>5, could serve as the potential source of the mineralization. The mineralized mylonite is a product of the granitic host. The Fe-Mg-Ca-P-Sr-Zr-V-Co-U may reflect mineralization fluid composition.

Keywords: Kanawa; Alteration; Uranium; Hematitic silica; Hematization

Introduction

Uranium is a mobile element, generally been classified as a lithophilic element which tends to accumulates at the later stage of magma crystallization. It may oxidize to hexavalent state as the complex uranyl (UO_2^{+2}), which permits extensive into two phases, the volatile and fluid phase. The mobility of uranium makes it difficult in search for economic deposits. Shallow accumulate of surface occurrence may portray false existence of buried deposits. The understanding of the fundamental geologic and geochemical processes that concentrate uranium and likely environment in which the ore occur is paramount important in the search for uranium in granitoids and associated rocks.

The uranium occurrences of Gubrunde horst have been worked for three decades but reasonable prospects have not yet been identified. There is a controversy on the rock hosting the uranium occurrences, for example Funtua et al. pointed out that the mineralized portion is altered/brecciated ryholite while according to Suh and Dada, identified such mineralized portion as reddish hematitic silica, a product of deformed granites [1-3]. This paper discusses the alteration patterns, uranium mineralization and the host rock characterization using the new data on mineralogy and geochemistry of the host rocks and mineralized zone.

Geological Setting

The Kanawa uranium mineralization is located at Gubrunde Horst, Gubrunde Horst is a major structure in the Hawal massif which is believed to be a product of Lower Cretaceous block faulting between two series of NE-SW strike border faults which is bounded to the north and west by basalts of Biu and Cretaceous sediment respectively [1,3,4]. The horst is underlain by migmatites, granite gneiss, mafic and intermediate plutonic rocks syntectonic equigranular and porphyritic granites (Figure 1). The oldest rock units in the region are the migmatites and granite-gneiss with emplacement ages ranges between 680-620 Ma and 620-590 Ma, respectively [1,5]. Mesozoic-Tertiary magmatism have recorded giving rise to alkaline rhyolites and transitional alkaline basalts [4,6]. These volcanic rocks forms part of the Burashika biomodal volcanic Group [4].

The Cretaceous-Tertiary sedimentary rocks comprising sandstone, siltstone, limestones, shales and claystones. The Bima Formation

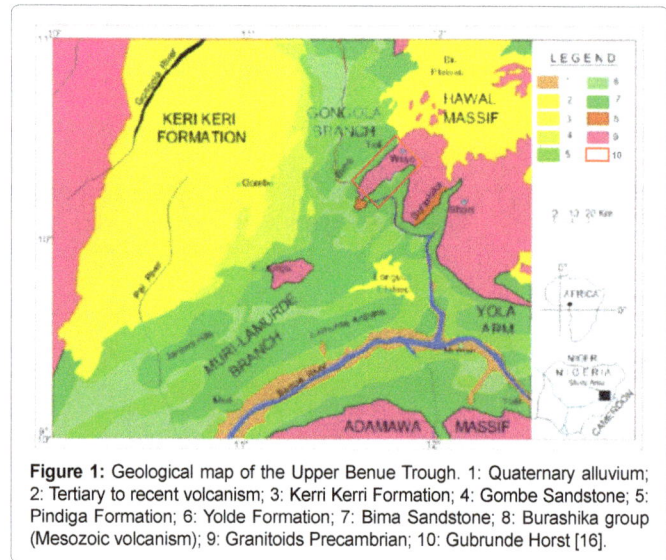

Figure 1: Geological map of the Upper Benue Trough. 1: Quaternary alluvium; 2: Tertiary to recent volcanism; 3: Kerri Kerri Formation; 4: Gombe Sandstone; 5: Pindiga Formation; 6: Yolde Formation; 7: Bima Sandstone; 8: Burashika group (Mesozoic volcanism); 9: Granitoids Precambrian; 10: Gubrunde Horst [16].

is the oldest lithologic unit occupying the base of the Cretaceous successions in the Northern Benue Trough; the sediments were mainly derived from juxtaposed basement rocks of older granites and gneisses and were deposited in a continental environment under widely varying conditions. The Yolde Formation lies conformably on the Bima Sandstone, represents the beginning of marine incursion into the Gongola Arm. The Turonian-Santonian Pindiga Formation conformably overlies the Yolde formation, which represents the full marine incursion in the Gongola arm [6,7]. The estuarine/deltaic Gombe Sandstone of Maastrichtian age overlies the Pindiga Formation

*Corresponding author: Saleh Ibrahim Bute, Department of Geology, Gombe State University, Nigeria; E-mail: salehbutee@yahoo.com which represents the youngest Cretaceous sediments in the Gongola Arm [5]. The Paleocene Kerri-Kerri Formation unconformably overlies the Gombe Sandstone and represents the only record of Tertiary sedimentation in the Gongola Arm [8].

Methodology

Sixteen (16) rock samples were selected for thin section and where cut in a plane normal to the mylonitic foliation. The slides were studied at the petrology laboratory of the Department of Geology, University of Ibadan. A total of thirty (30) samples were selected and analyzed using the ICP-MS method for the analysis. The samples were analyzed for major, minor, and rare earth elements (REE) at ACME Laboratories, Vancouver, Canada. Aggressive fusion techniques employing lithium metaborate/tetraborate fusion was chosen for the analyses, precision for major elements is better than 2% and for trace elements is better than 5%.

Field and Petrographic Studies

The Wuyo area is underlain by migmatites, medium grained granite, porphyritic granite, basalts and Bima sandstone (Figure 2).

Migmatites

The migmatite gneiss portrays banding of felsic and mafic minerals alignment. The felsic minerals are mainly quartz and plagioclase while the mafic are hornblende and biotite which shows evidence of deformation, where the biotite is stretched enclosing the ellipsoidal plagioclase. Mineral associations are polycrystalline ribbon quartz, minor K-feldspars, augen plagioclase porhyroblast, development of myrmekites at grain boundaries is observed. Hornblende with biotite flakes makes up the mafic band containing small amount of euhedral inclusion of accessory minerals (Figure 3a).

Granites

The granites are medium grained to porphyritic granites. The striking characteristics of porphyritic granites are its porphyritic texture consisting of microcline phenocryst. The medium grained granites are highly fractured blocks, blocks of medium grained granites are frequently found within the porphyritic granites implying the former were emplaced earlier than the latter (Figures 3b and 3c). Microscopically both the medium-grained and porphyritic granite display similar mineralogy (Table 1). They are composed of quartz, microcline, orthoclase, plagioclase with few mica shards, accessory minerals includes titanite, zircon, and apatite. Secondary alteration of micas to chlorite is observed. The feldspars are commonly deformed by transgranular fracturing.

Mylonites

The uranium occurrences is been hosted at the centre of nearly northerly trending sheared zone. This sheared zone is within the porphyritic granites which may have been derived from the precursor rocks and showed progressive increase in the intensity of deformation from the external limit (E-W) of the shear zones to the center. The mylonites contain stretched orthoclase and quartz porphyroblasts while the ellipsoid plagioclase are usually fragmented. The quartz grains portray stretched ribbon like features displaying undoluse extinction with sub-grains occurring as aggregate. The alteration observed moving E-S along N-S shear zone are kaolinization, silicification and hematization.

Basalts

There are several volcanic plugs in the study area; it contains olivines, labrodorite laths, augite and iron oxides. The olivines are slightly serpentinised, with corona (Figure 3d).

Alteration studies

The alteration of feldspars may produce free ions such as Si^{4+}, Al^{3+}, K^+, and Na^+ that can be easily remobilized together with the uranium. Evidence from the geochemical data, the inconsistency on the enrichment and depletion of alkalis, silica and iron oxide reflects

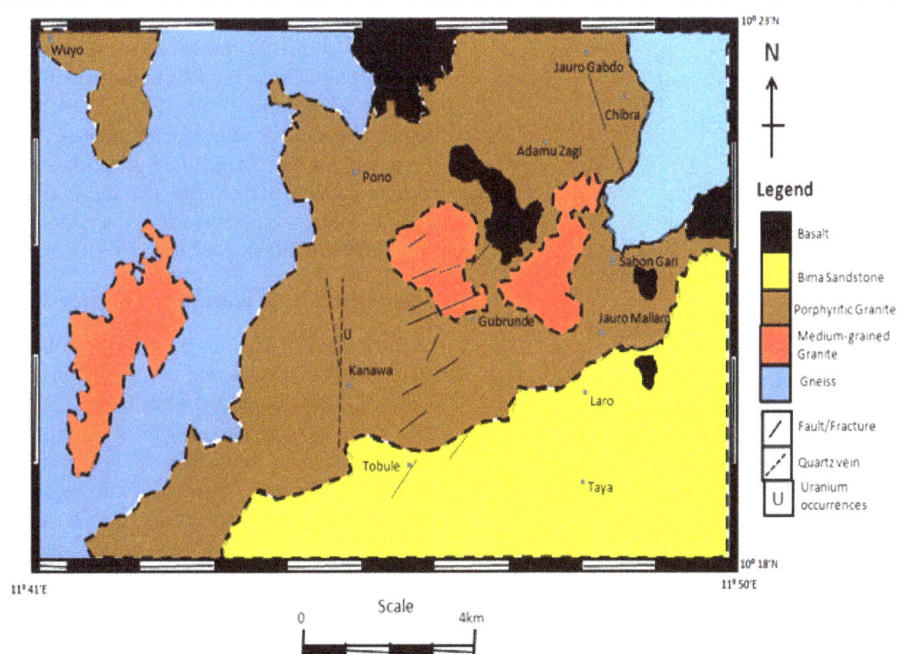

Figure 2: Geologic map of part of Wuyo-Gubrunde horst.

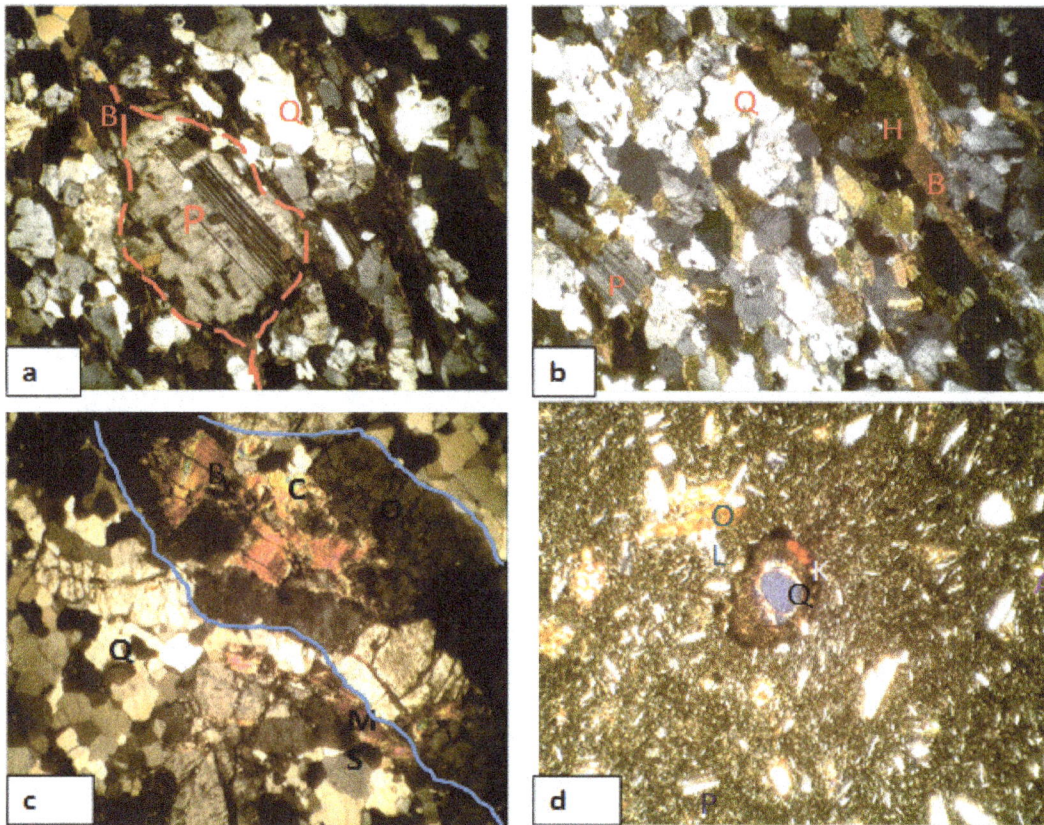

Figure 3: a) Augen plagioclase porphyroblast surrounded by stretched lenses of biotite crystals depicts plastic deformation; b) The banding character depicted by alignment of felsic and mafic minerals; c) Feldspar porphyroclasts depicting transgranular fracturing; d) Corona texture; large crystals of quartz was also observed. (Magnification 40X).

Metal	Gneiss	Porphyritic Granite	Medium grained Granite	Basalt
Quartz	30	30	35	5
K-feldspar	20	35	25	-
Plagioclase	15	13	25	30
Biotite	15	12	10	-
Muscovite	-	5	-	-
Hornblende	13	-	-	-
Olivine	-		-	20
Augite	-	-	-	10
Opaque	7	5	5	35

Table 1: Modal composition of the rock units in Wuyo-Gubrunde area.

remobilization of these elements. The mineralized mylonite is very poor in alkalis (<1), silica (<52), higher in Fe_2O_3=18%, Al_2O_3=15%, P_2O_5=2.4%, uranium mineralization is associated with desilification and ferruginization. The Ishikawa (alteration index) AI, where elements gained during chlorite and sericite alteration ($MgO+K_2O$) over the elements loss and gain ($Na_2O+CaO+MgO+K_2O$) was used as reference in this study (Table 2) [9]. The chlorite-carbonate-pyrite index (CCPI), Mg-Fe chlorite altered is typically developed where hydrothermal temperatures and water/rock ratios are at their highest level. Mg-Fe chlorite development commonly replaces plagioclase, K-feldspars which have been observed at mineralized portion on the thin section the dominant minerals is reddish iron oxide mineral (hematite) leading to the loss of alkalis which is observed by its low alkalis value of Na_2O (0.09-2.7 wt%) and K_2O (0.03-4.18 wt%) as shown in Figure 4. The

migmatites, basalt and ferruginised mylonite falls within the truncated triangular field (carbonate-sericite ± chlorite), where the AI varies from 3-52 and CCPI from 54-99.8. The porphyritic granites and silicified mylonites shows pervasive alteration typically exhibits AI>85 and 19<CCPI<60. The aplites falls in the field of least altered (Figures 4a and 4b). The relationship between Na_2O and the AI in the granitoids, the depletion pattern of sodium with increase AI is outlined (Figure 5). The trend shows a shift from albite-plagioclase field to the chlorite-sericite corners evidence of pervasive alteration. Also K_2O enrichment typically exhibits increase in AI which is dependent on the ratio of sericite to chlorite in the altered product (Table 3).

Petrogenesis of Kanawa granitoids

Two types of granites is been recognized by Chappell and White, each related to a particular orogenic belt [10]. I-type granite which is compositionally expanded while the S-type granite are compositionally restricted. Both granites have distinctive petrochemical characteristics which reflect the differences in the sources of magma. The I-type granites are derived from basic igneous source by remelting of deep seated igneous material or the mantle, while the S-type granite are derived from melting of meta sedimentary source materials. The distinguishing mineralogical, petrochemical and field relationship of different sources as proposed by Chappell and White are compared with those of Wuyo-Gubrunde granitoids in Table 4 [10]. Such comparison showed that the granitoids shows that they are of S-type characteristics and support the model of its derivation by partial melting of meta sedimentary rocks in the crust.

Rock	Granite gneiss		Ferruginized mylonite		Porphyritic granite			Silicified mylonite			Aplite	Basalt		
Sample	1	2	3	4	5	6	7	8	9	10	11	12	13	14
	n = 3	n = 3	n = 3	n = 3		n=3			n=3		n = 3			
SiO_2	52.86	64.35	51.28	50.83	65.02	67.44	74.97	61.77	80.37	79.6	66.32	51.55	51.2	51.5
TiO_2	2.22	1.14	2.59	2.63	0.62	0.36	0.3	0.17	0.12	0.12	0.48	2.59	2.64	2.61
Al_2O_3	15.76	14	15.28	14.79	18.74	17.87	14.02	9.81	11.25	12.5	16.04	14.77	14.89	14.77
Fe_2O_3	10.16	6.88	16.8	18	4.33	3.33	1.94	1.08	1.5	1.71	3.42	11.62	11.64	11.6
MnO	0.15	0.09	0.51	0.38	0.06	0.04	0.01	0.02	0.01	0.01	0.08	0.13	0.13	0.12
MgO	3.76	1.61	0.08	0.07	0.89	0.12	0.22	0.09	0.22	0.24	1.05	5.72	5.83	5.83
CaO	6.48	3.07	3.12	3.22	0.35	0.33	0.1	17.81	0.07	0.14	1.59	7.79	7.83	7.8
Na_2O	2.73	2.6	0.01	0.02	0.11	0.13	0.12	0.12	0.09	0.06	3.34	3.16	3.18	3.18
K_2O	3.73	4.62	0.03	0.09	3.57	3.92	4.21	4.18	3.55	1.8	5.83	1.45	1.42	1.45
P_2O_5	0.69	0.49	2.34	2.31	0.21	0.15	0.09	0.07	0.04	0.05	0.24	0.4	0.43	0.42
LOI	1	0.7	7.7	7.4	5.9	6.1	3.9	4.8	2.7	3.7	1.4	0.5	0.5	0.6
Total	99.57	99.61	99.75	99.77	99.81	99.77	99.84	99.89	99.88	99.95	99.78	99.7	99.7	99.7
Trace elements (ppm)														
Ni	21	<20	<20	<20	<20	<20	<20	<20	<20	<20	<20	121	122	121.5
Co	27.5	10.6	56.2	46.1	5.3	2.6	2.2	4.1	1.5	1.8	8.2	44.7	43.6	44.15
Sc	18	14	23	23	10	5	4	3	3	4	8	15	15	15
V	208	83	184	190	41	28	22	12	16	17	47	169	168	168.5
Rb	183.2	243.6	3.4	5.2	247.1	122.3	139.9	152.7	172.3	136.7	248.2	37.5	37.6	37.6
Sr	527.6	280.1	139.7	137.5	83.8	82.8	53.8	184.4	139.7	57	210.5	575.3	570.3	572.8
Ba	1065	894	306	196	442	986	865	495	703	131	504	368	372	372
Zr	577.4	756.6	501.2	482.7	355.6	367.2	196.3	85.8	63.7	81.2	316	192.5	192.4	192.4
Nb	36.6	33.1	45.5	44.6	20.2	19.8	9.5	6.6	6.4	10.5	19.3	37.7	38.9	38.3
Ta	2.1	3	2.7	2.6	1.5	1.4	0.5	0.4	0.6	1.1	1.5	2.4	2.4	2.4
Sn	4	12	2	2	12	5	4	2	4	5	13	3	3	3
Be	<1	3	9	2	6	4	1	2	5	3	8	3	<1	2
W	0.6	0.5	1.6	1.8	1.5	0.4	0.4	0.4	0.5	0.4	0.4	0.6	0.6	0.6
Cs	3.4	4.2	0.2	0.1	13.5	1.2	2.9	1.9	7	3.5	10.1	0.5	0.4	0.45
Ga	19.7	24.5	24.6	24	33.4	24.5	19.5	13.4	20.9	27.3	26.6	20.7	20	20.23
Y	43.5	42.4	70.6	70.2	37.3	20.4	9.7	27.9	7.9	24.8	21.9	22.4	22.15	22.5
Hf	14.3	19.6	11.6	10.9	10.7	10.1	5.5	2.7	2	2.6	9.8	4.7	4.7	4.7
Th	14.8	48.1	4.2	4.3	48.7	39.8	26.7	15.4	10.2	13.5	25.9	4.6	4.3	4.5
U	3.7	7.1	13.2	130.6	4.7	5.7	4.2	2.7	1.7	3.5	3.8	1.1	0.9	1
Th/U	4	6.78	0.31	0.31	10.26	7	6.35	5.7	6	3.86	6.83	4.18	4.78	4.45
Rb/Sr	0.34	0.87	0.03	0.04	2.95	1.48	2.6	0.83	1.23	2.4	1.17	0.07	0.07	0.07
AI	44.9	52.4	3.4	4.71	90.7	89.8	95.3	19.3	95.9	91.07	58.3	39.7	39.7	39.9
CCPI	68.3	54.1	99.8	99.4	58.7	46	33.3	21.4	32.1	51.18	32.8	79	79.1	79

Table 2: Major oxides compositions (wt %) and trace element (ppm) data of Wuyo-Gubrunde rocks [n=3-Number of samples analysed; Alteration Index (AI); Chlorite-carbonate-pyrite index (CCPI)].

Geochemistry of Kanawa uranium occurrence

Th/U ratio values of plutonic rocks ranges 4-6 reflects loss of uranium into the late-stage fluids, Th/U ratio is higher in porphyritic granites 7, the migmatites 5, aplite and basalt 5, 4 in silicified granites lowest in the mineralized zone 0.31. The oxidation of uranium is associated with loss in thorium which is clearly observed in this study, the mineralized portion is enriched with Fe_2O_3=18% and poor in Th=4.2 ppm.

The mineralized granitoids are S-type granites (Table 4) implying derived by remelting of pre-existing crustal material at a near depth of 20-30 km [11]. Remobilization of uranium from the surrounding host rock could be the likely source, U and Rb behave incompatibly in granitic melts; the kanawa uranium mineralization displays enrichment of U with loss in Rb (Figure 6) evidence of remobilization, magmatic sourced uranium on the hand shows positive correlation of Rb and U (Table 5) [12]. The variations in REE total values of the rocks (123-

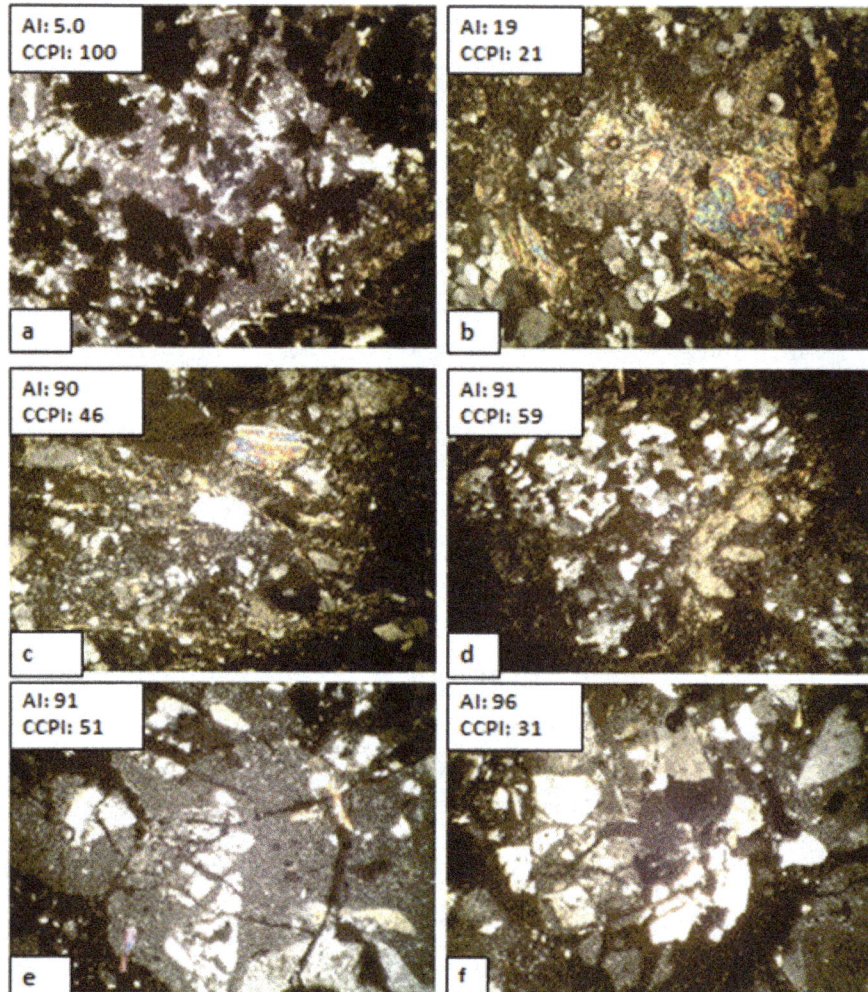

Figure 4: Photomicrograph of granitoids from Kanawa, displaying increasing intensity of hydrothermal alteration (Magnification 40X).

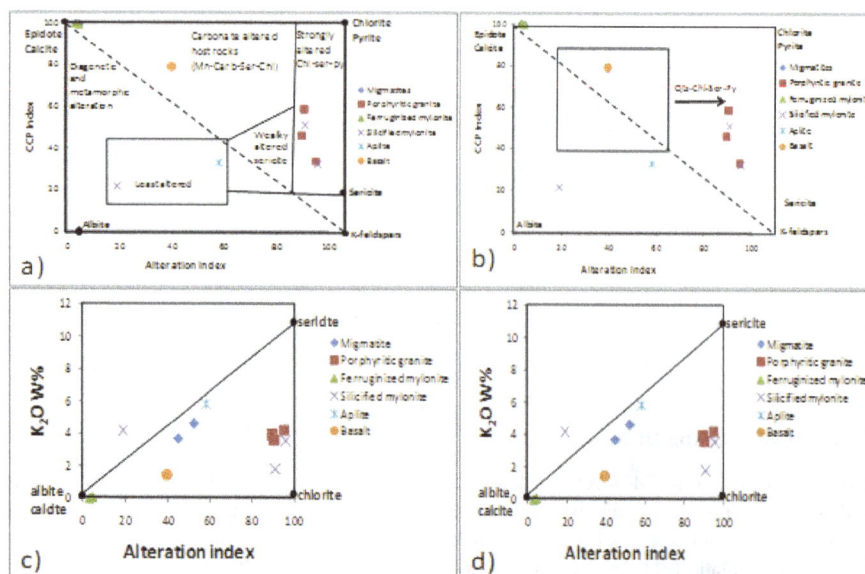

Figure 5: Alteration box plot of Kanawa alteration zones [9]; (a) and (b) showing the degree of alteration; Trends of AI with (c) Na$_2$O and (d) K$_2$O.

Rock	Granite gneiss		Ferruginised mylonite		Porphyritic granite			Silicified mylonite			Aplite	Basalt		
Sample	1	2	3	4	5	6	7	8	9	10	11	12	13	14
	n = 3	n = 3	n = 3	n = 3							n = 3			
La	80.6	149.3	62.1	61.8	74.6	92.1	59.8	27.6	33.5	36.4	87.4	27.9	29.1	28.5
Ce	178.1	296.3	150.4	151.6	141.2	170	112.4	69.9	52.6	56.7	166	54.6	57.9	56.3
Pr	21.12	32.7	18.19	17.86	15.55	18.6	12.2	6.19	6.12	6	21.16	6.49	6.78	6.64
Nd	83.2	113.8	86.8	82.4	53.6	65.9	42.6	23	21	19.8	84.4	28.1	29.1	28.6
Sm	14.87	19.31	18.97	18.14	11.02	10.66	6.94	5.4	3.61	3.29	19.8	6.42	6.67	6.55
Eu	3.55	2.84	5.45	5.22	1.17	1.47	1.23	1.71	0.68	0.63	2.37	2.12	2.19	2.15
Gd	13.2	13.98	19.42	18.99	9.15	7.26	4.68	5.4	2.33	3.59	17.05	6.27	6.74	6.5
Tb	1.83	1.9	2.74	2.74	1.42	0.93	0.56	0.82	0.31	0.63	2.33	0.95	0.98	0.97
Dy	8.89	9.17	14.19	14.09	7.4	4.41	2.45	3.95	1.4	3.71	11.75	4.8	4.95	4.87
Ho	1.64	1.56	2.69	2.74	1.35	0.73	0.4	0.75	0.26	0.77	1.85	0.89	0.86	0.87
Er	4.37	4.15	6.6	6.86	3.56	1.92	0.76	1.71	0.72	2.14	4.5	1.93	2.22	2.08
Tm	0.59	0.56	0.91	0.87	0.48	0.29	0.11	0.23	0.1	0.3	0.61	0.26	0.27	0.26
Yb	3.65	3.49	5.46	5.57	2.67	1.92	0.75	1.49	0.66	1.84	3.3	1.48	1.5	1.49
Lu	0.55	0.46	0.82	0.79	0.39	0.29	0.1	0.21	0.09	0.28	0.43	0.22	0.21	0.22
Total REE	416.2	649.5	394.7	389.7	323.6	376.5	245	148.4	123.4	136.1	423	142.4	149.5	145

Table 3: Rare earth element (ppm) data of Wuyo-Gubrunde rocks.

Criteria	I-type	S-type	Study area
Field	Massive, with little or no foliation contains mafic hornblende-bearing Xenoliths	Usually foliated. Contains meta sedimentary xenoliths. May be associated with regional metamorphism; more likely to be found near their source and shows evidence (migmatite, regional metamorphism)	Augen gneiss, migmatites, higly foliated, higly Faulted granites (sheared, altered and zoned)
Mineralogical	Horblende and biotite + accessory magnetite	Biotite, muscovite and corderite	Biotite, hornblende and very little muscovite
Chemical	High Na_2O>3.5% in felsic and >2.2% in more mafic types. Rich in Ca and poor Rb from the source region.	Low Na_2O; normally<3.5% in rocks with approximately 5% K_2O decreasing to >2.2% in rocks with about 2% K_2O. High Al, with high Rb in source rock.	Very low Na_2O (0.09-2.73%), K_2O (0.03-4.18%) Al_2O_3 (9.81-15.76) and Rb (3.4-243.6 ppm)
Ore association	Porphyry copper, Mo	Tin	Uranium

Table 4: Chappell and white I and S-type granite characteristics compared with the Wuyo-Gubrunde granitoids [10].

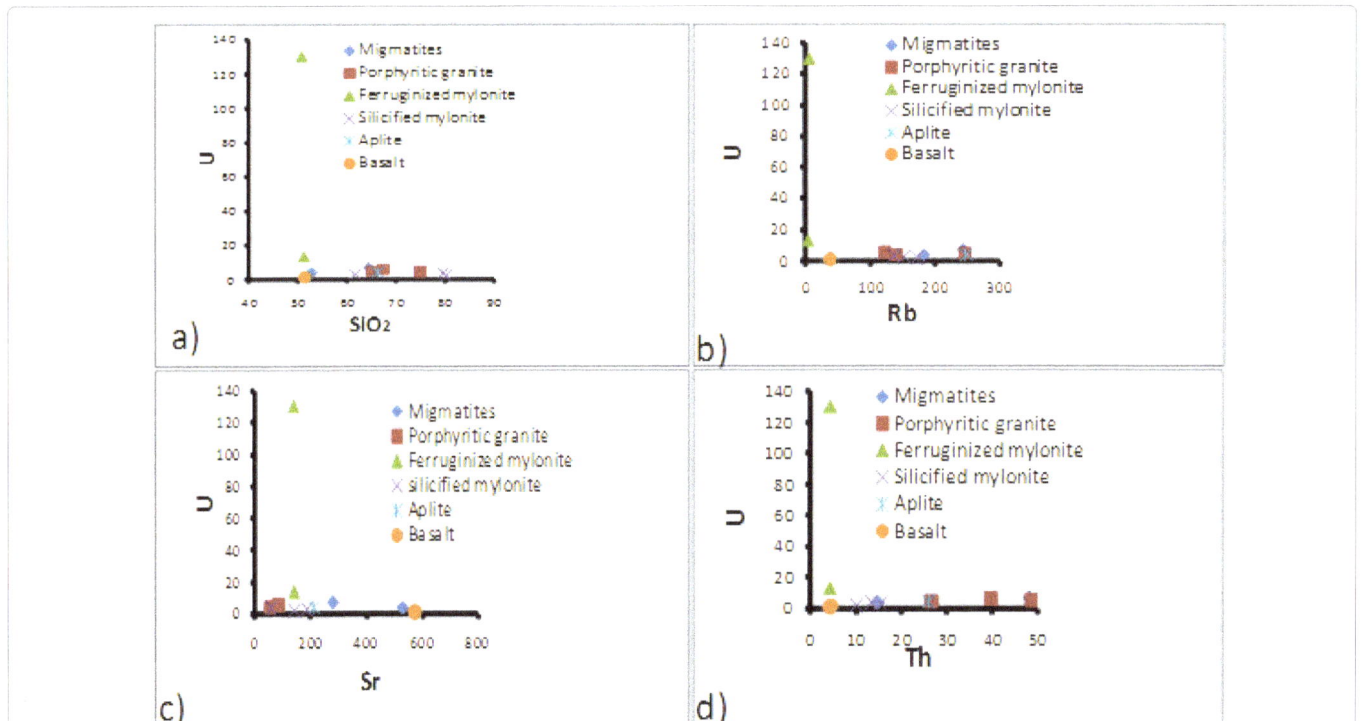

Figure 6: Harker-type variation plot of selected major and trace elements versus U for the Wuyo-Gubrunde granitoids.

Characteristics	Rössing model	Kanawa model
Lithology	Pegmatite-alaskite-gneiss; anatectic granite and migmatite	migmatite-gneiss, Porphyritic granite, Medium- grained granite, pegmatites, aplites and basalt
Derivation	reworked and recycled sialic crust	reworked crustal material
Initial Sr isotopes	>0.710	NA
Levels of emplacement	Catazonal	Catazonal
Tectonic stage	Syntectonic	Syntectonic
Age	commonly Protorezoic to early Paleozoic	Proterozoic-early Paleozoic (Pan-African age)
Tectonic setting	Orogenic	Oregenic
Metamorphic rock of country rocks	Middle-upper amphibolites	green-schist and unmetamorphosed

Table 5: Summary of principal similarities between the RÖssing and Kanawa uranium mineralization.

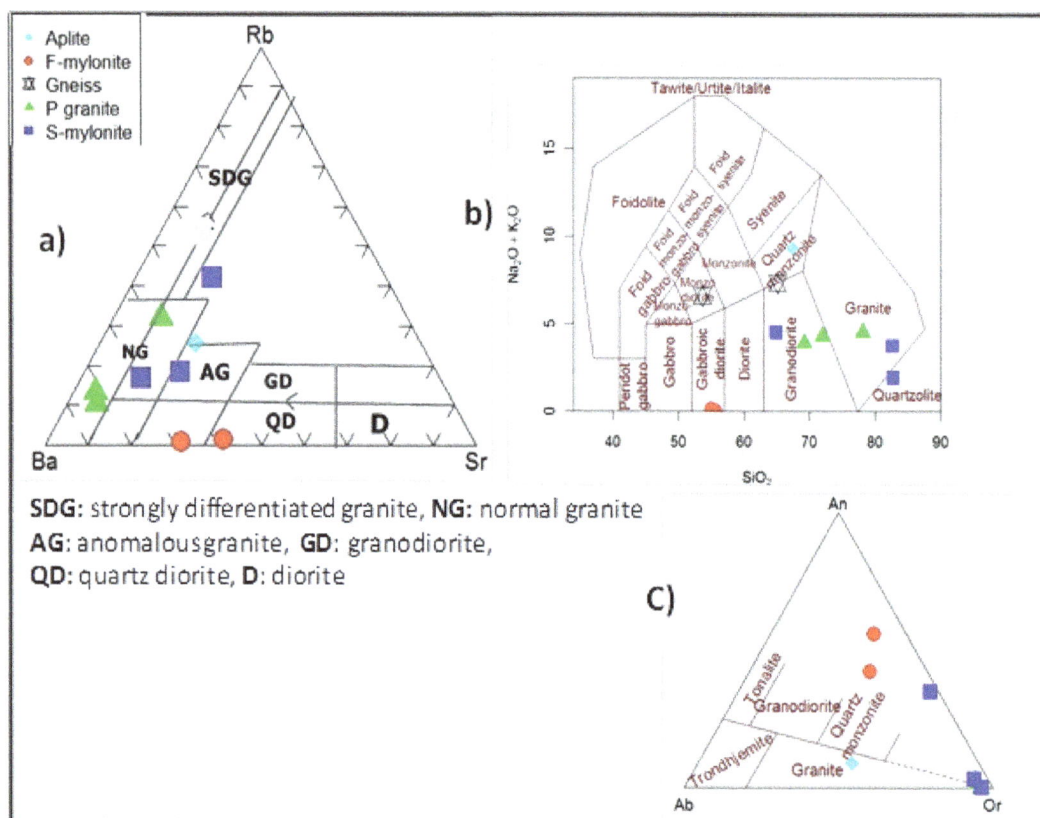

Figure 7: Plots of the analyzed granitoid rocks on; (A) Rb-Ba-Sr trivariant plot of Wuyo-Gubrunde granitoids [17]; (B) TAS plot of Wuyo-Gubrunde granitoids [14]; (C) An-Ab-Or ternary diagram [15].

649 ppm) most of the samples have lower value compared to the total abundances of acid and intermediate rocks (220-350 ppm) [13], suggest original magmatic signatures have been lost due to alteration/remobilizations.

The Rossing deposits of Southwest Africa (Namibia), which is considered to be an ideal example of anatectic remobilization of preexisting sialic crustal material is considered to be an excellent model similar to the Kanawa uranium occurrences (Table 4).

Constraints on the Genesis of the Rock Hosting Wuyo-Gubrunde Uranium

The mineralized portion is situated in the sheared/altered zone trending northerly, dipping almost vertical, bounded side by side by the porphyritic granite. It is fine grained, reddish tectonite. This host rock is been earlier reported as rhyolite [1], which is later argued to be a deformational product of the granites under oxidizing condition [3]. The plot of Wuyo-Gubrunde granitoids on Rb-Ba-Sr plot, most of the samples cluster in the field of normal-anormalous granites (Figure 7), the ferruginised granite that is mineralized fall in the field of granodiorites and quartz-diorites. Anomalous granite indicates that the rocks have suffered from metasomatism as a result of shearing, anoumalous granites may also shift in the field of granodiorite and quartz-diorite as a result of low Rb content, in this case the mineralized altered granite is depleted in Rb (3.4-5.2 ppm) may reflects Rb fractionation during metasomatism. On the TAS plot for volcanic rocks [14], most of the samples cluster in the field of dacite to rhyolites (Figure 7), but the mineralised altered ganite falls in the field of basalt due to low silica content (50.83-51.28 wt%), hence it is not rhyolitic. The granitoids can be classified by using normative An-Ab-Or teneray diagram of O'Connor [15]. All the granitoids fall in the field of granites except the ferruginised mylonite which shifts to the field of granodiorites.

Conclusion

A detailed field, petrographic and alteration studies of Kanawa area coupled with geochemistry of the host rock and uranium occurrences. Based on our results we characterize our study area:

1. The Kanawa granitoids are S-type granites, suggesting from partial melting of crustal materials.

2. The sheared zoned is highly altered. The alteration products are sericite, chlorite and hematite.

3. The uranium occurrences are epigentic derived by remobilization from the host rock to the sheared zones, probably through metasomatic process where feldspars have been replaced by U-Fe-Mg in an oxidized condition, the mineralizing fluids may be sourced from the numerous volcanic bodies around the area.

4. The mineralized mylonite is a product of the granitic host. The pathfinder elements of Kanawa uranium occurrences are U-Fe-Mg-Ca-P-Sr-Zr-V-Co.

References

1. Funtua II, Okujeni CD, Abaa SI, Elegba SB (1992) Geology and genesis of uranium mineralization at Kanawa Gubrunde horst, NE. Nigerian J Mining Geol 28: 171-177.

2. Funtua II, Okujeni CD, Elegba SB (1999) Preliminary note on the geology and genetic model of uranium mineralization in Northeastern Nigeria. J Mining Geol 35: 125-136.

3. Suh CE, Dada SS (1997) Fault rocks and differential reactivity of minerals in the kanawa violaine uraniferous vein, NE Nigeria. J Struct Geol 19: 1037-1044.

4. Ojo OM (1982) Geology and stream sediment geochemical survey of middle Gongola Basin (upper Benue trough) of Nigeria. Nigeria p: 386.

5. Carter JD, Barber W, Tait EA (1963) The geology of parts of Adamawa, Bauchi and Borno provinces in the Northeastern Nigeria. Geological Survey of Nigeria Bull 30: 106.

6. Popoff M (1984) Benue trough oblique rifting and the geodynamic evolution of the Gulf of Guinea. Revista Brasileira en Geociendas 18: 315.

7. Zaborski PM, Ugoduluwa F, Idornigie A, Nnbo P, Ibe K (1997) Stratigraphy and structure of the Cretaceous Gongola Basin, North Eastern Nigeria. Bull Cent Res Explor Product Elf 21: 153-185.

8. Dike EFC, Egbuniwe IG (1993) An uniformity between the Keri-Keri Formation and basement complex at Mainamiji, Bauchi State. J Mining Geol 29.

9. Large RR, Gemmell JB, Paulick H (2001) The alteration box plot: A simple approach to understanding the relationship between alteration mineralogy and lithogeochemistry associated with volcanic-hosted massive sulfide deposits. Eco Geol 96: 957-971.

10. Chappell BW, White AJR (1978) Granitoids from the Moonbi district, New England batholiths, Eastern Australia. J Geol Society Australia 25: 267-283.

11. Bute SI (2015) Geological and geochemical studies of rocks of part of Wuyo-Gubrunde horst, northeastern Nigeria: Implication on uranium occurrences.

12. Steenfelt A (1982) Uranium and selected trace elements in granites from the Caledonides of East Greenland. Mineral Mag 46: 201-210.

13. Haskin LA, Schmitt RA (1967) Rare-earth distributions. Researches in Geochemistry, John Wiley and Sons, New York.

14. Wilson M (1989) Igneous petrogenesis. Unwin Hyman, London.

15. O'Connor JT (1965) A classification for quartz-rich igneous rocks based on feldspars ratios. US Geol Surv Prof Paper 525B: B79-B84.

16. Haruna IV, Ahmed HA, Ahmed AS (2012) Geology and tectono-sedimentary disposition of the Bima sandstone of the upper Benue trough (Nigeria): Implications for sandstone-hosted uranium deposits. J Geol Mining Res 4: 168-173.

17. El Bouseily AM, El Sokkary AA (1975) The relation between Rb, Ba and Sr in granitic rocks. Chem Geol 16: 207-219.

20

Natural Vulnerability Estimate of Groundwater Resources in the Coastal Area of Ibaka Community, Using Dar Zarrouk Geoelectrical Parameters

Evans UF[1*], Abdulsalam NN[2] and Mallam A[2]

[1]Department of Sciences, Maritime Academy of Nigeria, Oron, Nigeria
[2]Department of Physics, University of Abuja, Nigeria

Abstract

The study is aimed at estimating the natural vulnerability of groundwater resources using Dar Zarrouk geoelectrical parameters. Hence, ground-based earth resistance measuring device (OYO McOHM resistivity meter, Model 2115A) and its accessories were deployed for the study. 15 Schlumberger vertical electrical soundings (VES) points distributed along three profiles within the Mbo plain area were conducted. The VES field data were interpreted manually on bi-logarithmic plot and then by applying the auxiliary point-partial resistivity curve matching technique. The interpretation was enhanced by sophisticated computer software (IPI2Win). The Dar Zarrouk geoelectrical parameters were deduced from the advanced interpreted results and use to produce contour maps. The lithology identified was of sands sequence of different grain sizes for aquifer zone. The absence of clay formation in the lithology log and low resistivity value (21.49 Ωm) at aquifer's depth within the VES 5 column suggested saline water ingress. The low range of overburden thickness lies between 0.3 and 4.6 m. The associated overburden longitudinal conductance was found to range from 0.000 to 0.019 Siemens, which was generally low. Therefore, suggesting that groundwater resources in the study area are vulnerable to contamination. Hence, aquifers in the study area were considered to be at high risk of contamination from environmental activities and saline water ingress. The information from study could help in the planning, monitoring and management of potable water supply and development in deltaic coastal aquifer system.

Keywords: Dar Zarrouk geoelectrical parameters; Lithology; Groundwater; Protection capacity; Aquifer

Introduction

Water resources in the entire Niger Delta are threatened by increasing exploration for oil and gas which has not only increased the demand for potable water due to population influx, but has also contaminated most of the surface water means. Groundwater resources are not free from contamination; however some local geologic formations have the potential to protect groundwater resources from contamination, which can occur naturally or artificially. The natural vulnerability expresses the sensitivity of aquifers to be adversely impacted upon by an imposed contaminant loads. The natural contamination of groundwater resources in coastal areas have been traced to the hydraulic connection between groundwater and seawater [1]. Seawater has the natural advantage of higher hydraulic pressure (due to higher mineral contents and sea level rise), therefore will always migrate into freshwater thereby contaminating the aquifers. At saline water ingress, groundwater system is overridden by a wedge of higher saline water that grades into seawater. The Dar-Zarrouk (D-Z) variables provide easily decipherable vision about the occurrence and distribution of fresh and saline water aquifers.

Freshwater aquifers contamination can also occur artificially by some anthropogenic activities such as navigation channels, drainage channels, agriculture, hysterical leakages of products from oil and gas installations such as pipelines and storage tanks and mobile vessels (oil tankers) and pumping of groundwater from coastal freshwater wells, etc. Contamination effect on groundwater resources can become very serious where the geologic materials are not capable of protecting the aquifers in an area [1]. Groundwater contamination makes an aquifer endorheic, this puts the populace at high risk of waterborne diseases, since some soil borne verses can survive for days in unsaturated zone, and eventually get to the saturated zone [2].

Resistivity of earth materials is widely influenced by a number of factors including salinity, moisture content, metallic mineral content, porosity, clay content, temperature and resistivity of pore fluid. Deviation of the resistivity of rock layer from that of the country rock indicates anomaly. Resistivity survey seeks to delineate anomalies, which are representations of the ground conditions. Ideally, in a clayey or saline environment, high resistivity anomalies are most probably indicators of potable water; while low resistivity anomalies would be the target for groundwater exploration in basement terrain.

Electrical resistivity data have a threefold intention: to identify external corrosion threat on buried facilities [3-6]; identify ore bodies, depth to bedrock and other geological processes [7] and for groundwater resources and environmental protection and management. The combination of thickness and resistivity of the overburden rock into single variables known as Dar Zarrouk parameters [8,9] can be used as a plinth for a proper estimation of the safety condition of aquifers as well as the protection of groundwater resources in any environment [6,10,11].

Researchers noted that groundwater is also susceptible to pollution from infiltration of leachate due to decomposed refuse from dump sites as well as leakage from septic tanks of various houses [12-14]. However, despite the large numbers of hydrogeological studies within the vicinity of the study area [15-17] only aquifer potential and groundwater reserve had been addressed using interpreted resistivity value of earth materials. Issues bordering on the safety of the aquifers within the study

***Corresponding author:** Evans UF, Department of Sciences, Maritime Academy of Nigeria, Oron, Nigeria, E-mail: egeosystems2@gmail.com

area are yet to receive adequate attention. It is for the purpose of safety of the rapid growing population witnessed in the study area that this investigation is carried out to ascertain the vulnerable level of aquifers in the area. Hence, the prime objective of this study to identify the local lithology and to determine if the local geologic formation has the natural capacity of protecting groundwater resources in the study area from contamination using Dar Zarrouk geoelectrical parameters.

Physiography, Geology and Hydrogeology

Akwa Ibom State lies entirely within coastal plain sand (otherwise known as the Benin Formation), which is a deltaic deposition environment and has a thickness in excess of about 800 m. The Benin formation extends from the west across the Niger Delta and southward beyond the present coastline. It is over 90 percent sandstone with silicon as the dominant element, though some places have minor shale intercalations. It is coarse grained, gravely, locally fine grained, poorly sorted, sub-angular to well-rounded, and bears lignite streaks and wood fragments. The Benin formation is thus partly marine, partly deltaic, partly estuarine and partly lagoonal and fluviolacustrine in origin [13]. Its age ranges from Miocene to recent. The terrain of the area is characterized by two types of land forms: highly undulating ridges and nearly flat topography. Various structural units (point bars, channel fills, natural levees, back swamp deposits and oxbow fills) are

particular within the formation indicating the variability of the shallow water depositional medium [18]. Stratigraphically, the coastal plain sand is overlain by recent alluvium and recent sediments and underlain by the Agbada formation. Its outcrop lateral equivalent is probably the Bende Amake sands (Figure 1). The coastal plain sand harbours prolific aquifers, which are the major source of potable water for domestic, agricultural, and industrial use in the study area [19].

Materials and Methods

The technique of the vertical electrical soundings with Schlumberger array was used for imaging the vertical variations in the ground electrical properties. The technique consists of a quite fast and versatile procedure of geophysical investigation. In addition, it is an extensively known technique to determine the location of the water table [9,15,16,20], whose information is essential for the elaboration of the piezometric surface map. The OYO McOHM resistivity meter (Model 2115A) was used to carry out 15 soundings along three profiles in July, 2016 in Ibaka community, Mbo Local Government Area. Electrodes were configured such that potential electrodes also called the inner electrodes (MN) were nested between the current electrodes (AB) also known as outer electrodes. The current electrodes separation was gradually adjusted (from 1 m to a maximum of 300 or 400 m depending on the terrain) to ensure the progressive penetration of

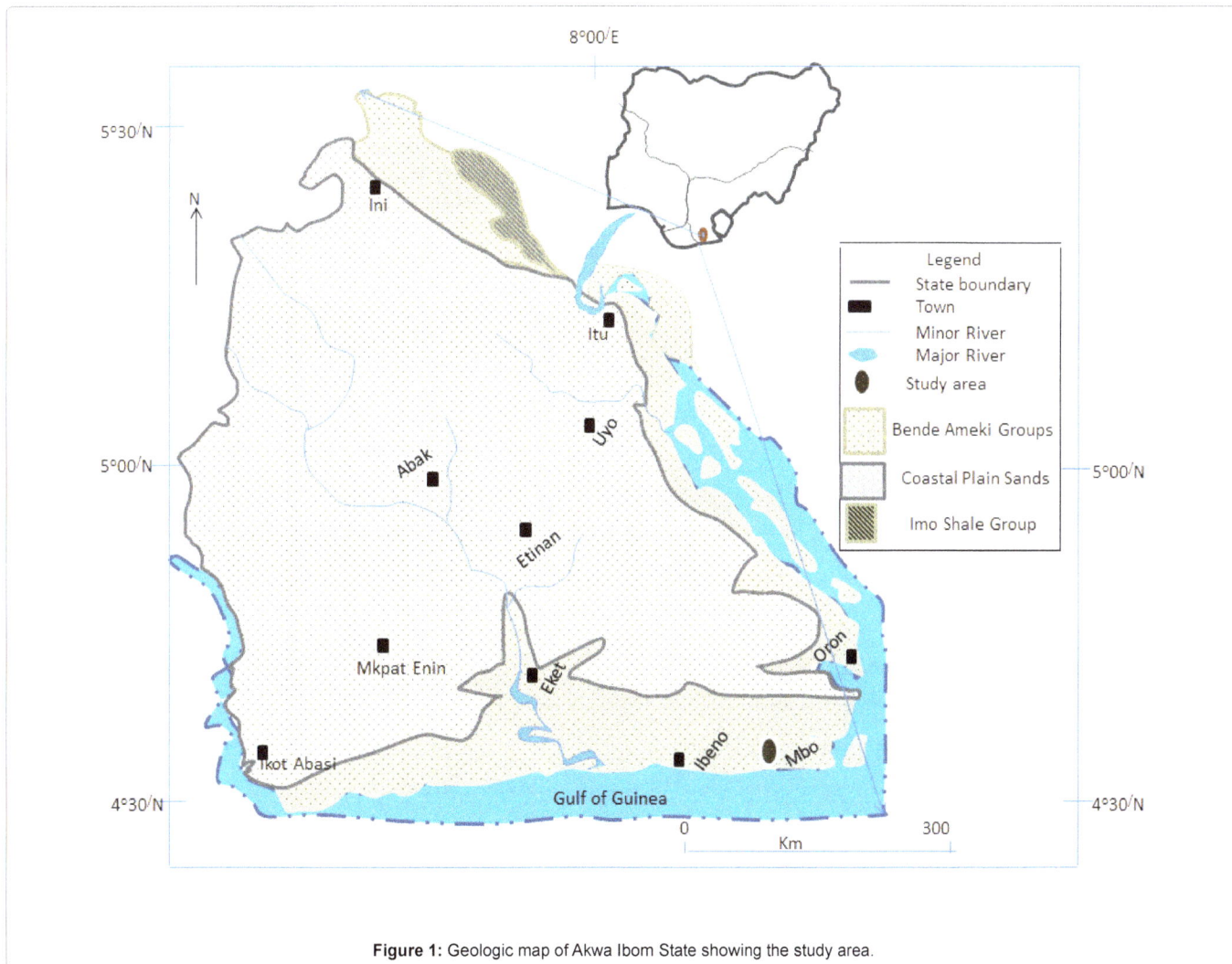

Figure 1: Geologic map of Akwa Ibom State showing the study area.

current into deeper layers. The potential electrodes (MN) separation was also adjusted from a minimum of 0.5 m to a maximum of 15 m. The adjustment was controlled such that MN separation did not exceed 1/5 of half AB separation in order to maintain measurable subsurface electrical potential.

The quantitative treatment of field data started with the conversion of measured resistance (R_a) to apparent resistivity (ρ_a) using a standard equation given as

$$\rho_a = \pi \left[\frac{\left(\frac{AB}{2}\right)^2 - \left(\frac{MN}{2}\right)^2}{MN} \right] * \frac{\Delta V}{I} \tag{1}$$

Data obtained were manually plotted on bi-logarithmic graph to produce apparent resistivity against half current electrodes spacing curves. The curves were manually smoothened to check resistivity curve discontinuities generated by different "MN" spacing of the same "AB" spacing induced by lateral inhomogeneity, anisotropy effect and other signatures arising from ambient noise on the resistivity curves [14,16,21]. Smoothened curves were subjected to curve matching procedure using master curves (Figure 2). Data obtained were inputted into the computer for inversion modeling interpretation using the IPI2Win computer software which generated models of the subsurface resistivity variation with depth. The inversion yielded resistivity models with fitting error between the observed and theoretical resistivity values of less than 10% and was considered to produce valued results. In addition, the computer processed and interpreted VES curves yielded geoelectric layer thickness as close as possible to the actual values of boreholes information in the study area (Figure 3). This was important because VES results are influenced by a number of factors such as salinity, lithology, water content, clay content and porosity [11,22].

The combination of subsurface resistivity and thickness into single parameter (Figure 4) gives rise to the Dar Zarrouk parameters deployed for the study. For a sequence of n horizontal stratified, homogeneous and isotropic layered earth model of resistivity ρ_i and thickness h_i, Orellana [20] and Zohdy [23] presented the combinations of the layer parameters (ρ_i and h_i) as

$$S_i = \sum_{i=1}^{n} \frac{h_i}{\rho_i} \tag{2}$$

and

$$T_i = \sum_{i=1}^{n} \rho_i * h_i \tag{3}$$

where Si and Ti represent longitudinal conductance (Siemens) and transverse resistance (ohm-m²) respectively. The combined resistance and thickness of earth layers was necessary because it checks some limitations, such as: heterogeneities effects, topographic effect and assumptions that beddings are horizontal associated with VES data interpretation. There was a need to correct for aquifers' resistivity in order to determine the hydrogeological characteristics of the underground water [24,25]. Hence, transverse resistance (T) was applied to reveal resistive layer confined between two or more conductive layers. While conductive layer sandwiched between two or more resistive layers was unraveled by its horizontal conductance (S). Therefore, analysis of Dar Zarrouk parameters was used as the basis for evaluating the groundwater properties such as aquifers' transmissivity, and groundwater resources protection. In order to determine the values for transmissivity, aquifers' hydraulic conductance were computed from

$$K = \varepsilon R_{aq}^{-0.93283} \tag{4}$$

where ε is an empirical constant given by 386.4 for coastal aquifers and R_{aq} is the aquifer's resistivity. According to Singh [24], Braga et al. [9] and Harb et al. [26] aquifer transmissivity is the product of its hydraulic conductivity and layer thickness (h_i), mathematically represented as

$$T_r = Kh \tag{5}$$

The transverse resistivity (ρ_T), longitudinal resistivity (ρ_L), pseudoanisotropy (λ) and the root mean square resistivity (ρ_{rms}) were determined using equations (6, 7, 8 and 9 respectively). This further helps in reducing ambiguities related to VES interpretation, which are mainly produced by principles of equivalence and suppression and cause intermixing in identifying depth limits for the electrical zones during interpretation.

$$\rho_T = \frac{T_i}{d_i} \tag{6}$$

$$\rho_L = \frac{d_i}{S_i} \tag{7}$$

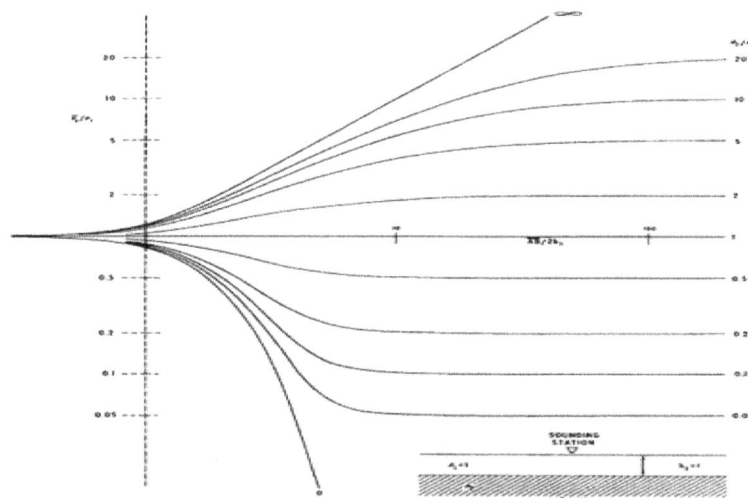

Figure 2: Two-layer master set of sounding curves for the interpretation of manually smoothened field curves [23].

$$\lambda_i = \left[\frac{T_i/d_i}{d_i/S_i} \right]^{\frac{1}{2}} \qquad (8)$$

$$\rho_{rms} = \left[\rho_T \, x \rho_L \right]^{\frac{1}{2}} \qquad (9)$$

Results and Discussion

The quantitative interpretation of VES provided geoelectrical information characterized by the values of resistivity and thickness (Table 1). Quantitatively, the correlation of geoelectrical interpreted

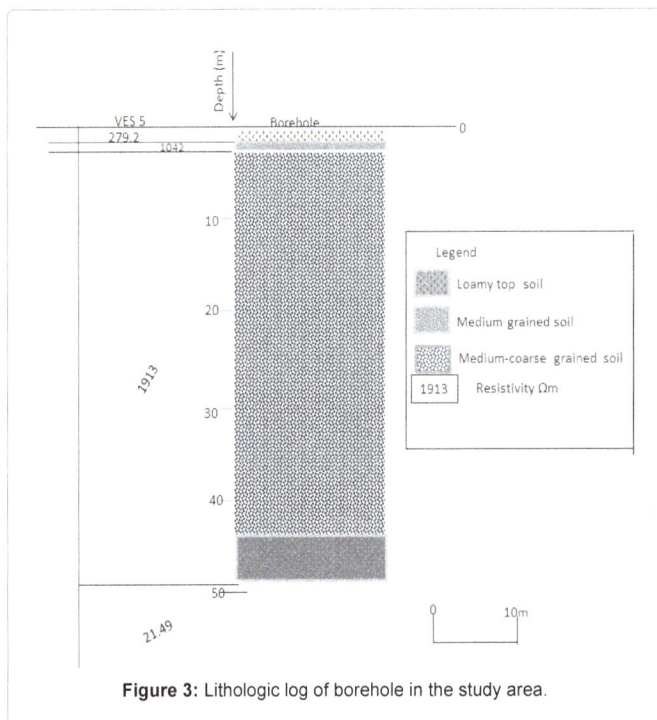

Figure 3: Lithologic log of borehole in the study area.

depth with borehole depth (near VES 5) produced a strong result (Figure 5) with a correlation coefficient of 0.98. This gave the researchers more confident that the VES data interpreted represents to a great extent, the true nature of the local geology of the study area.

The interpretation also determines a shallow depth water table (about 1.5 m), which was used to define the unsaturated sediments and saturated sediments for the study area as presented in Table 2. It is a geoelectrical model correlated with the local lithology in items of prevalence of rock type. The lithology succession which defines the stratigraphic sequence for the study area shows that the aquifers are unconfined. The combination of Tables 1 and 2 aided the understanding of the spatial variation of geoelectrical parameters which demarcated the fresh water bodies and envisaged the identification of saline water ingress. This informed the interpretation of the relatively low resistivity value (21.49 Ωm) observed at the 4th geoelectrical layer along VES 5 column as saline water ingress into the aquifer. The attribution of low resistivity at this depth to saline water ingress is because there were no records of clay layer formation at this depth in the study area.

This is further supported by the fact that, in sedimentary environment high resistivity may largely be associated with the presence of fresh water in porous medium aquifer, while low resistivity may be due to the presence of clay or brackish water [25,27]. The facies variations were identified to be sands of various grain sizes forming the saturation sediments.

Table 3 presents Dar Zarrouk (D-Z) geoelectrical parameters deduced for the study area. These parameters were used to characterize aquifers, delineate the groundwater potential zones, its lateral extent and to estimate the aquifer protective capacity in the area as well as assessing its recharge capability. The low resistivity zone of the map is assumed to be the area of saline water ingress from the Atlantic Ocean. Arising from Table 3, contour maps for longitudinal conductance (S), transverse resistance (T), transverse resistivity (ρ_t), longitudinal resistivity (ρ_l), pseudoanisotropy (λ) and root mean square resistivity (ρ_{rms}) were constructed.

Figure 6 presents groundwater zone for the study area showing isoresistivity (using rms values) and isopach maps of the saturated sediments. The root mean square (rms) value was preferred to the direct

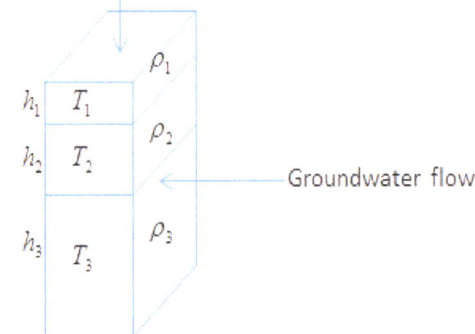
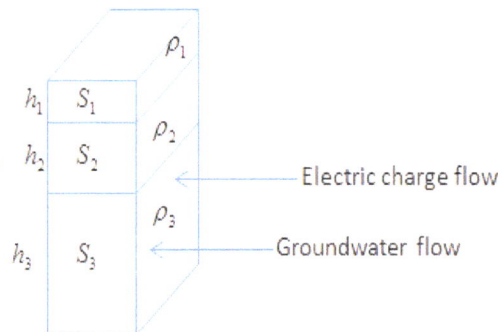

Figure 4: The theory and application for the combination of subsurface resistivity and thickness for two cases [24].

VES	N	ρ Ωm	ρ Ωm	ρ Ωm	ρ Ωm	d1 m	d2 m	d3 m	h1 m	h2 m	h3 m	% Er	Curve	Lat	Long
1	4	90.3	683	1390	368	0.5	12.5	42.4	0.5	12	29.9	1.1	AK	4.653	8.314
2	3	485	167	2383	-	2.3	4.73	-	2.3	2.5	-	2.2	H	4.633	8.305
3	4	154	475	4021	491	2.72	12.2	62.9	2.72	9.48	50.7	1.6	AK	4.645	8.302
4	3	398	285	5867	-	1.9	10.8	-	1.9	8.9	-	2.2	H	4.638	8.286
5	4	279.2	1042	1913	21.49	0.3	7	48.9	0.3	6.7	41.9	2.3	AK	4.663	8.251
6	3	42.4	2923	5579	-	0.8	7.9	-	0.8	7.2	-	2.3	A	4.68	8.242
7	3	510	4221	510	-	4.1	27.8	-	4.6	23.7	-	1.8	K	4.696	8.26
8	4	1010	4736	2787	1464	0.8	10.6	26.7	0.8	9.8	15.6	1.4	KH	4.703	8.245
9	4	1010	473	2847	1489	0.8	10.1	24.6	0.8	9.7	14	1.5	HK	4.684	8.263
10	3	867	6130	1587	-	3	18.6	-	3	15.5	-	1.9	K	4.687	8.264
11	3	53.1	1282	1586	-	0.5	19.1	-	0.5	18.6	-	2	A	4.711	8.26
12	3	42.4	1132	64.6	-	0.5	5.9	-	0.5	5.4	-	2.9	H	4.735	8.251
13	4	388	1596	8561	126	1	12.2	46.2	1	11.3	34	1.3	AK	4.804	8.219
14	4	365	1576	7581	117	1.1	14.5	47.3	1.1	13.3	32.8	1.6	AK	4.803	8.256
15	4	203.3	2532	629.4	845	0.5	1.34	9.58	0.5	0.8	8.2	4.5	AK	4.802	8.261

Table 1: Interpreted VES data for the subsurface studied.

Sediment	Lithology	Geoelectrical resistivity Ωm
Unsaturated sediments	Loamy top soil	42.4 ≤ ρ ≤ 1184
	Medium grain sand layer	167 ≤ ρ ≤ 6130
Saturated sediments	Medium-coarse grain sand layer	510 ≤ ρ ≤ 8561
	Medium grain sand layer	117 ≤ ρ ≤ 1489

Table 2: Identified lithology for the study area.

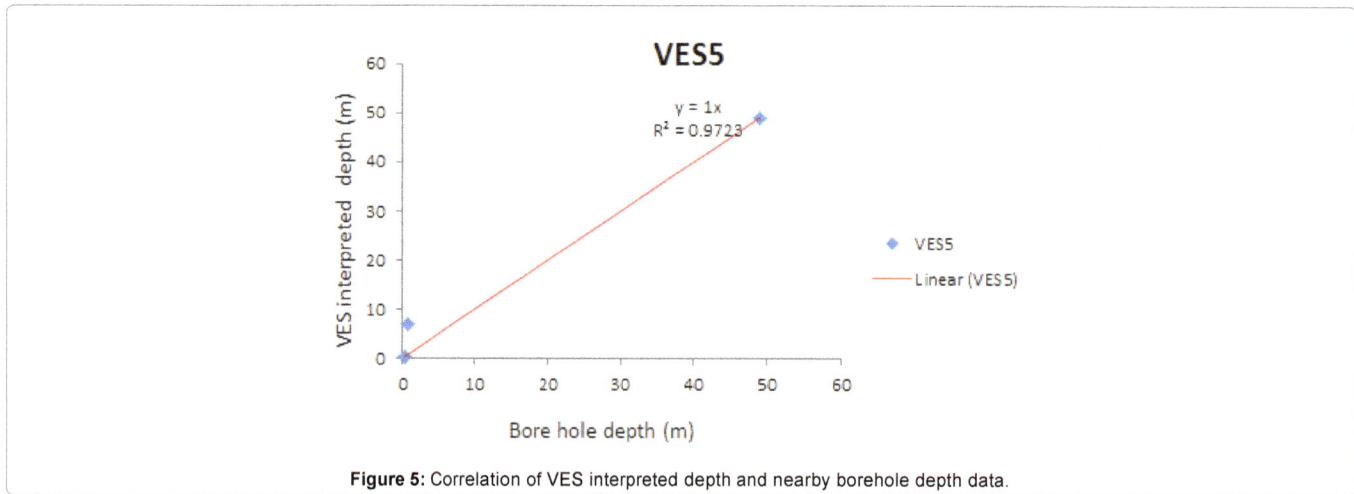

Figure 5: Correlation of VES interpreted depth and nearby borehole depth data.

VES	ρ (Ωm)	h(m)	S(Siemens)	T (Ωm²)	K(m/day)	Tr(m2/day)	ρ_T(Ωm)	ρ_L(Ωm)	λ	ρ_{rms}(Ωm)
1	1390	29.2	0.021	40588	0.452	13.198	957.26	2019.05	0.69	1390.2
2	167	2.5	0.015	417.5	3.263	8.158	88.27	315.33	0.53	166.8
3	4021	50.7	0.0126	203864.7	0.1678	8.508	48.81	4992.06	0.1	493.6
4	285	8.9	0.0312	2536.5	1.9819	17.639	234.86	346.15	0.82	285.1
5	1913	41.9	0.0219	80154.7	0.3355	44.219	1639.16	2232.88	0.86	1913.3
6	2923	7.2	0.0025	21045.6	0.2259	1.626	2664	3160	0.92	2901.4
7	4221	23.7	0.0056	100037.7	0.1604	3.801	3598.48	4964.29	0.85	4226.6
8	2787	15.6	0.0056	43477.2	0.2362	3.685	1628.36	4767.86	0.58	2786.4
9	2847	14	0.0049	39858	0.2317	3.244	1620.24	5020.41	0.57	2852.1
10	6130	15.5	0.0025	95015	0.1132	1.755	5108.33	7440	0.83	6164.8
11	1282	18.6	0.0141	23845.2	0.4874	9.066	1248.44	1354.61	0.96	1300.4
12	1132	5.4	0.0048	6112.8	0.5474	2.956	1091.57	1229.17	0.94	1158.3
13	8561	34	0.0034	291074	0.0929	3.046	6300.38	13288.24	0.67	9149.9
14	7581	32.8	0.0043	248656.8	0.402	5.829	5257.01	11000	0.69	7604.4
15	629.4	8.2	0.013	5161.08	0.9465	7.761	543.27	736.92	0.85	632.7

Table 3: Dar Zarrouk geoelectrical parameters computed for aquifer zone in the study area.

interpreted georesistivity because it further eliminates uncertainties from the interpreted results as well as helps in checking the certainty in the calculations of transverse resistivity (ρ_t) and longitudinal resistivity (ρ_l). It also justifies the consistency in the data.

The contour maps for transverse resistance, transmissivity, longitudinal conductance and anisotropy (Figure 7) clearly demonstrate the contour patterns of saline and fresh water aquifers over large regions with distinctly clear intermixing boundaries. This aided the

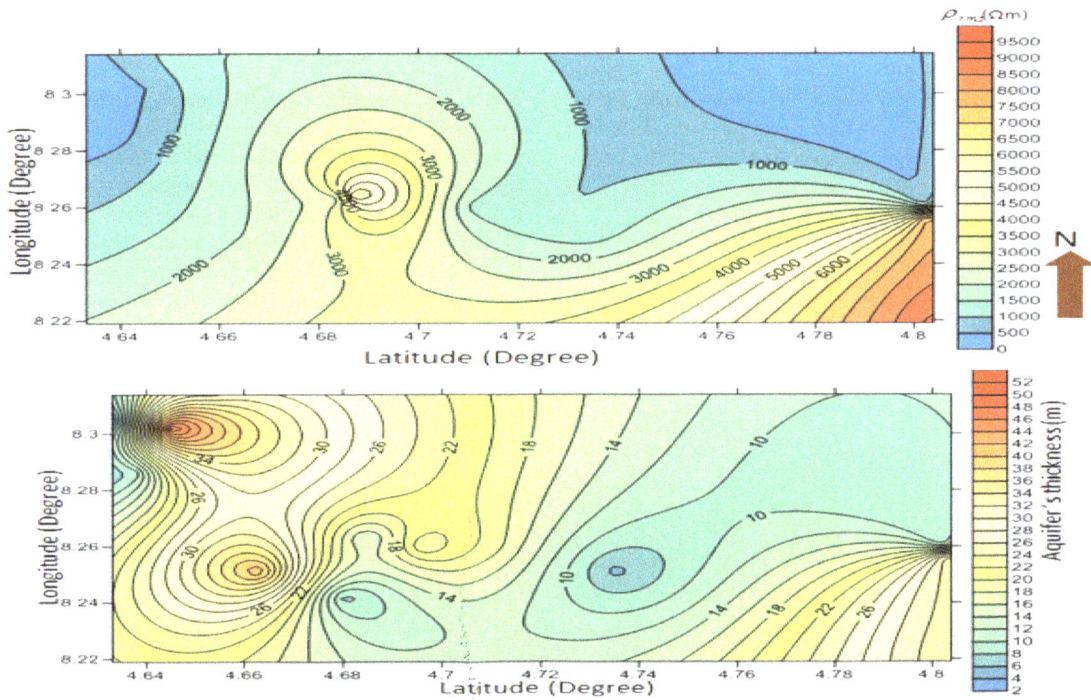

Figure 6: Contour map for groundwater zone showing aquifers rms resistivity at C.I. of 500 Ωm and thickness at C.I. of 2 m.

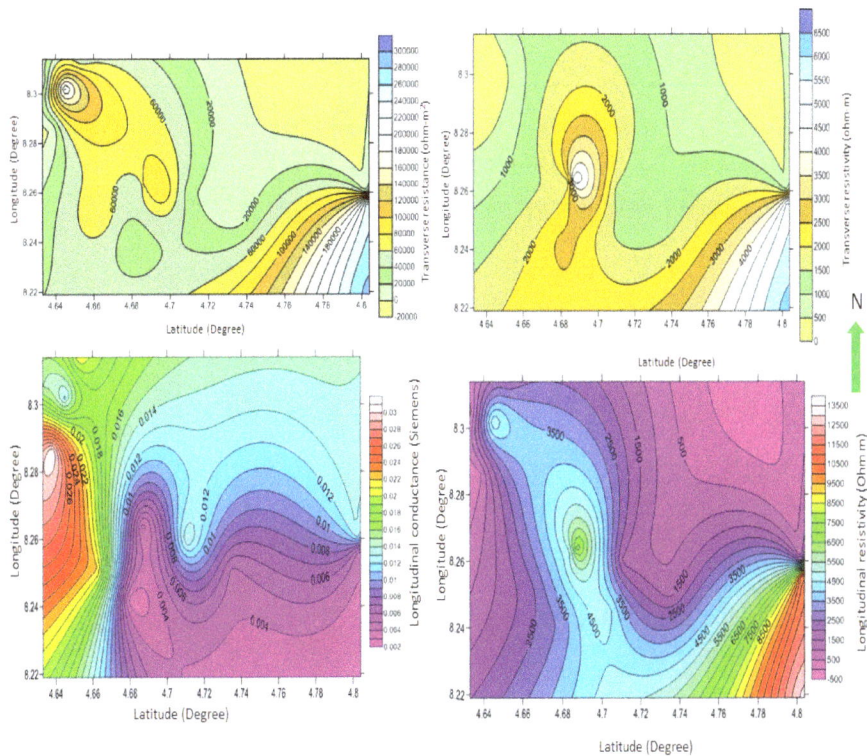

Figure 7: Contour maps of aquifer's D-Z geoelectrical properties for the study area.

identification of fresh water and saline water in the subsurface of the coastal environment studied. Besides, it provides useful evidences to overcome the problem of uncertainty, caused by resistivity data interpretation [24]. The maps of Dar Zarrouk geoelectrical parameters computed show that the transverse resistance (T) values were very high towards the Southeast part of the study area (Figure 7). Similar pictures are obtained for transverse resistivity and longitudinal resistivity maps, with the higher values directed towards the Southeast and these suggest the fresh water region. In sedimentary basins, high values of transverse resistance of phreatic zone is a pointer to aquifers interest, while the total longitudinal conductance of vadose layers best predicts the safety conditions of the aquifers [10,12].

The pseudoanisotropy (λ) of the aquifer zone for the study area calculated for every VES geoelectric column shows that all of the results gave value within the range 0.96 ≥ λ ≥ 0.1. The low values of anisotropy indicate that the rock constituting that aquifer's unit in the study area do not differ significantly. This is typical of sedimentary formation. Figure 8 is the anisotropy map of the aquifer unit in the study area.

To determine the protective capacity of the aquifer zone (saturated zone) by the overburden using D-Z geoelectrical parameter, Table 4 was constructed for every VES point. Considering the overburden longitudinal conductance range (0.1 ≤ S >10 Siemens) for weak excellent protective capacity published by researchers [6,9,10,28,29], the overburden protection potential could be considered as being proportional to the ratio of thickness to resistivity (longitudinal conductance).

The low range of overburden thickness (between 0.3-4.6 m) as well as the associated low values for overburden's longitudinal conductance (0.000-0.019 Siemens) suggested that aquifers in the study area are not naturally protected, therefore, are at probable high risk of contamination from environmental activities and saline water ingress. This result is in support of Orza and Panea [30] and Emmanuel [21]. They noted in their study on "estimation of the natural protective capacity of an aquifer system using surface geophysical" that low overburdens' longitudinal conductance is an indicator of low protective capacity of the aquifer by the overburden materials. Figure 9 is the map of the overburden protective nature determined for the area studied.

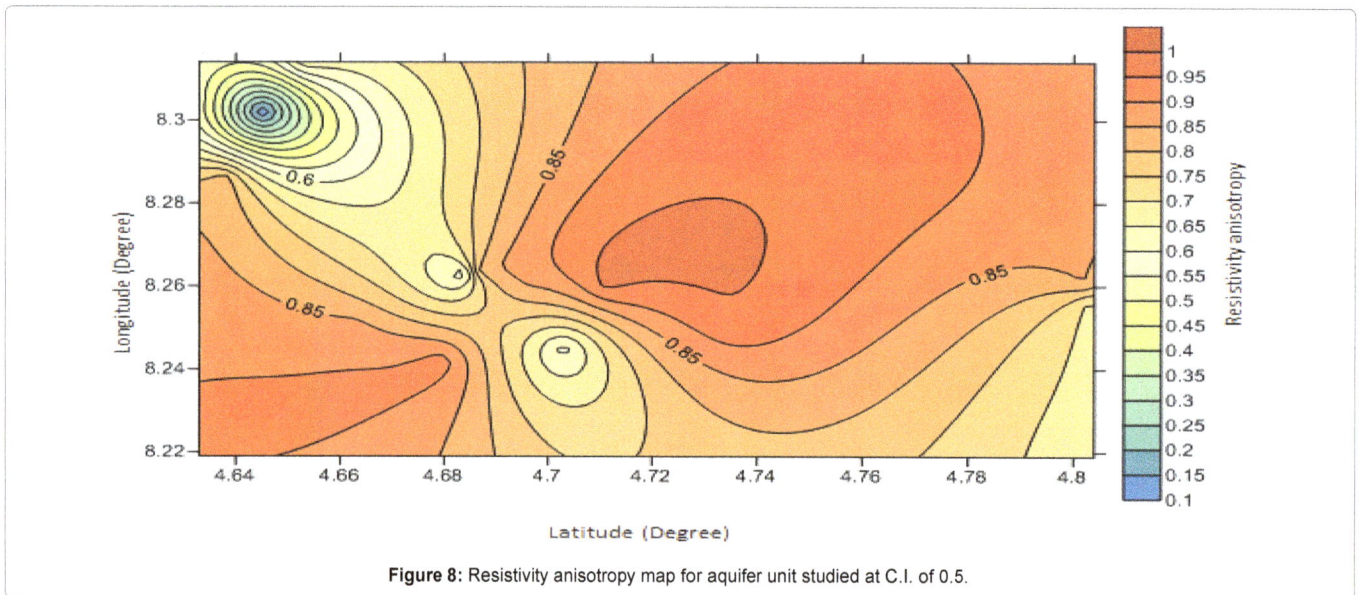

Figure 8: Resistivity anisotropy map for aquifer unit studied at C.I. of 0.5.

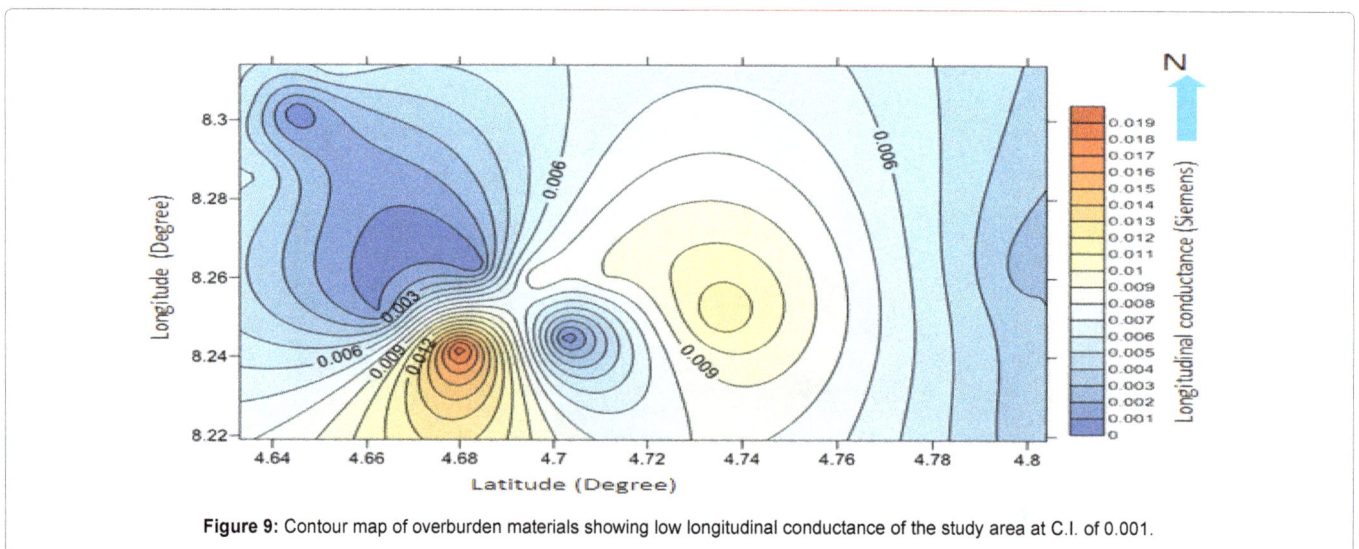

Figure 9: Contour map of overburden materials showing low longitudinal conductance of the study area at C.I. of 0.001.

VES	ρ(Ωm)	h₁(m)	Lat	Long	S(Siemens)	Remark
1	90.3	0.5	4.653	8.314	0.006	V
2	485	2.3	4.633	8.305	0.005	V
3	1184	0.8	4.645	8.302	0	V
4	398	1.9	4.638	8.286	0.005	V
5	279.2	0.3	4.663	8.251	0.001	V
6	42.4	0.8	4.68	8.242	0.019	V
7	510	4.6	4.696	8.26	0.009	V
8	1010	0.8	4.703	8.245	0	V
9	1010	0.8	4.684	8.263	0	V
10	867	3	4.687	8.264	0.003	V
11	53.1	0.5	4.711	8.26	0.009	V
12	42.4	0.5	4.735	8.251	0.012	V
13	388	1	4.804	8.219	0.003	V
14	365	1.1	4.803	8.256	0.003	V
15	203.3	0.5	4.802	8.261	0.002	V

Table 4: Dar Zarrouk parameter and groundwater resources vulnerability [V indicates vulnerability of groundwater resources to contamination].

Conclusion

The study was aimed at identifying lithology, estimating the relationship between Dar Zarrouk (D-Z) parameters and geoelectrical resistivity with groundwater resources protection. Geophysical technique involving the use of ground-based earth resistance measuring device was used to measure geoelectrical properties of the subsurface. Interpreted data shows that the aquifers delineated are unconfined and consist of sands of medium grain sizes. The high rms resistivity of the subsurface indicates opulent groundwater resources. Place with high transverse resistivity values were observed to produce high transmissivity and are potential areas were monitory wells can be drilled. This further explains the abundant of groundwater in most parts of the study area. However, the low values of the Dar Zarrouck variable (longitudinal conductance) are pointers to likely vulnerable location. For instance, the protective capacity of the overburden using the longitudinal conductance values which ranged between 0.000-0.019 Siemens, with overburden thickness that varies between 0.3 and 4.6 m, explain that the entire aquifer is at risk of contamination from environmental activities. Also, saline water ingress is most likely to occur at depth greater than 40 m within VES 5. This is due to the very low resistivity value (21.49 Ωm) without the presences of clay mineral within the depth of current penetration. The results show that the Dar-Zarrouk (D-Z) parameters provide a useful and confident solution in delineating the saline and fresh water aquifers. The identified aquifer's thickness, likely contamination of site, protection capacity could help in the planning, monitoring and management of potable water supply and development in deltaic coastal aquifer system.

References

1. Song S, Zemansky G (2012) Vulnerability of groundwater systems with sea level rise in coastal aquifers, South Korea. Environ Earth Sci 65: 1865-1876.

2. Rajsekhar K, Sharma PK, Shukla SK (2016) Numerical modeling of virus transport through unsaturated porous media. Cogent Geosci 2: 1-13.

3. Ismail AI, El-Shamy AM (2008) Engineering behaviour of soil materials on the corrosion of mild steel. Applied Clay Sci 10: 1-7.

4. Ekine AS, Emujakporue GO (2010) Investigation of corrosion of buried oil pipeline by the electrical geophysical methods. J Applied Sci Environ Manag 14: 63-65.

5. Okiwelu A, Evans U, Obianwu I (2011) Geoelectrical investigation of external corrosion of earth buried pipeline in the coastal area of Gulf of Guinea. J Am Sci 7: 221-226.

6. Evans UF, Okiwelu AA, Ude IA, Abima JO (2012) Estimation of corrosion induced flaw sizes on buried gas pipeline in the Nigerian sector of Niger Delta. Res J Environ Earth Sci 4: 264-269.

7. Corwin DL, Rhoades JD (1984) Measurement of inverted electrical conductivity profiles using electromagnetic induction. Soil Sci Soc Am J 48: 288-291.

8. Mailet R (1947) Fundamental equations for electrical prospecting. J Geophys 12: 528-556.

9. De Oliveira Braga AC, Fisho WM, Daurado JC (2006) Resistivity (DC) method applied or aquifer protection studies. Braz J Geophys 24: 573-581.

10. Henriet JP (1975) Direct applications of the Dar Zarrouk parameters in groundwater surveys. Geophys Prospec 24: 344-353.

11. Egbai JC, Iserhien-Emekeme RE (2015) Aquifer transmissivity Dar Zarrouk parameters and groundwater flow direction in Abudu, Edo State, Nigeria. Int J Sci Environ Technol 4: 628-640.

12. Okiongbo KS, Akpofure E, Odubo E (2011) Determination of aquifer protective capacity and corrosivity of near surface materials in Yenagoa City, Nigeria. Res J Applied Sci Environ Technol 3: 785-791.

13. Evans UF, George NJ, Omotosho OO (2010) Geoelectrical measurements for design of cathodic protection system to prolong the life span of Jetty. World J Applied Sci Technol 2: 217-223.

14. George NJ, Obianwu VI, Obot IB (2011) Estimation of groundwater reserve in unconfined frequently exploited depth of aquifer using a combined surficial geophysical and laboratory techniques in the Niger Delta, South-south, Nigeria. Pelagia Res Lib 2: 163-177.

15. Igboekwe MU, Akankpo AO (2012) Estimation of hydrogeological parameters for Michael Okpara University of Agriculture, Umudike, Southeastern Nigeria. Pac J Sci Technol 13: 455-461.

16. Abam TKS (1999) Dynamics and quality of water resources in the Niger Delta. Proceedings of the symposium of the International Union of Geodesy and Geophysics.

17. Edet A, Abdelaziz R, Merkel B, Okereke C, Nganje T (2014) Numerical groundwater flow modeling of the coastal plain sand aquifer, Akwa Ibom State, SE Nigeria. J Water Res Protect 6: 193-201.

18. Orellana E, Mooney HM (1966) Mastercurves for Schlumberger arrangement, Madrid.

19. Emmanuel CC (2013) Monitoring the quality of groundwater and leachate contamination near waste dump in coastal plain sand aqufer.

20. IPI2 Win V.2.1 users guide (2001) Computer software user guide catalog presented by Moscow State University.

21. Edet AE, Okereke CS (2002) Delineation of shallow groundwater aquifers in the coastal plain sand of calabar area using surface resistivity and hydrogeological data. J Afr Earth Sci 35: 433-441.

22. Zoddy AAR (1974) Use of Dar Zarrouk curves in the interpretation of verticals electrical sounding data. United States Geological Survey Bulletin pp: 1313.

23. Al-Yasi AI, Alridha NA, Shakir WM (2013) The exploitation of Dar-Zarrouk parameters to differentiate between fresh and saline groundwater aquifers of sinjar plain area. Iraqi J Sci 54: 358-367.

24. Kelly WE, Frohlich RK (1985) Relations between aquifer electrical and hydraulic properties. Ground Water 23: 10-16.

25. Singh UK, Das RK, Hodlur GK (2004) Significance of Dar-Zarrouk parameters in the exploration of quality affected coastal aquifer systems. Environ Geol 45: 696-702.

26. Harb N, Haddad K, Farkh S (2010) Calculation of transverse resistance to correct aquifer resistivity of groundwater saturated zones: Implications for estimating its hydrogeological properties. Lebanese Sci J 11: 105-115.

27. Oseji JO, Ujuanbi O (2009) Hydrogeophysical investigation of groundwater potential in Emu Kingdom, Ndokwa Land of Delta State, Nigeria. Int J Phys Sci 4: 275-284.

28. Word SH (1990) Resistivity and induced polarization methods USA investigations in geophysics, geotechnical and environmental geophysics. Society of Exploration Geophysics.

29. Akankpo O, Igboekwe MU (2011) Monitoring groundwater contamination using surface electrical resistivity and geochemical methods. J Water Res Protect 3: 318-324.

30. Orza RL, Panea I (2012) Estimation of the natural protective capacity of an aquifer system using surface geophysical. 74th EAGE Conference & Exhibition, Near Surface Geoscience Applications.

21

Geophysical Assessment of the Upper Dja Series Using Electrical Resistivity Data

Zoo Zame P, Mbida Yem*, Yene Atangana JQ and Ekomane E

Department of Earth Sciences, University of Yaounde I, Cameroon

Abstract

The upper Dja series consists of carbonates and shales deposits that date 580 ± 150 Ma. Petrography and mineral chemistry studies helped to differentiate this series into many sequences including massive limestone layers with calcite (CaO3) contents of about 30 to 42%. In order to determine the subsurface distribution of these sequences, a geophysical prospection campaign was carried out in which a total of 24 vertical resistivity soundings were recorded over a surface area of 9 km2. The processing and interpretation of data using the IX1D and OpendTect modeling tools permitted the distinction of two massive limestone layers in the upper Dja series. The first layer of about 10 to 35 m thick and outcrop in few places is characterized by resistivity ranging between 1110 and 2377 Ωm, while the second layer located beyond 50 m deep is separated by a very conductive clay stone layer with a capped thickness of 15 to 35 m. 3D modeling of the top and base of these formations indicates that, the whole Upper Dja series of about 190 m thick presents folded structure. These results can constitute a useful base of information in regards to a large scale economic study of the Upper Dja limestone series.

Keywords: Geophysical prospection; Upper Dja series; Vertical sounding; Limestone layer

Introduction

In Cameroon, limestone neoproterozoic metamorphism is essentially represented by the marble and travertine deposits of Bidzar and the Upper Dja series respectively (Figure 1). Other schisto-calcareous massifs have been indicated in the paleozoic to mesozoic trenches in Figuil, Moungo, Kompina and Logbadjeck. In the Dja series, the zones with the most important limestone deposit were revealed to be in relation with the sedimentary furrow of the Dja River and the Atog-Adjap outcrops [1-8]. These massifs are partially covered by the forest and the Dja phanerozoic basin.

The works of Gazel and Guiraudie and Van Houtte, presented the first observations as well as the early petrographic studies of these massifs [1,2]. More recent studies were carried out by Ntep et al., with aims to study the composition and their contents in calcium carbonate ores. In addition, these works consisted of deep drill cores, geological surveys, rock sampling, description of outcrops and pale environmental analysis [6].

In this study, the main questions were mostly concerned with the continuity and geometry of defined targets through drilling on a surface area of 9 km², with objective to constitute and put at the disposal of scientists and the industrial sector a database of useful information on the large scale evaluation of the upper Dja series of carbonates and perlite reserves.

Geological Context

The upper Dja carbonates series was dated using $^{207}P/^{204}P$ and $^{206}P/^{204}P$ isotopes, and the age obtained is 580 ± 150 Ma [6]. This series is hosted in a flexural basin of about 40 km wide. The research permit of this deposit is located between latitudes N 02°40' to 3° and Longitudes E 13° to 13°30', overlapped between a northern over flooded zone and a swampy southern zone. Up till date, most of the known limestone outcrops are found in the flooded (Figure 1).

Morphologically, in some places the very variable relief presents very steep valleys, whose unevenness can attain 30 m. On a geological point of view, the encountered formations are successively made up of:

- A base complex made up of migmatites, dolerites, diorites, granodiorites, amphibolites, granites and gabbro [9,10].

- The Mbalmayo-bengbis series which is essentially made up of chloritoschists [11].

- The upper Dja series made up of perlites, calcareous schist and clay stone [5,9].

- Phanerozoic formed by a recent sedimentary assembly, it is represented by conglomerates, siliceous sandstone, overlaid by a locally brecciated cuirassed laterite [12].

As well, control drilled boreholes F1 and F2 realized in the perimeters of the study area reveals that the limestone rich massif formations have variable thicknesses as follow [3,4]:

- At borehole F1, situated close to the SE2 and SE3 electrical soundings (Figure 2), limestone massifs extends from 58 m up to 142 m. Its overburden is made of clays and detritic formations from 14 to 58 m overlaid by laterite of about 9 m thick and recent colluvial material at the top of the borehole.

- At borehole F2 situated close to the SE19 electrical soundings, is made up of a clayey colluvial and lateritic cuirass cover on the first 15 meters, followed by an alternation of a massif and shaly limestone up to 103 m. A layer of pelite of about 16 m thick is found below the limestone alternation.

These data permitted the calibration of resistivity profiles by realizing controlled electrical soundings in line with some boreholes.

*Corresponding author: Mbida Yem, Department of Earth Sciences, University of Yaounde I, Cameroon, E-mail: yem04@yahoo.com

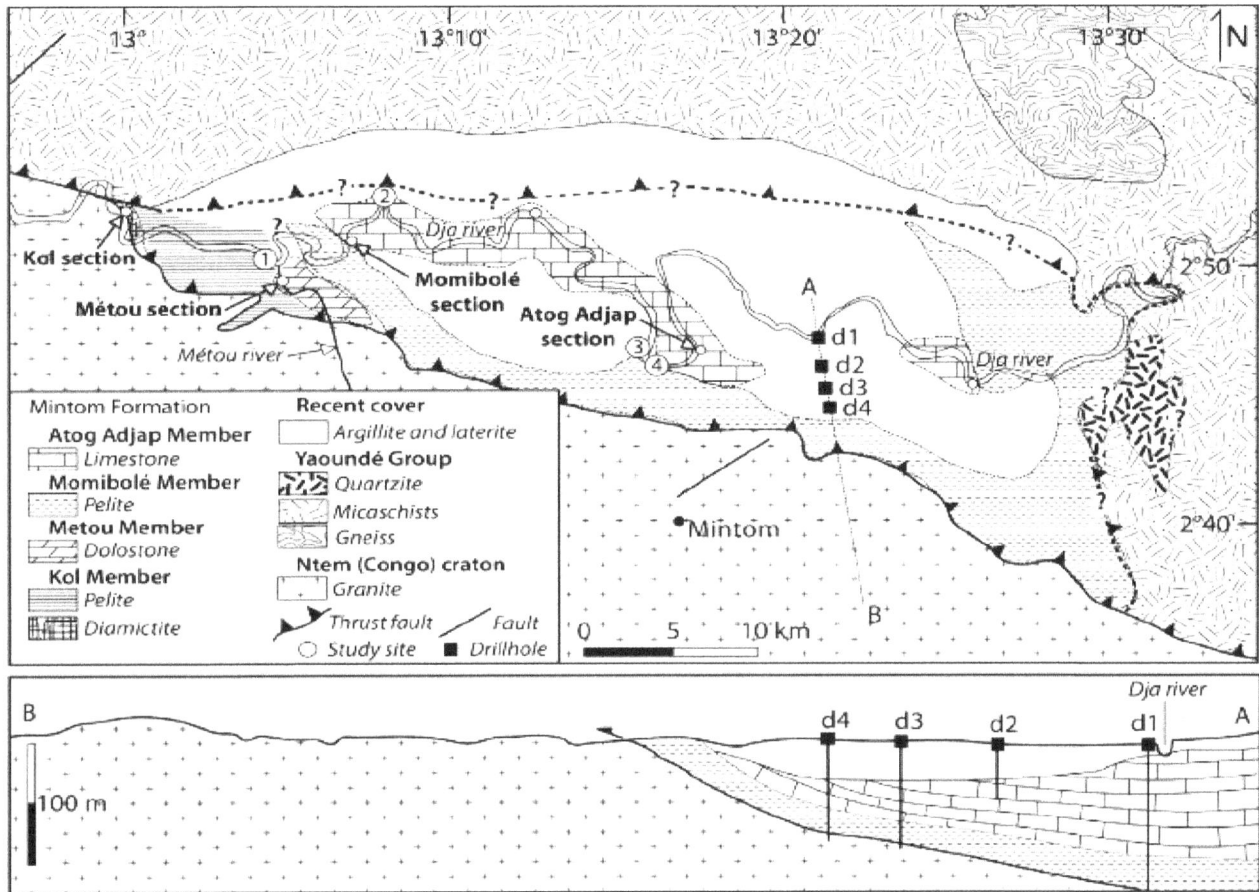

Figure 1: Geologic map of South Cameroon and the location of the study area [7].

Data and Methods

Geophysical methods represent important tools in the spatial characterization of geological formations [13]. Electrical resistivity mapping is an example of such well-developed methods and have been widely applied in the study of geological formations in the humid intertropical zone [14-17]. This method is particularly adapted in the study of subsurface discontinuity, marked by vertical and horizontal juxtaposition of geologic formations with well contrasted electrical properties. In vertical structures of homogenous milieu, where resistivity changes with depth, vertical electrical sounding has been identified to be the most adapted method. In the present study, this method was conducted following 24 profile lines with four over the core drilling boreholes (Figure 2). The Schlumberger electrode configuration with was use for the purpose to attain the maximum depth of the carbonate series as well as the top of the basement rock.

Vertical electrical sounding is a geophysical prospecting method for deep subsurface exploration [13,18,19]. Classically, the minimum size of subsurface anomalies investigated by vertical electrical sounding depends on the type of electrode array used as well as its physical dimension [13]. Theoretically, it has been shown that the Schlumberger configuration permits topographic corrections and it is very sensible to vertical resistivity variations [20,21]. In this study, the Schlumberger configuration was used with current electrode spacing varying from 12 to 500 m.

According to Abraham Bairu, the calculation of apparent resistivity values in the case of Schlumberger configuration is expressed by:

$\rho a = K \Delta V / I$

where: ΔV is the potential difference; I is the current intensity.

K is the geometric factor, which depends on the arrangement of electrodes.

For a AMNB electrode array, $K = \pi [(AB/2)^2 - (MN/2)^2]/MN$, with A-B being the current electrodes; M-N being the potential difference electrodes.

The spatial orientation and position of profile lines collected using this configuration is represented in Table 1. In order to determine geoelectrical sequences (Figure 3) over each sounding points, a software processing was done with the aid of the IX1D/RES2DINV codes [22].

Geoelectrical Correlations and Interpretations

Geoelectrical sequences

With reference to the overall vertical sounding curves, it is seen that, the underground cutting between the topographic surface (altitude ~ 640 m) and the bedrock substratum (altitude ~ 430 m) is made up of five (05) geoelectrical sequences, labeled: e1, e2, e3, e4 and e5 (Table 2).

Geoelectrical sequence e1

This sequence is situated at the top of the sedimentary section

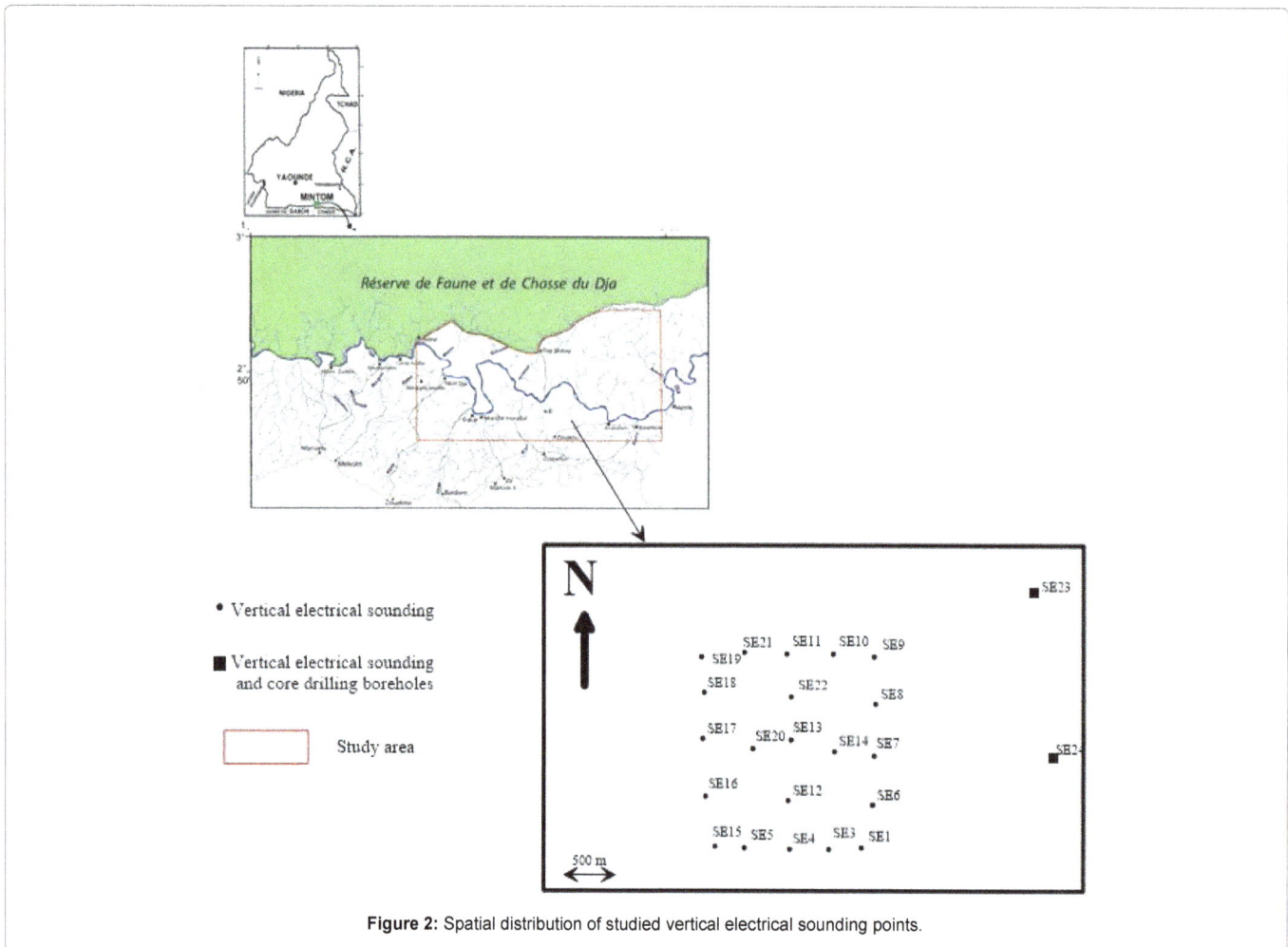

Figure 2: Spatial distribution of studied vertical electrical sounding points.

No.	Azimut	AB/2	X	Y	Z
SE1	E-W	150	E 13° 19' 35,4"	N02° 46' 00,7"	615
SE2	E-W	150	E13° 19' 30,6"	N02° 46' 00,6"	632
SE3	E-W	150	E13° 19' 12,5"	N02° 46' 00,3"	635
SE4	E-W	150	E13° 18' 49,9"	N02° 45' 59,7"	626
SE5	E-W	150	E13° 18' 23,2"	N02° 46' 01,2"	634
SE6	N-S	200	E13° 19' 37,0"	N02° 46' 22,2"	623
SE7	N-S	150	E13° 19' 37,6"	N02° 46' 48,5"	591
SE8	N-S	150	E13° 19' 39,4"	N02° 47' 15,0"	575
SE9	N-S	150	E13° 19' 40,6"	N02° 47' 39,4"	570
SE10	E-W	150	E13° 19' 15,2"	N02° 47' 41,2"	593
SE11	E-W	150	E13° 18' 47,7"	N02° 47' 41,4"	580
SE12	N-S	150	E13° 18' 49,9"	N02° 46' 24,3"	609
SE13	N-S	150	E13° 18' 51,1"	N02° 46' 56,6"	616
SE14	E-W	150	E13° 19' 15,4"	N02° 46' 50,1"	579
SE15	E-W	150	E13° 18' 06,4"	N02° 46' 01,9"	620
SE16	N-S	200	E13° 18' 00,6"	N02° 46' 27,3"	588
SE17	N-S	250	E13° 17' 59,5"	N02° 46' 57,5"	577
SE18	N-S	150	E13° 17' 59,9"	N02° 47' 21,3"	611
SE19	N-S	250	E13° 18' 00,7"	N02° 47' 32,1"	627
SE20	E-w	150	E13° 18' 28,1"	N02° 46' 51,9"	628
SE21	E-W	150	E13° 18' 24,3"	N02° 47' 41,3"	580
SE22	N-S	150	E13° 18' 51,0"	N02° 47' 19,4"	616
SE23 (F1)	E-W	250	E13° 21' 10,2"	N02° 48' 09,7"	609
SE24(F2)	E-W	150	E13° 21' 20,6"	N02° 46' 44,9"	566

Table 1: Characteristics of recording vertical electrical sounding.

and in some places, forms the cover terrain of e2. Details analysis of resistivity profiles (Figure 3) reveals that e1 thickness varied between 1 and 20 m. Its thickest section is observed over the sounding 18 (Figure 4a), while its thinnest portion is observed at the sounding points 11 and 21 (Figure 4b). In an electrical point of view, e1 is made up of two units as follow:

- U1: semi-resistive with a resistivity of 826 to 1068 Ωm.

- U2: very resistive with resistivity of 5011 to 13545 Ωm.

This differentiation clearly appears in the sounding point 6 (Figure 4c).

Geoelectrical sequence e2

Just as e1, e2 is also a continuous sequence and the general analysis of sounding curves show that, the thickness of e2 decreases from the western (Figure 4a) towards the eastern (Figure 4c) part of the study area. The resistivity values (1110-2377 Ωm) equally obtained from the inversion of resistivity data reveal that, the physical nature of the e2 turns to degrade from the South towards the North of the study area (Figure 4).

Geoelectrical sequence e3

In an electrical point of view, e3 is a conductive sequence whose resistivity vary between 181 and 746 Ωm. Some sounding curves obtained by resistivity inversion of the data collected in the North and

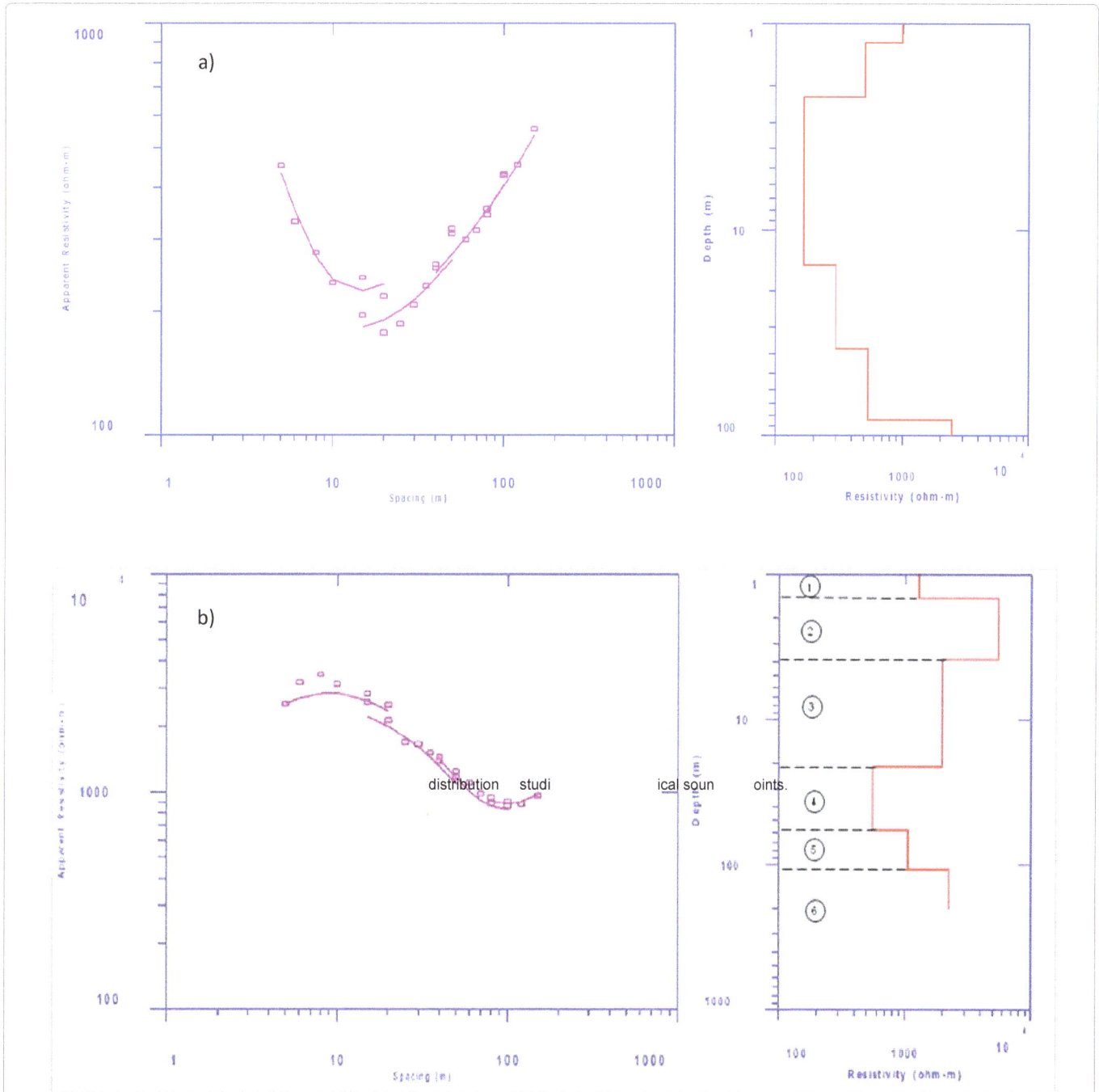

Figure 3: Inversed electrical resistivity sounding curves: a) resistivity profile type H. b) resistivity profile type Q.

Central parts of the study area (Figures 5a and 5b), show that e3 is made up of two units: U4 (conductive) and U5 (semi-conductive, Figure 5b). Contrary to e2, details analysis of the set of electrical pattern indicates that the thickness of e3 increases from the western (Figure 4a) towards the eastern (Figure 4c) of the said zone.

Geoelectrical sequence e4

As of e4, it is a very thick sequence in all the sounding points and appears as a resistive level with resistivity between 1056 to 2544 Ωm. Its thickness generally increases from the northern towards the southern part of the study area (Figures 5a-5c).

Geoelectrical sequence e5

This sequence is visible at the end of sounding profiles and appears as a deep semi-conductive terrain with an average resistivity of 1209 Ωm at the 14 sounding point (Figure 5b).

Resistivity Curves to Borehole Correlation

Resistivity records over cores boreholes points as been performed (Figure 5a). These records permitted to correlate sedimentary units and geoelectrical sequences. The average resistivity and thicknesses of the different layers are summarized is given on Table 2.

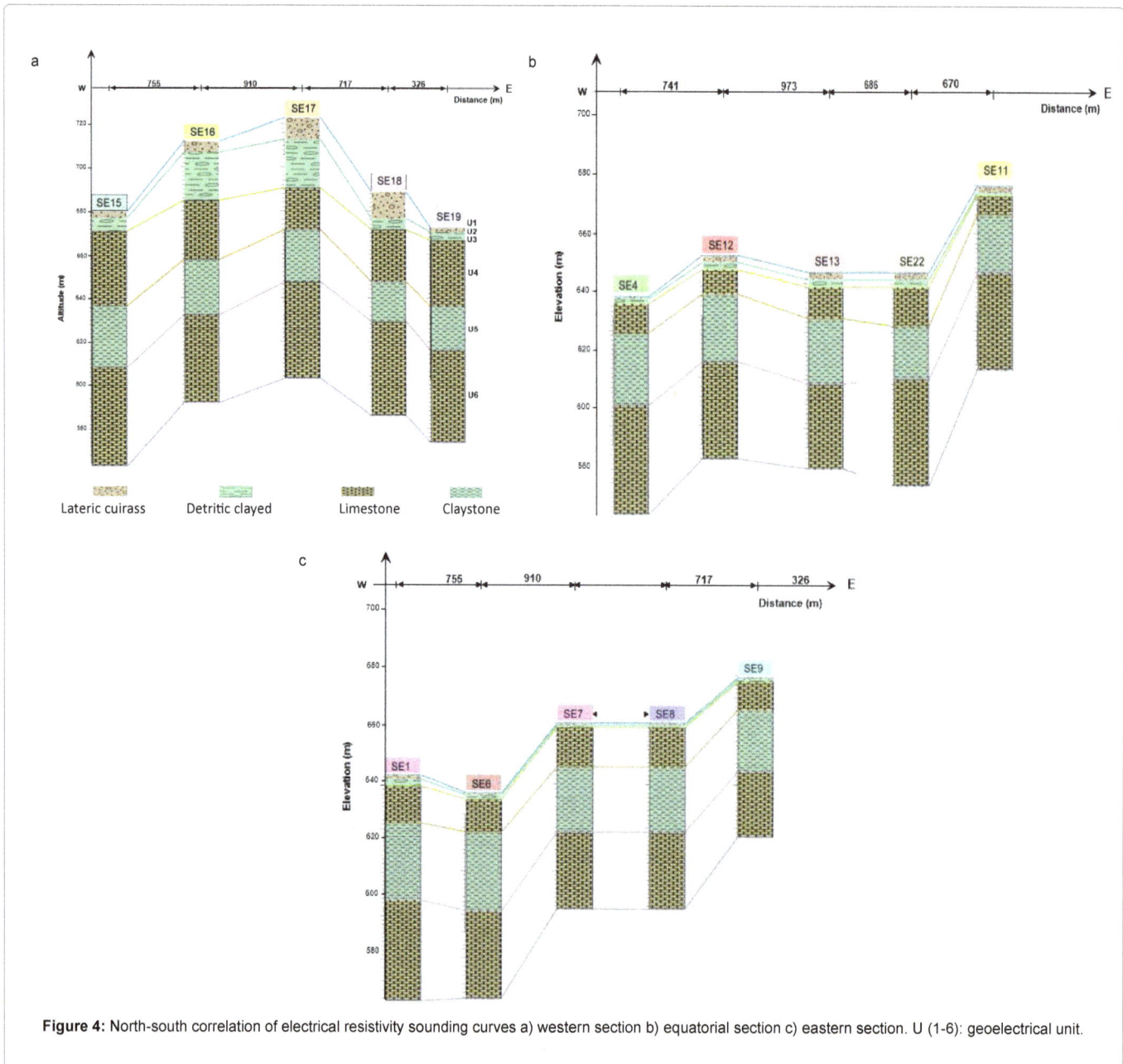

Figure 4: North-south correlation of electrical resistivity sounding curves a) western section b) equatorial section c) eastern section. U (1-6): geoelectrical unit.

Regarding e2 and e4 mineral ore bearing layers, the results of tridimensional modeling (Figure 6) indicate that their reserve can be estimate at 15343 m³ and 67989 m³ respectively.

Discussion and Conclusions

The results obtained show that the thickness of the ore bearing layers is almost constant (Figures 4 and 5). However, their variable geometry can either be due to post-deposition deformation or the original structure of the Dja Precambrian basin (Figure 5a). On the basis of geo-electrical parameters (Table 2), the Dja carbonate series is made up of seven petro physical units (U1, U2, U3, U4, U5, U6 and U7). The resistivity values associated carbonate ore bearing unit (Table 3) seems to be very continuous throughout the study area. These results are in good agreement with geo-electrical studies carried out in similar environments [23,24]. They also translate the relative homogeneity of the Dja carbonate ore bearing unit. Furthermore, the high resistivity values obtained from U3 at the sounding point 6 could have been associated to the very resistive nature of the subsurface capping layer (Figure 4c). While the low values observed at the sounding point 14 could be due to fluid circulation associated by N 55°E faulting. This observation was already made known by previous studies in the same study area by Ekomane [6]. In order to improve the results present on this paper, one of the solutions can be the densification of data grid acquisition or the use of 2D a multi-electrode resistivity acquisition.

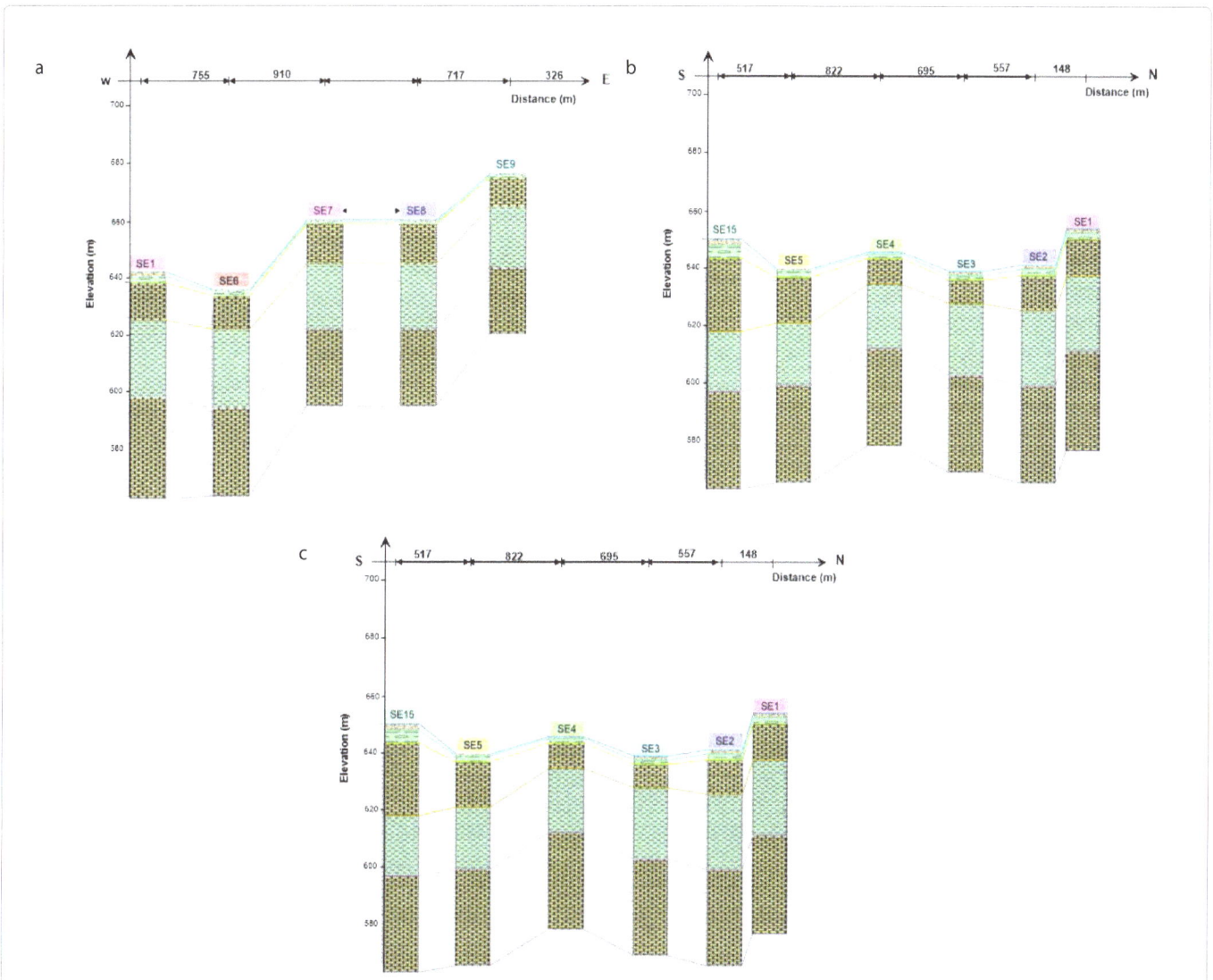

Figure 5: East-west correlation of electrical resistivity sounding curves a) southern section b) equatorial section c) northern section.

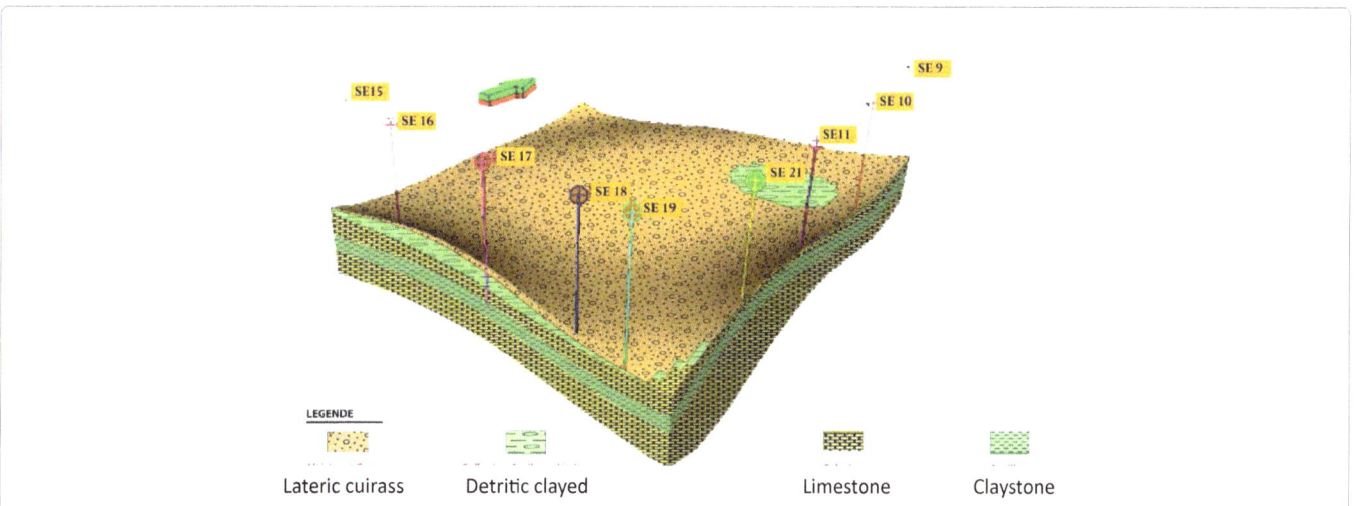

Figure 6: 3D correlation of electrical resistivity sounding curves.

Electrosequence	e1		e2	e3		e4	e5
Unit	U1	U2	U3	U4	U5	U6	U7
Resistivity (Ωm)	826-1068	5011-13545	1110-2377	181-440	225-746	1056-2544	1209
Thickness (m)	01-Dec	May-22	Nov-35	15-20	25-35	25-45	16-20
Lithology	Lateritic Cuirass	Clayey Colluvion	Massif Limestone	claystone	Shale	Massif limestone	Pelites

Table 2: Geoelectrical sequences and boreholes correlation.

e1	lateritic cuirass and detritic clayey deposits
e2	massif limestone of the Atog-Adjap unit
e3	claystone and shale
e4	massif limestone of the Metou unit
e5	alternation of pelite with a more or less carbonaceous deposits

Table 3: Recapitulation of the mapping ore bearing layers.

References

1. Gazel J, Guiraudie C (1955) Mission of the descent of Dja-Missions Geological 1 and 2: Unpublished reports. Cameroon.

2. Vanhoutte M, Salley P (1986) Mintom limestone reconstruction-mining research project, Southeast Cameroon. United Nations Development Program.

3. United Nations Development Program (UNDP) (1987) Mining research in South East Cameroon.

4. United Nations Development Program (UNDP) (1989) Review of the Mintom limestone geological survey and evaluation.

5. Vicat JP (1998) Review of the knowledge gained on the series of Dja (Cameroon), Nola (Central African Republic) and Sembe-Ouesso (Congo). Geosci in Cameroon. 1: 369-383.

6. Ekomane E (2010) Sedimentological and paleoenvironmental studies of the carbonate and pelitic rocks of the Mintom Formation (SE Cameroun).

7. Caron V, Ekomane E, Mahieux G, Moussango P, Ndjeng E (2010) The mintom formation (new): Sedimentology and geochemistry of a neoproterozoic, paralic succession in south-east Cameroon. J African Earth Sci 57: 367-385.

8. Caron V, Mahieux G, Ekomane E, Moussango P, Babinski M (2011) One, two or no record of late neoproterozoic glaciation in South-East Cameroon?. J African Earth Sci 59: 111-124.

9. Gazel J, Hourcq V, Nichelés M (1956) Geological map of Cameroon.

10. Vicat JP, Moloto-A-Kenguemba G, Pouclet A (2001) The granitoids of the proterozoic cover of the northern border of the Congo craton (south-eastern Cameroon and south-west of the Central African Republic), indicative of pre-Panafrican post-Kibarian magmatic activity. Earth Planet Sci 332: 235-242.

11. Nzenti JP, Barbey P, Macaudière J, Soba D (1988) Origin and evolution of the late Precambrian high-grade Yaounde gneisses (Cameroon). Precambrian Res 38: 91-109.

12. Censier C (1989) Sedimentary dynamics of a Mesozoic diamondiferous fluvial system. The formation of Carnot (Central African Republic).

13. Telford WM, Geldaart LP, Sheriff RE (1990) Applied Geophysics. 2nd edn, Cambridge Univ Press.

14. Albouy Y, Pion JC, Wackerman JM (1970) Application of electrical prospecting to the study of alteration levels. Cah Orstom Ser Geol 2; 161-170.

15. Robain H, Descloitres M, Ritz M, Yene Atangana JQ (1996) A multiscale electrical survey of lateritic soil system in the rainforest of Cameroon. J Appl Geophys 34: 237-253.

16. Ritz M, Robain H, Pervago E, Albouy Y, Camerlynch C (1999) Improvement to resistivity pseudo section modelling by removal of near surface inhomogeneity effect. Geophys Prospect 41; 85-101.

17. Beauvais A, Ritz MM, Parisot JC, Dukhan M, Bantsimba C (1999) Analysis of poorly stratified lateritic terrains overlying a granitic bedrock in west Africa, using 2D electrical resistivity tomography. Earth Planet Sci Lett 173: 413-424.

18. Keller GV, Frischknecht FC (1966) Electrical methods in geophysical prospecting. Pergamon Press, Oxford, pp: 519.

19. Loke MH (1999) Electrical imaging surveys for environmental and engineering studies. A practical guide to 2-D and 3-D surveys, Penang, Malaysia.

20. Storz H, Storz W, Jacobs F (2000) Electrical resistivity tomography to investigate geological structures of the earth's upper crust. Geophys Prospect 48: 455-471.

21. Dahlin T, Zhou B (2004) A numerical comparison of 2D resistivity imaging with ten electrode arrays. Geophy Prospec 52: 379-398.

22. Loke MH (1995) Res2Dmod ver 2.20a 2D Resistivity forward modelling. Malaysia.

23. Revil A, Cathles LM, Losh S, Nunn JA (1998) Electrical conductivity in shaly sands with geophysical applications. J Geophy Res 103: 23925-23936.

24. Revil A (2007) Final report on the mission of Tournemire. Large-scale electrical resistivity test at a test site.

Micropaleotological studies of Ewekoro Sediments Southwestern Nigeria

Oladosu YC* and Ogundipe OY

Department of Geology, Ekiti State University, P.M.B 5363, Ado, Ekiti State, Nigeria

Abstract

The study determines the micropaleontological studies of Ewekoro sediments in South Western region of Nigeria. The area of the study lies between latitudes 6°47N-6°53N and longitudes 335E-3°40E. Samples were collected from fleshly exposed surface of Ewekoro formation of Dahomey basin at an interval of 2 m. These samples were subjected to lithological description to determine the grain size, color, sorting also to standard micropaleontological analysis to recover the foraminifera which were used to determine the age, paleoenvironment of deposition and stratigraphic equivalence of the formation. The study showed that the limestone at the basal part of the sequence were sub-angular to sub-rounded shaped, poorly sorted and with yellowish brown color indicate disturbed environment. While the upper shale are fined grained, greenish in color, dark-grey and showing high fissility. The foraminifera recovered include, *Lenticulina degolyeri, globorotalia, pseudomonarrdii, Globigerina linaperta, Globigerina yeguaensis* and *cibicides* sp. This is occurrence of some planktonic foraminifera which are indicative of marine environment. Therefore, there must have been an incursion of marine water into the environment. However, the paleoenvironment of deposition ranged from inner neritic to middle neritic.

Keywords: Micropaleontology; Lithological; Stratigraphic; Grain size; Tectonics; Foraminifera; Environment

Introduction

A basin can be generally be defined as a depressional or shallow place and when filled or floored with different sort of sediments. It becomes a sedimentary basin. The various common rocks that in filled into a basin are: shales, limestones, clays, siltstone and sandstone. These sedimentary rocks are the end products of different sediments that had undergone the process of digenesis. The pre-processes that occur before diagnosis include: weathering, erosion, transportation of sediments, accumulation of sediment (deposition) the weathering due to overburden pressure and lithification. The Dahomey Basin is an upper Cretaceous to Pliocene sedimentary basin found on the continental margin of Guinea Coast of West Africa. It extends from Southwestern Ghana in the West through the Republic of Togo and Benin into the Southwestern flank in Nigeria (Niger Delta), as far the Okitipupa ridge [1,2]. In Nigeria, we have the south-eastern part of the basin extending from present day Ogun State in the Southwest to Agbabu-Okitipupa in the present day Ondo State. This basin lies approximately between longitude 1°E and 6°E and latitude 5N and 8°N. It is narrow and parallel to the coast line. About 40% of the basin contains 10,000-12,000 of Cretaceous and Tertiary rocks lie within Nigeria. The axis and the thickness of the sediment in the basin occur slowly to the west by faults and other tectonics structures associated with the land-ward extension of fracture zone [3]. This basin sequence, comprise mostly of dark-grey shale, grey-whitish limestone, with little of siltstone as described by Omatsola and Adegoke, Adegoke et al. and Billmam [4-6].

The area of study lies on the topographic sheet of Nigeria sheet 24 (North East) published in 1982 by Federal Geological Survey. The Ewekoro quarry which is approximately 60 km Northwest of Lagos latitude 648°N-635°N and longitude 335°E-330°E (Figure 1).

The Ewekoro quarry falls within the tropical rain forest of the sub-equatorial southwestern region, Nigeria. It experiences "both wet and dry" seasons in a year, which is a typical nature of Nigeria climate.

The average relative humidity of the study area is between 75-95% while the average mean annual rainfall and mean monthly temperature are about 1,500-2000 mm and 22°C-22.5°C respectively.

The Cretaceous sediments of Dahomey basin are partly cut off from the sediment of Niger Delta basin again supported ages assigned to the sequences, EwekoroAkinbo and other Tertiary formations range in age from Paleocene-middle to upper Eocene and the sediments are deposited in both marine and transitional environment. The formation of Ewekoro proved to be of economic importance as it being quarried for a production like cement by Lafarge (West African Portland Cement Company) Ewekoro and Dangote group Cement Company.

Several studies have been carried out on the Cretaceous-Eocene sediments of Dahomey basin particularly on micropaleontology, hydrogeology of Ewekoro shale and limestone. In early 1960's, the first work carried out by Parkinson on the post Cretaceous stratigraphy of southern Nigeria has made ways for more intensive work on the Ewekoro formation of the Western Dahomey basin. Other contributors are: Reyment, Coker, Deklsz, Eames and Fayose [7-11] have all provided valuable knowledge of the basin and it formations.

The following micro fossils so far found in the different sedimentary sequences by different authors like Reyment, Bergree, Adegoke et al. and Adegoke [12-15]; include *Globorotalia pseudo bulloides, G. Velaseoesis, G. acuta, Globigerina triloculinoides* assemblages, Reyment [13] also listed a typical Paleocene Ostracods fauna. There are also faunas like; molluscas, corals, crinoids, crustaceans, pelagic and planktonic foraminisfera [15]. These fauna assemblages have been very useful in dating the different sedimentary rocks of the Dahomey basin, as well as in the reconstruction of paleoenvironment of deposition.

Publications like Fayose, Folk, Kogbe, Frankle and Gemeraad

---

*Corresponding author: Oladosu YC, Department of Geology, Ekiti State University, P.M.B 5363, Ado, Ekiti State, Nigeria E-mail: yemidosu10@gmail.com

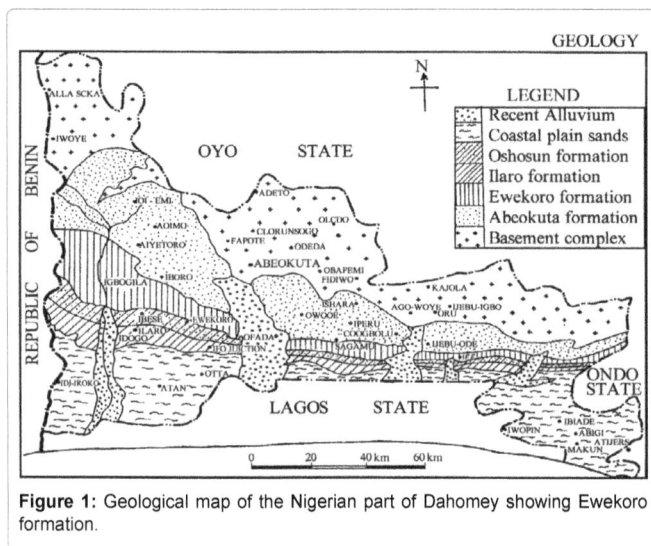

Figure 1: Geological map of the Nigerian part of Dahomey showing Ewekoro formation.

[16-20] who worked on the dolomitization of the Ewekoro formation has also been helpful. The tectonic evolution and Cretaceous stratigraphy of the Dahomey basin and the subsequently opening of the Atlantic Ocean by Omatsola and Adegoke [4]; had shed more light on the structural framework of the basin. Therefore, this study is aimed at presenting a vivid but comprehensive report of Ewekoro's shale and limestone in Dahomey basin from the foraminifera point of view. In the study, geological information of the sediments such as; age, depositional history, stratigraphic equivalence, textural and mineralogical characteristics, correlation and environmental deposition are documented based on micro fauna assemblages.

Material and Methods

The samples used for this study were obtained from a freshly exposed surface in Ewekoro quarry in Ogun state. Three samples were collected from freshly exposed surface to avoid picking weathered and contaminated sample. The sample were taken at an average interval of 2 m from the bottom of the well upwards to prevent stratigraphic leakages i.e. fragments of overlying strata will not contaminate the underlying strata. This fresh sample taken were immediately kept and sealed in polythene bags and labeled indicating the lithology. The samples collected were studied with the binocular microscope. The binocular microscope was used because of the physical properties of the samples are not visible on hand specimen. But, with the aid of the binocular microscope, the color, grain size, roundness, sorting etc. of each sample are determined. The process of sample preparation for micro paleontological study involved weighing, soaking, washing, sieving, drying and labeling of samples. About 40 g of each sample were weighed on the beam balance and observed in small tray under the binocular microscope. From this, the description on the lithology and the fossil content were made. Each of the weighed samples was crushed and soaked in hydrogen peroxide at room temperature for about 48 hours to disaggregate or dislodge the sediments from the fauna content present. After the time had elapsed, the hydrogen peroxide was decanted and the samples were again soaked in detergent were then washed, on at a time, under a jet of running water using the 63 μm and 73 μm sieve. The residue of each sieve was properly washed with water, placed in well-labeled plastic container and left to dry. Fossil were picked out from this dried sample at a magnification of 70X under the binocular microscope. The picked fossils of each sample were put in a labeled slide for a detailed identification and drawing.

Results and Discussion

Lithologic description

The Ewekoro shale and limestone has been studied using binocular microscope. The shale is dark-grey in colour, fine grained with high fossility; consist of carbonate stain, well sorted with shell fragments. The gluconitic shale is greenish in color, fine to medium grained, sub-angular and moderately sorted and not friable, while the limestone is whitish in color, consist of carbonate shell and quartz grains, sub-angular and moderately sorted and fine to medium grained. Table 1 show the lithologic description of the samples under binocular microscopes and this indicates that, the shale unit with aforementioned textural appearance shows it must have been transported from a very long distance before deposition in its present site. Attrition and abrasion capacity during the course of transportation are responsible for the sub-angular form of the limestone. The degree of sorting of the sediments which vary from moderately sorted to well-sorted is an indication that the sediments were deposited in quiet, low energy environment which allow sediments of similar size to accumulate together. While the basal limestone, were deposited in a disturbed environment.

Moreover, two major lithologies were recognized in the stratigraphic interval of 1-15 m of Ewekoro quarry. The two lithologies are the upper layers shale and the bottom layers limestone. The upper layers shales are very fossiliferous and the following foraminifera were recovered: *Lenticulina degolyeri*, *Globorotalia*, *Pseudotopilensis*, *Cibicides hampdenesis*, *Eponides pseudo elivatus*, *Globigerina angulisuturalis*. While the bottom layered limestone consist of *spirifer*, *cenoceras* sp., *pelecypoda*, calcareous broken shells. Also, in the recovery foraminifera *Globogerina yeguaensis*, *Globorotalia* sp., *Lenticulina degolyeri* and *cibicides* sp., are in abundance and also *Globorotalia* sp. are very common on all shale of the unit of the quarry. This foraminifera assemblage is similar to the assemblages of kalambina and Dange formations of the Ilumeden basin (Sokoto basin), which are dated Paleocene age.

Foraminifera assemblages

The samples collected from Ewekoro quarry at different intervals were prepared for foraminifera studies. Therefore, Table 2 reveals that the fauna recovered in the samples were identified based on their characteristic features (chamber shape and sculptures), which have varied appearance of the resistant outmost wall. The micro fauna are fairly high in abundance and diversity in the samples. Though, few samples are very poorly fossiliferous. The recoverable fauna are majorly planktonic foraminifera especially at the upper part of the sequence. This could be as a result of the depositional environment.

Geologic deduction from foraminifera assemblages

The various form of foraminifera recovered from samples collected at Ewekoro quarry of Dahomey Basin can be compared to be similar the

SAMPLE NUMBER	DEPTH (M)	LITHOLOGY	DESCRIPTION
S3	15	Shale	Fine grained with light grey in color, high fossility with carbonate stain, fine grained, well sorted.
S2	13	Glauconitic shale	Greenish in color, fine grained and not friable.
S1	12	Limestone	Whitish in color, presence of carbonate shale and quartz grains, sub-angular and moderately sorted, fine to medium grained.

Table 1: Lithologic description of samples under binocular microscopes.

Globorotalia sp, *Cibicides hampdenesis*, *Globigerina linaperta* of kerikeri formation of Chad basin [21]; which is dated Paleocene age. Also, the foraminifera assemblages of the Gambia, Kalambania and Dange units of Sokoto basin which include; *Globigerian trilocularis*, *Globigerina praebulloidis* [22]. They are all indicative of Paleocene age. Therefore, based on the written information, the rock samples studied for this work is here dated Paleocene age because of the strong similarity with work of Kogbe and Reyment [13,18].

Geological age

The geologic age of the sequence of Ewekoro formation of the Ewekoro quarry is dated Paleocene age (Figure 2). The age is characterized by the planktonic forams (*Globigerina* sp., *goborotalia* sp.) and bentonicforams (*Cibicides* sp., *Ammonia beccarii*) all dates Paleocene [15,23-25].

Foraminifera description systematics

Family: Globorotaliidae

Description: Test trachoids, earliest chambers often like *Globogerina* with a rough cancelled exterior, biconvex, dorsal side more or less flattened, ventral side strongly convex, wall calcareous,

SAMPLE NUMBER	DEPTH (M)	RECOVERABLE FORAMINIFERA
S3	15	*Cibicides* sp. *Globigerina yeguaensis*
S2	13	*Globigerina ciperroensis* *Lenticulina degolyeri* *Globorotalia* sp. *Globigerina yeguaensis* *Globigerina angulisuturalis* *Cubicdoides* sp.
S1	11	*Spirifer, Cenoceras* sp., *Pelecypoda Calcalreous*, broken shells

Table 2: Showing the recoverable foraminifera in different samples with depth in Ewekoro Quarry.

perforate frequently spinose in whole or in restricted area aperture large, opening into the umbilicus which is either open or particularly cover by a lip.

Occurrence: Cretaceous to recent

Family: Lagenidae

Genus: *Lenticuluna*

Description: Test similar to Robulus, tending to become uncoiled in the same species, the apertures radiate, at the peripheral angle, the silts equal.

Occurrence: Permian-recent

Family: Globigerinidae

Genus: *Globigerina*

Description: Test trichoid throughout, umbilicate, chambers in the young especially of the micropheric form in a flattened trachoid from like discorbis, usually smooth and the wall thin, later chambers globular, wall calcareous, thick and cancelled, in the well preserved especially pelargic specimens, clothed with long slender spines coming from the angles of the cancellation surface areas, the base of such areas with the pores of the wall; aperture large, opening into the umbilicus.

Occurrence: Cretaceous-recent

Family: Anomalinnidae

Genus: *Cibicides*

Description: Test plans convex, trochoid, usually attached by the flattened dorsal soidi, wall calcareous, coarsely perforated aperture pheripherical at the base of the chamber sometimes extending ventrally, but extension dorsally between the inner margin of the chamber and the previous whorl nearly the length of the chamber.

Occurrence: Jurassic, Cretaceous-recent

Paleo-environment of deposition

The Paleo-environment of the studied interval is generally marine.

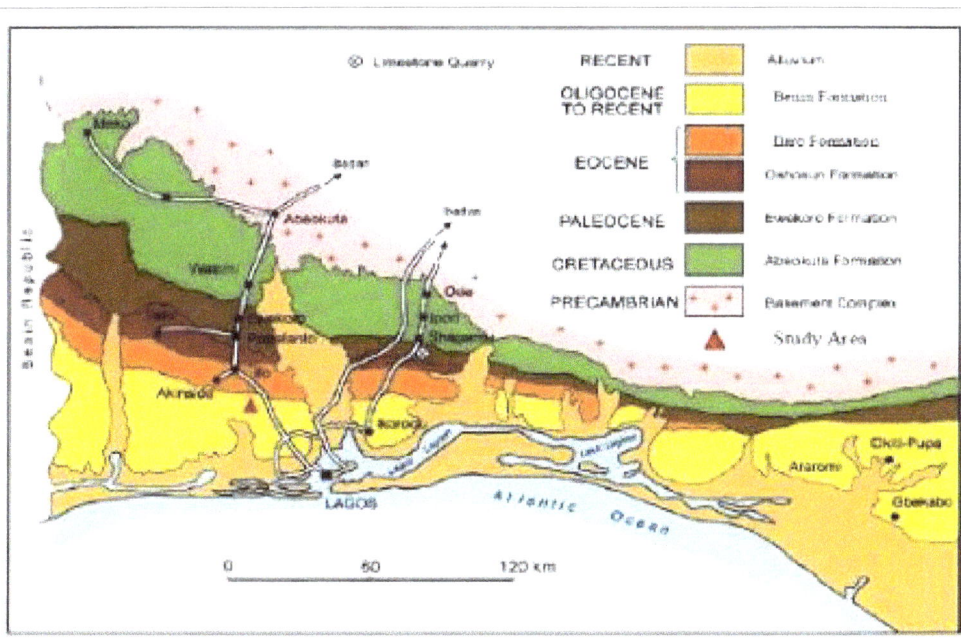

Figure 2: Geological map of the Nigerian part of the Dahomey embayment.

This characterized by the type of lithology (limestone and shale) and the derived bentonicforams. The lithological sequence varies from limestone at the base to the light grey shale at the top. The Paleobathymetry could vary slightly but all deposited in a marine environment.

Conclusion and Recommendation

The textural and lithological appearances of Ewekoro shale dark-grey colored with fine grained sizes and limestone, which is yellowish brown with sub-angular, sub- rounded shape. This indicates that the basal limestone of the sequence was deposited under disturbed condition while the shale sequence was deposited in a low energy environment. The mircopaleontological analysis carried out on the samples yielded diagnostic assemblages of foraminifera. The recoverable foraminifera include *Globigerina yeguaensis*, *Globigerina pseudotopilensis*, *Cibicides* sp., *Lenticulin adegolyeri*, and *Globigerina angulisuturalis*. All these microfauna are dated paleoenvironment of deposition is thought to have been deposited in a continental environment with a later incursion of marine water. The marine setting Deepings downward from the shale sequence at the top to the carbonate deposit at the bottom.

As a result of these, the following recommendations are hereby suggested to the government and to the various oil films:

1. The Ewekoro sediment is of high economic value and therefore the area should be explored in detail.

2. The presence of Benin formation and some other fossils and the occurrence of favorable rock types associated with petroleum may indicate the presence of crude oil.

References

1. Jones H, Hockey RD (1964) The geology of the part of south west Nigeria. Geol Surv Nig Bull 31: 1-87.

2. Antiloni P (1968) Eocene phosphate in Dahomey basin. J Min Geol 3: 17-23.

3. Adegoke OS et al. (1980) The bituminous of Nigeria. Nig Min Geosci 227-378.

4. Omatsola ME, Adegoke OS (1981) Tectonic evolution and stratigraphy of the dahomey basin. J Min Geol 18: 130-137.

5. Adegoke OS (1970) In: Dessauvagie TFJ, whiteman AJ (Eds.) Microfauna of the Ewekoro formation (patidence) of south-western Nigeria. African Geology University of Ibadan Press pp269-276.

6. Billmam HG (1976) Offshore stratigraphy and paleontology of the dahomey. Embayment proceeding 78th Africa micropaleontology College, Ile-Ife (University press).

7. Reyment RA (1964) Review of the Nigeria cretaceous Stratigraphy. J Min Geol 2: 61-80.

8. Coker SJL, Ejedawe JE (1987) Petroleum prospects benin basin Nigeria. J Min Geol 23: 27-42.

9. Deklsz, Chene J (1978) Biostratigraphical study of the borehole ojo -1 south west Nigeria. Rene de micropal 21: 123-139.

10. Eames Fe (1957) Eocene mollusca from Nigeria. A revision brit. Mus (nathist) bull 3: 25-70.

11. Fayose EA (1970) Stratigraphical paleontology of Afowo – 1 –well, south western Nigeria. J Min Geol 5.

12. Reyment RA (1963) Aspects of the geology of Nigeria. Ibadan University press, p: 133.

13. Reyment RA (1964) Quantitative Paleoecologic analysis of Ewekoro and oshosun formation of Western Nigeria. Geol Foren Stockh Forhand 86: 248-256.

14. Bergreen WA (1960) Paleocene biostratigraphy and planktonic foraminifera of Nigeria (West Africa). Proc.21st Interna Geol p41-55.

15. Adegoke OS (1969) Eocene stratigraphy of southern Nigeria. extract dememries du bureau durecherdedseologique Min 67: 23-40.

16. Fayose EA, Azeez LO (1972) Micropaleontological investigation of Ewekoro area, Southwestern Nigeria. Micropal 18: 369-385.

17. Folk RL (1974) Some aspects of recrystallization in ancient limestones. Dolomitiz Limes Diagen 13: 14-18.

18. KOGBE CA (1974) The Upper cretaceous Abeokuta formation of south west Nigeria. Nigerian Field 39: 4.

19. Frankle EJ, Cordry EA (1967) The Niger Delta oil province recent development onshore and offshore. Proc. 7th ward petrol congress 2: 195-309.

20. Lawal O (1982) Biostratigraphic palynologicqueet paleoenvironments des formation cretaceesdela haute Benue, Nigeria Mand oriental these cycle. Univnice p.219.

21. Kogbe CA (1976) Some paleocene corals from Ewekoro Southwest Nigeria. J Min Geol 13: 1.

22. Jardine S, Magloire J (1965) Palynology and Stratigraphic ducreater des basin du senegalitdecotedivoremem. Bur Rech Geol Min 32: 189-245.

23. Gemeraad JH, Hopping CA (1968) Palynology of tertiary sediment from tropical areas. Rev palaeobot palynol 6: 3-6.

24. Lawal O, Moullade M (1986) Paynological biostratigraphy of cretaceous sediments in the upper Benue basin. Rev Micropaleo 29: 61- 83.

25. Omatsola, Doust (1990) In: Edwards JD, Santogresi PA (eds.) Summary and conclusion. Divergent passive margin Basins. AAPG Bull 68: 390-394.

23

Petro-geochemistry, Genesis and Economic Aspects of Mafic Volcanic Rocks in the West and Southern Part of The Mamfe Basin (SW Cameroon, Central Africa)

Nguo Sylvestre Kanouo[1]*, Rose Fouateu Yongue[2], Tanwi Richard Ghogomu[2,3], Emmanuel Njonfang[4], Syprien Bovari Yomeun[2] and Emmanuel Archelaus Afanga Basua[5]

[1]*Mineral Exploration and Ore Genesis Unit, Department of Mines and Quarries, Faculty of Mines and Petroleum Industries, University of Maroua, Cameroon'*
[2]*Department of Earth Sciences, University of Yaoundé I, Cameroon*
[3]*Department of security, Quality and Environment, Faculty of Mines and Petroleum Industries, University of Maroua, Cameroon'*
[4]*Higher Teachers Training School, University of Yaoundé I, Cameroon*
[5]*Department of Earth Sciences, China University of Geosciences wuhan, China*

Abstract

Geologic prospecting, petrographic and geochemical analyses of mafic volcanic exposures in the west and southern part of the Mamfe Basin (SW Cameroon) distinguishes: basanites, picro-basalts, alkali basalts and tholeiitic basalts. They are relatively LREE-enriched, undersaturated, saturated or oversaturated due to presence or absence of normative nepheline, hypersthene or quartz. Basanites mainly form pillow-like lavas, and are aphyric or porphyritic. They have significant concentration of Ni (up to 387 ppm) and Ba (up to 436 ppm). These alkaline rocks cooled from less evolved mantle source magma. Picro-basaltic fragments exclusively found in the western part of the basin are Ni (up to 259 ppm) Ba (up to 2090 ppm) -enriched porphyritic, alkaline or subalkaline rocks. They also cooled from less evolved mantle source magma. Basalts form volcanoclasts, flow and dykes. They are aphyric or porphyritic, alkaline, transitional or subalkaline. Some of these rocks are Al-enriched. They crystallized from variably evolved mantle source magma within the Oceanic Island Basalt and Continental Rift Basalt tectonic settings.

Keywords: Cameroon; Mamfe basin; Mafic volcanic rock; Geologic prospecting; Petrography; Geochemistry; Alkaline; Tholeiitic

Introduction

Mafic volcanic rocks are melanocratic to mesocratic igneous rock which crystallized from a basaltic magma within the oceanic and continental crust [1-4]. Rocks from basaltic magma are varied with different petrographic, physical and geochemical features whose determination lead to their characterization and to a better understanding of their condition of formation [1,2,5-7]. They occur as aa, pillow, columnar or pahoehoe lava flow, or form dyke and sill which are easy to recognize in the field [2,7-9]. Some of these rocks are mineralized (e.g., host interesting PGE concentration: or can host gemstones as xenocrysts and in xenoliths sourced from other rocks by the ascending magma) [10-13]. Sulfide metallic mineralizations are associated with pillow basalts in the Yaeyama Central Graben, Southern Okinawa Trough, Japan [14]. Native iron is found in some basaltic rocks in Buhl basalts (Kassel Germany) and Siberian flood basalts [15-17]. FeO-Al_2O_3-TiO_2 rich rocks are found associated with transitional tholeiitic lava flows in the Tertiary Bana plutono-volcanic complex in the continental sector of the Cameroon Volcanic Line [18]. Peridotite xenoliths with significant Ni-Co concentrations are hosted in basaltic lavas in Kumba (SW region of Cameroon) [19]. Basaltic rocks can be mined as industrial mineral used for building and road construction [20,21]. Thus, mafic rocks have both scientific and economic interests.

The Mamfe Sedimentary Basin in the SW region of Cameroon (Figure 1) encloses mainly uncharacterized mafic flows, fragments, and dykes found in both the sedimentary part of the basin and some rocks of the basement [22]. These mafic rocks crop out in the west, south and south-east of the Manyu Division, which hosts the basin. Most of these rocks are recorded as basaltic with no specific identification. Although some of those rocks were petrographically characterized, information is limited for most of them. Wilson and Dumort presented a field reconnaissance report and constructed a map, respectively [23,24].

Kanouo presented field data and major element geochemical feature for mafic volcanic flow and volcano clasts found in the western part of the basin [22]. Much still has to be done to identify and characterize these rocks and verify if they have any interests in mining industries. Gem corundum and coarse zircon are found in detrital sediments (in Nsanaragati) at about 10 km east of the nearest basaltic flow (in Ekok) [22,25]. Although not yet confirmed, the presence of diamond within sedimentary clasts in the western part of Mamfe Basin is mentioned in Laplaine and Soba [26]. Studies carried out on corundum and zircon show that they are largely from magmatic crystallizations with some zircon being from kimberlitic origin [22,27-30]. Research studies carried out in South-East Asia, Australia and in many gem corundum fields in the world have shown that, alkali basalts can host corundum as xenocrysts or in xenoliths [31,32]. The origin of corundum found in the western part of the Mamfe Basin is still debated. During this research study a series geological prospecting was carried out in many volcanic terrains and localities found in the Mamfe Basin in order to locate gem or metallic host rocks.

In this paper, we present field data, petrographic and geochemical results for mafic volcanic rocks found in the west and southern part of the Mamfe Basin in order to characterize those rocks, understand the history of their genesis and present their economic aspects.

*Corresponding author: Nguo Sylvestre Kanouo, Department of Mines and Quarries, University of Maroua, Cameroon
E-mail: skdasse@gmail.com, sylvestrekanouo@yahoo.fr

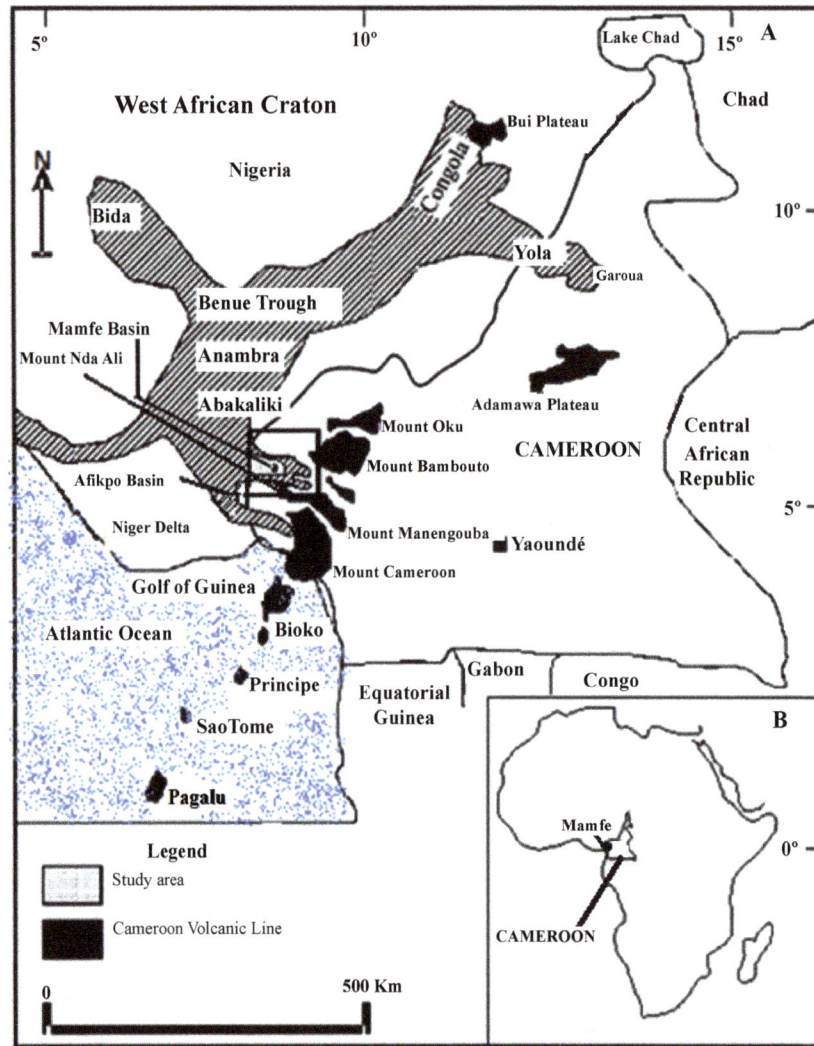

Figure 1: (A) and (B) location of the Mamfe sedimentary basin in the regional context [76,77].

Geography and geologic setting

The Mamfe Sedimentary Basin located between 5°30' to 6°00'N and 8°15' to 9°45' E underlies a coastal plain in Cameroon with low to slightly high relief whose heights range from 30 to 300 m [22,24]. It is locally bordered by high igneous terrains (e.g., Mount Nda Ali: 1200 m, Mount Mbinda: 1000 m, Nkogho hills: up to 600 m). The basin is regionally bordered by upland areas (Figure 1) (e.g., Mount Rumpi, Bambouto, Bamenda, Manengouba and Koupé), which are part of the Cameroon Volcanic Line [33,34]. The Mamfe Basin is administratively situated in the Manyu Division made up of four sub-divisions (Mamfe Center, Eyumojock, Upper Bayang, and Akwaya) and occupied by three main ethnic groups: the Kenyangs, Akwayas, and Ejagham [22]. The climate in this Division is hot and humid and consists of a rainy and a dry season modified by the deviation of the monsoon and the relief of Mount Cameroon [35]. The vegetation is dominantly that of the equatorial rain forest [22]. The drainage system is principally that of the Cross River (Figure 2) whose main source is found in Mount Bambouto [36]. The sources of its main tributaries the Munaya and Badi Rivers are at the Mount Rumpi and Nda Ali respectively [22].

The Mamfe Basin is an assigned Cretaceous age intracontinal sedimentary filled depression whose opening is related to the Atlantic Ocean and is one of the south eastern branches of the Benue Trough in Nigeria [24]. It is suggested to be genetically related to that of the Benue Trough as they both have similar structures [24,37]. The Mamfe Sedimentary Basin is filled by very fine to very coarse-grained immature and mature siliciclastic lithified sediments (e.g., conglomerates, sandstones, arkoses, shales and mudstones) [22,37-42]. These are locally associated with evaporites [23,24,43-45]. Part of these rocks (e.g., shales in Bachuo Ntia and Etoko) are good potential hydrocarbon sources [42]. Sedimentary rocks in the west of the Mamfe Basin are locally overlain by recent detritus enclosing gemstones of magmatic and metamorphic origin [22,28,30]. Lithified sediments in this basin are also locally overlain or cross-cut by basaltic, phonolitic, or trachytic rocks probably formed from post sedimentary volcanism [22,23,46,47]. These exposures are underlain by assumed Precambrian age basement made up of granites, gneisses, migmatites, syenites, and mica-schists (Figure 2) and undated gabbros, diorites, syenites, and monzonites at the Mount Nda Ali [22-24,46,48]. Part of the Precambrian rocks encloses polycrystalline, recrystallized and fractured quartz, and kinked biotite probably arising from post-crystallization cataclastic deformation [22].

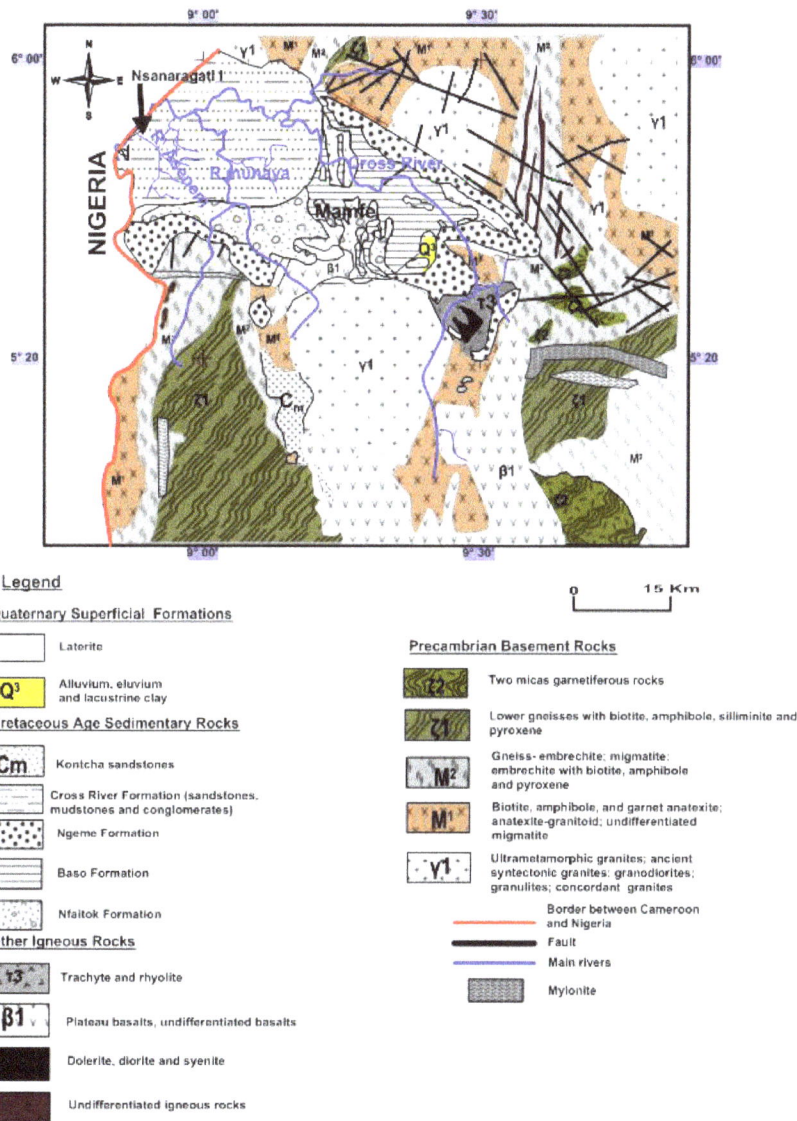

Figure 2: Sketch geological map showing Asenem River flowing on sedimentary formations in the western part of the Mamfe Basin; modified from Dumort [24] and Eyong [39].

Field work and analytical methods

With the aim to prospect, locate and characterize the mafic rocks and any associated gemstone sources and metallic mineralization, detailed geologic surveys were carried out in western and southern Mamfe basin. A total of ten villages (Ekok, Nsanaragati, Otu, Araru, Nkogho, Mbiofong, Babi, Ogurang, Ossing, and Kembong) were surveyed, outcrops were described and more than one hundred rock fragments were collected from flows, dykes, alluviums, colluviums and eluviums. The sampled rocks were macroscopically and microscopically characterized, and their geochemical features determined. The description of the outcrops included: outcropping mode; colour, size, and shape (for fragments); width, length, strike and dip (for a dyke or sill); nature of the surface, width, and height (for flows); presence of any xenoliths and/or xenocrysts; type and nature of host and underlying rock; and the relationship between the mafic rock and surrounding geologic formations. The macroscopic description of collected fragments was based on criteria presented in and Winter [1,2,7].

More than fifty thin sections were prepared at the Geology and Mining Research Institute (Yaoundé-Cameroon) and later characterized under the petrographic microscope (EUROMEX) at the Department of Earth Sciences of the University of Yaoundé I. Particular attention was paid to the presence of xenocryst during the macroscopic characterization. The characterization was based on parameters such as texture, colour of minerals, shape, weathering and mineral replacement, relief, extinction angle, interference colour as defined in Mackenzie et al., Mackenzie and Adams, Mackenzie and Guilford, Higgins and Beaux et al. [3,49-52].

Twenty samples were analyzed to determine their elementary abundance in China and South Africa. The major and minor elemental concentration in seven samples (NKK3, NKK4, NKK5, NKK7, KEG, TAE, OSG) was determined by an X-ray fluorescence spectrometer in China. The analytical procedures used to acquire these results are similar to those presented in Wu et al. [53]. Thirteen selected samples (e.g., NSI1, BAI3, BAI4, ARK16) were analyzed for their major, minor, trace and rare earth elements contents at ALS Chemex South Africa (Pty) Ltd, Johannesburg Gauten. These results were obtained by ICP-AES (Inductively Coupled Plasma-Atomic Emission spectroscopy) and ICP-MS (Inductively Coupled Plasma-Mass Spectroscopy). Major, minor (with Loss on Ignition) and selected trace element contents (e.g., Pb, Mo, Cd, As, Cu, Zn, Sc, and Ag) were determined by ICP-AES. A prepared sample (0.200 g) was added to lithium metaborate/lithium tetraborate flux (0.90 g), mixed well and fused in a furnace at 1000°C. The resulting melt was then cooled and dissolved in 100 ml of 4% nitric acid/2% hydrochloric acid. This solution was then analyzed by ICP-AES and the results were corrected for spectral inter-element interferences. Oxide concentration was calculated from the determined elemental concentration and the result was reported in that format. If required, the total oxide content was determined from the ICP analyte concentrations and loss on Ignition (L.O.I.) values. A prepared sample (1.0 g) was placed in an oven at 1000°C for one hour, cooled and then weighed. The percent loss on ignition was calculated from the difference in weight. Base metals can be reported with the ME-MS81 by a four acid digestion but not all elements were quantitatively extracted. A prepared sample (0.25 g) is digested with perchloric, nitric, hydrofluoric and hydrochloric acids. The residue was topped up with dilute hydrochloric acid and the resulting solution was analyzed by ICP-AES. Results were corrected for spectral inter-element interferences. Most of the trace elements (e.g., Ba, Cr, Ga, Hf, Ta, Th, U, V, Sn, Rb, Sr, Zr, and Y) and all the REE were quantified by ICP-MS. A prepared sample (0.200 g) was added to lithium borate flux (0.90 g), mixed well and fused in a furnace at 1000°C. The resulting melt was then cooled and dissolved in 100 ml of 4% HNO_3/ 2% HCl solution. This solution is then analyzed by inductively coupled plasma mass spectrometry. Results for Co and Mo were not obtained by this method.

Results

The mafic rocks are first classified using the analytical results. The named rocks are then described on field nature, macroscopic and microscopic petrography and major and trace element (including REE) results.

Nomenclature and field description

Nomenclature: The major element data as oxides (Table 1), were plotted on the total alkali versus silica diagram with a subalkaline-alkaline dividing line, for each rock's classification (Figure 3) [54]. A few plots lie in the sub-alkaline field and on the dividing line, while most have alkaline affinity. The major element compositions plotted on Le Bas et al. binary total alkali versus silica diagram (Figure 3) shows that the samples range from basanite to picro-basalt, and basalt [5].

Field description: Basanites crop out at the center of Ekok, Babi, and Nkogho. They form dark grey and massive lava flows with some being much more vesicular than others. At Ekok, two generation of flows include a vesicle-enriched flow underlying a less vesicular flow (Figure 4a and 4b). The contact between the flows is partly made up of well cemented pebble to cobble size sedimentary clasts (Figure 4b). The vesicle-enriched flow locally overlies a less consolidated conglomerate. The upper flow has a less developed top soil. In Nkogho, outcropping pillow lava partly overlies granitic bedrock. The surface of each flow is uneven and full of regular and irregular cracks similar to those of vesicle-enriched basanite at Ekok. Basanite at Babi occurs as scattered fragments within soils around some hills. Picro-basalts outcrops south west of Ekok and show variable shape with smooth blocky fragments overlying fine to coarse-grained sandstones (Figure 4c). Fragments are scattered in reddish clastic soil found on the only hill (up to 200

m high). Basalts form dykes, fragments (in alluvium, colluvium and/or eluvium) and/or flows in Nkogho, Babi, Araru, Nkogho, Kembong, Talangaye, Ossing, and Nsanaragati. A NE-SW trending vertical dyke more than 30 m length and up to 50 cm width cross-cuts granitic basement SE of Nkogho (Figure 4d). It is dark grey and massive with vesicles locally filled with white matter and is separated by a very thin dark layer, a probable product of a contact metamorphism. Just west and north, other basalts include volcanoclasts (colluviums and eluviums) associated with granitic fragments and sandstones in streams, while to the North West these clasts overlie granitic pegmatite and sandstone. Volcanoclasts also appear NW of Asenem River (in Nsanaragati). Variable shaped basalt fragments are widespread in Nkogho and Ossing, and locally found in Talangaye and Babi, while basaltic lava flow localities found in Ossing, Talangaye, Kembong, and Araru. They form a patch on granitic basement in Kembong, and overlie mica-schist in Araru, but contacts between the lava and surrounding rocks were not visible in Ossing and Talangaye.

Petrography: In Ekok, vesicles are less 10% in vesicle-depleted basanites (Figure 5a) and more than 30% in vesicle-enriched basanites

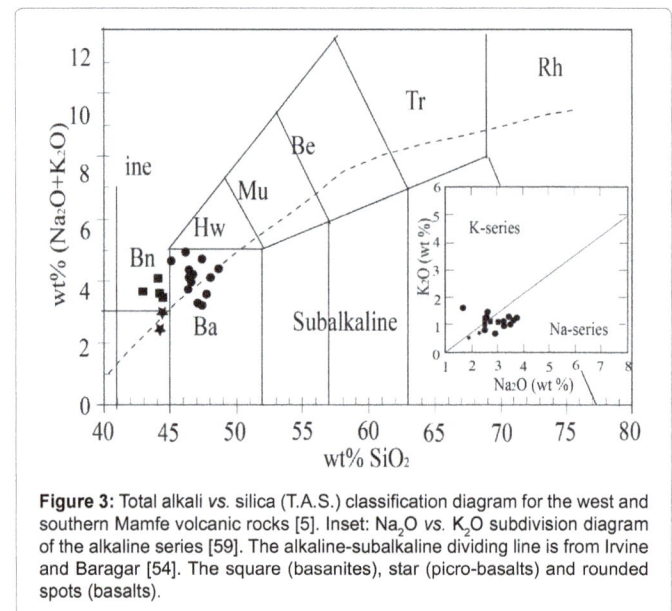

Figure 3: Total alkali *vs.* silica (T.A.S.) classification diagram for the west and southern Mamfe volcanic rocks [5]. Inset: Na_2O *vs.* K_2O subdivision diagram of the alkaline series [59]. The alkaline-subalkaline dividing line is from Irvine and Baragar [54]. The square (basanites), star (picro-basalts) and rounded spots (basalts).

Figure 4: Mafic volcanic rocks outcrops in the west and southern part of the Mamfe Basin (a and b: basanites in Ekok; c: picro-basalt in Ekok; d: basaltic dyke in Nkogho).

Sample composition	Basanites		Picro-basalts						Basalts											
(wt.%)	NKK3	BAI3	EKK1	EKK3	EKK6	EKK11	NSI2	NSI3	NKK1	NKK2	NKK4	NKK5	NKK6	NKK7	NKK10	ARK16	KEG2	TAE2	OSG	BAI4
SiO_2	42.95	44.1	44.4	44.2	44.4	44.2	46.4	46.3	46.4	46.7	46.14	46.36	48	47.4	48.6	47.4	47.05	45.06	46.58	47.7
TiO_2	3.55	2.54	2.27	2.31	2.28	2.27	2.32	2.32	2.74	2.65	3.27	3.15	2.49	3.2	2.74	0.8	2.73	3.08	2.94	3.54
Al_2O_3	12.4	12.4	13.05	13	13.5	13.3	13.45	13.3	13.85	13.7	13.48	13.56	13.6	13.97	14.7	15.35	15.35	13.81	13.32	14.05
Fe_2O_3*	13.97	12.55	12.6	12.4	11.85	12.05	12.15	11.95	12.8	12.7	13.75	13.32	12.9	12.98	10.75	12	12.43	13.32	13.57	13.5
MnO	0.19	0.18	0.2	0.18	0.18	0.16	0.19	0.18	0.16	0.16	0.17	0.17	0.16	0.19	0.17	0.22	0.16	0.18	0.25	0.17
MgO	10.78	13.15	11.4	11.35	10.2	11.3	9.58	10	8.97	8.5	8.05	8.2	8.45	7.07	7.06	8.2	5.79	8.38	7.63	5.49
CaO	9.06	8.64	9.98	9.87	10.05	9.64	9.45	9.28	9.22	9.29	9.31	9.24	9.08	9.32	9.91	10.5	9.19	9.51	9.54	9.07
Na_2O	2.48	3	2.47	2.53	2.29	1.87	3.17	2.54	2.98	3.14	3.72	2.93	3.19	3.55	3.44	1.66	2.49	3.39	2.59	2.89
K_2O	1.18	1.08	1	1.06	0.7	0.54	1.2	1.22	1.1	1.09	1.2	1.19	0.94	1.16	0.98	1.55	0.78	1.27	1.4	0.67
P_2O_5	0.69	0.69	0.57	0.6	0.59	0.52	0.55	0.55	0.56	0.55	0.52	0.55	0.48	0.58	0.5	0.08	0.51	0.74	0.62	0.45
Cr_2O_3	-	0.09	0.05	0.06	0.05	0.05	0.06	0.06	0.03	0.03	-	-	0.04	-	0.04	0.04	-	-	-	0.01
LOI	2.62	1.36	1.56	1.96	2.87	3.5	0.93	2.63	1.33	1.48	0.48	1.13	1.11	0.47	1.25	2.02	3.07	1.1	1.31	2.02
Total	99.87	99.79	99.56	99.53	98.99	99.42	99.47	100.35	100.15	100	100.09	99.8	100.46	99.89	100.16	99.82	99.55	99.84	99.75	99.56
Al_2O_3/TiO_2	3.49	4.88	5.74	5.63	5.92	5.86	5.78	5.73	5.05	5.17	4.12	4.3	5.46	4.37	5.36	19.19	5.62	4.48	4.53	3.97
Na_2O/K_2O	2.1	2.77	2.27	2.39	3.27	3.46	2.64	2.08	2.71	2.88	3.1	2.46	3.39	3.06	3.51	1.07	3.19	2.67	1.85	4.31
$(Na_2O+ K_2O)-(0.37*SiO_2-14.43)$	2.2	2.19	1.47	1.67	0.99	0.49	1.63	1.06	1.34	2.78	2.28	1.4	0.8	1.6	0.87	0.1	0.29	2.42	1.19	0.34
Mg#	56.51	63.82	60.37	60.65	59.17	61.23	56.7	58.5	54.13	52.98	49.64	50.9	52.45	47.84	52.51	53.5	43.96	51.07	48.63	40.64

CIPW normative Wt (%)

	NKK3	BAI3	EKK1	EKK3	EKK6	EKK11	NSI2	NSI3	NKK1	NKK2	NKK4	NKK5	NKK6	NKK7	NKK10	ARK16	KEG2	TAE2	OSG	BAI4
Quartz	-	-	-	-	-	-	-	-	-	0.97	-	-	-	-	-	-	12.73	2	8.22	3.13
Orthoclase	6.38	6.38	5.78	5.78	5.83	5.74	3.83	6.52	6.67	0.97	5.86	6.36	5.12	6.23	5.31	8.41	4.78	7.6	8.4	3.61
Albite	18.55	18.55	16.33	16.33	20.61	16.36	14.74	23.22	19.9	23.17	24.17	22.41	24.87	27.29	26.69	12.9	21.84	22.26	22.26	22.29
Anorthite	17.59	17.59	15.56	15.56	21.27	19.23	24.49	17.39	19.7	19.45	18.25	18.38	18.44	17.04	19.97	27.41	29.44	18.95	20.91	21.32
Nepheline	0.36	0.36	3.61	3.61	1.76	1.76	2.97	0.77	-	-	2.96	-	-	-	-	-	-	-	-	-
Diopside	15.39	15.39	15.27	15.27	17.92	17.59	13.13	17.7	15.47	15.2	16.32	15.83	15.89	17.1	17.71	16.15	4.02	14.97	11.84	13.71
Wollastonite	-	-	-	-	-	-	-	-	-	-	-	-	-	-	-	-	1.91	1.24	-	-
Hypersthene	-	-	-	-	6.81	4.24	13.16	-	8.02	4.52	1.69	4.4	10.25	1.17	3.5	10.04	-	0.21	-	11.01
Olivine	15.54	15.54	21.15	21.15	4	17.55	9.95	12.14	7.93	8.66	11.23	8.62	2.82	7.34	6.7	7.54	-	-	-	-
Magnetite	18.54	18.54	16.48	16.48	18.02	16.47	16.27	17.05	17.05	17.05	16.75	17.45	17.23	17.09	14.29	15.98	18.68	19.99	19.99	17.84
Ilmenite	6.17	6.17	4.37	4.37	4.25	4.02	4.02	4.05	4.08	4.78	4.58	5.41	4.36	5.52	4.77	1.4	5.37	5.92	5.67	6.13
Apatite	1.46	1.46	1.45	1.45	1.3	1.27	1.27	1.17	1.18	1.19	1.16	1.15	1.02	1.22	1.06	0.17	1.22	1.74	1.46	0.95

Table 1: Major element abundance (in wt. %) and CIPW norm in mafic volcanic rocks from the west and southern part of Mamfe Basin (Manyu Division, SW Cameroon). [Mg#=100 × (MgO/40.31)/(MgO/40.31)+Fe$_2$O$_3$ × 0.8998/71.85 × (1-0.15), assuming Fe$_2$O$_3$ / Fe$_2$O$_3$+FeO)=0.15. Recalculated to 100% anhydrous; total iron is expressed as Fe$_2$O$_3$; LOI: Loss 0n Ignition. Sample Location, EKK: Ekok; BAI: Babi; NSI: Nsanaragati; NKK: Nkogho; ARK: Araru; KEG: Kembong; TAE: Talangaye; and OSG: Ossing].

Figure 5: Hand specimens for mafic volcanic rocks from the west and southern part of the Mamfe Basin (a and b: basanites; c and d: picro-basalts; e and f: basalts).

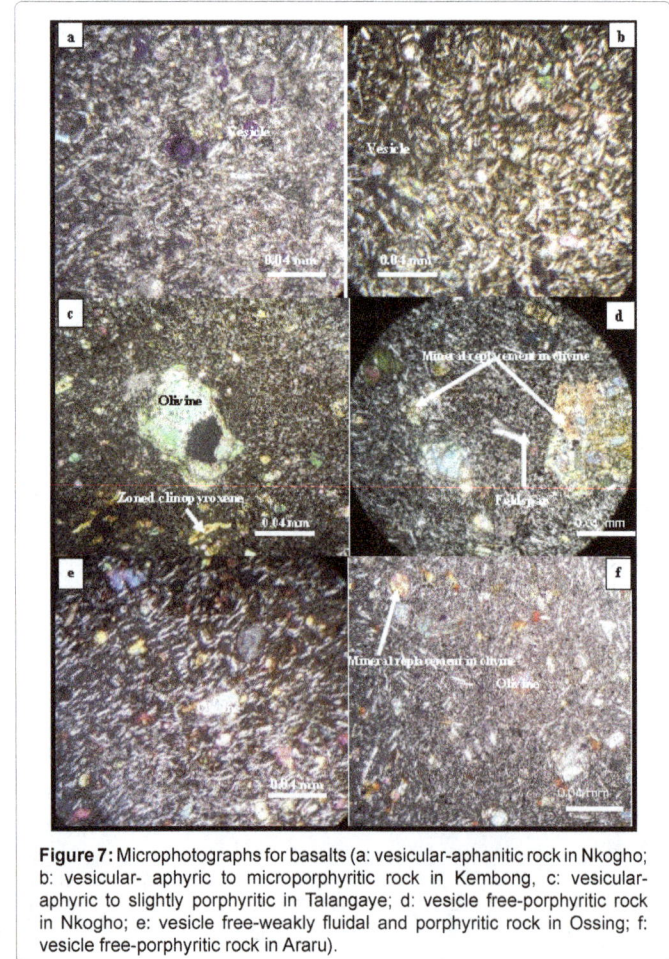

Figure 6: Microphotographs for basanite and picro-basalt (a and b: basanites; c: picro-basalt).

(Figure 5b). Vesicle diameter ranges from 0.5 mm to more than 1.0 cm (with a depth of up to 2 cm). In Nkogho, basanites are vesicle-depleted (<8% in each collected sample). Vesicle diameter in this rock is up to 5 mm, whereas the depth is below 2 mm. Basanites from Ekok are melanocratic, and aphanitic to slightly porphyritic. Plagioclase laths (Figure 6a) form 65% a very fine-grained groundmass, that includes olivine (17%) and clinopyroxene (8%) phenocrysts (anhedral to euhedral, >0.2 mm) and micro-phenocrysts (sub-hedral to euhedral, >0.01 mm) and opaque mineral (10%) grains (anhedral to subhedral, <0.01 to ≤ 0.03 mm in size). Basanite from Nkogho (Figure 6b) is melanocratic, aphanitic and composed of olivine, plagioclase, clinopyroxene and opaque minerals whose proportion is not easily estimated. Basanite from Babi has a very fine-grained groundmass dominated by plagioclase laths (up to 60%) and has a weakly fluidal structure. Olivine phenocryst (5%) and microphenocrysts (up to 25%), along with microlitic clinopyroxene and opaque minerals (10%) form the remainder of the rock. Picro-basalt is melanocratic when fresh (Figure 5c) and reddish-yellow when partly weathered. Some fragments enclose quartz fragments, and sparse olivine macrocrysts suggest a porphyritic texture (Figure 6c). Plagioclase laths (45%), dominant in the fine-grained groundmass, which hosts euhedral to anhedral fractured olivine (48%) phenocrysts (<1.7 mm) and microphenocrysts (0.03-0.07 mm), clinopyroxene (5%) are mostly sub-hedral microphenocrysts (up to 0.05 mm). Opaque minerals (3%) occur mostly as microlites (≤ 0.01 mm). Basalts are dark-grey, massive and very fine-grained. Some are vesicular while others are not (Figure 5e and 5f). Vesicular fragments are found in a dyke (south of Nkogho) and in patches within southern Kembong and Talangaye. Vesicular basalts from Nkogho are aphanitic (Figure 7a) and contain plagioclase (75%), olivine (10%), clinopyroxene (5%), opaque minerals (10%) and some irregular vesicles. Slightly aphyric to microporphyritic, vesicular fragments from Kembong (Figure 7b) have a similar mineralogy to those from Nkogho, but differ in the size and proportion of plagioclase laths and other materials. Plagioclase laths (70%) dominate in a fine-grained groundmass. Euhedral (four or six sided prisms) and anhedral (mostly sub-rounded) olivine (15%) and

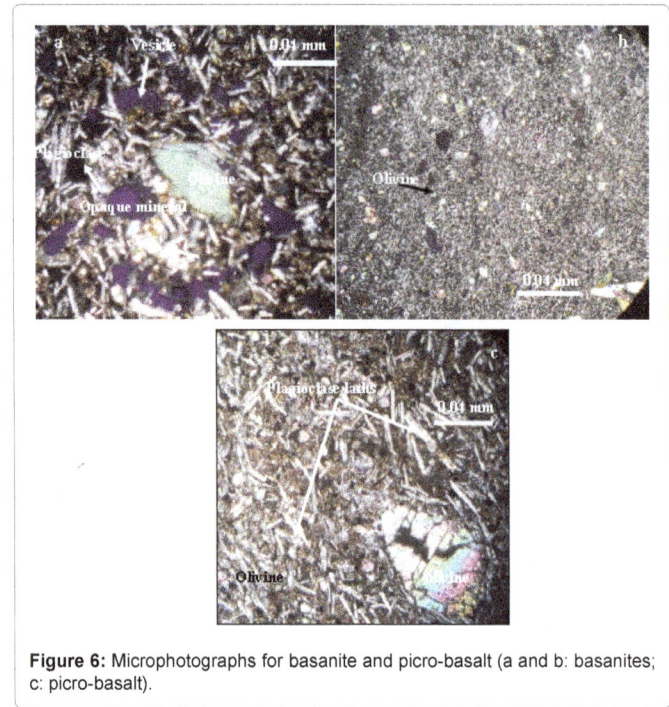

Figure 7: Microphotographs for basalts (a: vesicular-aphanitic rock in Nkogho; b: vesicular- aphyric to microporphyritic rock in Kembong, c: vesicular-aphyric to slightly porphyritic in Talangaye; d: vesicle free-porphyritic rock in Nkogho; e: vesicle free-weakly fluidal and porphyritic rock in Ossing; f: vesicle free-porphyritic rock in Araru).

pyroxene (10%) microphenocrysts (0.03-0.05 mm) are included within feldspar laths, while spotted vesicles and opaque mineral (sub-hedral to anhedral, <0.015 mm in size) complete the groundmass. The grain size of minerals differentiates basalts from Talangaye from those at Nkogho and Kembong. Talangaye basalts are aphyric to weakly porphyritic, with partly corroded olivine phenocrysts (5 wt. %; up to 0.3 mm), locally zoned clinopyroxene microphenocrysts (10%, up to 0.06 mm) and opaque minerals (15%) and vesicles disseminated in a very fine-grained groundmass dominated by plagioclase (70%) microlite (Figure 7c). Vesicle-barren basalts from fragments and flows in Nkogho, Ossing, Araru and Babi have similar compositions, but differ in proportions of their minerals. In Nkogho, the basalts are porphyritic (Figure 7d),

with olivine (20%) and clinopyroxene (5%) forming anhedral and euhedral (four, six or eight sided prisms) phenocrysts (0.1-0.3 mm), and/or microphenocrysts (0.03-0.08 mm) in a fine-grained plagioclase-rich (65%) groundmass that includes opaque mineral (10%). Some thin sections show microphenocrystic nepheline characterized by grey-white simple twinning. Brownish-red to yellow mineral replacement alters some olivine phenocrysts (Figure 7d), when observed under crossed-polar light. Basalts from Ossing (Figure 7e) are similar to those from Nkogho, but show a weakly fluidal structure, lack common mineral replacement and do not carry nepheline. Dominant plagioclase

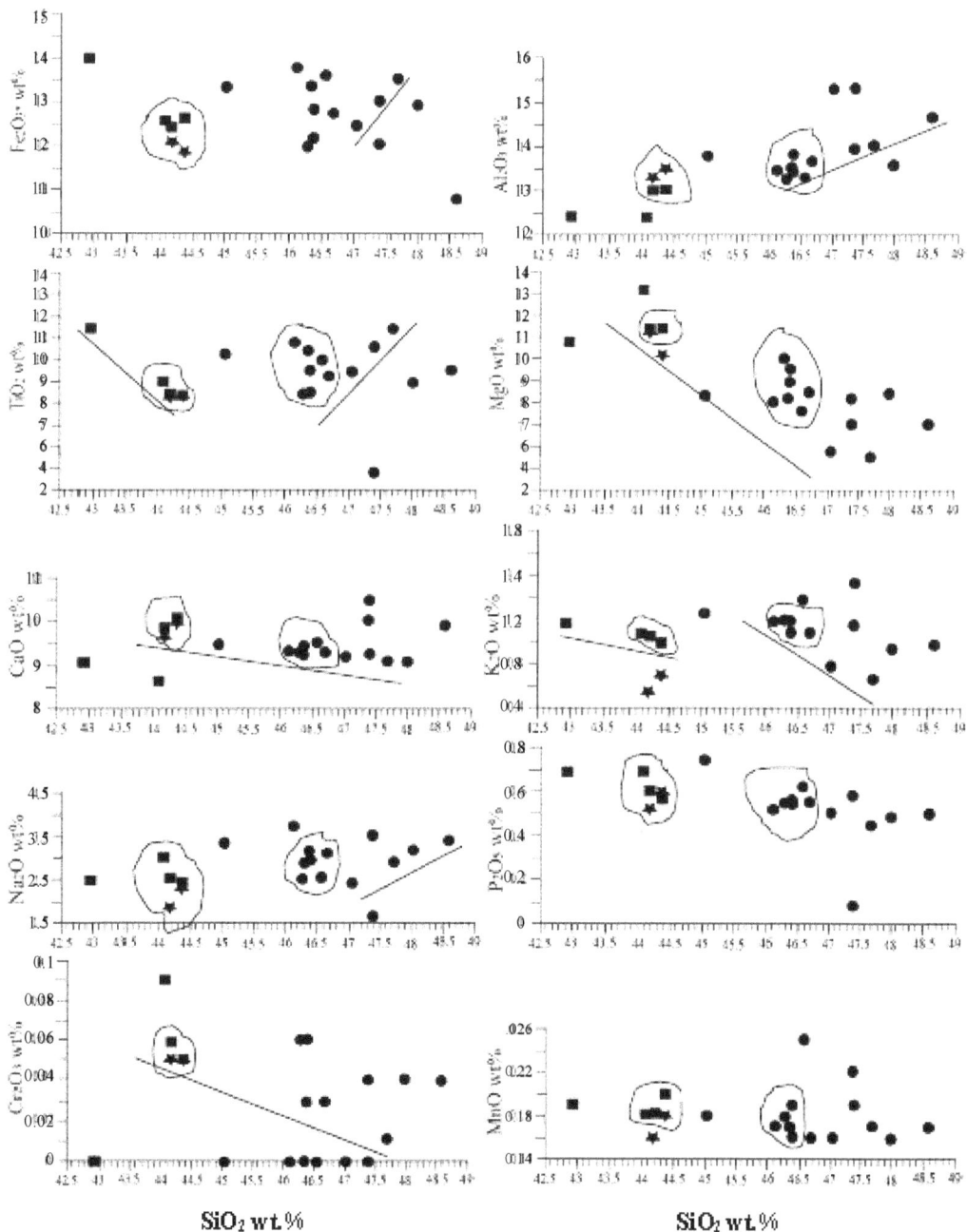

Figure 8: Variation of major and minor element oxides (in wt. %) in the west and southern Mamfe mafic volcanic rocks versus SiO$_2$. The line the shows either the decrease or increase in major and minor elements contents with that of SiO$_2$

(up to 80%) mostly occurs as microlite. Olivine (10%) forms anhedral to euhedral phenocrysts (up to 0.15 mm) and microphenocrysts (below 0.03 mm), clinopyroxene crystals (≤ 8%) are mostly subhedral microphenocrysts (≤ 0.03 mm), and opaque mineral (≤ 2%), are set within the feldspathic groundmass. In Babi and Araru (Figure 7f), basalts resemble Nkogho porphyritic basalts, but lack nepheline and fluidal structure. Cracks in olivine locally enclose yellowish products. Proportionally, olivine is estimated at 8%, plagioclase at 72%, clinopyroxene at 10% and opaque mineral at 10%.

Geochemistry

Major and minor element oxide (wt. %) and trace (including rare earth) element (ppm) concentrations in the analyzed rocks are presented in terms of rock types. The calculated CIPW norms are presented with the major and minor element values.

Major and minor element geochemistry: For twenty analyzed samples in Table 1, SiO_2 contents range from 42 to 48 wt. %: in basalts, from 42 to 44 wt. % in basanites; slightly over 44 wt. % in picro-basalts and from 45 to 48 wt. % in basalts. The TiO_2 contents (0.8-3.6 wt. %) show the highest values in basalts. The TiO_2 abundance in basalts distinguishes: low Ti-basalts (TiO_2 ≤ 0.8 wt. %), mid Ti-basalts (TiO_2 ≤ 2 wt. %, and high Ti-basalts (TiO_2 ≥ 2 wt. %) [55,56]. The SiO_2 versus oxide plots in Figure 8 lack major variation; although some plots vary in oxide content with the increase in SiO_2. General grouping of rocks suggest some genetic relationships.

The Al_2O_3 contents range from 12.4 to 13.6 wt. % in basanites, ≤ 13.5 wt. % in picro-basalts and 13.3 to 15.4 wt. % in basalts. Four samples (NKK10, BAI4, ARK16, and KEG2) have elevated alumina contents (>14 wt. %). The Al_2O_3 versus SiO_2 plot (Figure 8) defines both a low-Al group and high-Al group. The calculated Al_2O_3/TiO_2 ratio ranges from 3.95 to 19.90 with the highest ratio for ARK16. The Fe_2O_3 contents range from 11.8 to 14.0 wt. % in basanites, <12.1 wt. % in picro-basalts and from 10.7 to 13.8 wt. % in basalts.

The MgO abundances range from 10.0 to 13.2 wt. % in basanites, ≤ 11.3 wt. % in picro-basalts, and range from 5.4 to 11.0 wt. % in basalt. The basalts show MgO contents from 5.0 to 8.0 wt. % are within the range limit (2.0 to 8.0 wt. %) of Hagos et al. [6] for highly to moderately fractionated mafic rocks. The Mg number values range from 40 to 64, and most values being <55. The highest Mg values were found mainly in picro-basalt and basanite. The plotted oxides versus Mg# in Figure 9 show some decrease of SiO_2, TiO_2, and Fe_2O_3 with Mg# increase, and partly, the increase of Al_2O_3 and K_2O when Mg# increases. The CaO contents range from 9.0 to 10.1 wt. % (basanites), <11.0 wt. % (picro-basalts) and range from 9.1 to 10.5 wt. % in basalts.

The Na_2O contents in wt. %, range from 2.4 to 3.0 in basanite, ≤ 2.9 (picro-basalts) and 1.6 to 3.8 in basalts. The K_2O abundances are: <1.2 wt. % in basanites, ≤ 0.7 wt. % in picro-basalts and 0.6 to 1.6 wt. % in

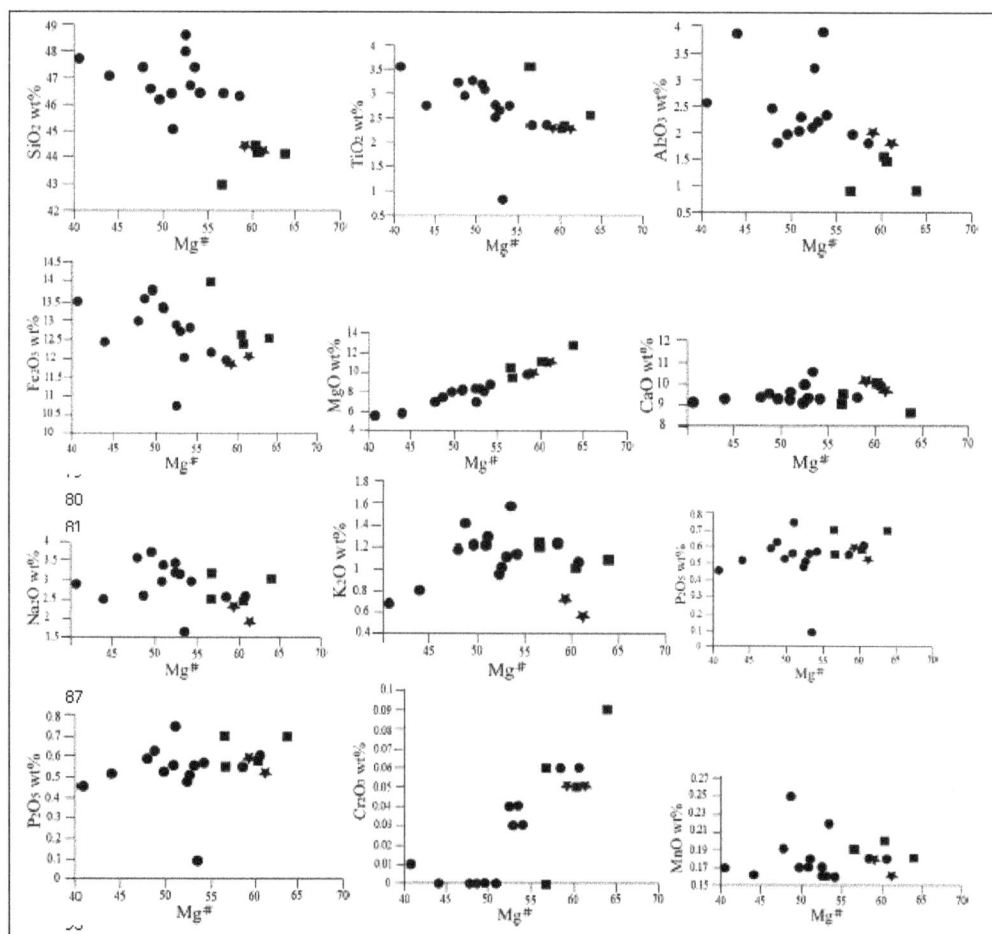

Figure 9: Variation of major element oxides (in wt. %) in the west and southern Mamfe mafic volcanic rocks versus Mg#.

basalts. The Na_2O/K_2O ratios range from 1.0 to 4.4, mostly being less than 2.2 and are lowest in sub-alkaline basalts from Kembong (KEG2), Araru (ARK16) and Babi (BAI4). The alkalinity index $(Na_2O+K_2O)-$ $(0.37*SiO_2-14.43)$ ranges from 0.1 to 2.8 with the lowest values in sub-alkaline picro-basalts and basalts. The sub-alkaline basalts are also termed tholeiitic basalts [2,57,58]. The west and southern Mamfe basaltic rocks mostly belong to the Na-series in a Na_2O versus K_2O binary diagram (Figure 3b) [59]. A few plot within the K-series or on the line separating the two fields.

The MnO contents are mostly between 0.16 to 0.19 wt. %, although EKK1, ARK16 and OSG are higher. The P_2O_5 contents generally range from 0.4 to 0.8 wt. %, except for a low value for ARK16 (0.08 wt. %). Basanites and a Kembong basalt (KEG2), shows high LOI (2.60 to 3.5 wt. %). Low LOI contents (0.46 to 2.02 wt. %) mainly prevail in picro-basalts and basalts.

The CIPW norms in Table 1 show most basanites are nepheline and olivine normative. Picro-basalts, most basalt (e.g., NSI3, NKK1 and NKK7) and a transitional basanite (EKK1) are hypersthene and olivine

normative. Some tholeiitic basalts (KEG2, TAE2, OSG and BAI4) are quartz and/or wollastonite normative, other alkali basalts are nepheline and olivine normative (NSI2 and NKK4).

Trace element geochemistry: Trace element abundances for thirteen selected samples in Table 2 and Figure 10 (few elements) show variations related to petrographic rock type. Figure 10a shows groupings of rocks that increase in SiO_2 contents, whereas Figure 10b partly shows variations in trace elements with increase in Mg#. Significant trace element ranges in the rock suites include Ba ($\leq$ 436 ppm: basanites; $\leq$ 2090 ppm: picro-basalts; and 173-800 ppm: basalts), Cr (342-616 ppm: basanites; 342: picro-basalts; 68-411 ppm: basalts); and Sr ($\leq$ 755 ppm: basanites; $\leq$ 625 ppm: picro-basalts; 142-722 ppm: basalts). Other important elements are Zr (170-230 ppm: basanites; 170 ppm: picro-basalts; 50-221 ppm: basalts), V ($\leq$ 222 ppm: basanites; $\leq$ 241 ppm: picro-basalts; 190-239 ppm: basalts), Zn ($\leq$ 120 ppm: basanites; $\leq$ 141 ppm: picro-basalts; 91-143 ppm: basalts) and Ni ($\leq$ 387 ppm: basanites; $\leq$ 252 ppm: picro-basalts; 52-193 ppm: basalts). Similar Zr contents (170 ppm) and other trace elements in basanites and picro-basalts from Ekok suggest an affinity between the two rock

Sample composition		Basanites		Picro-basalts				Basalts					
Trace elements (ppm)	BAI3	EKK1	EKK3	EKK6	EKK11	NSI2	NSI3	NKK1	NKK2	NKK6	NKK10	ARK16	BAI4
Ba	385	400	436	2090	519	706	609	358	332	473	797	173.5	218
Zr	230	170	170	170	170	220	210	190	170	160	160	50	190
V	202	214	222	234	241	198	198	201	206	191	199	274	328
Zn	114	118	120	141	124	128	124	134	134	126	140	92	142
Sr	755	726	658	625	588	721	693	597	556	505	561	142.5	411
Rb	21.6	21.4	24	14.7	11.1	28.2	31.2	22.1	23.2	21.3	13.2	76.6	6.8
Ga	19.9	20.2	21.2	21.2	21.8	21	20	22	20.8	21.9	24.1	16.9	24
Y	28.2	22.2	22.8	24.5	24.2	22.7	23.1	23	21.7	21.8	25.5	18.4	32
Nb	53.8	48.5	53.4	46.4	48	50.4	51.3	48	43.9	39.9	38.7	5.3	22.7
Ni	387	237	233	254	252	182	192	156	151	157	156	74	52
Co	49	54	49	66	57	51	54	45	48	46	61	51	44
Cu	56	63	62	64	65	54	54	59	59	54	61	30	57
Sc	18	21	21	22	22	19	18	18	17	18	19	36	26
Pb	5	8	11	15	11	9	8	5	9	7	6	14	4
Ta	3.1	2.7	3.1	2.8	2.7	3.2	2.9	2.8	2.7	2.3	2.7	0.3	1.3
Hf	5.5	4.4	4.5	4.5	4.4	5.6	5.1	4.9	5	3.9	4.5	1.8	5
Th	3.6	4	4.4	4	4.1	4.9	4.7	4.1	3.5	3.1	3.5	1.4	1.9
U	1	1	1.1	1.1	1	1.2	2.2	1	1	0.8	1	0.3	0.5
W	<1	1	1	<1	<1	<1	1	<1	<1	<1	<1	1	<1
Sn	3	1	2	2	2	2	1	2	2	2	2	2	2
Mo	1	2	2	2	2	1	2	1	1	1	2	<1	<1
Cd	<0.5	<0.5	<0.5	<0.5	<0.5	<0.5	<0.5	<0.5	<0.5	<0.5	0.7	<0.5	<0.5
Ag	<0.5	<0.5	<0.5	<0.5	<0.5	<0.5	<0.5	<0.5	<0.5	<0.5	<0.5	<0.5	<0.5
As	<5	<5	<5	5	<5	<5	<5	<5	<5	<5	<5	<5	<5
Cr	615.79	342.11	410.56	342.11	342.11	410.53	410.5	205.26	205.26	273.68	273.68	273.68	68.42
K	8965.6	8301.5	8799.57	5811.04	4482.8	9961.78	10128	9131.6	9048.6	7803.4	8135.46	12867	5562
Ti	15227	13609	13848.4	13668.6	13609	13908.4	13908	16426	15887	14928	16426.3	4796	21222
P	3011	2487.4	2618.3	2574.7	2269.2	2400.09	2400	2443.7	2400.1	2094.6	2181.91	349.11	1964
Rare earth elements (ppm)													
La	42.4	33	36.6	34.7	32.4	38	35.1	30.6	28.8	24.3	27.8	5.8	20.9
Ce	87.1	62.9	71.2	65.2	62.6	71.3	68.9	62.5	58	49.7	54.8	12.9	45.8
Pr	10.6	7.2	8.1	7.5	7.2	8.3	7.7	7.4	7	5.9	6.7	1.4	6
Nd	46.9	32.4	34.7	31.8	32	36.3	34.2	32.8	31.6	27.5	31.3	7.7	28.5
Sm	9.1	6.8	7.2	6.9	6.3	7.2	7	7	6.4	6.2	7.1	2.2	7
Eu	3.2	2.3	2.5	2.4	2.5	2.5	2.4	2.5	2.5	2.2	2.4	0.6	2.6
Gd	8.8	6.6	6.7	6.4	6.3	6.6	6.7	6.9	6.9	6.5	7	2.5	7.5
Tb	1.3	0.9	1	1	0.9	1	1	1	0.9	0.9	1	0.5	1.2
Dy	6	4.4	4.8	4.7	5	5.2	4.9	5.1	4.8	4.5	5.3	2.9	6.1
Ho	1.2	0.9	0.9	1	0.9	0.9	0.9	0.9	0.8	0.9	0.9	0.7	1.2

Er	2.8	2.2	2.4	2.3	2.4	2.4	2.1	2.2	2	2	2.3	2	3.3
Tm	0.4	0.2	0.3	0.3	0.3	0.3	0.3	0.3	0.2	0.3	0.3	0.3	0.4
Yb	2.1	1.7	2	1.6	1.7	1.8	1.6	1.5	1.4	1.5	1.6	2.2	2.5
Lu	0.3	0.2	0.3	0.3	0.3	0.2	0.2	0.2	0.2	0.2	0.3	0.3	0.4
Calculated ratios and anomalies													
Rb/Sr	0.029	0.029	0.036	0.024	0.019	0.039	0.045	0.037	0.042	0.042	0.024	0.538	0.017
Zr /TiO2	90.55	74.89	73.59	74.56	74.89	94.83	90.52	69.34	64.15	64.26	58.39	62.7	53.67
Y /Nb	0.524	0.458	0.427	0.528	0.504	0.45	0.45	0.479	0.494	0.546	0,659	3.49	1.41
Th/U	3.6	4	4	3.64	4.1	4.08	2.14	4.1	3.5	3.875	3.5	4.67	3.8
Zr/Hf	41.82	38.64	37.7	37.7	38.64	39.29	41.18	38.78	34	41.03	35.56	27.77	38
Nb/Ta	17,35	17,96	17,22	16,57	17,77	15,75	17,68	17,14	16,26	17,35	14,33	17,66	17,46
Ba/Nb	7.15	8.25	8.16	45.04	10.81	14.01	11.87	7.46	7.56	11.85	20.59	32.74	9.61
Zr/Nb	4.28	3.51	3.18	3.66	3.54	4.37	4.09	3.96	3.87	4.01	4.14	9.43	8.37
Nb/U	53.8	48.5	48.55	42.18	48	42	23.3	48	43.9	49.88	38.7	17.67	17.46
Th/La	0.085	0.121	0.12	0.115	0.013	0.129	0.134	0.134	0.122	0.128	0.126	0.241	0.091
Ba/La	9.08	12.12	11.91	60.23	16.02	18.58	17.35	11.7	11.53	19.47	28.67	29.91	10.43
Nb/La	1.269	1.47	1.459	1.337	1.481	1.326	1.462	1.569	1.524	1.642	1.392	0.914	1.086
Ce/Pb	17.42	7.86	6.47	4.35	5.9	7.91	8.61	12.5	6.44	7.1	9.13	0.92	11.45
(Ce/Yb)N	11.01	9.82	9.45	10.82	9.78	10.16	11.43	11.06	11	8.8	9.09	1.56	4.86
(La/Yb)N	13.77	13.24	12.48	14.79	13	13.89	14.96	13.91	14.03	11.05	11.85	1.8	5.7
(Tb/Yb)N	2.77	2.37	2.24	2.8	2.37	2.49	2.8	2.98	2.88	2.68	2.8	1.02	2.15
Eu/Eu*	0.74	0.74	0.75	0.77	0.84	0.74	0.74	0.79	0.84	0.81	0.77	0.69	0.88
LREE	140.1	103.1	115.9	107.4	102.2	117.6	111.7	100.5	93.8	79.9	89.3	20.1	72.7
MREE	68	48.1	51.1	47.5	47.1	52.6	50.3	49.2	47.4	42.4	47.8	13	45.6
HREE	14.1	10.5	11.7	11.2	11.5	11.8	11	11.2	10.3	10.3	11.7	8.9	15.1
∑REE	222.2	161.7	178.7	166.1	160.8	182	173	160.9	151.5	132.6	148.8	42	133.4
LREE /HREE	9.94	9.82	9.91	9.59	8.89	9.97	10.15	8.97	9.11	7.76	7.63	2.26	4.82

Table 2: Trace and rare earth element abundance for mafic rocks from the west and southern part of Mamfe Basin (Manyu Division, SW Cameroon) [The Eu/Eu* was calculated following similar method in Kamgang et al. Eu/Eu*=EuN / (SmN × NdN)1/2 with EuN, SmN, NdN being chondrite-normalized values of those elements [34]. Sample location, EKK: Ekok; BAI: Babi; NSI: Nsanaragati; NKK: Nkogho; ARK: Araru; KEG: Kembong; TAE: Talangaye; and OSG: Ossing].

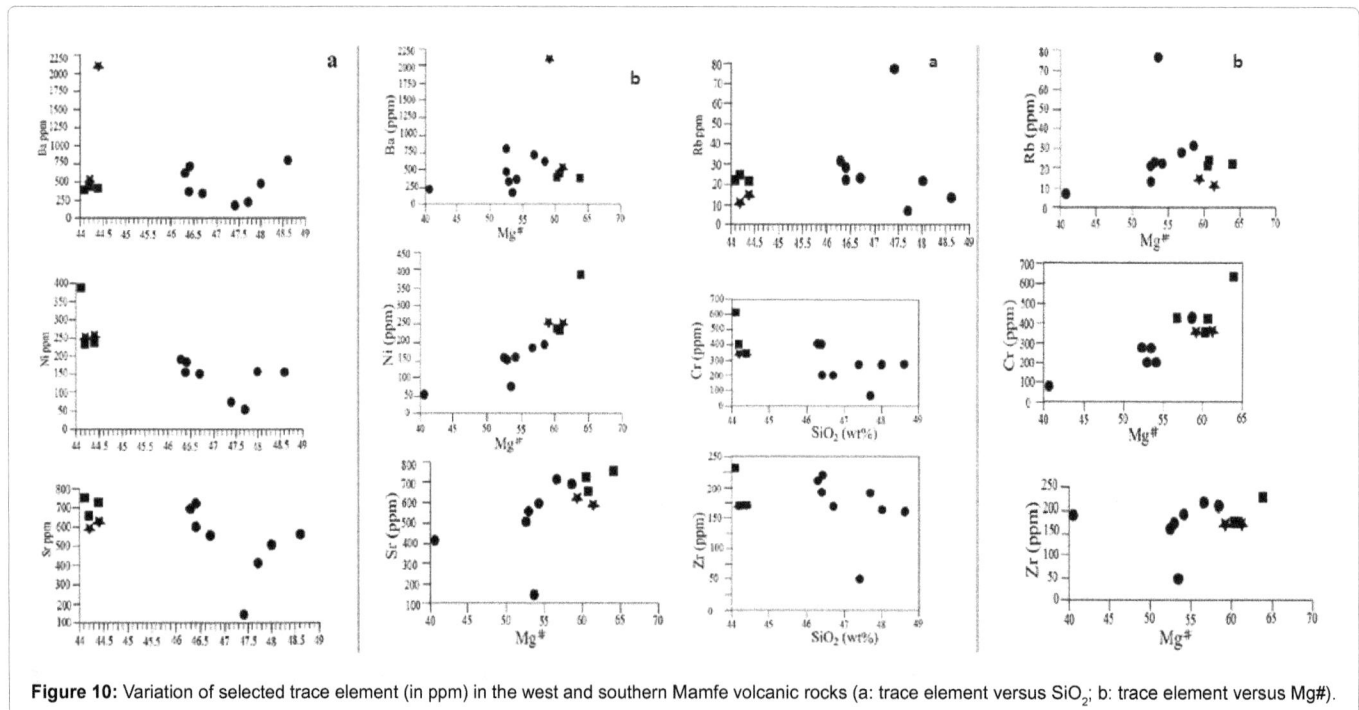

Figure 10: Variation of selected trace element (in ppm) in the west and southern Mamfe volcanic rocks (a: trace element versus SiO$_2$; b: trace element versus Mg#).

types, as in the trace element versus silicate plot diagrams (Figure 10a). The Zr/TiO$_2$ ratios vary from 53 to 95 with the lowest values found in NKK10, ARK16 and BAI4. Relatively higher Ni contents are recorded in picro-basalts and basanites, with the lowest values being among the sub-alkaline basalts. The Ni contents in the Mamfe mafic volcanic rocks generally exceed those of many mafic rocks of the Cameroon Volcanic

Line [34,58,60]. The relative high Ni content in the picro-basalts and basalts suggests enrichment of this element in the source magma and its incorporation into olivine.

Rubidium, Ga, Nb, Y, Co, Cu, and Sc abundances are generally low (<70 ppm), with Rb being relatively low in picro-basalts and tholeiitic basalts. Niobium content is lowest in basalt samples (ARK16:5.3 ppm, BAI4:20.7 ppm). The Rb/Sr ratios range from 0.01 to 0.53 with most values being <0.05 (within the limit:<0.07 in mantle peridotite [58]. The Y/Nb and Zr/Nb ratios range from ≤ 0.65 to >1.00 and 3.5 to 10.0, respectively. Lead, Hf, Th, Ta, U, W, Sn, Mo, Cd, Ag, and As contents are very low, ranging from <5 to ≤ 15 ppm. The Zr/Hf, Th/U and Nb/U ratios range from 27 to 42; 2 to 5 and 17 to 50 respectively. The Y/Nb versus Zr/Nb, Zr/Nb versus Zr/Y, and Zr versus Nb diagrams (Figure 11) discriminate between the alkali and transitional tholeiitic basaltic rocks suggesting lower to higher degrees of partial mantle melting were involved in the source magmas.

The primary mantle-normalized trace element patterns (Figure 12a and 12b) show weakly negative Th and strongly negative K anomaly in basanites and picro-basalts [61]. Thorium anomalies are weakly to strongly negative in basalts, whereas the K anomaly is generally weakly negative, except in ARK16 which is strongly positive (Figure 12c). Weakly positive Ba and weakly to strongly positive Pb anomaly feature in the basanites, picro-basalts and alkali basalts.

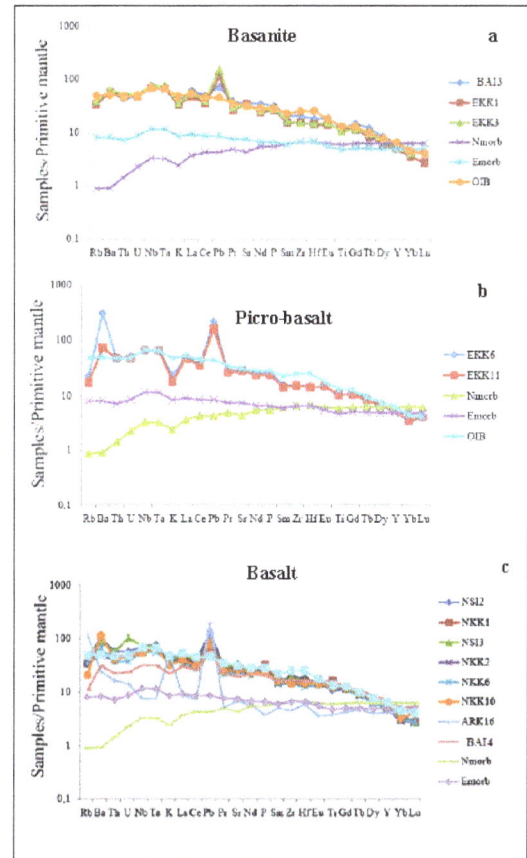

Figure 12: Primitive mantle-normalized multi-element patterns for west and southern Mamfe Basin volcanic rocks. a): basanite, b): picro-basalt, c): basalt. The normalizing values and the values of OIB, N-MORB and E-MORB are from Sun and McDonough [61].

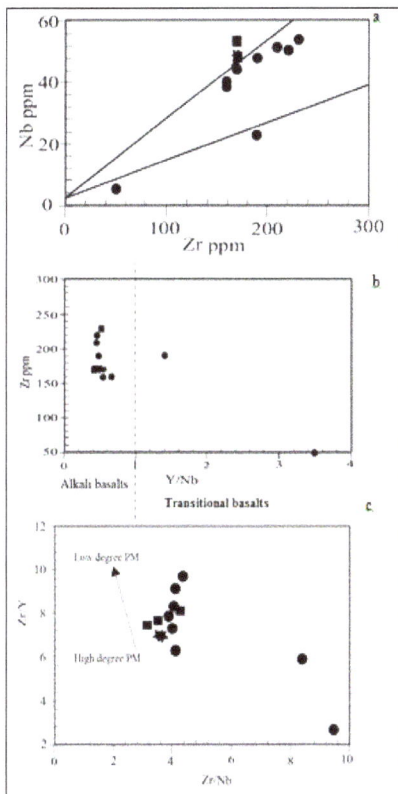

Figure 11: (a) Nb versus Zr plot diagram showing that basanites, picro-basalts and basanites from west and southern Mamfe Basin represent two different magma types. (b) Zr versus Y/Nb plot diagram distinguishing two basaltic groups. The boundary (by dashed) is from Pearce and Cann [64] (c) Zr/Y against Zr/Nb plot showing qualitative trend of compositions resulting from low versus high degrees of partial melting of the west and southern basaltic rocks (PM: primary mantle partial melting).

Rare earth element geochemistry: The REE suites for thirteen selected samples (Table 2) show some variation and the domination of light rare earth elements (ΣLREE:20-142 ppm) over middle (ΣMREE:13-68 ppm) and heavy rare earth elements (ΣHREE:8-16 ppm). The total REE ranges from 41 to 223 ppm are generally higher in basanites and picro-basalts. The LREE /HREE ratios range from 2 to 10, with the lowest values in sub-alkaline basalts. Values within the lighter REE in the rock suites include: La (≤ 42.4 ppm:basanites; ≤ 34.7:picro-basalt; 5-39 ppm: basalts), Ce (≤ 87.1 ppm: basanites; ≤ 65.2 ppm: picro-basalt; 12-72 ppm: basalts), Nd (≤ 46.9 ppm: basanites; ≤ 32 ppm: picro-basalt; 7-37 ppm: basalts). The Pr, Sm, Gd, and Dy contents (<11 ppm), are relatively low, while Eu, Ho, Tb, Er, Ho, Tm, Yb and Lu contents (<3.0 ppm) are very low. The calculated Eu anomaly as Eu/Eu* ranges from 0.6 to 10.0 with no marked Eu anomalies on the chondrite normalized REE patterns (Figure 13). The $(La/Yb)_N$ and $(Tb/Yb)_N$ ratios ranges from 1.8 to 15.0 and 1.0 to 3.0, respectively. The highest degree of LREE enrichment $(La/Yb)_N>10$ is predominantly found in basanites, picro-basalts and alkaline basalts. Part of the $(Tb/Yb)_N$ ratios lie within the range (1.89-2.45) for alkali basalts from Hawaii which are considered to have been generated in a garnet-bearing lherzolitic mantle [62].

The chondrite normalized REE patterns (Figure 13) generally show a distinction between basanites in Ekok and those in Babi and between alkaline basalts and some of the tholeiitic basalt (e.g., BAI4). Basanites

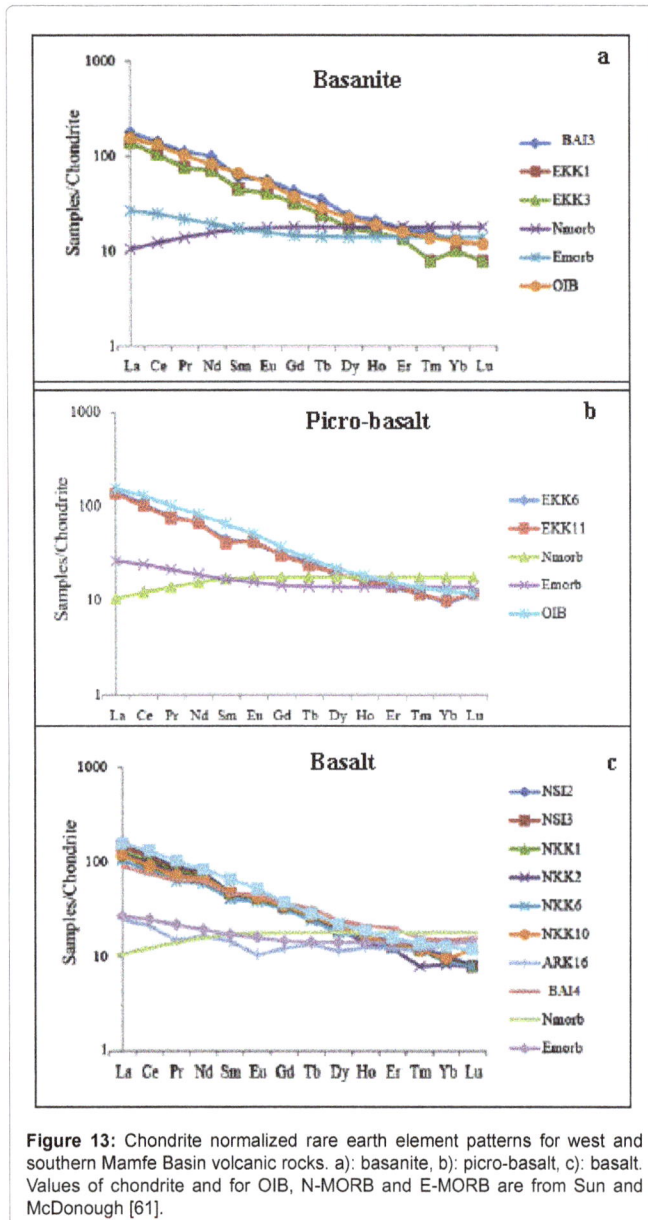

Figure 13: Chondrite normalized rare earth element patterns for west and southern Mamfe Basin volcanic rocks. a): basanite, b): picro-basalt, c): basalt. Values of chondrite and for OIB, N-MORB and E-MORB are from Sun and McDonough [61].

in Ekok show negative Tm and positive Yb anomalies, whereas no Tm and Yb anomalies are shown for basanite from Babi. The Yb anomaly is negative in picro-basalts (Figure 13b) and in some basalt (Figure 13c).

Distinct incompatible trace element characteristics of these rocks include: large variation of total rare earth element (ΣREE) contents (41-223 ppm) and trace and rare earth element ratios such as Nb/La (0.9-1.7) and Ba/Nb (7.1-46.0). Two other key parameters are: Ce/Pb=0.9-17.5 and Th/La=0.01-0.25.

Discussion

The obtained field, petrographic, and geochemical data are discussed separately to characterize each rock type, to help understand the overall petrogenetic history.

Field data and petrography: Basanites form variously vesiculated flows that show blocky to irregular jointing. They resemble pillow lava and lack rubbly base and top (common in 'aa' type flow), and also lack

columnar cooling joints. Pillow lavas indicate eruption into water or under ice [2,9]. The basanites probably cooled in water related to the overlying conglomeratic sedimentary deposits. The overlying flow (vesicle depleted in Ekok, Figure 4a and 4b) has a very smooth top, a feature found in pahoehoe flow [2,9,63]. Alternatively, the smoothness of the surface may be due to mechanical action of running water. A feature that supports the pahoehoe nature of this flow is the presence of vesicles in its inner part. The interior of pahoehoe lava is usually vesicular, often containing over 20% vesicles by volume. The pahoehoe nature of the basanite may typify cooling of low-viscosity basaltic lava [2]. The cemented conglomeratic clasts (Figure 4b) at the contact between the lower and upper flows suggest that they are products of two different magmatic eruptions. The hardened clasts are derived from the less consolidated surrounding conglomerates. The elongate parts of the vesicles may indicate either shearing during flow or later flattening, and the high concentration of vesicles towards the top of the lower basanite indicates a lava flow rather than a sill [9]. The aphyric basanites suggests rapid crystallization of melts lacking large suspended crystals and slightly porphyritic basanites a different stage of crystallization: when phenocrysts, grew in a magmatic chamber before rapid cooling at the surface produced a fine grained groundmass [1,2,7,9]. The more porphyritic basanites may signify some crystal fractionation and accumulation, as interpreted by Deshmukh [8].

Picro-basalts form blocky large fragments with some elongation, with visible smooth and prismatic surface features common in columnar flows [9]. Although picro-basalts can form in flows and dykes, it is difficult ascertain their exact origin in the Mamfe Basin. Topographically, these rocks are found at the hilly part in Ekok (at about 200 m), showing that the source was probably a volcano. This geomorphology (high topography) may suggest the presence of a local magma chamber under this hilly area. Geophysical survey is needed to confirm the existence of this magmatic chamber, as its crystallization history is not easily interpreted by simple observation of fragments at the surface. The vesicles represent gas bubbles that form when molten lava becomes supersaturated with volatiles on ascent and their absence in the picro-basalts suggests an origin from volatile-depleted magma. Xenocrystic quartz with thin contact metamorphism zones suggest it came from post sedimentary materials from the surrounding sandstones. Olivine and clinopyroxene macrocrysts and microcrysts, in a very fine-grained ground mass in the picro-basalts may suggest three stages of crystallization: macrocrysts (slow cooling in a magmatic chamber); microcrysts (shallow depth cooling) and groundmass (rapid cooling at the surface) [8].

In summary, basalts occur as volcanoclasts in SW Nkogho and Nsanaragati, as eluvial fragments in Talangaye and Babi, pillow-like lava in Kembong, Araru and Ossing, and as a dyke in south Nkogho. Volcanoclasts probably come from nearby volcanoes as most occur at the foot of hills (mainly in Nkogho). The presence of pillow lava suggests aqueous involvement and a basaltic dyke intruding granitic basement in Nkogho provides age control for the tectonic fracturing and volcanic events. Sedimentary rocks in the Mamfe Basin were affected fracturing and volcanic events as they were over-run by lava and are locally cut by unstudied mafic dykes. Vesicle-rich basalts (e.g., Nkogho dyke, Kembong flow) and vesicle-poor basalts (e.g., Talangaye and Babi fragments, Araru and Ossing flows) mark two groups of magmas of different volatile contents. Textural difference in aphyric or porphyritic basalts suggests different cooling histories in magma production. Modal nepheline in basalt from Nkogho typifies an alkaline and undersaturated nature [2,7]. Pseudomorphous mineral replacement of olivine (Nkogho basalt) by secondary mineral products such as serpentine and iddingsite suggests low-grade metamorphism or hydrothermal alteration of basalts by exposure to hot circulating fluids [2].

Geochemical features and classification: The major elements and CIPW norms data in Table 1 differentiate the basanites, picro-basalts and basalts. Among the basanites (ultrabasic with SiO_2 <45 wt. %) the Nkogho basanite has the lowest SiO_2 (<43 wt. %) and highest Fe_2O_3 (<13 wt. %) and the Babi basanite shows the highest MgO (<13 wt. %) and Mg# (~64) [1]. The basanites are nepheline-olivine normative rocks, but one transitional type (EKK1) is hypersthene-olivine normative. The picro-basalts are ultrabasic, but hypersthene-olivine normative [2]. The picro-basalts differ in the total alkali versus silica diagram of Irvine and Baragar, as one (EKK6) plots slightly in the alkaline field and another (EKK11) in the sub-alkaline field [54]. They show highest LOI values (2.9 to 3.5 wt. %), which may affect their precise chemical designations. In most of the Harker plots diagrams (Figures 8 and 9) the alkaline picro-basalt (EKK6) plots close to transitional basanite from Ekok. They may be products of allied magmas, as they have similar plots in many discriminating diagrams (Figures 10, 11b, 11c and 14). The basalts differ from the other rocks in higher SiO_2 (>45 wt. %, basic: Al_2O_3 (>13 wt. %), Na_2O (dominantly>2.90 wt. %) and lower MgO (≤ 10 wt. %) and Mg# (40-59) [1]. Some of those major element contents are within the ranges of basanite found in Mount Bamenda, Cameroon volcanic line basalts although the basanites there are consistently higher in Na_2O contents [34]. In the Irvine and Baragar binary diagram three groups of Mamfe basalts are distinguished: (1) alkaline basalts; (2) sub-alkaline basalts; and (3) those plotted on the alkaline-subalkaline dividing line [54]. The transitional nature of some is confirmed by Y/Nb versus Zr plots (Figure 11a) [64]. The CIPW norms distinguish four groups: (1) nepheline-olivine bearing basalts (NSI1 and NKK4); (2) olivine-hypersthene bearing basalts (e.g., NSI3, NKK1, NKK2, and ARK16); (3) hypersthene-quartz basalts (TAE and BAI4); and (4) quartz-wollastonite bearing basalts (KEG2 and OSG). Based on Gill, three groups of basalts can be separated on presence or absence of normative quartz or nepheline: (1) silica oversaturated basalts (with quartz and no nepheline); (2) silica-saturated (no quartz or nepheline); and silica undersaturated (with nepheline) [2]. By combining the plots in Irvine and Baragar diagrams with the CIPW norms, nepheline-olivine rocks and part of olivine-hypersthene basalts are alkaline, whereas the wollastonite-quartz-bearing basalts are all sub-alkaline (tholeiitic) [54]. Basalts that plot on the divide line separating the sub-alkaline and alkaline fields are hypersthene-olivine normative. Altogether, it is then possible to distinguish seven groups of Mamfe Basin basalts: alkaline nepheline olivine basalts; alkaline olivine-hypersthene basalts; alkaline hypersthene-quartz basalts; transitional hypersthene-olivine basalts; wollastonite-quartz tholeiitic basalts; and hypersthene-olivine tholeiitic basalts. The Al_2O_3 (>13 wt. %) content in tholeiitic basalts is greater than the values (10.88 and 12.49 wt. %) in Rajahmundry (India) tholeiitic basalts. The Al_2O_3 difference between the Mamfe tholeiitic basalts and those of Rajahmundry shows that cooling magma never had the same chemical enrichment, as most of the other elemental contents are not the same.

The Mamfe Basin mafic volcanic rocks mainly plot in the Ocean Island Basalt field using a FeOt, MgO and Al_2O_3 ternary diagram (Figure 14) [65]. A few plot within the continental field or on the line separating the Mid-Ocean Ridge and Ocean Island Basalt fields. These plots suggest that the studied rocks mainly formed from three different sources of magma. Some match geochemical features of Indian Ocean basalts described in Hagos et al. and differ to mafic volcanic rocks in Mount Bamenda [6,34]. The basalts erupted at mid-ocean ridges are olivine tholeiitic basalts characterized by high silica content (up to

50 wt. %) and very low K_2O content [2]. The geochemical features in tholeiitic basaltic rocks in this study differ to those of continental tholeiitic rocks presented in Ngounouno et al [57]. The Sr abundances in Mamfe basaltic rocks are similar to those of basaltic rocks in Mount Bamboutos (≤ 1963 ppm: which are low strontium basaltic rocks (LSrB) [66]. The extreme variation of Ba contents (173 to 2090 ppm) lead to distinction of two groups of Mamfe basaltic rocks: high Ba basaltic rocks (HBaBr) and low Ba basaltic rocks (LBaBr). The only HBaBr is a picro-basaltic sample (EKK6).

Petrogenetic and tectonic evaluation: Basalt magmas produced by melting in the Earth's mantle range in their geochemistry, and carry up a variety of deep inclusions. So their study can elucidate the nature of the underlying upper mantle and even aspects of the composition of the lower mantle [2]. Petrogenetic interpretation of basaltic suites rocks can be approached by evaluating the behavior of less mobile elements (e.g., Ti and Al) and certain key trace elements (Nb, Cr, Co, Zr, Y and some REE). In the nearby Cameroon volcanic line, basaltic studies analyzed Mg# values and the Ni and MgO contents [58,60]. The Mg# values in primitive upper mantle lavas range from 68 to 72 and are also characterized by high Ni (300-500 ppm), Cr (300-500 ppm) and Co (50-70 ppm) contents [67,68]. The Mg# values (40-64) for the studied Mamfe volcanic rocks are too low for the primitive limit defined by Green and O'Hara [67]. BAI3 is the sample closest to primary features with relatively high Mg# (64), Ni (387 ppm), Cr (616 ppm) and Co (49 ppm). Within the mafic rocks, the Mg# values (40-44) for part of quartz-wollastonite and quartz-hypersthene basalts are more evolved than the other mafic rocks (Mg#: 47-64). For the Mg# values for less evolved rocks (46.3-66.1), also fall short of primitive mantle basalts. A more evolved source for some of those rocks are supported by their low MgO (<9 wt. %) and Ni (<75 ppm). The source magmas for those more evolved rocks could be generated from fractionation processes in mantle-derived magmas [60]. The alkaline nature of the Mamfe mafic rocks is reflected in their enrichment in incompatible elements and high fractionation indexes, $(La/Yb)_N$ ranging from 11-15. These values lie within the range (10.62-22.80) in alkaline mafic volcanic rocks studied by Kamgang et al. [34]. The high $(La/Yb)_N$ indicating relative enrichment of LREE and depletion in HREE was considered

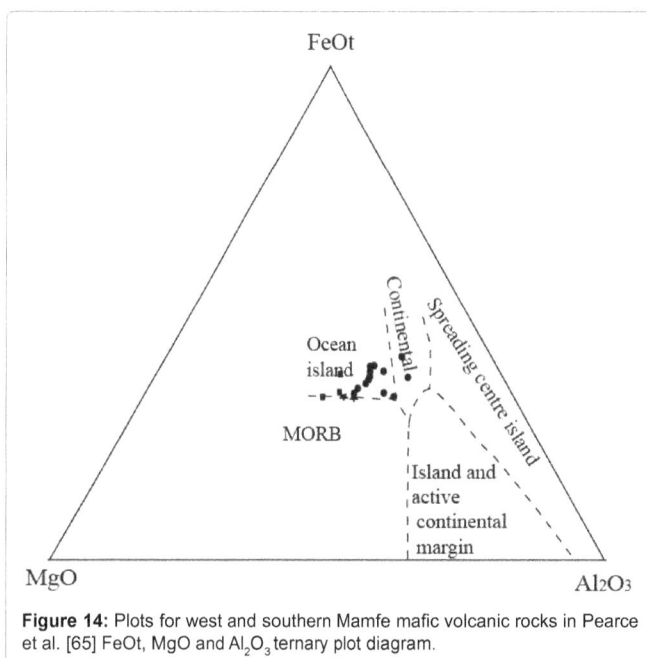

Figure 14: Plots for west and southern Mamfe mafic volcanic rocks in Pearce et al. [65] FeOt, MgO and Al_2O_3 ternary plot diagram.

to denote residual garnet in the source for these magmas and would likely apply also to the alkaline rocks from the Mamfe Basin. The other similarity is their $(Tb/Yb)_N$ >2.0, which are high and within the range limit for Bamenda Mountain mafic rocks 1.8-3.0 [34]. For Wang et al. high $(Tb/Yb)_N$ in mafic volcanic rock also characterizes the presence of residual garnet in the mantle source [69]. The calculated residual garnet proportions for the Mamfe rocks are shown in Figure 15. The high Ba contents mainly in Mamfe basanites and picro-basalts may stem from phlogopite in their magma sources based on the Zhao et al. and Ngounouno et al. studies [57,70].

The LILE, Nb, Zr, and Y contents in Ekok picro-basalt (EKK6) and basanites are similar. EKK6 may be a product from fractional crystallization in the magma source for the basanites. This is supported by the grouping and variation within plots in Figures 8-12. Figures 8-10, show grouping between Ekok basanites and EKK6. In a diagram (Figure 14), EKK6 and EKK3 (basanite) plot in Ocean Island Basalt and Mid Ocean Ridge Basalt fields. In general, crystal fractionation forms part of the processes that formed the range of mafic rocks within the Mamfe Basin. This reflected in the Harker diagrams, including trends in Ni and Cr contents and moderate Mg# values ($\leq$ 64).

The geochemical behavior of the High Field Strengths Elements (HFE) such as Nb, Ta, Zr and Hf makes it possible to explore the mantle sources by using Zr/Hf and Nb/Ta ratios which are less changeable during crustal contamination and crystal differentiation processes [71,72]. Ratios of Nb/Ta (17-18: basanites; 16-18: picro-basalts; 14-18:

basalts) and Zr/Hf (37-42: basanites; up to 38.4: picro-basalts; 27-42: basalts) for Mamfe mafic rocks in some case match those of primitive mantle (17.5 $\pm$ 2.0) and 36.27, respectively. This suggests the magmas for part of the basaltic rocks were derived from mantle sources without obvious crustal contamination. The ratios of La/Ce, Sm/Nd, Ce/Nb, Zr/Nb, of the basaltic rocks, however, differ from primitive mantle values (Table 4), being enriched in LREE and reflecting fertilized mantle sources [70]. Basalt ARK16 is depleted in Nb, Ti, Zr, Ta and Hf and enriched in Rb and has significant Sm, Ce, and Ba contents. This feature is similar to some mantle enclaves in south Hunan Province, China, where it was suggested to fingerprint mantle metasomatism [70]. The magma of this sample probably formed in a mantle containing metasomatic reactions. The Cr and Ni contents in Mamfe basanites, picro-basalts and part of alkaline basalts are higher than those of the tholeiitic basalts (e.g., ARK16). This suggests that the degree of partial mantle melting may be lower in the Cr-Ni relative enriched rocks and higher in the tholeiitic basalts [70]. The highly incompatible element ratios such as Zr/Nb, K/Nb, Ba/Th, Th/La, Ba/La and Ba/Nb (Table 4) are shown to be least susceptible to fractionation during partial melting, and are not significantly fractionated during limited degrees of low-pressure crystallization of OIB magmas [73]. Hence they could be useful indicators for basaltic end-member characterization of the Mamfe basaltic rocks. As presented in Figure 16 and Table 4, the mantle source magmas for alkaline, transitional and tholeiitic basaltic rocks from the Mamfe Basin are mainly influenced by the EM1 end-member in term of Nb/Th and Zr/Nb ratios, and by the HIMU-EM1 end-member in term of Zr/Y and Nb/Y ratios.

Basaltic volcanic rocks are erupted in a wide variety of tectonic environments (e.g., mid-ocean ridges, island arcs, back-arc basins, intra-plate oceanic islands, large igneous provinces and intra-continental rifts) [2]. Basaltic rocks in the western Mamfe Basin plot within the Ocean Island Basalt (OIB) and transitional fields (plots on the line separating OIB and Mid Ocean Ridge Basalt MORB) fields, whereas, the southern rocks are mainly OIB type and a few continental basalts. These OIB types can be distinguished by relative high contents of incompatible trace elements (e.g., Nb, Zr, Ta, Ba, Sr) and the relative low content of these elements in the continental basalts. The latter are probably post-rifting and post-Cretaceous volcanic products, formed after the likely Cretaceous opening of the W-E branch of the Benue Trough (Mamfe Basin). They are dominantly post-sedimentation as they locally overlie or cross-cut the sedimentary rocks in this basin.

Potential uses of the mafic rocks: Rocks such as the Mamfe basalts, picro-basalts, and basanites are mined and used as industrial mineral in civil engineering for construction of road and houses [20,21]. Some can host interesting concentration of elements (e.g., significant Al abundance: native and Mills; sulfide metallic deposits, or placer gemstones derived from xenocrysts or xenoliths [13,14,16,17,31,32,74]. Except for one significant concentration of Ba (2090 ppm in EKK6) and one elevated Ni value (387 ppm) most elements in the studied exposures are not sufficiently concentrated. No gemstone deposits or gem-host xenoliths were found. The above rocks, however, may have potential for some sources of industrial materials as building and road-constructing materials, or possibly aluminous sources in weathered form. The parent rock of lateritic bauxite deposits at Fongo-Tongo and Bangam in the western Cameroon is an aluminum-enriched basalt (Al_2O_3: up to 15.6 wt. %: Table 3) [74]. These Al-enriched basalts are always of interest for lateritic bauxite prospecting as they can produce accumulation of Al in soil profiles during intense weathering and

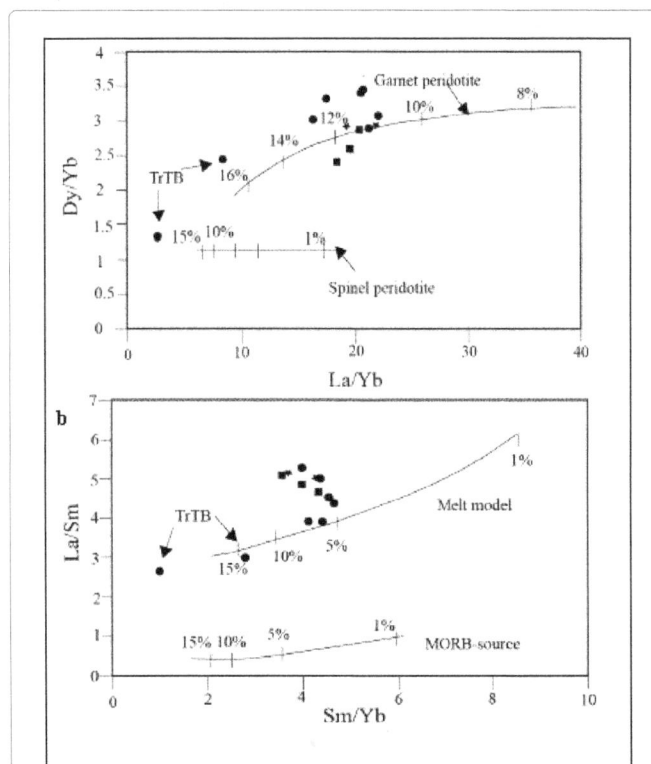

Figure 15: Plots for west and southern Mamfe volcanic rocks in two source discriminating diagrams. (a) La/Yb vs. Dy/Yb plot trend lines show degrees of partial mantle melting for garnet peridotite and spinel peridotite [78,79]. (b) Sm/Yb vs. La/Sm plot gives the melt plot trend lines are inverse batch partial mantle melting model and the depleted MORB source composition. TrTB (Transitional Tholeiitic Basalt) lie apart from the other plots and near higher degrees of melting generated from enriched source composition relative to N-MORB from Sun and McDonough [61].

Sample number and/or locality	Si2O	Ti2O	Al2O3	Fe2O3	MnO	MgO	CaO	Na2O	K2O	P2O5	Cr2O3	LOI
Fongo-Tongo	43.4	3.9	15.5	11.64	-	8.47	9.78	3.15	1.72	1.O9	-	-
Bangam	47.2	4.53	15,6	13,75	-	4.35	8.3	3.9	1.82	-	-	0.61
Araru (ARK16)	47.4	0.8	15.35	12	0.2	8.2	10.5	1.66	1.55	0.08	0.04	2.02
Kembong (KEG2)	47.05	2.73	15.35	12.43	0.16	5.79	9.19	2.49	0.78	0.51	-	3,07

Table 3: Comparing major element geochemical composition (in wt. %) for Al-enriched basalts from West Cameroon Fongo-Tongo and Bangam and south of the Mamfe Basin (Araru and Kembong).

Ratio	La/Ce	Sm/Nd	Ce/ N d	Zr/Nb	La/Nb	Ba/ N b	Ba/Th	Rb/Nb	K/Nb	Th/Nb	Th/La	Ba/La
PM(7)	0.384	0.325	1.35	14.8	0.94	9	77	0.91	323	0.117	0.125	9.6
Crust(7)				16.2	2.2	54	124	4.7	1341	0.44	0.204	25
EMI(7)				5.3-11.5	0.86-1.19	11.4-17.8	103-154	0.88-1.17	213-432	0.105-0.122	0.107-0.128	13.2-16.9
EMII(7)				4.5-7.3	0.89-1.09	7.3-11.0	67-84	0.59-0.85	248-378	0.111-0.157	0.122-0.163	8.3-11.3
Study rocks												
Basanites	0.487-0.525	0.194-0.210	1.85-2.06	3.1-4.3	0.68-0.79	7.1-8.3	100-110	0.40-0.50	164-172	0.066-0.083	0.085-0.121	9.0-12.2
Picro-basalts	0.518-0.532	0.197 -0.217	2.05-2.23	3.1-3.7	0.68-075	10.8-45.1	130-523	0.23-0.32	93-125	0.086	0.013-0.115	16.0-60.3
Basalts	0.456-0.535	0.198-0.306	1.60-2.02	3.5-9.5	0,61-1,09	7.4-32.8	87-228	0.30-14.45	190-2428	0.072-0.264	0.091-0.241	10.4-30.0

Table 4: Trace element characteristics of different mantle end-members and basaltic rocks from the west and southern part of Mamfe Basin (Manyu Division, SW Cameroon).

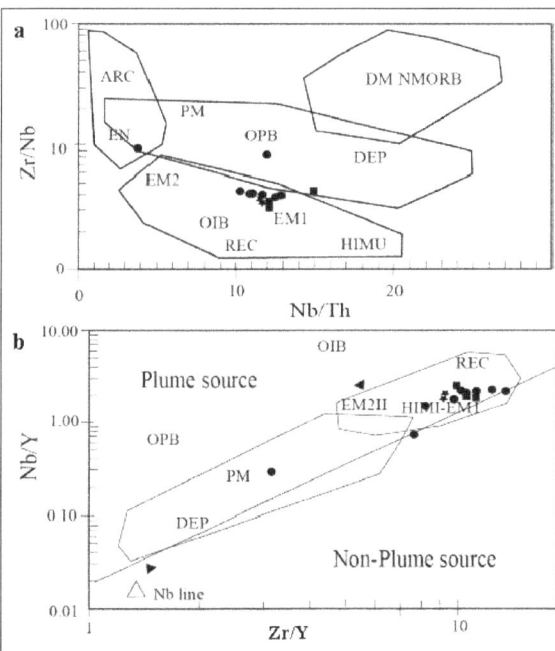

Figure 16: Plots for west and southern basaltic rocks relative to the mantle compositional components (filled star) and fields for basalts from various tectonic settings as defined by Weaver [73] and Condie [80]. (a) Nb/Th–Zr/Nb and (b) Zr/Y-Nb/Y (PM: primitive mantle; DM: shallow depleted mantle; HIMU: high mu (U/Pb) source; EM1 and EM2: enriched mantle sources; ARC: arc related basalts; N-MORB: normal ocean ridge basalt; OIB: oceanic island basalt; DEP: deep depleted mantle; EN: enriched component; REC: recycled component; OPB: oceanic plateau basalt.

element leaching within tropical climate and plateau topography. The Al_2O_3 content (15.35 wt. %) in ARK16 and KEG2 is very close to values found in Fongo-Tongo and Bangam Al-enriched basalts, but are less than values 17.98-20.81 wt. %: for high-alumina basalt in Sant'Antioco Island, SW Sardinia, Italy [75]. ARK16 and KEG2 through chemical weathering within tropical climatic conditions and plateau topography may pay future geochemical prospecting of Al concentration within lateritic soil overlying those basalts for potential bauxite developments.

Diamond occurrences were noted in western Mamfe Sedimentary Basin [26]. Geochemical analyses on coarse-grained detrital zircons show some zircon crystals had kimberlitic affinities [22,27,29,30]. The studied picro-basalts (ultra-basic and mantle origin) are exclusively found on high land (up to 200 m of height) in western Mamfe Basin [2]. The presence of these rocks shows a possible cooling of a mantle ultrabasic rock in this locality. Although not directly an indicator for kimberlite, these mantle source rocks may carry diamond as xenocrysts or in kimberlitic xenoliths showing the existence of kimberlite at shallow depth [2]. Detailed geochemical and geophysical prospecting for kimberlitic or other diamantiferous host rock indicators may lead to discovery of such rocks in that hilly topography.

Conclusions

The west and southern Mamfe mafic volcanic exposures are alkaline, transitional and sub-alkaline rocks including basanites, picro-basalts, and basalts mainly formed during post-sedimentary and tectono-volcanic events. Basanites are dominantly OIB-like undersaturated rocks formed from less evolved mantle source magma. These rocks host the highest nickel concentration. Picro-basalts are alkaline to subalkaline saturated rocks also formed from less evolved mantle source magma. They can host significant concentration of Ba and Ni. Some of these rocks are a product of fractionation of the alkali basalt and basanite source magmas.

Basalts are alkaline and transitional undersaturated rocks, or subalkaline, saturated and oversaturated, rocks formed from variably evolved mantle source magmas mostly representing Ocean Island type sources and include some continental rift basalt types. Some basalt has higher Al contents and has potential for developing bauxite during intense chemical weathering.

Acknowledgements

Special thanks for the two anonymous reviewers whose comments helped to improve this manuscript. We extend our gratitude to Professor Chen Shouyu (China University of Geosciences, Wuhan), who funded part of the major geochemical analysis.

References

1. Le Maitre RW, Streckeisen A, Zanettin B, Le Bas MJ, Bonin B, et al. (2002) Igneous rocks: A classification and glossary of terms. (2nd edn), Cambridge University Press p: 252.

2. Gill R (2010) Igneous rocks and processes: A practical guide. Wiley Blakwell, A John Wiley and Sons Ltd Publication p: 366.

3. Beaux FJ, Fogelgesang FJ, Agard P, Boutin V (2011) Atlas de Géologie Pétrologie. Dunod Paris, p:144.

4. Tchuimegnie NNB, Kamgang P, Chazot G, Agranie A, Bellon H, et al. (2015) Age, geochemical characteristics and petrogenesis of Cenozoic intraplate alkaline volcanic rocks in the Bafang region, West Cameroon. J African Earth Sci 102: 218-232.

5. Le Bas MJ, Le Maitre RW, Streckeisen A, Zanettin B (1986) A chemical classification of volcanic rocks based on the total alkali-silica diagram. J Petrol 27: 745-750.

6. Hagos M, Koeberl C, Kabeto K, Koller F (2010) Geochemical characteristics of the alkaline basalts and the phonolite-trachyte plugs of the Axum area, northern Ethiopia. Aus J Earth Sci 103: 153-170.

7. Winter DJ (2010) Principles of igneous and metamorphic petrology. (2nd edn) Pearson new international production p: 745.

8. Deshmukh SS (1988) Petrographic variations in compound flows of Deccan Trap and their signifance. Memoir Geological Society of India 10: 305-319.

9. Coe LA (2010) Geological field techniques. Wiley Backwell p: 337.

10. Shukla PN, Bhandari N, Das A, Shukla AD, Ray JS (2001) High iridium concentration of alkaline rocks of Deccan and implications to K/T boundary. J Earth Syst Sci 110: 103-110.

11. Crocket J, Paul D, Lala TJ (2013) Platinum-group elements in the Eastern Deccan volcanic province and a comparison with platinum metals of the western Deccan. J Earth Syst Sci 122: 1035-1044.

12. Seifert W, Rhede D, Tietz O (2008) Typology, chemistry and origin of zircon from alkali basalts of SE Saxony (Germany). N Jb Miner Abh 184: 299-313.

13. Sutherland FL, Coenraads RR, Abduryim A, Meffre S, Hoskin PWO, et al. (2015) Corundum (sapphire) and zircon relationships, lava plains gem fields, NE Australia: Integrated mineralogy, geochemistry, age determination, genesis and geographical typing. Mineral Mag 79: 545-581.

14. Watanabe M, Hoshino K, Shiokawa R, Takaoka Y, Fukumoto H, et al. (2006) Metallic mineralization associated with pillow basalts in the Yaeyama Central Graben, Southern Okinawa Trough, Japan. JAMSTEC Report of Res Develop 3: 1-8.

15. Medenbach O, ElGoresy A (1982) Ulvöspinel in native iron-bearing assemblages and the origin of these assemblages in basalts from Ovifak, Greenland, and Bühl, Federal Republic of Germany. Contribution to Mineral Petrol 80: 358-366.

16. Taylor LA, Day JMD, Goodrich CA, Howarth GH, Pernet-Fisher JF, et al. (2014) Metallic-Fe deposits in basalts: Siberia, Greenland, and Germany. International Mineralogical Association, Johannesburg, South Africa.

17. Mills S (2015) Mineralogy and petrology of unique native-iron basalts from northern Siberia. J Undergraduate Res at the University of Tennessee 6: 181-195.

18. Kuepouo G, Sato H, Tchouankoue PJ, Murata M (2008) FeO*-Al$_2$O$_3$-TiO$_2$ rich rocks of the Tertiary Bana igneous complex, West Cameroon. Resour Geol 59: 69-86.

19. Sababa E, Ndjigui DP, Ebah Abeng AS, Bilong P (2015) Geochemistry of peridotite xenoliths from the Kumba and Nyos areas (southern part of the Cameroon Volcanic Line): Implications for Au-PGE exploration. J Geochem Explor 152: 75-90.

20. Gomes LR, Rodrigues EJ (2006) Physical characterization and weathering of the basaltic rocks of the Paraná Basin, Brazil. IAEG 559: 1-9.

21. Danciu C, Buia G (2009) Classification and characterization of basalts of Branisca and Donra-Romania, for capitalization. Recent Adv Industries and Manufacturing Technologies pp: 64-69.

22. Kanouo SN (2014) Geology of the Western Mamfe Corundum Deposits, SW Region Cameroon: Petrography, geochemistry, geochronology, genesis, and origin. Univ de Yaounde I p: 225.

23. Wilson D (1928) Notes on the geology of the Mamfe Division, Cameroon SW province. Occasional papers. Geologic Survey of Nigeria n° 6.

24. Dumort JC (1968) Geological map of recognition of Cameroon at scale 1/500000 sheet Douala-West, with explanatory note. National Printers, Yaounde Cameroon p: 69.

25. Kanouo SN (2008) Geological study of the sapphire mineralization indexes in the southern part of the Mamfe Sedimentary Basin. Univ de Yaoundé I p: 91.

26. Laplaine L, Soba D (1967) Report of the geological service for the years 1965-1966-1967, prospecting of sapphires in the Cretaceous basin of Mamfe.

27. Kanouo SN, Zaw K, Yongue FR, Sutherland LF, Meffre S, et al. (2012a) U-Pb zircon age constraining the source and provenance of gem-bearing late Cenozoic detrital deposit, Mamfe Basin, SW Cameroon. Resour Geol J 62: 316-324.

28. Kanouo SN, Zaw K, Yongue FR, Sutherland FL, Meffre S, et al. (2012b) Detrital mineral morphology and geochemistry: methods to characterize and constrain the origin of the Nsanaragati blue sapphires, south-western region of Cameroon. J Afr Earth Sci 70: 18-23.

29. Kanouo SN, Yongue FR, Ekomane E, Njonfang E, Ma C, et al. (2015) U-Pb ages for zircon grains from Nsanaragati Alluvial Gem Placers: its correlation to the source rocks. Resour Geol J 65: 103-121.

30. Kanouo SN, Ekomane E, Yongue FR, Njonfang E, Zaw K, et al. (2016) Trace elements in corundum, chrysoberyl, and zircon: Application to mineral exploration and provenance study of the western Mamfe gem clastic deposits (SW Cameroon, Central Africa). J Afr Earth Sci 113: 35-50.

31. Sutherland FL, Schwarz D, Jobbins EA, Coenraads RR, Webb G (1998) Distinctive gem corundum from discrete basalt fields: a comparative study of Barrington, Australia, and west Pailin, Cambodia, gem fields. J Gemol 26: 65-85.

32. Sutherland FL, Graham IT, Pogson RE, Schwarz D, Webb GB, et al. (2002) The Tumbarumba basaltic gem field, New South Wales: In relation to sapphire-ruby deposits of eastern Australia. Records Aus Museum 54: 215-248.

33. Fitton JG (1980) The Benue trough and the Cameroon line a migrating rift system in West Africa. Earth Planet Sci Letters 51: 132-138.

34. Kamgang P, Chazot G, Njonfang E, Ngongang TBN, Tchoua MF (2013) Mantle sources and magma evolution beneath the Cameroon volcanic line: Geochemistry of mafic rocks from the Bamenda Mountains (NW Cameroon). Gondwana Res 24: 727-741.

35. Suchel JB (1972) The distribution of rainfall and rainfall patterns in Cameroon, Talence.

36. Olivry JC (1986) Rivers and Rivers of Cameroon, Paris. Hydrographic Monograph, Orstom p: 734.

37. Ajonina HN, Ajibola OA, Bassey EC (2001) The Mamfé basin, SE Nigeria and SW Cameroon: A review of the Basin filling model and tectonic evolution. J Geosci Society Cameroon 1: 24-25.

38. Bassey CE, Ajonina HN (1997) Petrology of the upper member (Cenomanian) of the Mamfe formation, Mamfe embayment, Southwestern Cameroon. 15th Annual International Conference p: 40.

39. Eyong TJ (2003) Lithostratigraphy of the Mamfe cretaceous basin. South West Province of Cameroon-West Africa p: 256.

40. Eseme E, Littke R, Agyingi MC (2006) Geochemical characterization of a cretaceous black shale from the Mamfe basin, Cameroon. Petrol Geosci 12: 69-74.

41. Njoh OA, Nforsi MB, Datcheu JN (2015) Aptian-late cenomanian fluvio-lacustrine lithofacies and palynomorphs from Mamfe basin, Southwest Cameroon, West Africa. Int J Geosci 6: 795-811.

42. Njoh OA, Njie SM (2016) Hydrocarbon source rock potential of the lacustrine black shale unit, Mamfe basin, Cameroon, West Africa. Earth Sci Res 6: 217-230.

43. Le Fur Y (1965) Special report on corundum research. Archives Office of Geological and Mineral Research p: 69.

44. Eseme E, Agyingi MC, Foba-Tendo J (2002) Geochemistry and genesis of brine emanations from Cretaceous strata of the Mamfe basin, Cameroon. J Afr Earth Sci 35: 467-476.

45. Eyong TJ, Wignall P, Fantong YW, Best J, Hell VJ (2013) Paragenetic sequences of carbonate and sulphide minerals of the Mamfe basin (Cameroon): Indicators of palaeofluids, palaeo-oxygene levels and diagenetic zones. J Afr Earth Sci 86: 25-44.

46. Njonfang E, Moreau C (1996) The mineralogy and geochemistry of a subvolcanic alkaline complex from the Cameroon line, the Nda Ali massif, South-West Cameroon. J Afr Earth Sci 22: 113-132.

47. Kangkolo R (2002) Aeromagnetic study of the Mamfe basalts of Southwestern Cameroon. J Cameroon Acad Sci 2: 173-180.

48. Regnoult JM (1986) Cameroon Geological Synthesis, Ministry of Mines, Sodexic Yaounde p: 119.

49. Mackenzie SW, Donaldson HC, Guilford C (1982) Atlas of igneous rocks and their texture. British Library Catalogue in Publication Data p: 154.

50. Mackenzie SW, Adams EA (1994) A colour atlas of rocks and minerals in thin section. Manson publishing Ltd p: 98.

51. Mackenzie SW, Guilford C (1998) Atlas for rock-forming minerals in thin section. British Library Catalogue in Publication Data p: 106.

52. Higgins DM (2006) Quantitative and textural measurement in igneous and metamorphic petrology. Cambridge University press p: 273.

53. Wu WY, Li C, Xu JM, Xiong QS, Fan GZ, et al. (2016) Petrology and geochemistry of metabasalts from the Taoxinghu ophiolite, central Qiangtang, northern Tibet: Evidence for a continental back-arc basin system. Aust J Earth Sci 109.

54. Irvine TN, Baragar WRA (1971) A guide to the chemical classification of the common rocks. Canadian J Earth Sci 8: 523-548.

55. Simonov VA, Mikolaichuk AV, Safonova IYU, Kotlyarov AV, Kovyazin SV (2014) Late Paleozoic-Cenozoic intra-plate continental basaltic magmatism of the Tienshan-Junggar region in the SW Central Asian Orogenic Belt. Gondwana Res 27: 1646-1666.

56. Manikyamba C, Ganguly S, Santosh M, Saha A, Lakshminarayana G (2015) Geochemistry and petrogenesis of Rajahmundry trap basalts of Krishna-Godavari Basin, India. Geosci Frontiers 6: 437-451.

57. Ngounouno I, Déruelle B, Guiraud R, Vicat PJ (2001) Tholeiitic and alkaline magmatism of the Cretaceous half-grabens of Mayo Oulo-Léré and Babouri-Figuil (northern Cameroon-South of Chad) in the domain of continental extension. Académie Sci 333: 201-207.

58. Kuepouo G, Tchouankoue PJ, Nagao T, Sato H (2006) Transitional tholeiitic basalts in the tertiary Bana volcano-plutonic complex, Cameroon Line. J Afr Earth Sci 45: 318-332.

59. Middlemost EAK (1975) The basalt clan. Earth Sci Review 11: 337-364.

60. Nkouandou OF, Ngounouno I, Déruelle B, Ohnenstetter D, Montigny R, et al. (2008) Petrology of the Mio-Pliocene volcanism to the North and East of Ngaoundéré (Adamawa, Cameroon). CR Géosci 340: 28-37.

61. Sun S, McDonough W (1989) Chemical and isotopic systematics of oceanic basalts: Implications for mantle composition and processes. Geological Society of London Special Publications 42: 313-345.

62. Frey FA, Garcia MO, Wise WS, Kennedy A, Gurriet P, et al. (1991) The evolution of Mauna Kea volcano, Hawaii: Petrogenesis of tholeiitic and alkalic basalts. J Geophys Res 96: 14347-14375.

63. Allaby M (2008) Dictionary of earth sciences. (3rd edn) Oxford University Press p: 663.

64. Pearce JA, Cann JR (1973) Tectonic setting of basic volcanic rocks determined using trace element analysis. Earth Planet Sci Lett 19: 290-300.

65. Pearce TH, Gorman BE, Birkett TC (1977) The relationship between major element geochemistry and tectonic environment of basic and intermediate volcanic rocks. Earth Planet Sci Lett 36: 121-132.

66. Marzoli A, Piccirillo ME, Renne RP, Bellieni G, Iacumin M, et al. (2000) The Cameroon volcanic line revisited: Petrogenesis of continental basaltic magmas from lithospheric and asthenospheric mantle sources. J Petrol 41: 87-109.

67. Green DH, O'Hara MJ (1971) Compositions of basaltic magmas as indicators of conditions of origin: Application to oceanic volcanism. Philosophica Transactions of the Royal Society of London 268: 707-725.

68. Jung S, Masberg P (1998) Major and trace element systematics and isotope geochemistry of Cenozoic mafic volcanic from the Vogelsberg (Central Germany): Constraints on the origin of continental alkaline and tholeiitic basalts and their mantle sources. J Volcanol Geotherm Res 86: 151-177.

69. Wang K, Plank T, Walker JD, Smith EI (2002) A mantle melting profile across the Basin and range, SW USA. J Geophy Res 107.

70. Zhao Z, Bao Z, Zhang B (1998) Geochemistry of the Mesozoic basaltic rocks in southern Hunan Province. Science in China 41: 103-112.

71. Pfander JA, Münker C, Stracke A, Mezger K (2007) Nb/Ta and Zr/Hf in ocean island basalts: Implications for crust-mantle differentiation and the fate of niobium. Earth Planet Sci Lett 254: 158-172.

72. Huang H, Niu Y, Zhao Z, Hei H, Zhu D (2011) On the enigma of Nb-Ta and Zr-Hf fractionation: A critical review. J Earth Sci 22: 52-66.

73. Weaver BL (1991) The origin of ocean island basalt end-member compositions: Trace element and isotopic constraints. Earth Planet Sci Letters 104: 381-397.

74. Hiéronymus B (1973) Mineralogical and geochemical study of bauxite formations in Western Cameroon. Cahier Orstom Série Géologie 5: 97-112.

75. Conte MA, Palladino MD, Perinelli C, Argenti E (2010) Petrogenesis of the high-alumina basalt-andesite suite from Sant'Antioco Island, SW Sardinia, Italy. Per Mineral 79: 27-55.

76. Benkhelil J (1989) The origin and evolution of the cretaceous Benue trough (Nigeria). J Afr Earth Sci 8: 251-282.

77. Maluski H, Coulon C, Popoff M, Baudin P (1995) 40Ar/ 39Ar chronology, petrology and geodynamic setting of Mesozoic to early Cenozoic magmatisim from the Benue Trough, Nigeria. J Geologic Society of London 152: 311-326.

78. Thirlwall FM, Upton BGJ, Jenkins C (1994) Interaction between continental lithosphere and Iceland plume-Sr-Nd-Pb isotope geochemistry of tertiary basalts, NE Greenland. J Petrol 35: 839-879.

79. Bogaard PJF, Worner G (2003) Petrogenesis of basanitic to tholeiitic volcanic rocks from the Miocene Vogelsberg, Central Germany. J Petrol 44: 569-602.

80. Condie KC (2005) High field strength element ratios in Archean basalts: A window to evolving sources of mantle plume. Lithos 79: 491-504.

A Study of Earthquakes in Bangladesh and the Data Analysis of the Earthquakes that were generated In Bangladesh and Its' Very Close Regions for the Last Forty Years (1976-2016)

Md. Abdullah Al zaman[1]* and Nusrath Jahan Monira[2]

[1,2]Department of Physics, University of Chittagong, Bangladesh

Abstract

Bangladesh is a south Asian developing country which is used to struggle with various natural disasters and the earthquake is one of them. Bangladesh is of the most earthquake venerable countries of the world. Here we have tried to discuss about the risks of earthquakes in Bangladesh and the historical earthquakes that occurred in Bangladesh and its surrounding regions with some information. After that we have analyzed the earthquakes that were generated in Bangladesh and it's very close regions(between 20.35° N to 26.75°N Latitude and 88.03° E to 92.75° E Longitude) for the last forty years. We have observed that under the area of concern most of the earthquake occurred were not devastating but the occurrences of those small magnitude earthquakes have been increasing significantly.

Keywords: Bangladesh; Earthquake; Tectonic plates; History; United states geological survey

Introduction

Bangladesh, a geographically and geologically important country of south Asian region with an area of 1, 47,610 square kilometers, extends 820 kilometers north to south and 600 kilometers east to west, located between 24°0'0" N latitude and 90°0'0" E longitude, is often visited by many devastating natural disasters like flood, droughts, tropical cyclones, tornados, thunderstorms, excessive rainfalls, tidal bores, intense summer heat etc. On average Bangladesh is affected by the cyclones 16 times in a decade which are originated from the Bay of Bengal. But an even more lethal natural disaster "earthquake" has been threatening Bangladesh in the recent past. Though Bangladesh is located in a moderately risky territory and the density of population here and the infrastructural condition are been always a matter of concern here. As there is no technology developed to predict the time of occurring, makes earthquake a far devastating one. Recent earthquake in several parts of the world and the destruction occurred, clearly showing that Bangladesh has to take necessary steps to face the earthquake at any time in future.

Bangladesh is surrounded mostly by India, with some portion by Myanmar and the Bay of Bengal to the south. To the north of Bangladesh are the Himalayas, the world's largest mountain range. The fate of Bangladesh is vastly dependent on the Himalayas. Three major rivers the Brahmaputra, Ganges and Meghna-that were originated from the Himalayas and its neighbour mountains flow across the country. These gigantic rivers and their branches deposit huge amounts of mud and sand. All these elements formed the world's largest delta. The total area of Bangladesh can be divided into two regions. The first one is the broad deltaic plain covering 85% of land area of Bangladesh and the remaining one is the small hilly regions in some definite places. Bangladesh is one of the rainiest territories of this planet. During the monsoon the deltaic plain is frequently visited by the floods. But Bangladesh is not shaped by just rivers and flooding. It's also shaped by the incidents occurring beneath its surface, where tectonic plates are continuously changing their position. Bangladesh sits where three tectonic plates meet (Figure 1). These are the Indian plate, the Eurasian plate and the Burmese plate. In fact the whole Indian subcontinent is lying on the junction of the enormous Indian plate and Eurasian Plate. But living under the influences of third tectonic plate i.e. the Burmese plate and it's junction with the Indian plate makes Bangladesh one of the most tectonically active regions in the world.

Current condition of the tectonic plates related to Bangladesh

After detaching from the supercontinent Gondwana 110 million years ago, the Indian plate started its journey towards north and collided with Eurasian plate about 50 million years ago during late cretaceous period [1]. It was moving 20 centimeters per year before colliding, which was fastest among all other tectonic plates. The collision between the Indian plate and the Eurasian plate along the boundary between today's India and Nepal formed the Tibetan Plateau and the mighty Himalaya Mountains which are still rising [2]. Currently the Indian plate is moving in north east at a speed of approximately 6 cm per year.

The Eurasian plate's history is far older than relatively young Indian plate. It's the result of several collisions by many small cratons at different times in the past. It's a slow moving plate compared with the others. The whole Eurasia sits on the Eurasian plate. Currently the Eurasian plate is moving north at 2 centimeters per year. The Indo-Burman Mountain ranges mark the boundary between the Indian and Eurasian plates.

The Burmese plate located in the Southeast Asia is a minor tectonic plate or Micro-plate. It's in a highly tectonically active region. In the past it was a part of the Eurasian plate. It was separated from the Eurasian plate when the Indian plate collided with Asia. It's now surrounded by the Indian, Sunda, Australian and Eurasian plates. These plates gave the Burmese plate a quite strange shape due to convergences. It's a very slow moving plate and there is debate over the plate's motion.

---

*Corresponding author: Md. Abdullah Al zaman, Department of Physics, University of Chittagong, Bangladesh, E-mail: proyashzaman1@gmail.com

Figure 1: Geographical location of Bangladesh in terms of tectonic [2].

The Indian Plate is sub-ducting beneath the eastern facet of the Burma plate formed the Sunda trench causing it to move 46mm per year. The devastating 2004 earthquake and tsunami in Sumatra occurred along the boundary between the Indian and Burmese plate. [3]

Fault zones of Bangladesh

Bangladesh is surrounded by a number of tectonic blocks responsible for many earthquakes in the past. Calcutta, Assam, Tripura are the three very earthquake prone regions that are joined to Bangladesh in the borders in the Northern, Western and North-Eastern part respectively. If we consider the tectonics and geology, five major faults are significant for the occurrences of devastating earthquakes and these are

- Bogra Fault Zone

- Tripura Fault Zone

- Shilong Plateau

- Dauki Fault Zone

- Assam Fault Zone

Bogra fault is a normal or gravity fault (Table 1). It is very close to the Bogra town and Jamuna River. It might be associated with flexure of the basin along its western margin. It was active in Palaeogene and Neogene times. Movement along the Bogra fault led to the deposition of a large amount of sedimentary pile within the Bogra graben [4].

Sl. No.	Fault zone	Maximum magnitude
1	Bogra fault zone	7.0
2	Tripura fault zone	7.0
3	Shilong plateau	7.0
4	Dauki fault zone	7.3
5	Assam fault zone	8.5

Table 1: The maximum magnitude of the earthquakes that can be produced from the fault zones.

Tripura is one of the states of India surrounded by Bangladesh and two other states Mizoram and Assam of India. The area is surrounded by Koplili fault; Kaladan fault etc. which have produced many earthquakes. The Tripura-Naga orogenic belt is a zone of highly faulted tertiary deposits which has witnessed earthquakes of moderate magnitudes [5].

Shillong plateau is characterized as a seismically active and geologically complex region located on the collision boundary between

Indian and Eurasian plate in the Meghalaya state of India. The general altitude of the Plateau is about 1,500 m. The plateau is composed of the Precambrian Metamorphic rocks and the Tertiary and Quaternary deposits are limited on the southern foothills of the Shillong Plateau indicating the successive uplift Of the Shillong Plateau and the process started from the Pliocene time [6]. The 1897 Ms. 8.0 Great Assam earthquake is well-known as a historic earthquake that occurred below the Shillong Plateau. The Shillong plateau presently behaves like a rigid body tied to the Indian Shield at a velocity of 46.5 ± 1 mm/a toward N 51° E. [7,8]

Dauki fault zone is a 300 km long north dipping reverse fault along the Meghalaya-Bangladesh border and inferred to go through the southern margin of Shillong plateau (Figure 2). It has a major role in deforming the surrounding areas. The Dauki fault is believed to be active in the past and it is most likely the fault associated with the magnitude >7 earthquake in Sylhet (Shilchar) known as Cachar earthquake (10 January 1869) [8]. Though it is inactive in the recent times still it is considered as one of the major threats for Bangladesh for the occurrence of devastating earthquakes (Figure 2) [7].

But there are more reasons to worry. A recent study reveals the existence of subduction zone of about 250 kilometers that can produce an earthquake of magnitude 8.2-9.0 [9]. After setting two dozen ground-positioning (GPS) instruments linked to satellites, capable of tracking tiny ground motions and analyzing the ten years of data the scientists have shown that eastern Bangladesh and a bit of eastern

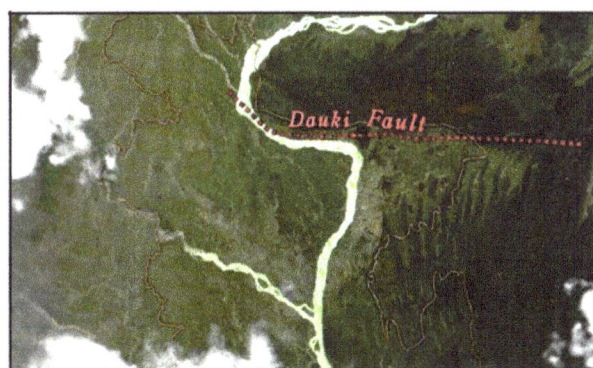

Figure 2: Dauki fault zone [2].

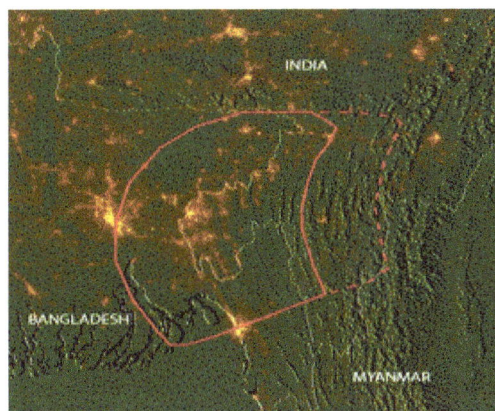

Figure 3: The most probable quake zone. Solid line representing the area under influence and dashed line represents a scenario in which the slip might take place along a separate fault [10]

India are pushing diagonally into western Myanmar at a rapid clip-46 millimeters per year or about 1.8 inches. After combining with the existing GPS data from India and Myanmar, the measurements show that much of the resulting strain has been taken up by several known, slowly moving surface faults in Myanmar and India. But the rest of the movement-about 17 millimeters, or two-thirds of an inch per year is shortening the distance from Myanmar to Bangladesh which is inferred as a subduction process going on. This shortening of distance is building pressure and that also a few kilometers below the surface and this process is going for a long time. The most uncomfortable reason is Dhaka, one the most densely populated territories and also the capital of Bangladesh is also under the range of this zone (Figure 3) [10].

Major Earthquakes in Bangladesh and its Surrounding Regions

When we talk about the history of earthquake in Bangladesh or Indian subcontinent there are always lack of evidences. We even don't have specific information about the earthquakes occurred around 500 years ago except some historical evidences which are not enough to specify the exact intensity or magnitude of the earthquakes. There are evidences of terrible earthquakes felt in Sylhet, Chittagong and Dhaka in 1548, 1642, 1663, 1762, 1765, 1812, 1865, 1869 but we do not have any information about their magnitudes. At most we can deal with their impact. It is reported that the 1663 earthquake in Assam and Sylhet lasted about half an hour. The 1765 earthquake was so devastating that it raised the Coast of Foul Island by 2.74 m and the northwest coast of Chedua island by 6.71 m above sea level and also caused a permanent submergence of 155.40 sq. km near Chittagong. The earthquake took 500 lives in Dhaka. [11] We have much clear data of earthquakes occurred more than two hundred years in this region. Here are some major earthquakes with some information.

Cachar earthquake

10th January 1869: This earthquake occurred in the Sylhet region (Silchar). The epicenter of the earthquake was 250 kilometers away from the current capital city Dhaka. According to the estimation by Braseys and Douglas the magnitude of this earthquake was 7.39 [12]. The Dauki fault is believed to be responsible for this earthquake. The earthquake was not a single shock rather it lasted, on and off and some shocks lasted for about five minutes. Most of the houses were down. But the number of casualties is not reported [8,11,13].

Bengal Earthquake

14th July 1885: The 1885 Bengal Earthquake, also known as Manikganj Earthquake had the magnitude 7.0. It's possible epicenter was at Kudalia in Saturia (Manikganj) which is 170 kilometers away from the capital Dhaka. This event was generally associated with the deep-seated Jamuna Fault. The earthquake was so strong that it was felt by the people of Bihar, Sikkim, Manipur (India) and Burma (Myanmar).

Destructions of Buildings and losses of lives were reported from Dhaka, Mymensingh, Sherpur, Pabna etc. There is probability of recurrence of this kind of earthquake in this region from 2015-2020 [8,14,15].

Meghalaya Earthquake

10 January, 1889: There is not much information about this earthquake occurred in 1889. The possible epicenter of this earthquake was Jaintia Hills in Meghalaya State of India. The magnitude of the earthquake was 7.5. It affected the Sylhet town and its surrounding regions. No losses of lives were reported [11].

Great Indian Earthquake

12th June 1897: It is also known as the Shilong Plateau earthquake. The magnitude of that earthquake was 8.0 [8]. The earthquake raised the northern edge of the plateau about 10 meters. The epicenter was 230 kilometers away from Dhaka. More than thousand people died in that event and most of the buildings in the affected region were damaged [11,13].

The Srimangal Earthquake

18th July 1918: This earthquake's epicenter was at Srimangal in Moulavi-Bazar (Sylhet) which is about 150 kilometers away from Dhaka with a magnitude of 7.6. The earthquake occurred in the afternoon and for this the losses of lives was not reported. The earthquake was felt in Myanmar and North East of India. The brick built houses in Srimangal were seriously damaged but minor effects were observed in Dhaka [11,16].

Meghalaya Earthquake

9th September 1923: This earthquake with a magnitude of 7.1 shook the south of Meghalaya, west Bengal (India) and Bangladesh in the morning. It caused heavy damages in Mymensingh, Cherrapunji and Guwahati. It was also felt in Chittagong and Barisal [13].

The Dubri Earthquake

3rd July 1930: The epicenter of this earthquake was in Dubri, Assam with a magnitude of 7.1. It shook Assam, west Bengal and Bangladesh early in the morning. Heavy damages occurred in Assam, many people were injured but fortunately there were no losses of lives as it occurred early in the morning. This earthquake was followed by six major aftershocks of magnitude 6. The eastern part of Rangpur district in Bangladesh was the worst sufferer of that earthquake. [11,13].

The Assam Earthquake

15th August 1950:One of the largest earthquakes of 20th century with a magnitude of 8.7 killed about 1500 people in India. Heavy damages were observed. It also shook Bangladesh, Myanmar and a part of China but no significant damages were reported in those regions. [11,13].

The Bay of Bengal Earthquake

11th August 2009: The epicenter of that earthquake was located at the North Andaman Islands of the Bay of Bengal and seacoast of Myanmar with a magnitude of 7.5. It was strongly felt from Dhaka but fortunately no heavy damages occurred [11].

The Myanmar Earthquake

24th August 2016:The epicenter of this earthquake was in 25 kilometers west of Chauk in Myanmar with a magnitude of 6.8. It was strongly felt in Chittagong and Dhaka. 3 people died in Myanmar but in Bangladesh, no casualties were reported but 20 people were seriously injured [17].

With the help of Google map and GPS Geoplaner (A web based application that provides several GIS and GPS utilities) we have generated an image that is showing the approximate location of the epicenters of the earthquakes (Figure 4).

Data Analysis

Now we are discussing about the earthquakes those generated in between 20.35° N to 26.75° N latitude and 88.03° E to 92.75° E longitude for the last forty years (1976-2016). The area we have chosen covers the area of Bangladesh and it's very nearly regions. The necessary data are taken from United States Geological Survey (USGS). The data contains the depth of the epicenters, magnitude of the earthquakes, time of the event occurred and some others related data. We have found 284 earthquakes occurred in that concerning area (Figure 5).

Results

The diagrams below are showing the characteristics of those earthquakes. Here we have represented 283 earthquakes with the depth of their epicenters and their magnitudes. We have also tried to represent the number of earthquakes occurred in every 10 years in this area from 1976-2016 (Figures 6-9) [20].

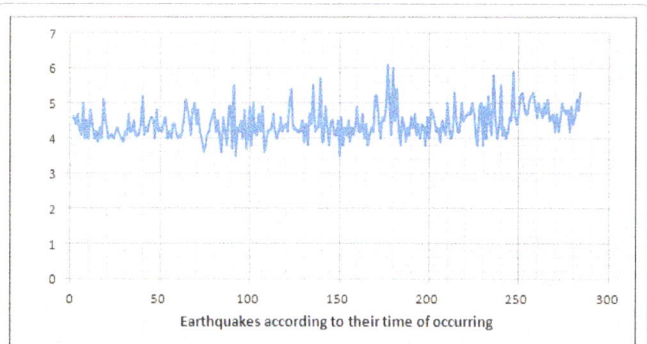

Figure 4: Epicenters (approximately) of the historical earthquakes in Bangladesh and its surrounding regions. Here A, B, C etc. are representing earthquakes, e.g. "A" signifies "The Cachar Earthquake (1869)" [19].

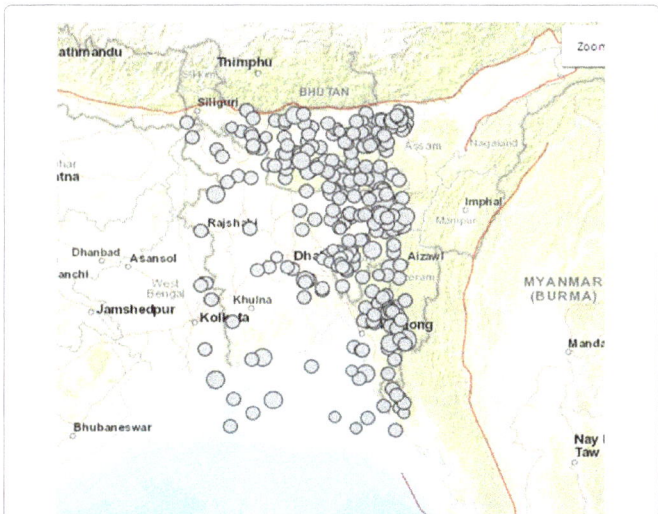

Figure 5: Earthquakes that were generated in the area of concern. The grey circles signifying the epicenters of the earthquakes and the size of the circle show the strength of the earthquake. The solid red lines are representing the plate boundary [18].

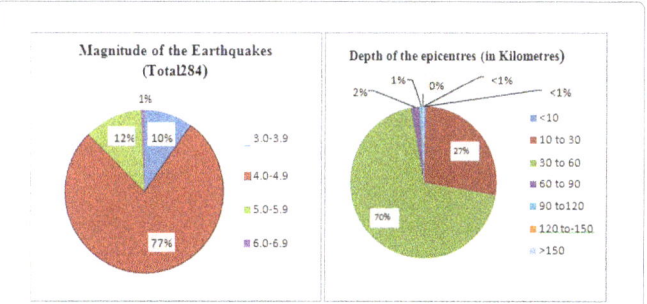

Figure 6: Earthquake magnitude vs. Earthquake according to their time of occurrence (i.e. 1 for the earthquake on 2016-11-15 or no. 291 for the earthquake on 1976-06-23).

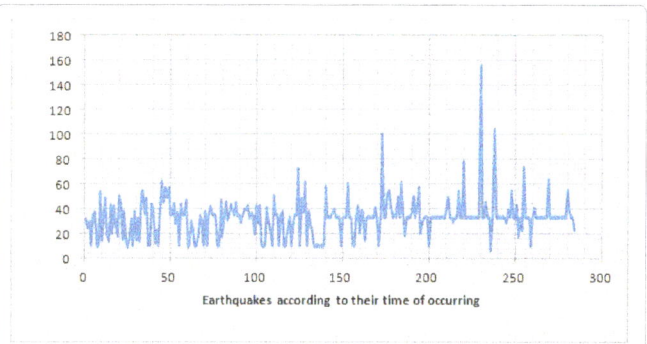

Figure 7: Depth of the epicenters vs. Earthquakes according to their time of occurrence (i.e. 1 for the earthquake on 2016-11-15 or 291 for the earthquake on 1976-06-23.)

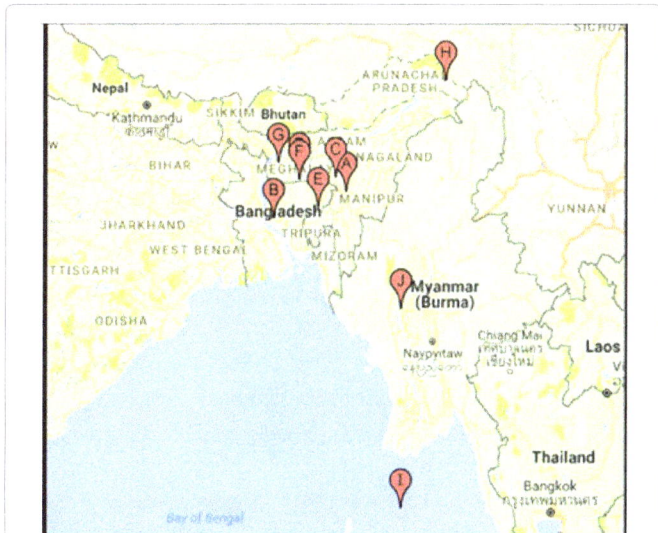

Figure 8(a): Earthquakes of different Magnitudes.   Figure 8(b): Depths of epicentres of the earthquakes

Figure 8: Magnitude and depth of the earthquake for the last forty years.

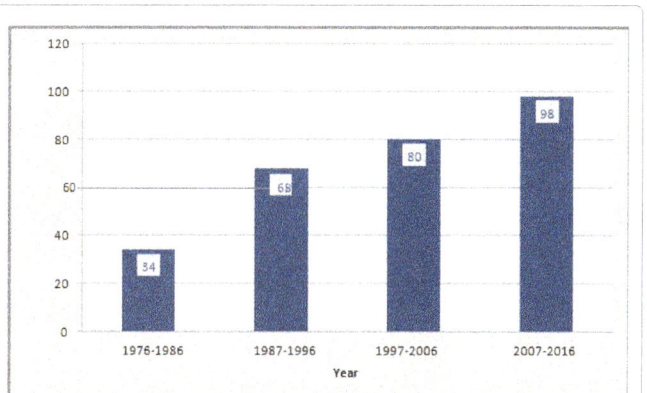

Figure 9: Frequency of the earthquakes.

Conclusion

We have analyzed the data of the earthquakes occurred in the last forty years in Bangladesh and it's very close regions (20.35° N to 26.75°N latitude and 88.03° E to 92.75° E longitude). From Figure 6 and 8a we have seen that most of the earthquakes occurred in the area of concern were not devastating. We have only two cases when the magnitude of the earthquake exceeded the 6.0 mark. Most of the earthquakes were ranged between magnitudes 4.0 to 4.9 (77%) which is not harmful enough and only 12% of them were ranged between magnitudes 5.0 to 5.9. Figures 7 and 8b have shown that depth of the epicenters varied between 8 kilometers to 155 kilometers beneath the surface of the earth. But in most cases (70%) the depth of the epicenters were ranged between 30 to 60 kilometers. About 27% of the earthquakes were generated 10-30 kilometers beneath the earth surface. We have only a single case when the epicenter is 155 kilometers beneath the earth surface. From Figure 9 we can see that the number of earthquakes generating in this area is increasing and for the last ten years it has increased significantly if we had taken more data of the earthquake for more twenty years in the past we might have a clearer picture about this.

This above analysis indicates that the earthquake occurred in the area of concern were not devastating. But it actually does not show the real picture of the risk of Bangladesh about the earthquakes. From Figure 4 suggests that if we had taken a larger area for study then few of the devastating earthquakes in the past would have come to our calculation as the earthquake prone zones are not so far from our area of concern. Besides we have studied only the history of forty years of earthquakes in this region. Forty years are not enough as compared to time of recurrence of a devastating earthquake in most cases.

The newly discovered subduction zone, the geological position and the historical earthquakes clearly indicates that the government of Bangladesh should put much emphasize on earthquakes and how to minimize its damage. Disaster management system should be more developed to cope with the requirement. The building codes should be followed by the citizens and the government should try to increase the awareness about the safety rules of earthquakes among the mass people.

Acknowledgement

The data and the Figure 5 are taken from the website of United states Geological survey's website. This agency provides data about the earthquakes. The link of the data: https://on.doi.gov/2t1DCfJ

References

1. Prakash K, Yuan X, Ravi K, Rainer K, Chadha RK (2007) The rapid drift of the Indian tectonic plate. Nature 449: 894-897.

2. Earthquake risk in Bangladesh (2013) American museum of natural history.

3. http://tectonicsofasia.weebly.com/burma-plate.html

4. Ali H (1998) Earthquake database and seismic zoning of Bangladesh. Bangkok 59-74.

5. Geological survey India

6. http://en.banglapedia.org/index.php?title=Shillong_Plateau

7. Morino M, Kamal ASM, Muslim D, Ali RME (2011) Seismic event of the Dauki Fault in 16th century confirmed by trench investigation at Gabrakhari Village, Haluaghat, Mymensingh, Bangladesh. J Asian Earth Sci 42: 492-498.

8. Bilham R (2000) Earthquakes in India and the Himalaya: Tectonics, geodesy and history. Annals Geophys 47:

9. Stickler S, Mondal DR, Akhtar SH (2017) Locked and loading megathrust linked to active subduction beneath the Indo-Burman Ranges. Nature Geosci 9: 615-618.

10. https://phys.org/news/2016-07-giant-quake-lurk-bangladesh.html

11. http://en.banglapedia.org/index.php?title=Earthquake

12. Ambraseys NN, Douglas J (2004) Magnitude calibration of North Indian earthquakes. Geophys J Int 159: 165-206.

13. http://aasc.nic.in/course%20material/Disaster/documentation%20on%20past%20disasters.pdf

14. https://sites.google.com/site/indiaquake/bengal-earthquake-of-1885

15. http://archive.thedailystar.net/2005/05/28/d505281501106.htm

16. https://www.scribd.com/document/37862838/Srimangal-Earthquake-8July-1918

17. https://www.theguardian.com/world/2016/aug/24/myanmar-struck-by-6-8-magnitude-earthquake

18. United States Geological Survey Agency (USGS).

19. http://www.geoplaner.com/

20. https://www.tes.com/lessons/uwZfS4kXsljS5g/earthquake-risk-in-bangladesh

Preliminary Magnetostratigraphic and Isotopic Dating of the Ngwa Formation (Dschang Western Cameroon)

Benammi M[1]*, Hell JV[2], Bessong M[2], Nolla D[2], Solé J[3] and Brunet M[1,4]

[1]*Institut de Paléoprimatologie, Paléontologie Humaine: Évolution et Paléoenvironnements (IPHEP), UMR-CNRS 7262, Bâtiment B35, 6 rue Michel.Brunet, 86022 Poitiers Cedex, France*
[2]*Institut de Recherches Géologiques et Minières du Cameroun, BP 4140, Yaoundé, Cameroun*
[3]*Universidad Nacional Autónoma de México, Instituto de Geología Dept. de Geoquímica Cd. Universitaria, Coyoacán 04510 México DF*
[4]*Collège de France, Chaire de Paléontologie humaine, 11 Place Marecelin Berthelot, 75231 Paris cedex 05*

Abstract

A magnetostratigraphic study has been carried out to constrain the age of the volcano-sedimentary Ngwa formation in the eastern part of the Dschang region. A stratigraphic section of about 80 meters thick corresponding to 26 sites has been sampled, and it is composed mainly of fine-grained sandstones, clays, lignite, volcanic sediment and tuffs. A magnetic study conducted on 56 samples shows one or two components of magnetization carried either by titanomagnetite, magnetite and Fe-sulphide. The section that was sampled shows one normal polarity and one reversed polarity. In the lower part of the section, a K-Ar radiometric dating was performed on the plagioclase minerals isolated from the tuffs level situated about 15 meters above the lignite seam, and gave an age of 20.1 ± 0.7 Ma. Constrained by this age, the observed polarity zones can be readily correlated with chrons C6An.1n-C6An.1r of the GPTS. This study suggests that the age of the lignite is comprised between 20.04 Ma and 20.21 Ma. The mean direction of the characteristic remanent magnetization documents a counterclockwise vertical axis rotation of about 8° with respect to the expected Lower Miocene direction derived from the Africa polar wander curve.

Keywords: Cameroon; Ngwa formation; Magnetostratigraphy; Lignite; Lower Miocene

Introduction

Most studied and well known continental basins from Cameroon are Mesozoic intracontinental basins such as Mamfe [1,2], Mbere Djerem [3,4], Babouri Figuil [5], Mayo Oulo Lere [6-8], Hamakoussou [9,10] and Mayo Oulo-Léré [6].

However, little attention has been paid to the continental Cenozoic sediments. Some fragmentary sediment lithofacies were identified at Ngwa Village (eastern Dschang) as well as some enclaves of sedimentary rocks randomly distributed in a volcanic complex [11]. However, very little is known about the features of these sedimentary rocks, their age or provenance. Research work started in 1925 in the Dschang area to assess the extent of the lignite deposits [12]. After several hundred meters of gallery and the test use of the lignite as a fuel, the exploitation was stopped because of the bad quality of the lignite. Capponi [12] suggested a Tertiary age for the Dschang lignite in comparison with lignite known in southern part of Nigeria. The lignite deposit is known also North-east of Bamenda, the strata there are vertical and overlaid by clays which are the product of basalt alteration. The lignite seams found in Dschang and Bamenda are commonly brownish to black in color and vary in thickness from a few centimeters to a maximum of 1.8 m. They are thinly laminated and contain fossils of leaves, fruits and wood fragments. Magnetic polarity sequences with K-Ar dating result will the calibration of the Ngwa Formation to the geomagnetic polarity time scale (GPTS). In this paper, we present magnetostratigraphic results of the Early Miocene section of the Ngwa Formation and its correlation to the GPTS. This study is of great importance, since it is the first one carried out in a Neogene continental formation of Cameroon.

Introduction to geological setting

The study was carried out in the area of Ngwa, a small locality in the eastern part of the Dschang region on the southern slope of Mount Bambouto belonging to the Cameroon Volcanic Line (CVL) (Figure

1). The CVL represents a 1600 km long chain of Cenozoic volcanic and sub-volcanic complexes that straddles the continent ocean boundary and extends from the Gulf of Guinea to the interior of the African continent [13]. It constitutes a major tectonic feature in Central Africa characterized by a zone of fault-bounded horsts and grabens that extend N30°E. These structures are thought to be induced by a network of combined faults, related to an intra-plate sliding system of high extension [14]. Its continental part is represented by the major volcanic massifs of Mounts Cameroon, Rumpi, Manengouba, Bambouto and Oku and the volcanic necks and plugs of the Kapsiki plateau and Benue Valley [15]. Mount Bambouto is the third largest volcano of the CVL after Mounts Cameroon and Manengouba [16]. According to Kagou Dongmo et al. [17], the geological history of Mount Bambouto is divided into three stages: 1) The pre-caldera stage corresponds to the building of the main shield volcano, between 21 and 16 Ma. It was mainly effusive and basaltic; 2) Collapse of a large caldera initiated the second stage characterized by the extrusion of ignimbritic trachytes, between 16 and 11 Ma [18]; 3) The post-caldera and third stage consists of the pouring out of intra-caldera and adventive basaltic flows, and of extrusions of phonolitic domes, between 9 and 4.5 Ma [13] (Table 1).

The Dschang region, on the southern slope of Mount Bambouto in the West Cameroon Highlands, forms the northern edge of the

*Corresponding author: Benammi M, Institut de Paléoprimatologie, Paléontologie Humaine: Évolution et Paléoenvironnements (IPHEP), UMR-CNRS 7262, Bâtiment B35, 6 rue Michel.Brunet, 86022 Poitiers Cedex, France, E-mail: mbenammi@univ-poitiers.fr

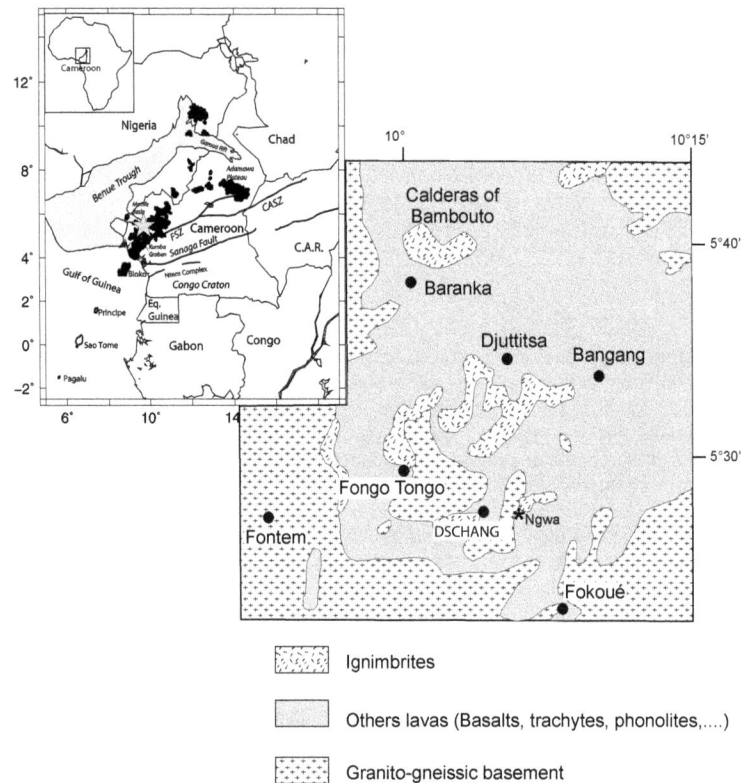

Figure 1: Sketch map of the central portion of the Cameroon Volcanic Line in West Equatorial Africa (see inset), FSZ, Foumban shear zone; CASZ, Central African shear zone. Geological sketch map of the Bambouto Mountains indicating the location of the study area [15].

Sample	Mineral	%K	^{40}Ar* (moles/g)	%^{40}Ar*	Age (Ma)
Cameroon	Plagioclase	0.984	3.45×10^{-11}	68.3	20.1 ± 0.7

Table 1: K-Ar datting of the volcanic tuff of Ngwa formation.

Mbo plain (Figure 1). It is characterized by various volcanic products covering the basement granitoids [11]. The basement rocks in the Dschang region consist of Neoproterozoic granite-gneisses, Late Proterozoic granitoids intruded within the granite gneisses and gabbroic dykes that crop out in the two previous units. The basement rocks are partly covered by a very thick layer of volcanic deposits derived from Mount Bambouto. Volcanic activity at Mount Bambouto originated as a complex sequence of basalt, trachyte and phonolite lava flows and ignimbrites [13,15,18,19].

The most important sedimentary rock outcrop is at the Ngwa locality in the eastern part of the Dschang region (Figure 1), and a number of small lignite deposits are encountered. The Ngwa Formation consists of a sequence of coarse-grained sandstone, brown and dark coloured clays and carbonaceous shale intercalated within lignite seams of continental origin. A sedimentological study was conducted by Kenfack et al. [11] who highlighted four lithofacies, and the volcanic sediment was related to volcanic activity at Mount Bambouto during the Miocene. This led to the deposition of volcano-clastic sediment which occurs at the base of the sedimentary sequence that unconformably overlys the Precambrian basement (Figure 5).

Description of the Ngwa section

The sedimentary succession exposed in the Ngwa corresponds to 80 m of lacustrine and volcano-clastic sediments belonging to the Ngwa Formation. The lower portion of the sequence is represented by an interstratified succession of conglomerate, clay, sand, volcanic sediment and lignite, all of variable thickness and texture. The conglomerate beds in the lower part lay on the granite substratum, and are composed of chert, quartz, lithic, and laterite clasts (Figure 5 and Figure 6A). The clasts don't exceed 5 cm in diameter. The conglomerate beds occur at distinct layers, but it is commonly present at the bases of sandstone units. Lignite is developed as stratified lignitic clays and marls or as massive lignite (Figure 5 and Figure 6B). They commonly contain well preserved macrofloral remains as the lignitic clays (leaves, fruits and wood). The absence of rooting structures indicates an allochthonous origin of macrofloral material. The sandstones are massive, and the bases of the beds are non-erosive. A fine grained yellowish tuff layer (about 2 m thickness) occurs above the basal conglomerate. This layer is called "Bamilékite" according to Capponi [12], and it marks a break slope in the section (Figure 6A). The overlying bed (nearly 40 m thickness) of the section is dominated by volcano-clastic sediment (Figure 5). All the section is covered by laterized soil. The soil cover is composed of massive unconsolidated and porous reddish brown laterized sandy clay.

Materials and Methods

Paleomagnetic sampling and sample analysis

A total of 29 cores (Figure 5) were drilled in the field from 19 levels with a portable gasoline powered drill and oriented in situ with

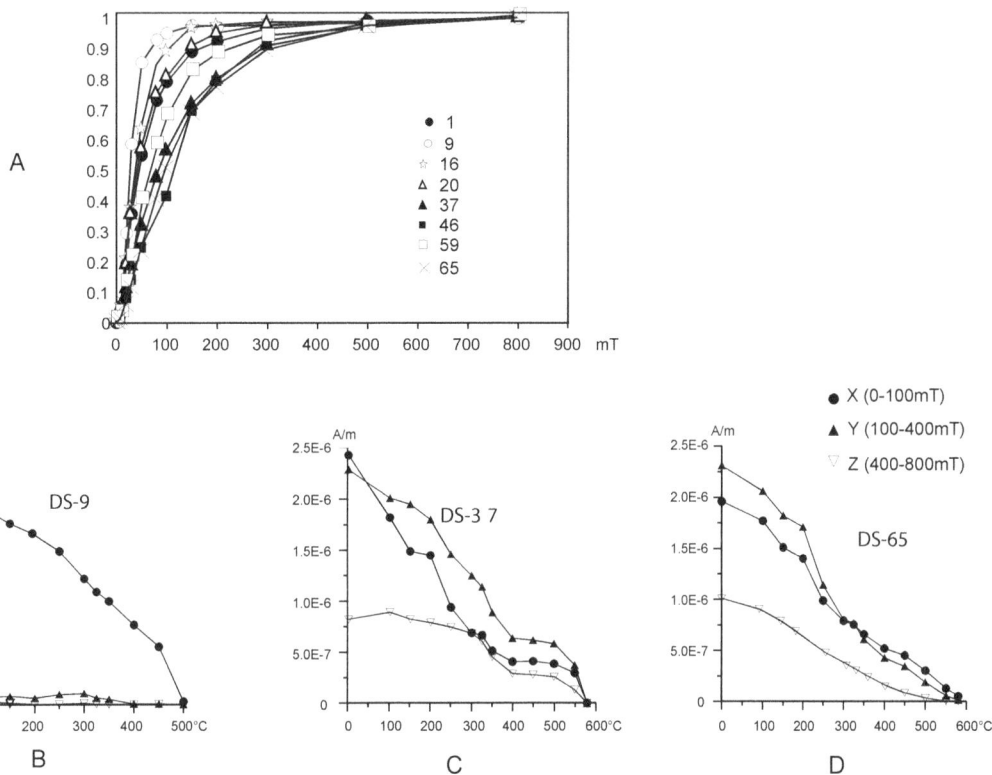

Figure 2: A) Progressive acquisition of an isothermal remanent magnetization in field of 800 mT. B-D): Thermal demagnetization of a composite IRM acquired on three orthogonal axes.

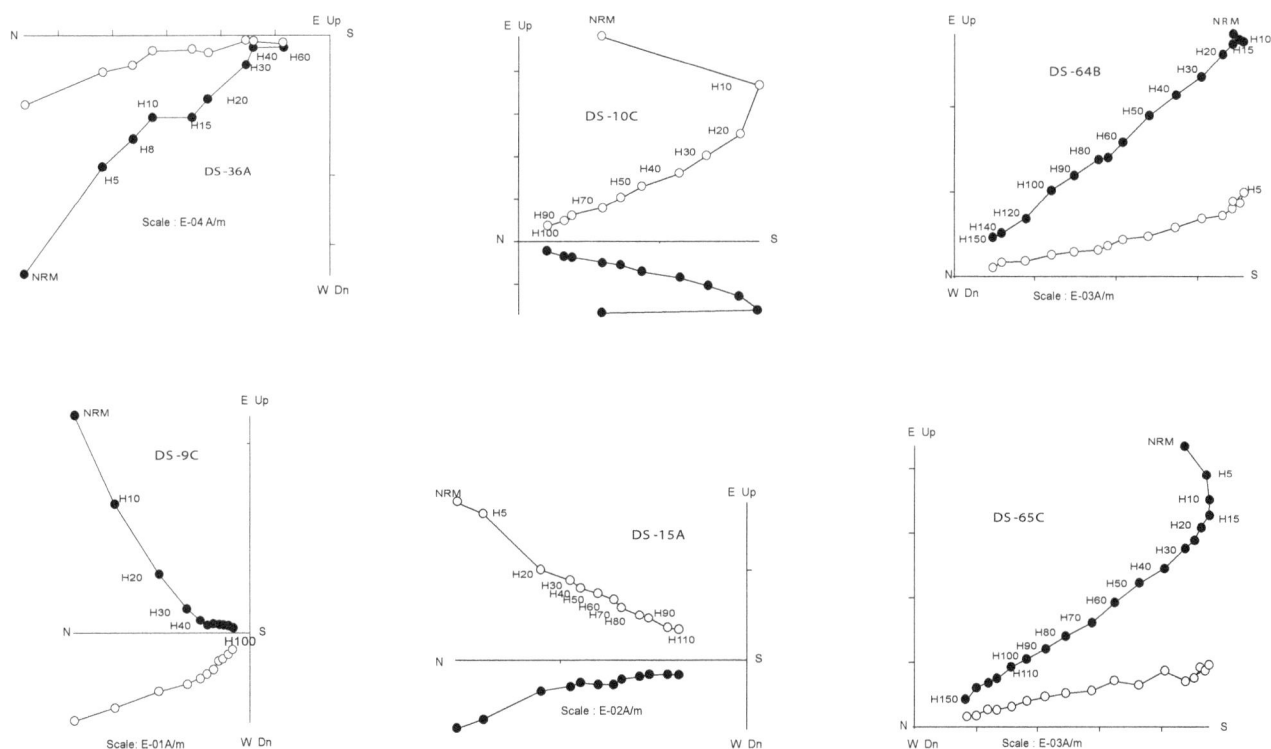

Figure 3: Orthogonal vector plots of AF demagnetization of representative samples. Solid circles indicate horizontal component; open circles indicate vertical component. Demagnetization steps in mT. NRM is the natural remanent magnetization.

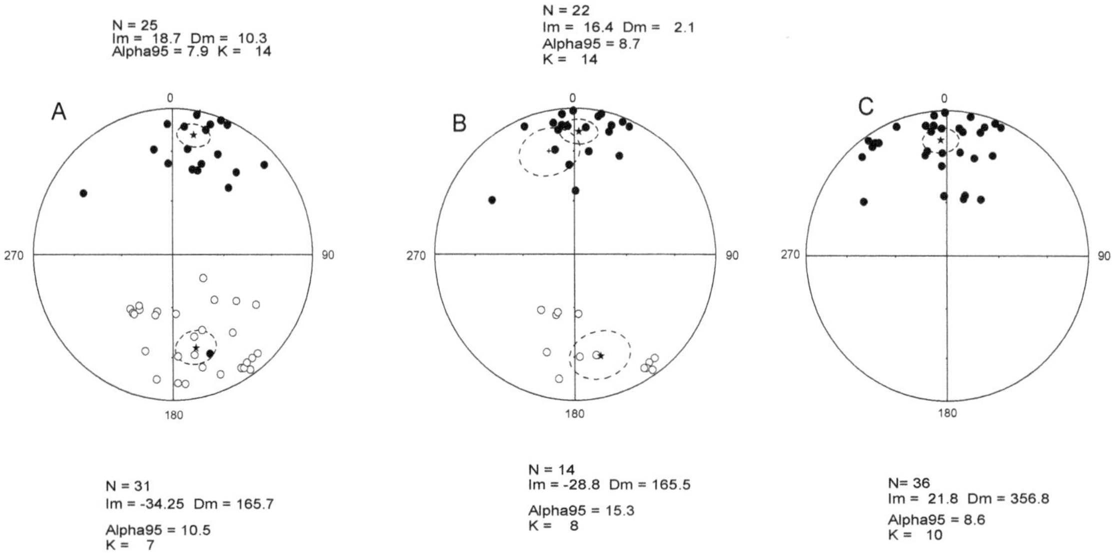

Figure 4: Equal-area stereographic projection of characteristic directions (normal and reversed polarities) isolated from samples. Circle around mean direction represents 95% confidence limit. A: mean direction for all normal and reversed samples. B: mean direction of normal and reverse drilled samples appear to be antipodal, and the reversal test is positive. C: mean direction of all drilled site (normal and reversed polarity) are plotted on the same hemisphere.

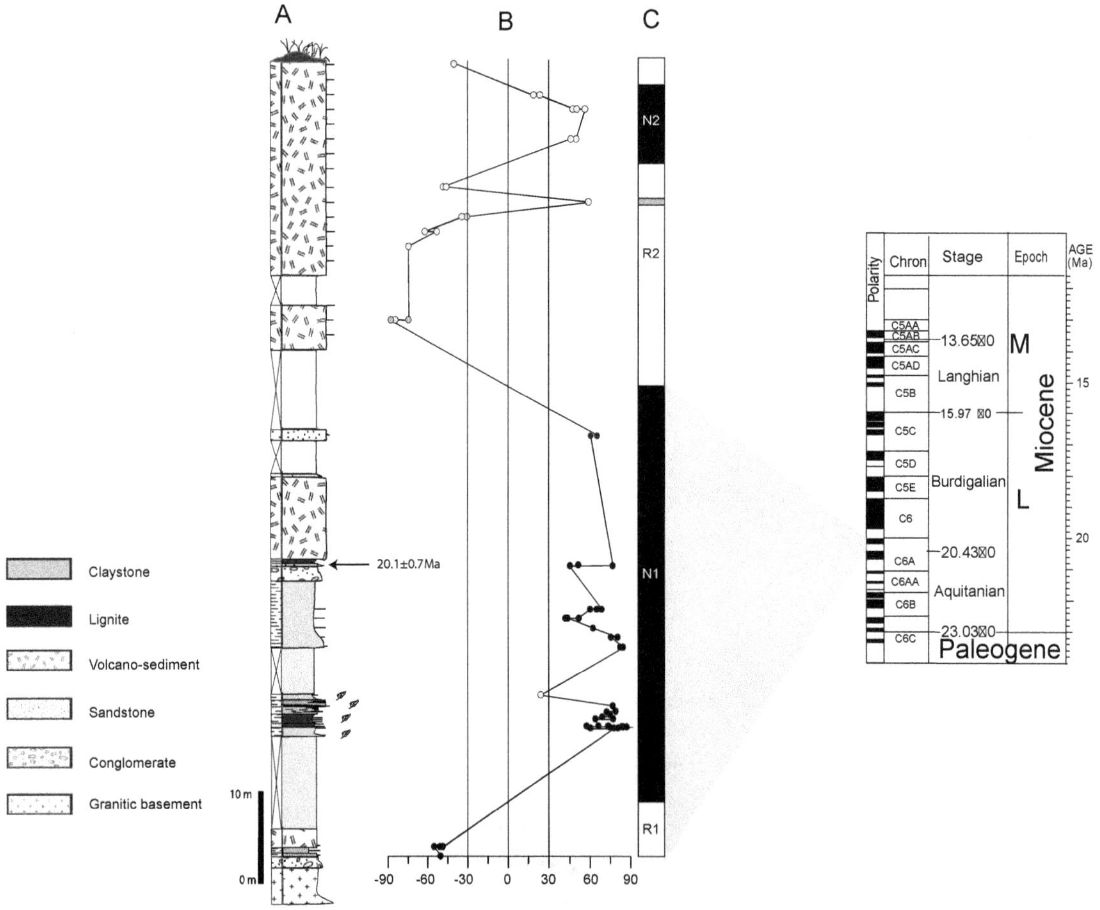

Figure 5: Lithology, magnetostratigraphy, and proposed correlation with the geomagnetic polarity time-scale. A: section and stratigraphic position of sampling sites. B: Positive and negative virtual geomagnetic pole (VGP) latitudes represent normal and reverse polarities respectively, gray dots correspond to hand bloc samples (C). D: Portion of the GPTS [27]. C: Polarity column (black/white bars represent normal/reverse polarity).

Figure 6: A) View of the lower part of the section with sandstone, conglomerate and grained yellowish tuff. B) View of the lignite layer, black dots indicates the paleomagnetic sampling. C) Volcanic tuff used for K-Ar dating.

a magnetic compass. Most drilled sites correspond to the lower part of the section. One or two orientated hand samples (blocks) were collected from the other 17 sites after digging to remove a lateritic soil that covers the formation. At each stratigraphic horizon described herein, a flat face was shaved on the in situ samples with a hand rasp, and then the strike and the dip of the face were measured with a Topochaix compass. Of the 17 originally sampled sites, 7 were lost during cutting or transport. The sampled lithology includes sandstones, tuff, clays and shale. One fresh sample about 15 m above the lignite seams was retained for radiometric determinations (Figure 6C).

Samples were analyzed in the paleomagnetic laboratory at the iPHEP (Institut de Paléoprimatologie, Paléontologie Humaine: Evolution et Paléoenvironnements) at the University of Poitiers. Remanent magnetization was measured with a JR6 magnetometer combined with stepwise thermal or alternating field demagnetization in a magnetically shielded room. To better constrain the magnetic mineralogy, we studied the acquisition of isothermal remanent magnetization (IRM), and then the stepwise thermal demagnetization of three-axis differential IRM following the method of Lowrie [20]. The specimens were subjected to stepwise thermal demagnetization in steps up to 600°C. The IRM was determined with a pulse electromagnet. Thermal demagnetization was done with a magnetic measurement thermal demagnetizer (*MMTD80*) shielded furnace. Progressive thermal demagnetization was carried out, in steps of 30°C to 40°C, from 100°C, until either the magnetization intensity fell below the noise level or the direction became erratic. The majority of specimens were submitted to stepwise alternating field (AF) demagnetization in increments of 5-10 mT, up to a maximum field of 150 mT using a Molspin Ltd. high-field shielded demagnetizer. Characteristic magnetization components were isolated by applying the method of Kirschvink [21] to vector segments composed of at least

more than seven points with a maximum angular deviation of less than 15. Data resulting from AF demagnetization were plotted on orthogonal vector plots [22]. To determine characteristic magnetic directions, principal components analysis was carried out on all samples. These paleomagnetic directions were then analyzed, using Fisher statistics, to determine site mean declinations, mean inclinations, and associated precision parameters.

Radiometric dating

K-Ar analysis on tuff sample was conducted at the Instituto de Geología, UNAM (Univesidad Nacional Autonoma de Mexico). Plagioclase was obtained by crushing, sieving, selection of the best fraction (300-400 μm), washing, and separation by magnetic methods (Isodynamic Frantz separator). The K content was measured by XRF on 50 mg aliquots using a specific regression for measuring K in K-Ar samples [23]. Analytical precision was better than 2%. Duplicate samples weighing between 6 and 8 mg were degassed under high vacuum at ~150°C for twelve hours before analysis in order to reduce atmospheric contamination. Argon was extracted by complete sample fusion using a 50W CO_2 laser defocused to 1-3 mm of diameter. The evolved gasses were mixed with a known amount of 38Ar spike and purified with a cold finger immersed in liquid nitrogen and two SAES getters in a stainless steel extraction line. Measurements were done in static mode with an MM1200B mass spectrometer using electromagnetic peak switching controlled by a Hall probe. Analytical precision on ^{40}Ar and ^{38}Ar peak heights was better than 0.2%, and better than 0.5% on ^{36}Ar. The data was calibrated with internal standards and the international reference materials LP-6 and HD-B1 biotites. All ages were calculated using the constants recommended by Steiger and Jäger. A detailed description of the procedure and calculations can be seen in Solé [24].

Results

Magnetic properties and characteristic directions

A set of rock magnetic experiments was conducted to characterize and identify the magnetic mineralogy of the main lithologies. We first analyzed the acquisition of IRM (Isothermal Remanent Magnetization) up to 800 mT and its subsequent thermal demagnetization. Following the procedure described by Lowrie [20], magnetic fields of 1, 0.4 and 0.12 T were successively applied to each of the three perpendicular directions prior to thermal demagnetization. The IRM acquisition curves (Figure 2A) show a broad range of coercivities. The initial increase of magnetization up to 100-150 mT indicates the presence of low coercivity minerals. Saturation was not achieved at 800 mT, which indicates the presence of high coercivity minerals.

Thermal demagnetization shows that the low field (0.12 T) component is dominant and decreases up to a maximum unblocking temperature of approximately 580°C indicating magnetite as magnetic mineral (Figure 2B). The high coercivity component is present, and is unblocked at a maximum temperature of 350°C. This unblocking temperature is consistent with maghemite or Fe-sulfides as carrier. In Figure 3C and 3D, the first drop appears on the soft and medium components between 300°C and 350°C, indicating the existence of magnetic mineral with soft coercivity, probably corresponding to low-Ti titanomagnetite. The second drop is observed at 580°C indicating the presence of magnetite (Figures 2C and 2D). The harder components, less than 40% of the total IRM (Figure 2C), decrease regularly up to temperature of 300-350°C and suggest the presence of a Fe-sulphide (pyrrhotite or greigite).

The natural remanent magnetization displays moderately high values starting at 76×10^{-6} A/m in sandstone levels, and reaching up to $\sim 6 \times 10^{-3}$ A/m in the volcanic tuff; the majority of samples fall within the 1 to 10×10^{-3} A/m interval. The visual inspection of orthogonal demagnetization diagrams (Figure 3) reveals that the total NRM is generally composed of one or two components. For almost all of the samples, the individual characteristic remanent magnetization (ChRM) was clearly isolated. Nearly all samples are contaminated by a small secondary component, which is removed entirely completely between 10 and 40 mT and is therefore of recent origin. Characteristic remanent magnetization is generally removed at AF demagnetization between 10-40 mT and 100-150 mT (Figure 3). Demagnetization vector plots are generally of good quality, and in most cases, characteristic directions can be reliably determined.

The stable component, referred to as the characteristic remanent magnetization component (ChRM), shows both polarities: northerly declination with positive inclination and southerly declination with negative inclination (Figure 4). The mean direction was calculated separately for the normal polarity (Dec=10.3°; Inc=18.7°; k=14; K95=7.9) and for the reverse polarity (Dec=165.7°; Inc=-34.3; k=7; K95=10.5) (Figure 4A). The angular distance between the mean normal and reverse directions is 25.1° ± 9.6 and the 95% confidence circles about them do not overlap. Applying McFadden and McElhinny's reversal test (1990), this angle is greater than the critical angle at the 95% confidence level, so the reversal test is negative. The tow directions are not antipodal probably due to the strong overprint of the present day field which may not have been removed completely or due to the directions obtained from the blocs (may-be the blocs have been moved before sampling). The bedding attitudes of the formation do not allow the fold test to be applied because of the horizontal bedding of the strata. To limit the biasing effect we removed the bloc samples

from the data set, the mean direction calculated for the normal polarity is (Dec=2.1°; Inc=16.4°; k= 14; K95=8.7) and for the reverse polarity is (Dec=165.5°; Inc=-28.8; k=8; K95=15.3) (Figure 4B). The α95 confidence cones overlap when plotted on the same hemisphere, and yield a class C positive reversals test [25] and indicates primary of the characteristic remanence. When all the sites mean directions (normal and reversed polarity) are plotted on the same hemisphere (Figure 4C), the mean declination is 356.8°, which indicates that since the time of deposition the Ngwa Formation has rotated 8° counter-clockwise. This rotation is probably related to the reactivation of the shearzone in the central part of the CVL which the Ngwa formation belongs to.

Magnetostratigraphic correlation and discussion

The ChRM directions of each sample were converted into virtual geomagnetic pole (VGP) latitude [26] with respect to the combined mean direction pole at 5.43 N, 10.08 E site coordinates. Based on VGP latitude, each level was assigned a polarity (Figure 5), and the data yielded the following results. The lower three sites representing 4 m thick have reverse polarity R1. Above these, and after a gap of about 10 m, the section continues with 30 m of normal polarity, followed by 26 m of reversed polarity. The last normal polarity N2 is represented by two sites that are 4 m thick. The question remains as to which part of the reversal time scale this series of reversals belongs to. The only independent mean of correlation available at the present time is the radiometric age. The K-Ar dates obtained from mineral concentrates yield an age of 20.1 ± 0.7 Ma on the tuff layer situated some 15 m above the lignite seam (Table 1). If we consider that remanent magnetization of the section we sampled is primary, we can compare our data with the geomagnetic polarity timescale GPTS. The chron C6An.1n in the lower part of the Burdigalian stage lasts 0.17 my; it lies between 20.04 and 20.21 Ma [27], which means that our normal polarity N1 can be correlated with chron C6An.1n, and the reversed polarity R1 can be correlated with chron C6An.1r. If we exclude the data obtained from the hand bloc samples, the proposed correlation does not change.

Conclusion

This study has provided paleomagnetic and rock magnetic results of the Ngwa Formations and indicates that the dominant detrital magnetic minerals are titanomagnetite, magnetite and Fe-sufides. The characteristic magnetization is primary based on the positive reversal test, mean polarity direction counter-clockwise in respect with expected directions for the Lower Miocene of Africa. Magnetostratigraphic analysis and K-Ar age demonstrate that the lignite of Ngwa Formation can be correlated with chron C6An.1n and has an age comprised between 20.0 and 20.2 Ma. Using the magnetostratigraphic data and age estimates for the duration of the magnetic polarity chrons, we have, for the first time determined the age of the lacustrine lignite of Ngwa Formation.

Acknowledgements

We sincerely thank the IRGM Authority for providing logistic and technical assistance with field work, as well as Mathieu Schuster, Guy Franck, and Xavier Valentin. The research reported in this paper was supported by ANR-09-BLAN-0238-02 program. We are very grateful to Teodoro Hernández Treviño for assistance in mineral separation

References

1. Petters SW, Okereke CS, Nwajide CS (1987) In: Matheis G, Schandelmerer H (Eds.) Geology of the Mamfe rift, southeastern Nigeria. Current research in African earth sciences, Balkema, Rotterdam p299-302.

2. Nguimbous-Kouoh JJ, Takam Takougang EM, Nouayou R, Tabod C, Manguelle-

Dicoum E (2012) Structural Interpretation of the Mamfe Sedimentary Basin of Southwestern Cameroon along the Manyu River Using Audiomagnetotellurics Survey. ISRN Geophys 413042: p. 7.

3. Njike PR, Eno BSM, Ndjeng E, Hell JV, Tsafack JPF (2000) Contexte tectonogénique de la mise en place des basins sédimentaires camerounais du Sud au Nord. J Geosci Soc Cameroon GSAf12: Geo-environmental catastrophes in Africa. Abstract pp: 96-97.

4. Touatcha MS, Richard NNP, Salah MM, Said DA, Emmanuel EG (2010) Existence of "late continental" deposits in the Mbere and Djerem sedimentary basins (North Cameroon): Palynologic and stratigraphic evidence. J Geol Min Res 2: 159-169.

5. Dejax J, Michard JG, Brunet M, Hell J (1989) Empreintes de pas de dinosauriens dates du Crétacé inférieur dans le bassin de Babouri-Figuil (Fossé de la Bénoué, Cameroun). N Jb Géol Paläont Abh 178: 85-108.

6. Brunet M, Dejax J, Brillanceau A, Congleton J, Downs W, et al. (1988) Mise en évidence d'une sédimentation précoce d'âge Barrémien dans le fossé de la Bénoué en Afrique occidentale (Bassin du Mayo Oulo Léré, Cameroun), en relation avec l'ouverture de l'Atlantique Sud. CR Acad Sci Paris Série 306: 1125-1130.

7. Ndjeng E, Brunet M (1998) Modèles d'évolution géodynamique de deux bassins de l'Hauterivien-Barrémien du Nord-Cameroun: Les bassins de Babouri-Figuil et du Mayo Oulo-Léré (Fossé de la Bénoué). Géoscience au Cameroun, Presse Univ. Yaoundé I pp: 163-165.

8. Stendal H, Toteu SF, Frei R, Penaye J, Njel UO, et al. (2006) Derivation of detrital rutile in the Yaounde region from the Neoproterozoic Pan-African belt in southern Cameroon (Central Africa). J African Ear Sci 44: 443-458.

9. Dejax J, Brunet M (1996) Les flores fossiles du bassin d'Hama-koussou, Crétacé inférieur du Nord-Cameroun: corrélations biostratigraphiques avec le fossé de la Bénoué, implications paléogéographiques. Géologie de l'Afrique et de l'Atlantique Sud. Actes Colloques Angers. Bulletin Centres Recherches Exploration-Production Elf-Aquitaine 16: 145-173.

10. Bessong M, El Albani A, Hell JV, Fontaine C, Ndjeng E, et al. (2011) Diagenesis in Cretaceous Formations of Benue Trough in the Northern Part of Cameroon: Garoua Sandstones. World J Eng Pure Appl Sci 1: 58-67.

11. Kenfack PL, Tematio P, Kwekam M, Gabriel N, Njike PR (2011) Evidence of a Miocene volcano-sedimentary lithostratigraphic sequence at Ngwa (Dschang Region, West Cameroon): Preliminary analyses and geodynamic interpretation. J Petrol Tech Alter Fuels 2: 25-34.

12. Capponi A (1945) Le lignite de Dschang. Bull Soc Et Camerounaises 7: 75-86.

13. Marzoli A, Piccirillo EM, Renne PR, Bellieni G, Iacumin M, et al. (2000) The Cameroon volcanic line revisited: Petrogenesis of continental basaltic magma from lithospheric and asthenospheric mantle sources. Oxford University Press pp: 87-109.

14. Déruelle B, Moreau C, Nkoumbou C, Kambou R, Lissom J, et al. (1991) In: Kampunzu AB, Lubala RT (Eds.) The Cameroon Line: a review. Magmatism in Extensional Structural Settings: The Phanerozoic African Plate. Springer-Verlag, Berlin pp: 274-327.

15. Nono A, Njonfang E, Kagou Dongmo A, Nkouathio DG, Tchoua FM (2004) Pyroclastic deposits of the Bambouto volcano (Cameroon Line, Central Africa): evidence of an initial strombolian phase. J African Ear Sci 39: 409-414.

16. Dongmo AK, Wandji P, Pouclet A, Vicat JP, Cheilletz A, et al. (2001) Evolution volcanologique du mount Manengouba (ligne du Cameroun), nouvelles données pétrographiques. CR Acad Sci Paris Série 333: 155-162.

17. Dongmo AK, Nkouathio D, Pouclet A, Bardintzeff JM, Wandji P, et al. (2010) The discovery of late Quaternary basalt on Mount Bambouto: Implications for recent widespread volcanic activity in the southern Cameroon Line. J African Ear Sci 57: 96-108.

18. Youmen D, Schmincke HU, Lissom J, Etame J (2005) Données géochronologiques: Mise en évidence des différentes phases volcaniques au Miocène dans les Monts Bambouto (Ligne du Cameroun). Sci Technol Dev 11: 49-57.

19. Nkouathio DG, Dongmo AK, Bandintzeff JM, Wandji P, Bellon H, et al. (2008) Evolution of volcanism in graben and horst structure along the Cenozoic Cameroon Line (Africa): implications for tectonic evolution and mantle source composition. Mineral Petrol 94: 287-303.

20. Lowrie W (1990) Identification of ferromagnetic minerals in a rock by coercivity and unblocking temperature properties. Geophys Res Lett 17: 159-162.

21. Kirschvink JL (1980) Least-squares lines and plane the analysis of paleomagnetic data. Geophys J Int 62: 699-718.

22. Zijderveld JDA (1967) In: Collinson DW, Creer KM, Runcorn SK (Eds.) AC demagnetization of rocks: Analysis of results, Methods in Paleomagnetism: Amsterdam, Netherlands, Elsevier pp: 254-286.

23. Solé J, Enrique P (2001) X-ray fluorescence analysis for the determination of potassium in small quantities of silicate minerals for K-Ar dating. Analytica Chimica Acta 440: 199-205.

24. Solé J (2009) Determination of K-Ar ages in milligram samples using an infrared laser for argon extraction. Rapid Commun Mass Spectrom 23: 3579-3590.

25. McFadden PL, McElhinny MW (1990) Classification of the reversal test in palaeomagnetism: Geophys J Int 103: 725-729.

26. Opdyke ND, Channell JET (1996) Magnetic Stratigraphy, Academic Press, San Diego, California, USA p. 345.

27. Gradstein FM, Ogg JG, Smith AG (2004) A Geologic Time Scale 2004, Cambridge, UK, Cambridge University Press. p. 589.

Permissions

All chapters in this book were first published in JGG, by OMICS International; hereby published with permission under the Creative Commons Attribution License or equivalent. Every chapter published in this book has been scrutinized by our experts. Their significance has been extensively debated. The topics covered herein carry significant findings which will fuel the growth of the discipline. They may even be implemented as practical applications or may be referred to as a beginning point for another development.

The contributors of this book come from diverse backgrounds, making this book a truly international effort. This book will bring forth new frontiers with its revolutionizing research information and detailed analysis of the nascent developments around the world.

We would like to thank all the contributing authors for lending their expertise to make the book truly unique. They have played a crucial role in the development of this book. Without their invaluable contributions this book wouldn't have been possible. They have made vital efforts to compile up to date information on the varied aspects of this subject to make this book a valuable addition to the collection of many professionals and students.

This book was conceptualized with the vision of imparting up-to-date information and advanced data in this field. To ensure the same, a matchless editorial board was set up. Every individual on the board went through rigorous rounds of assessment to prove their worth. After which they invested a large part of their time researching and compiling the most relevant data for our readers.

The editorial board has been involved in producing this book since its inception. They have spent rigorous hours researching and exploring the diverse topics which have resulted in the successful publishing of this book. They have passed on their knowledge of decades through this book. To expedite this challenging task, the publisher supported the team at every step. A small team of assistant editors was also appointed to further simplify the editing procedure and attain best results for the readers.

Apart from the editorial board, the designing team has also invested a significant amount of their time in understanding the subject and creating the most relevant covers. They scrutinized every image to scout for the most suitable representation of the subject and create an appropriate cover for the book.

The publishing team has been an ardent support to the editorial, designing and production team. Their endless efforts to recruit the best for this project, has resulted in the accomplishment of this book. They are a veteran in the field of academics and their pool of knowledge is as vast as their experience in printing. Their expertise and guidance has proved useful at every step. Their uncompromising quality standards have made this book an exceptional effort. Their encouragement from time to time has been an inspiration for everyone.

The publisher and the editorial board hope that this book will prove to be a valuable piece of knowledge for researchers, students, practitioners and scholars across the globe.

List of Contributors

Olusiji Samuel Ayodele
Department of Geology, Ekiti State University, P.M.B. 5363, Ado-Ekiti, Nigeria

Muthamilselvan A, Srimadhi K, Nandhini R, Pavithra P, Balamurugan T and Vasuki V
Centre for Remote Sensing, Bharathidasan University, Trichy, India

Mohamed Mhmod, Liu Hai Yan, Liu Cai and Feng Xuan
College of Geo-Exploration Science and Technology, Jilin University, Changchun 130026, China

Amal Kumar Ghosh, Virendra Kumar Sharma and Rajeev Kumar Singh
Bhagwant University, Ajmer, Rajasthan-305004, India

Monia Aloui and Chedly Abbes
Faculty of Sciences of Sfax, Road Soukra, Sfax 3038, Tunisia

Yannick Bleuzen
Society of Mining and Industrial Engineering, ZA Brushes 2, 9 Rue de la Hyssop 37270 Larcay, Tunisia

Elhoucine Essefi
National Engineering School of Sfax, University of Sfax, Tunisia

Sandeep Meshram and Sunil P Khadse
College of engineering-PUNE, Maharastra, India

Victor OM, Ude AE and Valeria AC
Department of Geological Sciences, Nnamdi Azikiwe University, Awka, Nigeria

Yusoff AH and Mohamed CAR
School of Environmental and Natural Resource Sciences, University Kebangsaan Malaysia, 43600 Bangi, Selangor, Malaysia

Nyakundi ER and Ambusso WJ
Kenyatta University, Department of Physics, Nairobi, Kenya

Githiri JG
Jomo-Kenyatta University of Agriculture and Technology, Department of Physics, Nairobi, Kenya

Peng K Hong and James M Dohm
The University Museum, The University of Tokyo, 7-3-1 Hongo, Bunkyo-ku, Tokyo 113-0033, Japan

Seiji Sugita
Department of Earth and Planetary Science, The University of Tokyo, 7-3-1 Hongo, Bunkyo-ku, Tokyo 113-0033, Japan

Hideaki Miyamoto
The University Museum, The University of Tokyo, 7-3-1 Hongo, Bunkyo-ku, Tokyo 113-0033, Japan
Planetary Science Institute, 1700 East Fort Lowell, Tucson, AZ 85719-2395, USA

Takafumi Niihara
The University Museum, The University of Tokyo, 7-3-1 Hongo, Bunkyo-ku, Tokyo 113-0033, Japan
Lunar and Planetary Institute, Universities Space Research Association, 3600 Bay Area Boulevard, Houston, TX 77058, USA

Kenji Nagata and Masato Okada
Department of Complexity Science and Engineering, The University of Tokyo, 5-1-5 Kashiwanoha, Kashiwa, Chiba 277-8561, Japan

Kasi Viswanadh Gorthi
Department of Civil Engineering, JNTUH College of Engineering Hyderabad (Autonomous), Hyderabad, India

Mohan Babu M
Department of Civil Engineering, Sri Venkateswara College of Engineering and Technology (Autonomous), Chittoor, Andhra Pradesh, India

Oluwadare OA, Osunrinde OT, Abe SJ and Ojo BT
Department of Applied Geophysics, Federal University of Technology, Nigeria

Nwankwoala HO
Department of Geology, University of Port Harcourt, Port Harcourt, Nigeria

Orji MO
Department of Petroleum Engineering and Geosciences, Petroleum Training Institute, Effurun, Warri, Delta State, Nigeria

Mbah Victor O, Onwuemesi AG, Aniwetalu and Emmanuel U
Geophysical Science Department, Nnamdi Azikiwe University, Awka, Nigeria

Ndam Noupayou JR and Mvondo Ondoua J
Laboratory of Engineering Geology and Alterology, Department of Earth Sciences, Faculty of Science, University of Yaounde 1, PO Box: 812, Yaounde, Cameroon

Kouassy Kalédjé PS
Laboratory of Engineering Geology and Alterology, Department of Earth Sciences, Faculty of Science, University of Yaounde 1, PO Box: 812, Yaounde, Cameroon
Department of Mining and Geological, Sub-Regional Bilingual University of Mining, Sciences, Technology, Management and Professional Training, PO Box: 863, Yaounde, Cameroon

Djomou Djomga PN
Laboratory of Material and Inorganic Industrial Chemistry, Department of Applied Chemistry, National School of Agro-Industrial Sciences (ENSAI), University of Ngaoundéré, PO Box: 455 Ngaoundere, Cameroon

Onimisi M
Department of Earth Sciences, Kogi State University, Anyigba, Nigeria

Abaa SI
Department of Geology and Mining, Nasarawa State University, Keffi, Nigeria

Obaje NG
Department of Geography and Geology, Ibrahim Babangida University, Lapai, Nigeria

Sule VI
Department of Physics, Kogi State University, Anyigba, Nigeria

Mohamed K Salah
Department of Geology, American University of Beirut, Lebanon

Ahmed Abd El-Al
Civil Engineering Department, Najran University, Saudi Arabia

Geology Department, Al Azhar University, Egypt

Abdel-Hameed AT
Geology Department, Tanta University, Egypt

Owoyemi OO
Department of Geology, Kwara State University, Malete, Nigeria

Adeyemi GO
Department of Geology, University of Ibadan, Nigeria

Saleh Ibrahim Bute
Department of Geology, Gombe State University, Nigeria

Evans UF
Department of Sciences, Maritime Academy of Nigeria, Oron, Nigeria

Abdulsalam NN and Mallam A
Department of Physics, University of Abuja, Nigeria

Evans UF
Department of Sciences, Maritime Academy of Nigeria, Oron, Nigeria

Abdulsalam NN and Mallam A
Department of Physics, University of Abuja, Nigeria

Oladosu YC and Ogundipe OY
Department of Geology, Ekiti State University, P.M.B 5363, Ado, Ekiti State, Nigeria

Nguo Sylvestre Kanouo
Mineral Exploration and Ore Genesis Unit, Department of Mines and Quarries, Faculty of Mines and Petroleum Industries, University of Maroua, Cameroon'

Rose Fouateu Yongue and Syprien Bovari Yomeun
Department of Earth Sciences, University of Yaoundé I, Cameroon

Tanwi Richard Ghogomu
Department of Earth Sciences, University of Yaoundé I, Cameroon
Department of security, Quality and Environment, Faculty of Mines and Petroleum Industries, University of Maroua, Cameroon'

Emmanuel Njonfang
Higher Teachers Training School, University of Yaoundé I, Cameroon

Emmanuel Archelaus Afanga Basua
Department of Earth Sciences, China University of Geosciences wuhan, China

Md. Abdullah Al zaman
Department of Physics, University of Chittagong, Bangladesh

Nusrath Jahan Monira
Department of Physics, University of Chittagong, Bangladesh

Benammi M
Institut de Paléoprimatologie, Paléontologie Humaine: Évolution et Paléoenvironnements (IPHEP), UMR-CNRS 7262, Bâtiment B35, 6 rue Michel.Brunet, 86022 Poitiers Cedex, France

Hell JV, Bessong M and Nolla D
Institut de Recherches Géologiques et Minières du Cameroun, BP 4140, Yaoundé, Cameroun

Solé J
Universidad Nacional Autónoma de México, Instituto de Geología Dept. de Geoquímica Cd. Universitaria, Coyoacán 04510 México DF

Brunet M
Institut de Paléoprimatologie, Paléontologie Humaine: Évolution et Paléoenvironnements (IPHEP), UMR-CNRS 7262, Bâtiment B35, 6 rue Michel.Brunet, 86022 Poitiers Cedex, France
Collège de France, Chaire de Paléontologie humaine, 11 Place Marecelin Berthelot, 75231 Paris cedex 05

Index